Western North Atlantic Palaeogene and Cretaceous Palaeoceanography

Geological Society Special Publications

Series Editors

Special Publication reviewing procedures

The Society makes every effort to ensure that the scientific and production quality of its books matches that of its journals. Since 1997, all book proposals have been refereed by specialist reviewers as well as by the Society's Publications Committee. If the referees identify weaknesses in the proposal, these must be addressed before the proposal is accepted.

Once the book is accepted, the Society has a team of series editors (listed above) who ensure that the volume editors follow strict guidelines on refereeing and quality control. We insist that individual papers can only be accepted after satisfactory review by two independent referees. The questions on the review forms are similar to those for *Journal of the Geological Society*. The referees' forms and comments must be available to the Society's series editors on request.

Although many of the books result from meetings, the editors are expected to commission papers that were not presented at the meeting to ensure that the book provides a balanced coverage of the subject. Being accepted for presentation at the meeting does not guarantee inclusion in the book.

Geological Society Special Publications are included in the ISI Science Citation Index, but they do not have an impact factor, the latter being applicable only to journals.

More information about submitting a proposal and producing a Special Publication can be found on the Society's web site: www.geolsoc.org.uk

It is recommended that reference to all or part of this book should be made in one of the following ways.

KROON, D., NORRIS, R. D. & KLAUS, A. (eds) 2001. *Western North Atlantic Palaeogene and Cretaceous Palaeoceanography*. Geological Society, London, Special Publications, **183**.

RÖHL, U., OGG, J. G., GEIB, T. L. & WEFER, G. 2001. Astronomical calibration of the Danian time scale. *In*: KROON, D., NORRIS, R. D. & KLAUS, A. (eds) 2001. *Western North Atlantic Palaeogene and Cretaceous Palaeoceanography*. Geological Society, London, Special Publications, **183**, 163–183.

GEOLOGICAL SOCIETY SPECIAL PUBLICATION NO. 183

Western North Atlantic Palaeogene and Cretaceous Palaeoceanography

EDITED BY

DICK KROON
(University of Edinburgh, UK)

R. D. NORRIS
(Woods Hole OI, USA)

and

A. KLAUS
(Ocean Drilling Program, USA)

2001
Published by
The Geological Society
London

THE GEOLOGICAL SOCIETY

The Geological Society of London was founded in 1807 and is the oldest geological society in the world. It received its Royal Charter in 1825 for the purpose of 'investigating the mineral structure of the Earth' and is now Britain's national society for geology.

Both a learned society and a professional body, the Geological Society is recognized by the Department of Trade and Industry (DTI) as the chartering authority for geoscience, able to award Chartered Geologist status upon appropriately qualified Fellows. The Society has a membership of 9099, of whom about 1500 live outside the UK.

Fellowship of the Society is open to persons holding a recognized honours degree in geology or a cognate subject, or not less than six years' relevant experience in geology or a cognate subject. A Fellow with a minimum of five years' relevant postgraduate experience in the practice of geology may apply for chartered status. Successful applicants are entitled to use the designatory postnominal CGeol (Chartered Geologist). Fellows of the Society may use the letters FGS. Other grades of membership are available to members not yet qualifying for Fellowship.

The Society has its own Publishing House based in Bath, UK. It produces the Society's international journals, books and maps, and is the European distributor for publications of the American Association of Petroleum Geologists (AAPG), the Society for Sedimentary Geology (SEPM) and the Geological Society of America (GSA). Members of the Society can buy books at considerable discounts. The Publishing House has an online bookshop (http://bookshop.geolsoc.org.uk).

Further information on Society membership may be obtained from the Membership Services Manager, The Geological Society, Burlington House, Piccadilly, London W1V 0JU (E-mail: enquiries@geolsoc.org.uk; tel: +44 (0) 207 434 9944).

The Society's Web Site can be found at http://www.geolsoc.org.uk/. The Society is a Registered Charity, number 210161.

Published by The Geological Society from:
The Geological Society Publishing House
Unit 7, Brassmill Enterprise Centre
Brassmill Lane
Bath BA1 3JN, UK
Orders: Tel. +44 (0)1225 445046
Fax +44 (0)1225 442836
Online bookshop: http://bookshop.geolsoc.org.uk

British Library Cataloguing in Publication Data

A catalogue record for this book is available from the British Library.

ISBN 1-86239-078-9

Typeset by Alden Multimedia, Westonzoyland, UK

Printed by The Alden Press, Oxford, UK

Distributors

USA
AAPG Bookstore
PO Box 979
Tulsa
OK 74101-0979
USA
Orders: Tel. +1 918 584-2555
Fax +1 918 560-2652
E-mail bookstore@aapg.org

Australia
Australian Mineral Foundation Bookshop
63 Conyngham Street
Glenside
South Australia 5065
Australia
Orders: Tel. +61 88 379-0444
Fax +61 88 379-4634
E-mail bookshop@amf.com.au

India
Affiliated East-West Press PVT Ltd
G-1/16 Ansari Road, Daryaganj,
New Delhi 110 002
India
Orders: Tel. +91 11 327-9113
Fax +91 11 326-0538
E-mail affiliat@nda.vsnl.net.in

Japan
Kanda Book Trading Co.
Cityhouse Tama 204
Tsurumaki 1-3-10
Tama-shi
Tokyo 206-0034
Japan
Orders: Tel. +81 (0)423 57-7650
Fax +81 (0)423 57-7651

Geological Society Special Publication No. 183

Western North Atlantic Palaeogene and Cretaceous Palaeoceanography

ERRATUM

The folding figure between pages 10 and 11 has been bound with the wrong Norris *et al.* paper. It should be with the Norris, Klaus & Kroon paper starting on page 23.

Contents

IUBER, B. T. & ERBACHER, J. Cretaceous-Palaeogene ocean and ›ical North Atlantic 1

ƘROON, D. Mid-Eocene deep water, the Late Palaeocene Thermal ope mass wasting during the Cretaceous–Palaeogene impact 23

. J. & COBABE, E. A. Deposition of sedimentary organic matter in y the geochemistry and petrography of high-resolution samples, ıtlantic 49

No extinctions during Oceanic Anoxic Event 1b: the Aptian– record of ODP Leg 171 73

graphic subdivision and correlation of upper Maastrichtian Coastal Plain and Blake Nose, western Atlantic 93

B. T. The Maastrichtian record at Blake Nose (western North r global palaeoceanographic and biotic changes 111

-HUERTAS, M., KROON, D., SMIT, J., PALOMO-DELGADO, I. & ɔf the Cretaceous–Tertiary boundary at Blake Nose (ODP Leg 131

-HUERTAS, M., PALOMO-DELGADO, I. & SMIT, J. K–T boundary ODP Leg 171B) as a record of the Chicxulub ejecta deposits 149

T. L. & WEFER, G. Astronomical calibration of the Danian time 163

. D. Biostratigraphic implications of mid-latitude Palaeocene– rom Hole 1051A, ODP Leg 171B, Blake Nose, western North 185

UIS, H. & WILLIAMS, G. L. Mid- to Late Eocene organic-walled P Leg 171B, offshore Florida 225

. North Atlantic climate variability in early Palaeogene time: ty study 253

NORRIS, R. D. Orbitally forced climate change in late mid-Eocene B): evidence from stable isotopes in foraminifera 273

tion and removal during the Late Palaeocene Thermal Maximum: ary treatment of the isotope record at ODP Site 1051, Blake Nose 293

nental implications of palygorskite clays in Eocene deep-water central Atlantic 307

317

Cretaceous–Palaeogene ocean and climate change in the subtropical North Atlantic

RICHARD D. NORRIS[1], DICK KROON[2], BRIAN T. HUBER[3] & JOCHEN ERBACHER[4]

[1]*Department of Geology and Geophysics, Woods Hole Oceanographic Institution, Woods Hole, MA 02543, USA*

[2]*Department of Geology and Geophysics, University of Edinburgh, Grant Institute, West Mains Road, Edinburgh EH9 3JW, UK*

[3]*Department of Palaeobiology, Smithsonian Institution, MRC:NHB 121, National Museum of Natural History, Washington, DC 20560, USA*

[4]*Bundesanstalt für Geowissenschaften und Rohstoffe, Stilleweg 2, 30655 Hannover, Germany*

Abstract: Ocean Drilling Program (ODP) Leg 171B recovered continuous sequences that yield evidence for a suite of 'critical' events in the Earth's history. The main events include the late Eocene radiolarian extinction, the late Palaeocene benthic foraminiferal extinction associated with the Late Palaeocene Thermal Maximum (LPTM), the Cretaceous–Palaeogene (K–P) extinction, the mid-Maastrichtian event, and several episodes of sapropel deposition documenting the late Cenomanian, late Albian and early Albian warm periods. A compilation of stable isotope results for foraminifera from Leg 171B sites and previously published records shows a series of large-scale cycles in temperature and $\delta^{13}C$ trends from Albian to late Eocene time. Evolution of $\delta^{18}O$ gradients between planktic and benthic foraminifera suggests that the North Atlantic evolved from a circulation system similar to the modern Mediterranean during early Albian time to a more open ocean circulation by late Albian–early Cenomanian time. Sea surface temperatures peaked during the mid-Cretaceous climatic optimum from the Albian–Cenomanian boundary to Coniacian time and then show a tendency to fall off toward the cool climates of the mid-Maastrichtian. The Albian–Coniacian period is characterized by light benthic oxygen isotope values showing generally warm deep waters. Lightest benthic oxygen isotopes occurred around the Cenomanian–Turonian boundary, and suggest middle bathyal waters with temperatures up to 20 °C in the North Atlantic. The disappearance of widespread sapropel deposition in Turonian time suggests that sills separating the North Atlantic from the rest of the global ocean were finally breached to sufficient depth to permit ventilation by deep waters flowing in from elsewhere. The Maastrichtian and Palaeogene records show two intervals of large-scale carbon burial and exhumation in the late Maastrichtian–Danian and late Palaeocene–early Eocene. Carbon burial peaked in early Danian time, perhaps in response to the withdrawal of large epicontinental seas from Europe and North America. Much of the succeeding Danian period was spent unroofing previously deposited carbon and repairing the damage to carbon export systems in the deep ocean caused by the K–P mass extinction. The youngest episode of carbon exhumation coincided with the onset of the early Eocene Warm Period and the LPTM, and has been attributed to the tectonic closure of the eastern Tethys and initiation of the Himalayan Orogeny.

Cretaceous and Palaeogene marine deposits provide the opportunity to study Earth system processes during partly to entirely deglaciated states. Certain key intervals are marked by rapid climate change and massive carbon input. These intervals are the Late Palaeocene Thermal Maximum (LPTM) and Oceanic Anoxic Events (OAEs) in early Aptian–Albian time and at the Cenomanian-Turonian boundary. These time intervals are particularly significant to current earth science objectives because focused research has the potential to considerably improve the understanding of the general dynamics of the Earth's climate during rapid perturbation of the carbon cycle.

Deep ocean sections that are continuously cored provide the high-resolution records to document complex sequences across unique

From: KROON, D., NORRIS, R. D. & KLAUS, A. (eds). 2001 *Western North Atlantic Palaeogene and Cretaceous Palaeoceanography*. Geological Society, London, Special Publications, **183**, 1–22.
0305-8719/01/$15.00

events in the geological record that show global biogeochemical variations. These records are needed to register major steps in climate evolution: burial of excess carbon, changes in global temperatures, nutrient cycling, etc. Stable isotope records are fundamental in showing the sequence of events and the amplitude of change in the system. Most Cretaceous stable isotope records are based on bulk carbonate. Here, we present a compilation of known planktonic and benthic foraminiferal stable isotope records to highlight long-term trends in surface and deep ocean palaeoceanography. We present this stable isotope record to provide a palaeoceanographic context to the 14 papers presented in this Special Publication. Several portions of the stable isotope record were compiled from Ocean Drilling Program (ODP) Leg 171B results. This ODP Leg was dedicated to drill Cretaceous–Palaeogene sequences at Blake Nose, northwestern Atlantic, in 1998. The scientific crew on the *JOIDES Resolution* decided to compile a set of papers based on drilling results, in a Special Publication of the Geological Society of London. The focus would be western Atlantic Cretaceous–Palaeogene palaeoceanography and the results are now in front of you. Before we present the compilation of stable isotope results, we would like to introduce you to ODP Leg 171B drilling results.

Ocean Drilling Program Leg 171B drilling

ODP Leg 171B was designed to recover a series of 'critical boundaries' in the Earth's history during which abrupt changes in climate and oceanography coincide with often drastic changes in the Earth's biota. Some of these events, such as the Cretaceous–Palaeogene (K–P) extinction and the late Eocene tektite layers, are associated with the impacts of extraterrestrial objects, such as asteroids or meteorites, whereas other events, including the benthic foraminifer extinction in the late Palaeocene and mid-Maastrichtian events, are probably related to intrinsic features of the Earth's climate system. Three of the critical intervals, early Eocene, the Cenomanian–Turonian boundary interval and the late Albian, are characterized by unusually warm climatic conditions when the Earth is thought to have experienced such extreme warmth that the episodes are sometimes described as 'super-greenhouse' periods. The major objectives of Leg 171B were to recover records of these critical boundaries, or intervals, at shallow burial depth where microfossil and lithological information would be well preserved, and to drill cores along a depth transect where the vertical structure of the oceans during the boundary events could be studied. The recovery of sediments characterized by cyclical changes in lithology in continuous Palaeogene or Mesozoic records would help to establish the rates and timing of major changes in surface and deep-water hydrography and microfossil evolution.

Accordingly, five sites were drilled down the spine of the Blake Nose, a salient on the margin of the Blake Plateau where Palaeogene and Cretaceous sediments have never been deeply buried by younger deposits (Fig. 1). The Blake Nose is a gentle ramp that extends from *c.* 1000 to *c.* 2700 m water depth and is covered by a drape of Palaeogene and Cretaceous strata that are largely protected from erosion by a thin veneer of manganiferous sand and nodules. We recovered a record of the Eocene and Palaeocene epochs that, except for a few short hiatuses in mid-Eocene time, is nearly complete. Thick sequences through the Maastrichtian, Cenomanian and Albian sequences have allowed us to create high-resolution palaeoclimate records from these periods. The continuous expanded records from Palaeogene and Cretaceous time show Milankovitch-related cyclicity that provides the opportunity for astronomical calibration of at least parts of the time scale, particularly when combined with an excellent magnetostratigraphic record and the presence of abundant calcareous and siliceous microfossils. Our strategy was to drill multiple holes at each site as far down as possible to recover complete sedimentary sequences by splicing multisensor track (MST) or colour records.

Lithostratigraphy and seismic stratigraphy of Blake Nose

The sedimentary record at Blake Nose consists of Eocene carbonate ooze and chalk that overlie Palaeocene claystones as well as Maastrichtian and possibly upper Campanian chalk (Fig. 2). In turn, Campanian strata rest unconformably upon Albian to Cenomanian claystone and clayey chalk that appear to form a conformable sequence of clinoforms. A short condensed section of Coniacian–Turonian nannofossil chalks, hardgrounds and debris beds is found between Campanian and Cenomanian rocks on the deeper part of Blake Nose. Aptian claystones are interbedded with Barremian periplatform debris, which shows that the periplatform material is reworked from older rocks. The entire middle Cretaceous and younger sequence rests on a Lower Cretaceous, and probably Jurassic,

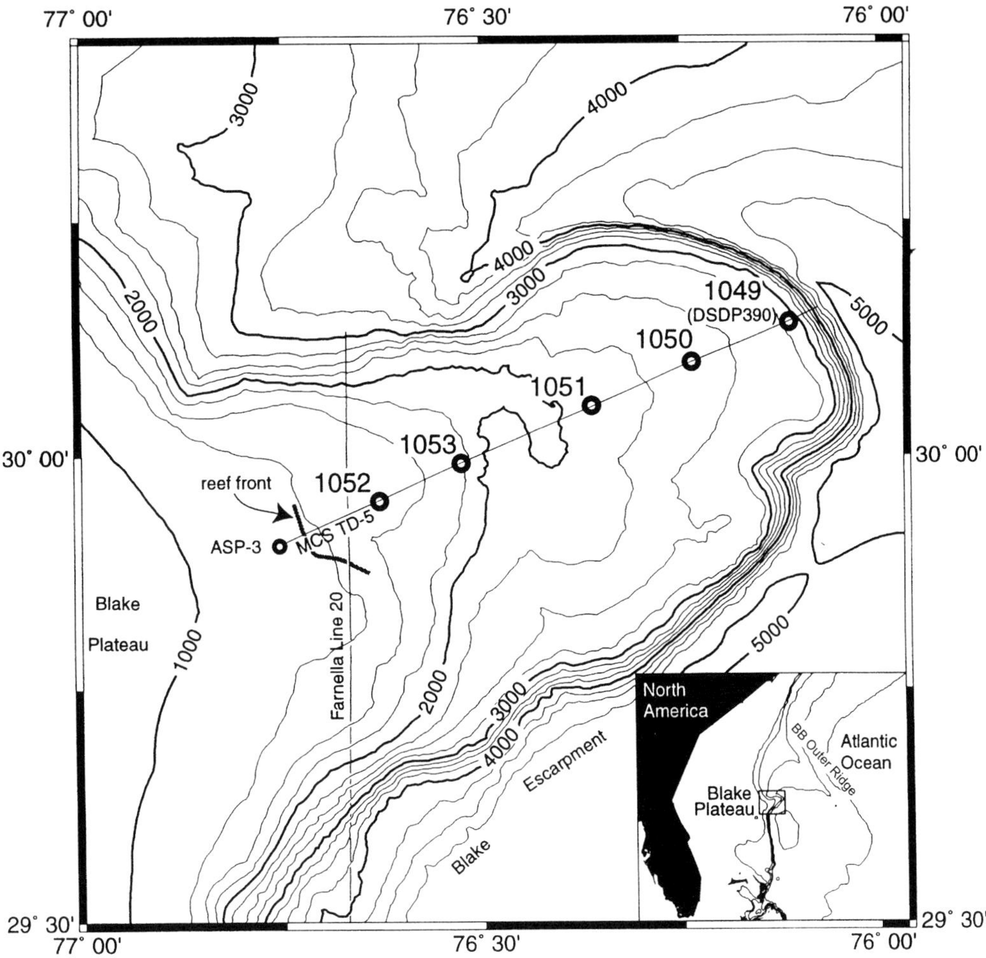

Fig. 1. Location of Blake Nose showing ODP Leg 171B borehole sites and location of Multichannel seismic profile TD-5. BB Outer Ridge, Blake–Bahama Outer Ridge.

carbonate platform that is more than 5 km thick in the region of Blake Nose (Shipley *et al.* 1978; Dillon *et al.* 1985; Dillon & Popenoe 1988).

Seismic records show the presence of buried reef build-ups at the landward end of the Blake Nose (Fig. 3). Fore-reef deposits and pelagic oozes, built seaward of the reef front, rest on relatively flat-lying Barremian shallow-water carbonates and serve largely to define the present bathymetric gradient along Blake Nose (Benson *et al.* 1978; Dillon *et al.* 1985; Dillon & Popenoe 1988). Single-channel seismic reflection data (SCS) lines collected by the *Glomar Challenger* over Deep Sea Drilling Project (DSDP) Site 390 and our reprocessed version of multichannel seismic reflection (MCS) line TD-5 show that more than 800 m of strata are present between a series of clinoforms that overlap the reef complex and the sea bed. ODP Leg 171B demonstrated that most of the clinoform sequence consists of Albian–Cenomanian strata and that a highly condensed sequence of Santonian–Campanian rocks is present in places between the lower Cenomanian and the Maastrichtian sequences (Norris *et al.* 1998). The Maastrichtian section is overlapped by a set of parallel, continuous reflectors interpreted as being of Palaeocene and Eocene age that become discontinuous updip. Most of the Eocene section is incorporated in a major clinoform complex that reaches its greatest thickness down dip of the Cretaceous clinoforms.

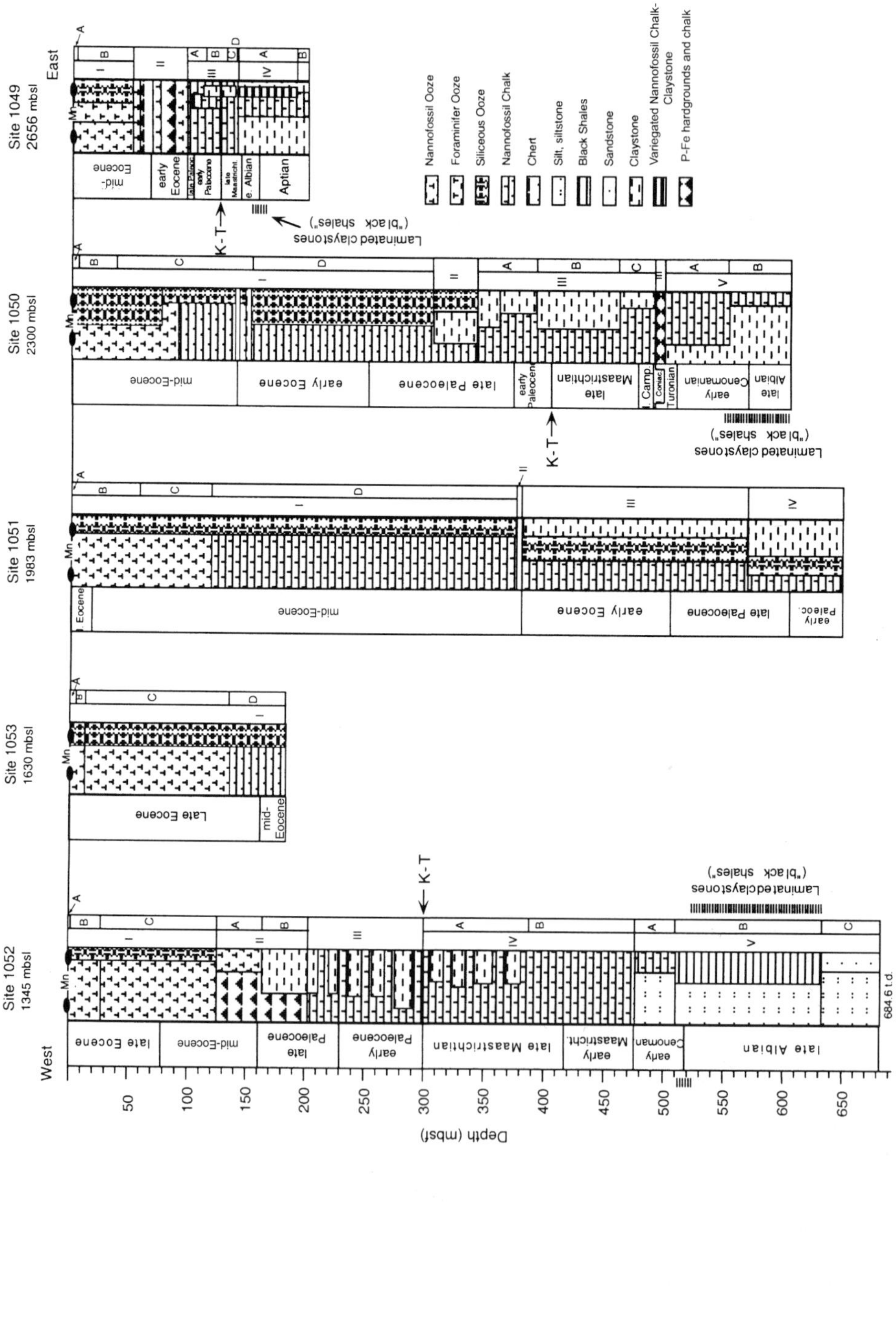

Fig. 2. Simplified lithostratigraphy of ODP Leg 171B sites (after Norris *et al.* 1998).

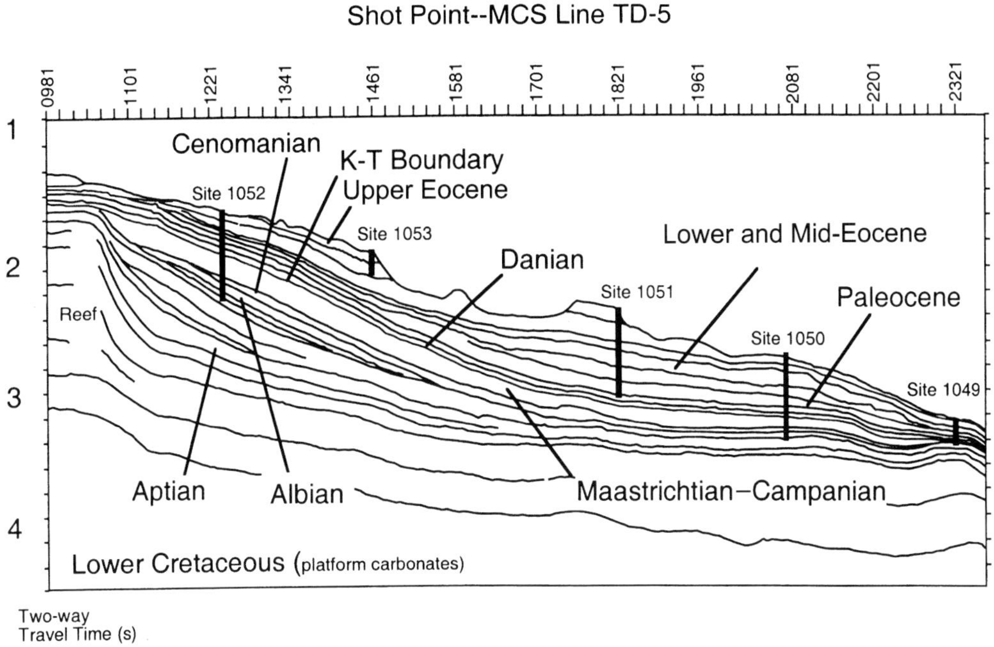

Fig. 3. Schematic interpretation of multichannel seismic profile TD-5 showing ODP Leg 171B drill sites.

Early Albian black shale event (OAE 1b c. 112 Ma)

ODP Leg 171B recovered upper Aptian–lower Albian sediments at Site 1049 consisting of green, red, tan and white marls overlying Barremian and Aptian pelletal grainstones and carbonate sands. A similar succession is present at DSDP Sites 390 and 391 drilled during DSDP Leg 44 as well as in DSDP holes drilled on the Bahama platform. The distal equivalents of these sediments are present at DSDP Site 387 on the Bermuda Rise, where they are light grey limestones interbedded with chert and green or black claystone deposited at water depths of over 4 km (Tucholke & Vogt 1979).

On Blake Nose, the multicoloured claystones contain a single, prominent laminated black shale bed (sapropel) about 46 cm thick (Fig. 4). The sapropel is laminated on a millimetre scale and contains pyrite nodules and thin stringers of dolomite and calcite crystals. Total organic carbon content ranges from *c.* 2 wt % to over 11.5 wt % and has a hydrogen index typical of a type II kerogen (Barker *et al.* this volume). A bioturbated interval of marl about 1 cm thick occurs in the middle of the black sapropel. The lower contact of the sapropel is gradational into underlying green nannofossil claystone. In contrast, the upper contact is sharp below bioturbated olive green nannofossil claystone. The planktonic foraminifer assemblage (characteristic of the upper *Hedbergella planispira* Zone) together with the nannoflora (representing biozone CC7c) suggest an early Albian age and a correlation with OAE 1b.

Ogg *et al.* (1999) estimated sedimentation rates during and after deposition of the OAE 1b sapropel. Spectral analysis of physical property records suggests that the dominant colour cycles (between red, green and white marls) are probably related to the eccentricity and precession cycles and yield average sedimentation rates of *c.* 0.6 cm ka^{-1} across the OAE. Sapropel deposition persisted for at least *c.* 30 ka (Ogg *et al.* 1999). This duration for OAE 1b is probably an underestimate, as it is common for organic matter in sapropels to be partly removed once oxic conditions return to the sea floor (e.g. Mercone *et al.* 2000). Therefore, the OAE 1b sapropel may originally have been thicker and represented a longer period of disaerobic seafloor conditions than is at present the case.

By any measure, the OAE 1b sapropel represents a remarkably long interval of disaerobic conditions. Sapropels formed during Plio-Pleistocene time in the Mediterranean basins were deposited on time scales of no more than a few thousand years (e.g. Mercone *et al.* 2000). The best studied example of OAE 1b in Europe is the Niveau Paquier in the Southeast France Basin. There, the upper Aptian–lower Albian sequence

Fig. 4. Conspicuous beds recovered during ODP Leg 171B. (**a**) Close-up of the K–P boundary in Hole 1049A showing the spherule bed that is interpreted as ejecta material from Chicxulub. (**b**) Close-up of the lower Albian black shale that is time equivalent to Oceanic Anoxic Event 1b. (**c**) Condensed section separating sediments of Campanian age (above) and Santonian age (below), from Hole 1050C.

is characterized by numerous black shales, all of which, with the exception of OAE 1b, are not present at Blake Nose. On the basis of the ecology and distribution of benthic foraminifers, Erbacher *et al.* (1999) demonstrated the synchroneity of OAE 1b between Blake Nose and France.

The distribution of the foraminifera was extensively studied across OAE 1b by Erbacher *et al.* (1999) and Holbourn & Kuhnt (this volume). Sediments deposited before OAE 1b contain a low-diversity fauna of opportunistic phytodetritus feeders: foraminifera that feed on the enhanced carbon flux to the ocean floor. This unique fauna replaced a highly diverse fauna and is indicative of the large environmental changes leading to the OAE 1b. The finely laminated black shale itself contains a very impoverished microbiota indicating disaerobic conditions at the sea floor. Although faunal turnovers have been found to be associated with the OAEs, Holbourn & Kuhnt (this volume) concluded from their foraminiferal distributional study that there were no foraminiferal extinctions associated with OAE 1b.

Erbacher *et al.* (in press) showed new stable isotope data on foraminifera from early Albian OAE 1b. Those workers demonstrated that the formation of OAE 1b was associated with an increase in surface-water runoff, and a rise in surface temperatures, that led to decreased bottom-water formation and elevated carbon burial in the restricted basins of the North Atlantic. The stable isotope record has similar features as the Mediterranean sapropel record from the Pliocene–Quaternary period inasmuch as there is a large negative shift in $\delta^{18}O$ of planktic foraminifera in both instances that probably reflects the freshening and perhaps temperature rise associated with the onset of sapropel formation. However, the geographical extent of the OAE 1b is much larger and its duration is at least four times longer than any of the Quaternary sapropels. Hence, current results suggest that OAE 1b was associated with a pronounced increase in surface-water stratification that effectively restricted overturning over a large portion of Tethys and its extension into the western North Atlantic.

Mid-Cretaceous sea surface temperatures and OAE 1d and 2

A thick section of Albian–Cenomanian continental slope and rise sediments were recovered in Holes 1050C and 1052E on Blake Nose. In Hole 1052E, 215 m of black and green laminated claystone, minor chalk, limestone and sandstone were recovered, representing the uppermost Albian and lowermost Cenomanian interval. Sandstones recovered at the base of the hole give way to dark claystones with laminated intervals and thin bioturbated limestones higher in the sequence. A partly correlative sequence was recovered in Hole 1050C where the section starts in latest Albian time and continues through most of Cenomanian time. The Turonian, Santonian and Coniacian periods are represented in a highly condensed sequence of multi-coloured chalk just above the last black shales and chalks of the Cenomanian sequence.

The upper Albian–lower Cenomanian strata in Hole 1052E preserve a biostratigraphically complete sequence from calcareous nannofossil Zone CC8b to CC9c (planktonic foraminifer biozones *Rotalipora ticinensis* to *R. greenhornensis*). The time scales of Gradstein *et al.* (1995) and Bralower *et al.* (1997a) and the biostratigraphy from Site 1052 suggest the sequence records *c.* 8 Ma of deposition between *c.* 94 and *c.* 102 Ma. Sedimentation rates drop off dramatically in the upper *c.* 25 m of the Cenomanian sequence about 98 Ma. The remainder of the sequence was deposited within a 2 Ma interval at sedimentation rates of about 9–10 cm ka^{-1}. The termination of these high sedimentation rates coincides with a lithological change from black shale to chalk deposition, probably reflecting the sea-level rise in early Cenomanian time.

The section across the Albian–Cenomanian boundary in Hole 1052E is correlative with OAE 1d. Black and green shales become increasingly well laminated and darker in colour approaching the Albian–Cenomanian boundary and are accompanied by interbeds of white to grey limestone. High-resolution resistivity (Formation Microscanner; FMS) logs from Hole 1052E demonstrate that the OAE is not a single event, but represents a gradual intensification of the limestone–shale cycle that terminated abruptly at the end of the OAE (Kroon *et al.* 1999; Fig. 5).

Cyclostratigraphy suggests there is a low-frequency cycle (seen best in the unfiltered FMS log) with a wavelength of *c.* 9–10 m that is expressed in cycles of increasing and then decreasing FMS amplitude. In turn, the 10 m cycle consists of bundles of 1.8–2 m cycles. Filtering the FMS record (that is sampled at better than 1 cm resolution) with a 70 cm–20 m window reveals these dominant cycles clearly. There is a still higher frequency cycle about 10–15 cm wavelength that is particularly well expressed in the claystone interbeds and are expressed by high FMS resistivity. These high-resistivity claystones contrast with the low resistivity of more carbonate-rich beds, including the limestone beds that

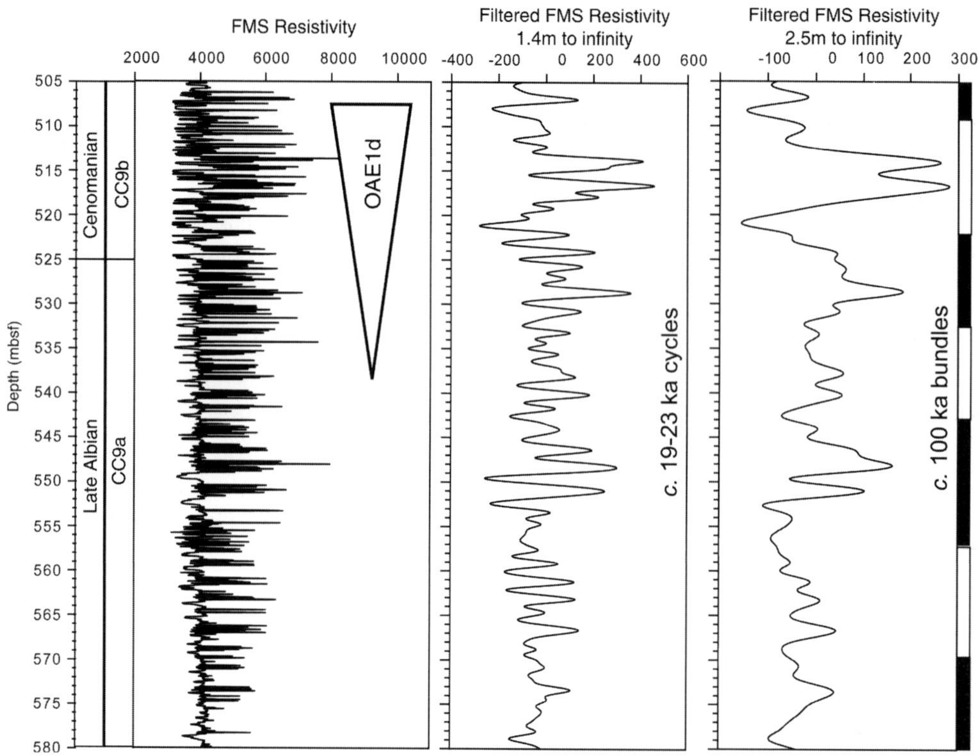

Fig. 5. High-resolution resistivity (FMS) log of ODP Hole 1052E (modified after Kroon *et al.* 1999). The interval shown is characterized by black shale formation paced by orbital cycles. The black shales are characterized by resistivity maxima. Lower frequencies of the Milankovitch spectrum can be observed in the upper part of the record. The record has been filtered (two panels to the right) to emphasize the low-frequency cycles. We interpret the 10 m cycle to be an expression of the eccentricity cycle (100 ka) and the 2 m cycle to be the precessional band (~21 ka).

are prevalent in the interval around the Albian–Cenomanian boundary.

The ratios of the 10 and ~ 2 m cycles are about right to represent the 100 ka and *c.* 21 ka bands of orbital precession (Fig. 5). The higher-frequency cycles do not correspond neatly to orbital bands. We suggest that either our initial guess is wrong that the longer-wavelength cycles represent eccentricity or differential compaction around the limestone stringers has altered the relative spacing of cycles sufficiently to obscure the duration of the high-frequency cycles. However, if we assume that the low-frequency cycle is in the eccentricity band, we can estimate an average sedimentation rate of *c.* 9.0–10.0 cm ka^{-1}. The whole interval represented by the increasing cycle amplitude to the termination of the OAE (between *c.* 575 and 506 mbsf (metres below sea floor)) represents a little over a 750 ka interval of which the peak of OAE 1d lasts for *c.* 400 ka of earliest Cenomanian time.

Norris & Wilson (1998) have shown that the planktonic foraminifera found in the shale beds are extremely well preserved and record an original stable isotopic signature of the oceans across OAE 1d. They found that the subtropical North Atlantic had sea surface temperatures on average warmer than today at between 30 and 31 °C. Temperatures peaked at about the level (*c.* 510 mbsf) where the maximum amplitude of the FMS cycle is also recorded. Apparently the increase in intensity of the OAE was directly mirrored by a rise in surface temperatures. Temperature and the vertical thermal gradient both collapsed near the termination of the OAE, suggesting a fundamental change in ocean circulation and stratification near or at the end of OAE 1d.

The Cenomanian–Turonian boundary interval was investigated at Site 1050 at Blake Nose by Huber *et al.* (1999). This important interval is characterized by one of the major Cretaceous

Anoxic Events (OAE 2). Although the sequence at Site 1050 is not entirely complete, Huber *et al.* (1999) were able to measure stable isotopes of unaltered calcareous planktonic and benthic foraminifera across the Cenomanian–Turonian transition. One of the astonishing results is the massive warming of middle bathyal temperatures based on benthic oxygen isotope results. Bathyal temperatures were already rather high (15 °C) before the boundary but rose to about 19 °C within OAE 2. This deep-water warming does not seem to be mirrored by sea surface water temperatures and therefore (Huber *et al.* 1999) concluded that most heat during OAE 2 time, a super-greenhouse event, was transported via the deep ocean. The warming may have been responsible for the extinction of deeper-dwelling planktonic foraminiferal genera such as *Rotalipora.*

Mid-Maastrichtian extinctions and palaeoceanographic events

The calcareous nannofossil stratigraphy shows that the Maastrichtian sediments, although slumped in parts, are biostratigraphically complete (Self-Trail this volume). The mid-Maastrichtian interval is characterized by a series of important biological and palaeoceanographic events. Extinction of deep-sea inoceramid bivalves and tropical rudist bivalves occurred at the same time with geochemical shifts in deep-sea biogenic carbonates and a pronounced cooling of high-latitude surface waters. It is not clear how all these events are related. Blake Nose middle Maastrichtian sediments, although complicated in places by slumping and coring gaps, yield conclusive evidence in the form of stable isotopes from foraminifera and extinctions concerning the subtropical palaeoceanographic evolution of the area. In contrast to cooling at high-latitude sites, Blake Nose surface waters show a temperature rise of about 4 °C (MacLeod & Huber this volume). Blake Nose results highlight regional mid-Maastrichtian differences. The benthic foraminiferal oxygen isotopes do not show a marked shift in mid-Maastrichtian time in contrast to southern Ocean and Pacific stable isotope records. Therefore, MacLeod & Huber (this volume) excluded build up of ice during the course of Maastrichtian time as the force behind the chain of mid-Maastrichtian events, although the details of mid-Maastrichtian palaeoceanography and cause of extinctions remain enigmatic.

Cretaceous–Palaeogene boundary impact

The ODP Leg 171B recovered a conspicuous Cretaceous–Palaeogene (K–P) boundary interval at Blake Nose (Fig. 4). At the deepest Site 1049 three holes were drilled through the K–P interval. The boundary layer, mostly consisting of green spherules, was interpreted to be of impact origin and ranges in thickness from 6–17 cm in the three holes at Site 1049. The largely variable thickness suggests reworking of the ejecta material down slope after deposition. Martínez-Ruiz *et al.* (this volume *a* & *b*) have shown that the green spherules represent the diagenetically altered impact ejecta from Chicxulub. Martínez-Ruiz *et al.* (this volume *b*) described the now predictive sequence of meteorite debris and associated chemistry across the K–P boundary at Blake Nose: the impact-generated coarse debris or tektite bed is followed by fine-grained pelagic ooze of the earliest Danian period rich in iridium.

One of the remarkable features of the pre-impact sediments at Blake Nose is deformation and large-scale slope failures probably related to the seismic energy input from the Chicxulub impact, some of it clearly induced before the emplacement of the ejecta from the impact (Norris *et al.* 1998; Smit 1999; Klaus *et al.* 2000). Mass wasting as a response to the impact occurred at a large scale. Correlation between Blake Nose cores and seismic reflection data indicates that the K–P boundary immediately overlies seismic facies characteristic of mass wasting (Klaus *et al.* 2000). Sediments 1600 km away from the Chicxulub impact on the Bermuda Rise were 'shaken and stirred'. Norris *et al.* (this volume) found that the seismic reflector associated with mass wasting at the K–P boundary is found over nearly the entire western North Atlantic basin, suggesting that much of the eastern seaboard of North America had catastrophically failed during the K–P impact event. Norris *et al.* (unpubl. data) made a survey of rise and abyssal sediment cores off North America, Bermuda and Spain, and concluded that indeed mass wasting may well have disrupted pelagic sedimentation at the K–P boundary in all those places.

The biostratigraphy of the K–P interval at Blake Nose is typical of an Atlantic seaboard deep-sea K–P section (Norris *et al.* 1999). Ooze immediately below the spherule bed contains characteristic late Maastrichtian planktonic foraminifera and nannofossils. The ooze above the spherule bed contains abundant extremely small Palaeocene planktic foraminifers in addition to large Cretaceous foraminifera. Norris *et al.* (1999) argued that the large Cretaceous planktonic foraminifera found in Darian sediments have been reworked. Small specimens of the same Cretaceous species are hardly present in the post-impact ooze, which implies that some form of sorting has changed the usual size distribution of planktonic foraminifera, which tends to be dominated by small individuals in pelagic sediments. The important conclusion is that the impact ejecta exactly coincided with the biostratigraphic K–P boundary and the planktonic foraminiferal extinction was caused by the Chicxulub impact.

Late Palaeocene Thermal Maximum

Drilling at Blake Nose recovered a continuous sequence of the Palaeocene–Eocene transition at a relatively low-latitude site. The upper Palaeocene section at Site 1051 consists of greenish grey siliceous nannofossil chalk that exhibits a distinctive colour cycle between 23–29 cm wavelength in Palaeocene time and *c.* 1 m wavelength in latest Palaeocene and early Eocene time. A distinctive bed of angular chalk clasts, now deformed by compaction, occurs at about the level at which the increase in colour cycle wavelength is seen. Norris & Röhl (1999) found that the LPTM as defined by a *c.* 2–3‰ $\delta^{13}C$ excursion and the appearance of a distinctive 'excursion fauna' of planktic foraminifera occurs just above the chalk breccia. Furthermore, Katz *et al.* (1999) found the extinction of an assortment of cosmopolitan benthic foraminifera in site 1051 that are also known to have become extinct at the LPTM in other regions around the world. Hence, we are confident that we recovered a section typical of the LPTM on Blake Nose.

Bains *et al.* (1999), Katz *et al.* (1999) and Norris & Röhl (1999) have interpreted the chalk breccia horizon, and step-like features in the $\delta^{13}C$ anomaly as strong evidence that the LPTM was associated with rapid release of buried gas hydrates. Oxidation of released methane to CO_2 would have caused a rapid increase of this greenhouse gas in the atmosphere and thus resulted in warming of the planet. Bains *et al.* (1999) showed that the onset of the carbon isotope anomaly occurred in a series of three pronounced drops in $\delta^{13}C$ separated by plateaux (Fig. 6). They interpreted the steps and intervening plateaux in $\delta^{13}C$ as intervals of rapid methane outgassing separated by intervals where the rate of carbon burial roughly balanced its rate of release into the ocean and atmosphere. Katz *et al.* (1999) showed that the chalk clast breccia can be reasonably interpreted as a debris flow produced by methane hydrate destabilization and venting during the initial phases of the LPTM.

The continuous sequence of Blake Nose across the Late Palaeocene Thermal Maximum (LPTM) displays very pronounced cyclicity of the sediments shown in spectral reflectance records at Site 1051. The carbon isotope anomaly within C24r coincides with a major shift in lithology and cyclicity of various physical property records including sediment colour, carbonate content and magnetic susceptibility. Norris & Röhl (1999) used the cyclicity to provide the first astronomically calibrated date for the LPTM (*c.* 54.98 Ma) and a chronology for the event itself using the Milankovitch induced cyclostratigraphy at Site 1051. Kroon *et al.* (1999) showed that the LPTM may be part of a long-term climatic cycle with a wavelength of 2 Ma by using the downhole gamma-ray log (Fig. 7). The LPTM or carbon anomaly coincides exactly with one of the gamma-ray maxima, possibly showing increased influx of terrigenous clay or less dilution by relatively reduced carbonate content.

Dickens (this volume) reported the results of a numerical modelling study designed to evaluate the potential range for the mass of carbon released. Using the $\delta^{13}C$ record from Site 1051 as the target site, he used a simple box model to assess the implications of variations in the mass and/or isotopic composition of the primary carbon fluxes and reservoirs. Dickens (2000*b*) concluded that a massive release of 1800 Gt of methane hydrate explains best the carbon isotope excursion.

Palaeocene–Eocene climate variability

The Eocene sequence consists largely of green siliceous nannofossil ooze and chalk. The upper part of the sequence is typically light yellow, and the colour change to green sediments below is sharp. This colour change is diachronous across Blake Nose and probably relates to a diagenetic front produced by flushing the sediment with sea water. Planktonic foraminifers, radiolarians and calcareous nannofossils are well preserved throughout most of the middle Eocene sequence, but calcareous fossils are more overgrown in the lower middle Eocene and lower Eocene sequence. A distinct feature of the Eocene sequence is a high number of vitric ash layers that were found at each site. These ash layers can be correlated across and serve as anchor points within the highly cyclical (at the Milankovitch scale) sequences. One of the conspicuous dark upper Eocene layers at Site 1053 contains nickel-rich spinels, which indicates that this layer contains potentially extraterrestrial material as a consequence of the Chesapeake Bay impact (Smit, pers. comm.). The Palaeocene and lower Eocene sediments are relatively clay rich compared with the middle and upper Eocene deposits. The

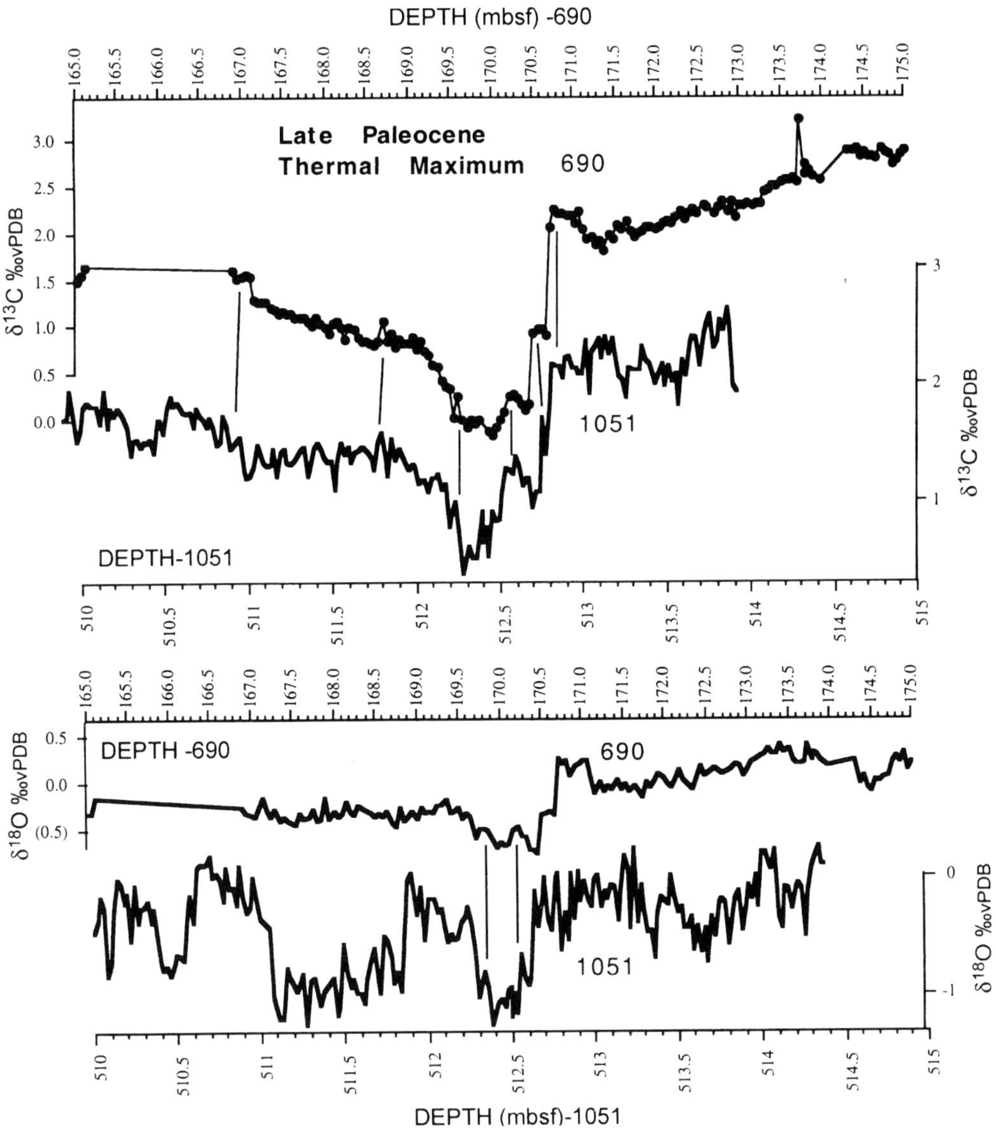

Fig. 6. High-resolution (*c.* 1–2 cm spacing) stable isotope records of bulk carbonate across the LPTM at ODP Site 1051 (Blake Nose, western North Atlantic) and ODP Site 690 (Southern Ocean) after Bains *et al.* (1999). Noteworthy features are the close correlation that can be achieved during the LPTM despite the considerable difference in location of the two sites, and the set of 'steps' leaving to the most negative $\delta^{13}C$ during the onset of the LPTM. These 'steps' suggest that ^{12}C was introduced into the biosphere in a series of rapid pulses separated by intervals of approximate balance between outgassing and carbon burial. Bains *et al.* (1999) proposed that the 'steps' reflect massive failure of methane hydrate reservoirs.

upper Palaeocene sequence contains chert or hard chalk, and preservation of most fossil groups is moderate to poor. The lower Palaeocene sediment is typically an olive green, clay-rich nannofossil chalk or ooze. Calcareous microfossils are typically very well preserved, whereas siliceous components are absent.

A number of excellent biostratigraphic studies have developed from the Blake Nose Palaeogene records. The preservation of the mid- and late Eocene calcareous fossils is very good and the preservation of the siliceous fossils is adequate to construct a detailed biostratigraphy, one of the first for late Palaeocene and early Eocene time.

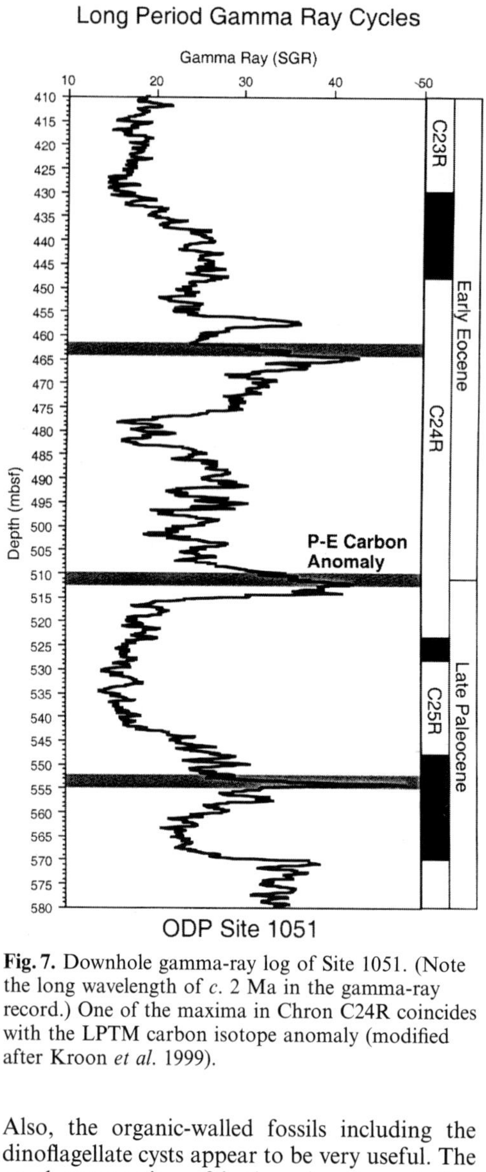

Fig. 7. Downhole gamma-ray log of Site 1051. (Note the long wavelength of *c.* 2 Ma in the gamma-ray record.) One of the maxima in Chron C24R coincides with the LPTM carbon isotope anomaly (modified after Kroon *et al.* 1999).

Also, the organic-walled fossils including the dinoflagellate cysts appear to be very useful. The good preservation of both carbonate and siliceous microfossils led to the first well-integrated biostratigraphy of mid-latitude faunas and floras (Sanfilippo & Blome this volume). The sedimentary sequence is unique in containing well-preserved radiolarian faunas of late early Palaeocene to late mid-Eocene age. Sanfilippo & Blome (this volume) have documented 200 radiolarian biohorizons. Middle to Upper Eocene sediments contain dinocyst assemblages characteristic of warm to temperate surface waters (van Mourik *et al.* this volume). The absence of known cold-water species shows there was no influence of a northern water mass in the area of Blake Nose and thus subtropical conditions prevailed throughout mid–late Eocene time.

One of the main objectives for drilling Blake Nose was to recover complete sequences of Palaeogene sediments by drilling multiple holes. The recovery of continuous sequences characterized by cyclical changes in lithology would help to establish the rates and timing of major changes in surface and deep-water hydrography. Röhl *et al.* (this volume) have presented a study on the use of Milankovitch forcing of sediment input, expressed in the periodicity of iron (Fe) concentrations, to calculate elapsed duration of the Danian stage. A new astronomical Danian time scale emerged from counting obliquity cycles at Sites 1050C and 1001A (Caribbean Sea), which surprisingly appeared to be the dominant frequency in the record. Röhl *et al.* elegantly made use of the Fe records observed in the cores and various downhole logs to fill in for drilling gaps. The results from both drilling sites were neatly similar. Analysis of the cyclostratigraphy both from the cores and downhole logs appears to be successful in the older parts of the stratigraphy and useful in calibrating magnetic polarity zones.

Wade *et al.* (this volume) have presented the first detailed benthic and planktonic foraminiferal oxygen isotope curves in combination with spectral reflectance records from late mid-Eocene time. Those workers found clear evidence of Milankovitch cyclicities at Site 1051 (Fig. 8; Wade *et al.* this volume). At Site 1051 in the top 30 m, the colour records document a dominant cyclicity with wavelengths of 1.0–1.4 cycles m^{-1} forced by precession (Fig. 8), whereas the stable isotope records demonstrate all frequencies of the Milankovitch spectrum. Changes in the $\delta^{18}O$ records of the surface-dwelling planktonic foraminifera are astonishingly large. Wade *et al.* (this volume) concluded that upwelling variations may be responsible for these large variations resulting in large sea surface temperature variations. Sloan & Huber (this volume) in a modelling study also found that changes in wind-driven upwelling and in continental runoff on a precessional time scale should be observed in regions of the central North Atlantic.

Another subject relates to the distribution of clay minerals. One of the most exciting aspects is that the origin of palygorskite clay minerals found in lower Eocene pelagic sediments in the western Central Atlantic may be authigenic (Pletsch this volume). Of particular interest and greater potential significance is the idea that authigenic palygorskite formation required the presence of relatively saline bottom waters, such as the elusive 'Warm Saline Bottom Water'

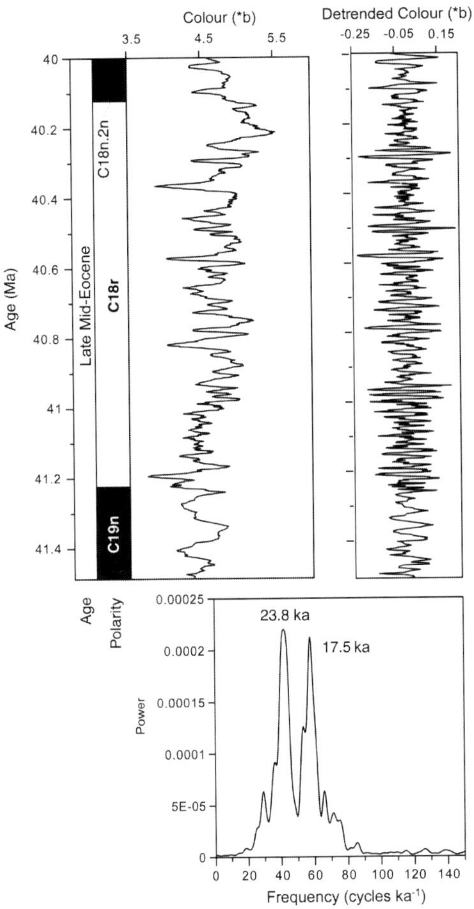

Fig. 8. Cyclostratigraphy for late mid-Eocene time from ODP Site 1051. The colour cycle has been assigned an age scale based on magnetic polarity datum points and detrended with a 50 point moving average (right panel) to draw out the high-frequency cycle periods. Spectral analysis of the resulting detrended record is shown below and suggests that the main cycles correspond approximately to the precession bands.

(WSBW) of Brass *et al.* (1982). As such the distribution of palygorskite could become a palaeo-watermass tracer of WSBW.

A Cretaceous–Palaeogene stable isotope record

Palaeoclimate records of the Mesozoic and Cenozoic eras display a long-term trend from extremely warm conditions during the mid-Cretaceous optimum to cooler climates of late Palaeogene and Neogene time. The Neogene climate record is reasonably well known on a time scale of *c.* 50–100 ka resolution or better, based on extensive surveys of benthic and planktonic stable isotope stratigraphies from all the major ocean basins save the Arctic. However, Palaeogene and Cretaceous climates are currently poorly resolved owing to the scarcity of high deposition rate sections and the frequent poor preservation of microfossils upon which various faunal and geochemical climate proxies are based. Consequently, most of our highly resolved records of Cretaceous and Palaeogene climate come from stable isotope analyses of bulk carbonates. The uncertain role of diagenesis in these isotopic data has made it difficult to draw firm conclusions about many aspects of ocean climate and circulation.

To improve the situation we have combined stable isotope records from Leg 171B sites with previously published datasets to illustrate long-term trends in $\delta^{18}O$ and $\delta^{13}C$ for the surface and deep ocean (Fig. 9). These data include records of planktonic foraminifera that record the most negative $\delta^{18}O$ as well as benthic species to show the vertical $\delta^{18}O$ and $\delta^{13}C$ gradients in the oceans. Currently, there are insufficient data to determine which of the features of these isotope records reflect global trends rather than regional trends for the interval before Campanian time. However, Campanian and younger trends have been duplicated in a number of localities suggesting that the basic patterns are of global significance. Trends in mid-Cretaceous records have been illustrated in several low- to moderate-resolution datasets from both deep-sea and outcrop sections (e.g. Clarke & Jenkyns 1998; Stoll & Schrag 2000). Our foraminifer isotopic records broadly mirror patterns in the bulk sediment stable isotope data. It is likely that parts of the Albian and Cenomanian records will ultimately be found to differ in amplitude and absolute values from deep Pacific or Indian Ocean records. The North Atlantic was partly a silled basin during Albian and Cenomanian time and was part of the Tethys seaway, suggesting that it may have been filled with intermediate waters or even surface waters from elsewhere in the world's oceans.

Cretaceous climatic optimum

Benthic $\delta^{18}O$ in early Albian time was broadly similar to that during the Cretaceous and early Palaeogene time. However, Atlantic intermediate waters (*c.* 1500 m water depth at ODP Site 1049) approached the same $\delta^{18}O$ values as planktonic foraminifera during early Albian time. The convergence of planktonic and benthic $\delta^{18}O$

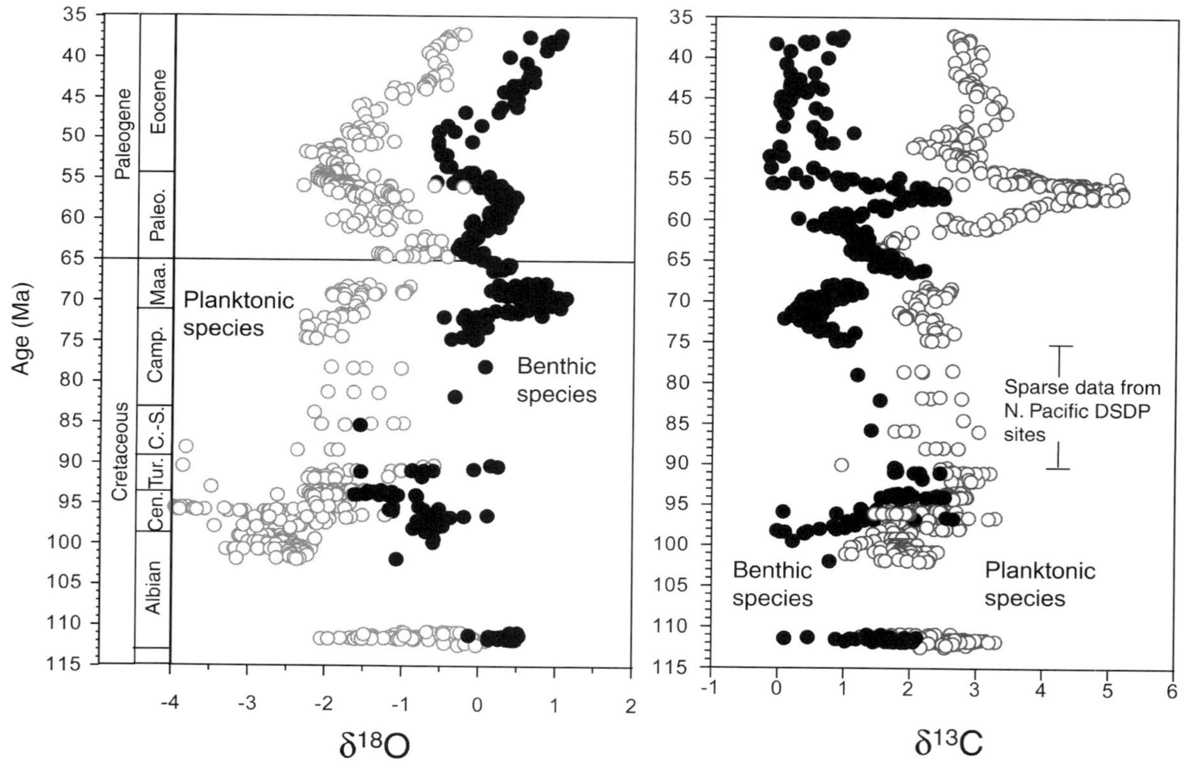

Fig. 9. Compilation of stable isotope records for planktonic foraminifera (○) and benthic foraminifera (●) showing long-term climate trends in the Atlantic and global ocean. Not all the trends shown are strictly global as some, such as the mid-Maastrichtian $\delta^{18}O$ increase in benthic foraminifera, are not observed in the Atlantic and others, such as the various trends in Albian and Cenomanian time may include features unique to the silled basins of the Atlantic. Early Albian data from ODP Hole 1049C (Blake Nose, western North Atlantic; Erbacher *et al.* 2000), late Albian–early Cenomanian data from ODP Site 1052 (Blake Nose, western North Atlantic; Norris and Wilson 1998), Cenomanian–early Turonian data from ODP Hole 1050C (Blake Nose, western North Atlantic; Huber *et al.* 1999, unpubl. data); latest Cenomanian data from DSDP Site 144 (equatorial western Atlantic; Norris *et al.* unpubl. data); Turonian–early Campanian datasets from various North Pacific DSDP sites (mostly DSDP Sites 463, 305 and 311 (Douglas & Savin 1973, 1975, 1978; Barrera 1994)); Maastrichtian and late Campanian data from DSDP Site 463 (equatorial Pacific; Barrera & Savin 1999); Palaeocene data from DSDP Site 384 (northwest Atlantic; Berggren & Norris 1997), and late Palaeocene and Eocene data from ODP Site 865 (north central Pacific; Bralower *et al.*, 1995).

records suggests that the North Atlantic basin was either filled with overturning surface waters or that the planktonic foraminifera chosen for analysis grew in subthermocline waters. We think it most likely that the occasional similarity in deep and surface $\delta^{18}O$ reflects variability in the intensity of surface stratification, as the $\delta^{18}O$ gradient swings between *c.* 0.5‰ and nearly 2‰ during OAE 1b (*c.* 112 Ma). Hence, the isotopic data suggest that the western North Atlantic was probably filled with moderately high-salinity waters not unlike the modern Mediterranean Basin that were occasionally more strongly stratified by runoff from the adjacent continents or by inflow of low-salinity waters from adjacent ocean basins such as the Arctic or the Pacific. Indeed, tectonic reconstructions show that the North Atlantic could have been hydrographically restricted from the deep Indo-Pacific by shallow sills across the Central American Seaway and the myriad of elevated plateaux and tectonic terranes in the Tethys Seaway (Hay *et al.* 1999).

Our isotopic data (Erbacher *et al.* 2000) suggest that the western North Atlantic behaved like the Plio-Pleistocene Mediterranean basins. There, the modern vertical temperature gradient can be as low as 3–4 °C and salinities of *c.* 36‰ increase $\delta^{18}O$ of planktonic and benthic foraminifera by as much as +1‰ over the open eastern North Atlantic (Miller *et al.* 1970). The potential for unusually high salinities in the North Atlantic during early Albian time makes it difficult to estimate absolute sea surface temperatures (SST). However, heavy $\delta^{18}O$ values from other records (Huber *et al.* 1995; Clarke & Jenkyns 1999) may suggest that temperatures during early Albian time were lower than during any other period in the mid-Cretaceous.

We have no foraminifer $\delta^{18}O$ data from mid-Albian time. Indeed, a survey of existing DSDP and ODP sites suggests that mid-Albian time is very poorly represented everywhere in the deep oceans, or where present, has little calcareous fossil material suitable for stable isotopic analysis. The situation changed by the late Albian, where isotopic results from ODP Hole 1052E show that both planktonic and benthic $\delta^{18}O$ were nearly the most negative ratios seen in the last 100 Ma. Only around the Cenomanian–Turonian boundary interval can lighter isotopic values be observed. The large vertical $\delta^{18}O$ gradients between planktonic and benthic foraminifera suggest that by late Albian time, the North Atlantic had ceased to be an extension of a silled Tethys Seaway with an estuarine circulation and had developed deeper marine connections to other deep ocean basins.

Norris & Wilson (1998) inferred that SST reached at least 30–31 °C during late Albian and earliest Cenomanian time (*c.* 98–102 Ma) in the western North Atlantic. These high temperatures were maintained or even raised further during parts of the Cenomanian. For example, planktonic foraminifera from DSDP 144 (9° N) in the equatorial western North Atlantic have average $\delta^{18}O$ values of −3.9‰, equivalent to temperatures of *c.* 32–34 °C using estimates from standard palaeotemperature equations, modern salinity, and assumption of an ice-free world. Notably, benthic $\delta^{18}O$ also peaks in upper Cenomanian time, reaching ratios over 1‰ more negative than that seen at any point in the deep oceans during the Cenozoic. Intermediate waters may have been unusually warm near the C–T boundary (Huber *et al.* 1999). Alternatively, tectonic barriers to exchange with other ocean basins may have become sufficiently restrictive that the North Atlantic was filled with thermocline waters or upper intermediate waters flowing in over shallow sills.

Our data (Huber *et al.* unpubl. data) from ODP Site 1050 illustrate that the mid-Cretaceous thermal optimum was not uniformly warm or stable. A pronounced increase in both planktonic and benthic $\delta^{18}O$ occurred in the mid-Cenomanian during the *Rotalipora reicheli* Zone (*c.* 97 Ma). Stoll & Schrag (2000) recorded a similar event in $\delta^{18}O$ of bulk limestone and marl in outcrop sections from Spain and Italy. They suggested on the basis of the magnitude of the event and its abrupt onset, that it records glaciation and a shift in whole ocean $\delta^{18}O$ as a result of ice build-up. A similar event is present in our data from the middle Cenomanian sequence in ODP 1050 and the $\delta^{18}O$ analyses of bulk carbonates from the Indian Ocean (Clarke & Jenkyns 1999) and Tethys (Stoll & Schrag 2000). Although the widespread occurrence of the $\delta^{18}O$ maxima suggests a global shift in $\delta^{18}O$, such as that produced by ice volume changes, intermediate water palaeotemperatures at this site and in the southern South Atlantic (Huber *et al.* 1995) were none the less higher (>11 °C) than would be expected if there were a significant volume of polar ice unless the North and South Atlantic basins were isolated from high-latitude sources of deep water.

The possible existence of large ice volume changes during the peak of mid-Cretaceous warmth raises questions about mechanisms that regulate ice growth and decay. Huber *et al.* (1995) showed that latitudinal thermal gradients were unusually low during intervals of the mid-Cretaceous thermal optimum. Therefore, it is hard to understand how ice growth could begin when high-latitude SSTs were nearly as high as equatorial temperatures. The abrupt $\delta^{18}O$ swings in Cenomanian time suggest that there may be threshold effects that can dramatically alter ocean circulation and temperature at the warm end of the climate spectrum, in much the same way that feedback systems such as moisture balance and runoff operate to abruptly change boundary conditions in Pleistocene time at the cool end of the climate spectrum.

Campanian–Maastrichtian refrigeration

We have limited stable isotope data for foraminifera from late Turonian to mid-Campanian time, an interval of *c.* 15 Ma. Data are contradictory for this interval. Huber *et al.* (1995) showed that DSDP Site 511 in the South Atlantic was bathed with warm waters from Cenomanian to Coniacian time and SST did not begin to fall appreciably until early Campanian time. Barrera (1994) presented a handful of measurements from North Pacific DSDP sites that also suggest peak temperatures were maintained until some time in the Santonian period. However, $\delta^{18}O$ data of bulk carbonates from Indian Ocean DSDP sites (Clarke & Jenkyns 1999) suggest that the decline in temperatures occurred during late Turonian time. The planktonic foraminifera show a tendency towards heavier isotopic values during the course of the Turonian as suggested by the dataset from ODP Site 1050 (Huber *et al.* unpubl. data), although some extreme light isotopic values can still be seen. Results from Site 1050 suggest that planktonic $\delta^{18}O$ was similar to or more positive than that in late Campanian time and suggest that the surface waters were significantly cooling during Turonian time, although highly variable. One could conclude from the planktonic isotope record that the mid-Cretaceous climatic optimum was over by *c.* 90 Ma if not earlier, although the benthic isotope data show that a persistent fall towards heavier values occurred in Campanian time. More benthic isotope data are needed to document the exact timing of the end of the mid-Cretaceous climatic optimum.

Barrera & Savin (1999) published Campanian and Maastrichtian data for planktonic and benthic $\delta^{18}O$ from a number of sites around the world. Most records show the long-term trend to more positive $\delta^{18}O$ in both surface and deep waters and an abrupt step in this trend about 71 Ma. Miller *et al.* (1999) interpreted the step increase in foraminifer $\delta^{18}O$ to reflect an increase in ice volume and glacioeustatic sea-level lowering. MacLeod & Huber (this volume) have shown that the positive $\delta^{18}O$ deflection in benthic foraminifera is not nearly so pronounced in benthic records from the North Atlantic. Their data suggest either that the $\delta^{18}O$ shift observed elsewhere is not a glacial step or that the North Atlantic record is biased by the introduction of an unusually warm deep water mass that overprints the glacial increase in $\delta^{18}O$. Notably, the size of the mid-Maastrichtian increase in $\delta^{18}O$ in planktonic foraminifera is about a third to half the amplitude of the change in benthic foraminifera even in the Pacific. Therefore, it seems likely that at least half of the mid-Maastrichtian $\delta^{18}O$ shift is due to cooling or an increase in salinity of deep waters.

Cretaceous–Palaeogene boundary and Danian climate

The Cretaceous–Palaeogene (K–P) boundary is preceded by a rise in global ocean $\delta^{13}C$ and bottom-water temperatures starting in mid-Maastrichtian time (*c.* 71–72 Ma) that culminates in early Danian time (Fig. 9). The overall rise in benthic $\delta^{13}C$ and deep-water temperatures probably reflects carbon burial and CO_2 sequestration (Zachos *et al.* 1989; Stott & Kennett 1990). Sea-level fall may have played a role in carbon burial through the formation of large coal swamps with the retreat of epicontinental seas in late Maastrichtian time. The deep-sea $\delta^{13}C$ record is also influenced by a general decrease in inter-basin $\delta^{13}C$ gradients near the end of Maastrichtian time that reduced the isotopic contrast between relatively young deep waters in the North Atlantic and relatively old deep waters in the Pacific and Indian Ocean (e.g. Corfield & Norris 1996, 1998; Barrera & Savin 1999). Frank & Arthur (1999) explained the reduction of interbasinal $\delta^{13}C$ gradients by suggesting that the opening of deep passages between the North and South Atlantic played a key role in ventilating the deep North Atlantic during mid-Maastrichtian time. None the less, the North Atlantic continued to maintain a distinctive young deep and intermediate watermass through the mid-Palaeocene (Corfield & Norris 1996, 1998).

Bottom temperatures stayed the same or rose from mid-Maastrichtian time to the K–P boundary, whereas surface water temperatures fell until the last 200 ka of the Cretaceous period. The general decline in SST may be related to the withdrawal of epicontinental seas (Frank & Arthur 1999) or to the sequestration of carbon and CO_2 during the rise in global $\delta^{13}C$ in late Maastrichtian time. The deep North Atlantic and deep waters elsewhere in the oceans display distinctly different $\delta^{18}O$ throughout late Maastrichtian time. The North Atlantic was consistently *c.* 2 °C warmer or less saline than the other deep basins, a contrast that was maintained through early Palaeocene time (e.g. Corfield & Norris 1996). Relatively high deep-water temperatures in the North Atlantic may reflect the presence of young deep waters conditioned by overflow from Tethyan basins much like the conditioning of modern North Atlantic deep water by Mediterranean outflow waters.

The early Danian period (64–65 Ma) is best known as a time of tremendous turnover in marine pelagic ecosystems following the Cretaceous–Palaeogene mass extinction (D'Hondt & Keller 1991; Gerstel *et al.* 1987; Jablonski & Raup 1995; Keller 1988; MacLeod 1993; Olsson *et al.* 1992; Smit 1982). The extinction eliminated *c.* 95% of planktonic foraminifer species and had a profound effect on other members of the plankton. The extent of the devastation of the pelagic ecosystem is reflected by the extended (*c.* 3 Ma) collapse of the carbon pump and the vertical $\delta^{13}C$ gradient in the oceans (e.g. D'Hondt *et al.* 1998).

The recovery was also associated with large-scale changes in marine climate and carbon-cycle dynamics. There is a general decrease in global ocean $\delta^{13}C$ that may reflect reduced productivity and carbon burial (Shackleton & Hall 1984). In addition, the vertical $\delta^{18}O$ gradient was greatly reduced for an interval of almost 3 Ma in early Danian time coincident with the reduction in the vertical $\delta^{13}C$ gradient. The declines in vertical $\delta^{18}O$ and $\delta^{13}C$ gradients are both likely to be partly related to reorganization of biotic communities brought on by the end-Cretaceous mass extinction. A reduction in the vertical $\delta^{18}O$ gradient is perhaps best explained by the widespread extinction of surface-dwelling species of planktonic foraminifera during the K–P mass extinction and a tendency for palaeoceanographers to analyse species that grew mostly in thermocline waters during early Danian time.

At the K–P boundary, the mass extinction resulted in a dramatic 1‰ decrease in surface ocean $\delta^{13}C$ with little or no change in deep-water $\delta^{13}C$. The absence of any large negative shift in $\delta^{13}C$ of deep waters strongly suggests that changes in the vertical $\delta^{13}C$ gradient are due to the extinction of surface ocean biota rather than a change in the global $\delta^{13}C$ reservoir (Hsü *et al.* 1982; Zachos & Arthur 1986; Zachos *et al.* 1989; D'Hondt *et al.* 1998). Some records show an inversion of the surface-to-deep $\delta^{13}C$ gradient (Hsü & McKenzie 1985; Zachos & Arthur 1986) that has been attributed to biomass burning (Ivany & Salawitch 1993) but could also reflect measurement artifacts due to low carbonate content in the boundary interval (e.g. Shackleton 1986).

Following the collapse of planktonic and benthic $\delta^{13}C$ gradients during the mass extinction, global $\delta^{13}C$ continues to rise. Maximum benthic $\delta^{13}C$ of *c.* 2.2‰ is recorded less than 100 ka above the K–P boundary and is succeeded by a long-term decline in $\delta^{13}C$ over the next 4 Ma. Hence, it appears that the process of carbon burial that led to the Maastrichtian rise in benthic $\delta^{13}C$ was reversed just after the K–P mass extinction and set in motion a long-term interval of unroofing previously deposited carbon. We suggest that the change from net carbon burial to net erosion may reflect the final draining of epicontinental seas and the onset of weathering of coal and organic shales deposited during late Cretaceous time. By the end of the Danian (*c.* 61 Ma) benthic $\delta^{13}C$ had returned to a ratio very similar to that achieved in mid-Maastrichtian and late Albian time, and very close to that later reached during the early Eocene.

Palaeocene–Eocene climate trends

One of the most striking features of the early Cenozoic and Cretaceous stable isotope records is the dramatic positive shift in $\delta^{13}C$ of both planktonic and benthic foraminifera in late Palaeocene time. Benthic foraminifer $\delta^{13}C$ increased by *c.* 2‰ and planktonic foraminifer $\delta^{13}C$ increased by *c.* 3‰ between 61 and 58 Ma. Part of the increase in $\delta^{13}C$ of planktonic foraminifera is probably due to the re-evolution of photosymbiosis after the K–P mass extinction (e.g. Corfield & Norris 1998). However, the increase in $\delta^{13}C$ of benthic species is seen throughout the oceans (e.g Miller *et al.* 1987; Corfield & Cartlidge 1992), suggesting that it represents a reservoir effect of burying large quantities of organic carbon. It is not at all clear where all this carbon was deposited, as there are few large oil or coal reservoirs of late Palaeocene age. The relatively rapid decline of $\delta^{13}C$ of both benthic and planktonic foraminifera between 58 and 55 Ma suggests that much of the carbon buried during Palaeocene time was exhumed by the end of the Palaeocene epoch. Beck *et al.* (1998) have suggested that much of the carbon deposited during late Palaeocene time may have accumulated in Tethyan basins that were subsequently uplifted and eroded during the initial stages of the Himalayan Orogeny.

The peak of the Palaeocene $\delta^{13}C$ increase coincides with the most positive benthic foraminiferal $\delta^{18}O$ ratios in the Palaeogene period. A modest increase in $\delta^{18}O$ of planktonic foraminifera is also present, suggesting cooling of both surface and deep waters during the late Palaeocene 'carbon isotope maximum'. We speculate that CO_2 drawdown associated with organic carbon burial may have been responsible for the fall in ocean temperatures. Evidence for snowmelt and cool interior climates in the Rocky Mountains (e.g. Dettman & Lohmann 2000) as well as palaeobotanical estimates of mean annual temperature (e.g. Wing 1998) suggest that cooling in late Palaeocene time occurred over the continental interiors as well as the oceans. The late Palaeocene cool phase was succeeded by a *c.* 3 Ma interval of increasingly warm conditions leading up to the Late Palaeocene Thermal Maximum (LPTM) at *c.* 55 Ma.

The LPTM occurred at the point where deep-water temperatures and SST had nearly reached the highest levels in Cenozoic time. Zachos & Dickens (1999) and Katz *et al.* (1999) have proposed that the LPTM occurred because deep-water temperatures exceeded a threshold level above which methane hydrates began to catastrophically destabilize and contribute to a runaway greenhouse effect. However, Bains *et al.* (1999) pointed out that the long-term warming trend in latest Palaeocene time is highly aliased and that detailed $\delta^{18}O$ records from both foraminifera and fine fraction carbonate display no significant warming trend for at least 200 ka before the LPTM. Hence, it is unclear whether the million-year scale drift to higher temperatures has anything to do with the LPTM. Other proposed mechanisms to initiate the LPTM include changes in deep-water circulation inspired by tectonics (Beck *et al.* 1998), volcanism (Eldholm & Thomas, 1993; Bralower *et al.* 1997*b*) or slope failure (Bains *et al.* 1999; Norris & Röhl 1999).

Dickens (1999) suggested that sedimentary carbon reservoirs act as 'capacitors' in the global carbon cycle that store and release greenhouse gases to modulate global climate. This theory supposes that methane dissociation events, as proposed for the LPTM, are common in the geological record (Dickens 2000*b*). Indeed, the example of the LPTM has spawned a resurgence of interest in the Cretaceous and Palaeogene climate record and has led to the discovery of other large perturbations in the carbon cycle. Large negative $\delta^{13}C$ anomalies have been identified with the onset of several Oceanic Anoxic Events (OAEs) in Cretaceous time (Jenkyns 1995; Wilson *et al.* 1999). Likewise, analysis of benthic foraminifer assemblages and isotopic anomalies provides strong hints that there may be other events like the LPTM in late early Eocene and mid-Palaeocene time (Thomas & Zachos 1999). Short, but intense, $\delta^{13}C$ anomalies are of great interest as 'natural experiments' in the biological and climatological effects of transient perturbations of the carbon cycle. The increasing evidence that there may be several LPTM-like events offers the exciting opportunity to compare and contrast these events to better evaluate the palaeoceanographic context, trigger, duration and biological effect of large-magnitude changes in carbon reservoirs of whatever cause. There is also the possibility that some of the $\delta^{13}C$ events represent large emissions of greenhouse gases and can provide natural analogues for modern greenhouse warming.

The early Eocene benthic $\delta^{18}O$ record shows that the Earth was essentially warmer than today. That is to say, the high-latitude oceans were much warmer than today, raising the global mean temperature (e.g. Zachos *et al.* 1994), and the Earth was characterized by the absence of large ice sheets. None the less, maximum sea surface temperatures in the tropics were not necessarily higher than modern temperatures. Most planktonic isotope records show that low-latitude early Eocene sea surface temperatures were lower than at present (Boersma *et al.* 1987; Zachos *et al.* 1994; Bralower *et al.* 1995). This is not well understood. Either heat diffused into the deep ocean or the stable isotope records of the planktonic foraminifera do not record the surface, or have been affected by diagenesis. Wade *et al.* (this volume) have shown that massive upwelling in late mid-Eocene time influenced low-latitude sea surface temperatures. Upwelling at Milankovitch periodicities may explain why sea surface temperature estimates were low in the low latitudes. Towards mid-Eocene time increases in $\delta^{18}O$ can be seen in both the planktonic and benthic foraminiferal records. These trends indicate cooling of the Earth (Kennett & Shackleton 1976) starting near the early Eocene–mid-Eocene transition and continuing in a series of steps to the Eocene–Oligocene boundary.

Conclusions

A number of outstanding post-cruise studies resulted from ODP Leg 171B drilling at Blake Nose in the North Atlantic concerning biostratigraphy, palaeoceanography–palaeoclimatology and meteorite impacts. Here, we have summarized the highlights of these results, which appear in this volume. In addition, we have combined the new planktonic and benthic foraminiferal stable isotope records of ODP Leg 171B with previously published datasets to elucidate the main features of the Cretaceous–Palaeogene oceans. The dataset is still sparse and additional data are needed to confirm the trends described here. Nevertheless, it represents the first long-term stable isotope record solely based on planktonic and benthic foraminifers and gives a unique overview of palaeoceanographic changes and their probable causes during the early stage of the evolution of the Atlantic Ocean.

References

BAINS, S., CORFIELD, R. M. & NORRIS, R. D. 1999. Mechanisms of climate warming at the end of the Palaeocene. *Science*, **285**, 724–727.

BARKER, C. E., PAWLEWICZ, M. & COBABE, E. A. 2000. Deposition of sedimentary organic matter in black shale facies indicated by the geochemistry and petrography of high-resolution samples, Black Nose, western North Atlantic. *This volume*.

BARRERA, E. 1994. Global environmental changes preceding the Cretaceous–Tertiary boundary: early–late Maastrichtian transition. *Geology*, **22**, 877–880.

BARRERA, E. & SAVIN, S. 1999. Evolution of late Maastrichtian marine climates and oceans. *In*: BARRERA, E. & JOHNSON, C. (eds) *Evolution of the Cretaceous Ocean–Climate System*. Geological Society of America Special Papers, **332**, 245–282.

BECK, R. A., SINHA, A., BURBANK, D. W., SERCOMBE, W. J. & KAHN, A. M. 1998. Climatic, oceanographic, and isotopic consequences of the Palaeocene India–Asia collision. *In*: AUBRY, M.-P., LUCAS, S. & BERGGREN, W. A. (eds) *Late Palaeocene–early Eocene Climatic and Biotic Events in the Marine and Terrestrial Records*. Columbia University Press, New York, 103–117.

BENSON, W. E., SHERIDAN, R. E. *et al.* (eds) 1978. *Initial Reports of the Deep Sea Drilling Project*, **44**. US Government Printing Office, Washington, DC.

BERGGREN, W. A. & NORRIS, R. D. 1997. Biostratigraphy, phylogeny and systematics of Palaeocene trochospiral planktonic foraminifera. *Micropalaeontology*, **43** (Supplement 1), 1–116.

BOERSMA, A., PREMOLI SILVA, I. & SHACKLETON, N. J. 1987. Atlantic Eocene planktonic foraminiferal palaeohydrographic indicators and stable isotope palaeoceanography. *Palaeoceanography*, **2**, 287–331.

BRALOWER, T. J., FULLAGAR, P. D., PAULL, C. K., DWYER, G. S. & LECKIE, R. M. 1997*a*. Mid-Cretaceous strontium isotope stratigraphy of deep-sea sections. *Geological Society of America Bulletin*, **109**, 1421–1442.

BRALOWER, T. J., THOMAS, D., ZACHOS, J. *et al.* 1997*b*. High resolution records of the late Palaeocene thermal maximum and circum-Caribbean volcanism: is there a causal link? *Geology*, **25**, 963–966.

BRALOWER, T. J., ZACHOS, J. C., THOMAS, E. *et al.* 1995. Late Palaeocene to Eocene palaeoceanography of the equatorial Pacific Ocean: stable isotopes recorded at Ocean Drilling Program Site 865, Allison Guyot. *Palaeoceanography*, **10**, 841–865.

BRASS, G. W., SOUTHAM, J. R. & PETERSON, W. H. 1982. Warm saline bottom water in the ancient ocean. *Nature*, **296**, 620–623.

CLARKE, L. J. & JENKYNS, H. G. 1999. New oxygen isotope evidence for long-term Cretaceous climatic change in the Southern Hemisphere. *Geology*, **27**, 699–702.

CORFIELD, R. M. & CARTLIDGE, J. E. 1992. Oceanographic and climatic implications of the Palaeocene carbon isotope maximum. *Terra Nova*, **4**, 443–455.

CORFIELD, R. M. & NORRIS, R. D. 1996. Deep water circulation in the Palaeocene Ocean. *In*: KNOX, R. W., CORFIELD, R. M. & DUNAY, R. E. (eds) *Correlation of the Early Palaeogene in Northwest Europe*. Geological Society, London, Special Publications, **101**, 443–456.

CORFIELD, R. M. & NORRIS, R. D. 1998. The oxygen and carbon isotopic context of the Palaeocene–Eocene Epoch boundary. *In*: AUBRY, M.-P., LUCAS, S. & BERGGREN, W. A. (eds) *Late Palaeocene–Early Eocene Climatic and Biotic Events in the Marine and Terrestrial Records*. Columbia University Press, New York, 124–137.

DETTMAN, D. L. & LOHMANN, K. C. 2000. Oxygen isotope evidence for high-altitude snow in the Laramide Rocky Mountains of North America during the Late Cretaceous and Palaeogene. *Geology*, **28**, 243–246.

D'HONDT, S. & KELLER, G. 1991. Some patterns of planktonic foraminiferal assemblage turnover at the Cretaceous–Tertiary boundary, *Marine Micropaleontology*, **17**, 77–118.

D'HONDT, S., DONAHAY, P., ZACHOS, J. C., LUTTENBERG, D. & LINDINGER, M. 1998. Organic carbon fluxes and ecological recovery from the Cretaceous–Tertiary mass extinction. *Science*, **282**, 276–279.

DICKENS, G. R. 1999. Back to the future. *Nature*, **401**, 752–755.

DICKENS, G. R. 2000*a*. Carbon addition and removal during the late Palaeocene Thermal Maximum: basic theory with a preliminary treatment of the isotope record at ODP Site 1051, Blake Nose. *This volume*.

DICKENS, G. R. 2000*b*. Methane oxidation during the late Palaeocene Thermal Maximum. *Bulletin de la Société Géologique de France* (in press).

DILLON, W. P. & POPENOE, P. 1988. The Blake Plateau Basin and Carolina Trough. *In*: SHERIDAN, R. E. & GROW, J. A. (eds) *The Atlantic Continental Margin, Volume I-2: The Geology of North America*. Geological Society of America, Boulder, CO, 291–328.

DILLON, W. P., PAULL, C. K. & GILBERT, L. S. 1985. History of the Atlantic continental margin off Florida: the Blake Plateau Basin. *In*: POAG, C. W. (ed.) *Geologic Evolution of the United States Atlantic Margin*. Van Nostrand Reinhold, New York, 189–215.

DOUGLAS, R. G. & SAVIN, S. M. 1973. Oxygen and carbon isotope analyses of Cretaceous and

Tertiary foraminifera from the central North Pacific. *In*: WINTERER, E. L., EWING, J. I. *et al.* (eds) *Initial Reports of the Deep Sea Drilling Project*, **17**. US Government Printing Office, Washington, DC, 591–605.

DOUGLAS, R. G. & SAVIN, S. M. 1975. Oxygen and carbon isotope analyses of Cretaceous and Tertiary foraminifera from Shatsky Rise and other sites in the North Pacific Ocean. *In*: LARSON, R. L., MOBERLY, R. *et al.* (eds) *Initial Reports of the Deep Sea Drilling Project*, **32**. US Government Printing Office, Washington, DC, 509–520.

DOUGLAS, R. G. & SAVIN, S. M. 1978. Oxygen isotopic evidence for the depth stratification of Tertiary and Cretaceous planktonic foraminifera. *Marine Micropalaeontology*, **3**, 175–196.

ELDHOLM, O. & THOMAS, E. 1993. Environmental impact of volcanic margin formation. *Earth and Planetary Science Letters*, **117**, 319–329.

ERBACHER, J., HEMLEBEN, C., HUBER, B. T. & MARKEY, M. 1999. Correlating environmental changes during Albian oceanic anoxic event 1B using benthic foraminiferal palaeoecology. *Marine Micropalaeontology*, **38**, 7–28.

ERBACHER, J., HUBER, B. T., NORRIS, R. D. & MARKEY, M. in press. Increased thermohaline stratification as a possible cause for a Cretaceous oceanic anoxic event. *Nature*.

FRANK, T. & ARTHUR, M. A. 1999. Tectonic forcings of Maastrichtian ocean–climate evolution. *Palaeoceanography*, **14**, 103–117.

GERSTEL, J., THUNELL, R. C. & EHRLICH, R. 1987. Danian faunal succession: planktonic foraminiferal response to a changing marine environment, *Geology*, **15**, 665–668.

GRADSTEIN, H. M., AGTERBERG, F. P., OGG, J. G., HARDENBOL, J., VAN VEEN, P., THIERRY, J. & HUANG, Z. 1995. A Triassic, Jurassic and Cretaceous time scale. *In*: BERGGREN, W. A., KENT, D. V., AUBRY, M.-P. & HARDENBOL, J. (eds) *Geochonology, Time Scales and Global Stratigraphic Correlation*. SEPM Special Publications, **54**, 95–126.

HAY, W. W., DECONTO, R., WOLD, C. N. *et al.* 1999. Alternative global Cretaceous palaeogeography. *In*: BARRERA, E. & JOHNSON, C. (eds) *Evolution of the Cretaceous Ocean–Climate System*. Geological Society of America Special Papers, **332**, 1–47.

HOLBOURN, A. & KUHNT, W. 2001. No extinctions during Oceanic Anoxic Event 1b: the Aptian–Albian benthic foraminiferal record of ODP Leg 171. *This volume*.

HSÜ, K. J. & MCKENZIE, J. A. 1985. 'Strangelove' ocean in the earliest Tertiary. *In*: SUNDQUIST, E. T. & BROECKER, W. S. (eds) *The Carbon Cycle and Atmospheric CO_2: Natural Variations Archean to Present*. American Geophysical Union, Washington, DC, 487–492.

HSÜ, K. J., MCKENZIE, J. A., WEISSERT, H. *et al.* 1982. Mass mortality and its environmental and evolutionary consequences. *Science*, **216**, 249–256.

HUBER, B. T., HODELL, D. A. & HAMILTON, C. P. 1995. Middle–Late Cretaceous climate of the southern high latitudes: stable isotopic evidence for minimal equator-to-pole thermal gradients. *Geological Society of America Bulletin*, **107**, 1164–1191.

HUBER, B. T., LECKIE, R. M., NORRIS, R. D., BRALOWER, T. J. & COBABE, E. 1999. Foraminiferal assemblage and stable isotope change across the Cenomanian–Turonian boundary in the subtropical North Atlantic. *Journal of Foraminiferal Research*, **29**(4), 392–417.

IVANY, L. C. & SALAWITCH, R. J. 1993. Carbon isotopic evidence for biomass burning at the K–T boundary. *Geology*, **21**, 487–490.

JABLONSKI, D. & RAUP, D. M. 1995. Selectivity of end-Cretaceous marine bivalve extinctions. *Science*, **268**, 389–391.

JENKYNS, H.C. 1995. Carbon-isotope stratigraphy and palaeoceanographic significance of the lower Cretaceous shallow-water carbonates of Resolution Guyot, Mid-Pacific Mountains. *In*: WINTERER, E. L., SAGER, W. W., FIRTH, J. V. & SINTON, J. M. (eds) *Proceedings of the Ocean Drilling Program Scientific Results*, **143**. Ocean Drilling Program, College Station, TX, 99–104.

KATZ, M. E., PAK, D. K., DICKENS, G. R. & MILLER, K. G. 1999. The source and fate of massive carbon input during the latest Palaeocene thermal maximum. *Science*, **286**, 1531–1533.

KELLER, G. 1988. Extinction, survivorship and evolution of planktonic foraminifera across the Cretaceous/Tertiary boundary at El Kef, Tunisia. *Marine Micropalaeontology*, **13**, 239–263.

KENNETT, J. P. & SHACKLETON, N. J. 1976. Oxygen isotope evidence for the development of the psychrosphere 38 Myr ago. *Nature*, **260**, 513–515.

KLAUS, A., NORRIS, R. D., KROON, D. & SMIT, J. 2000. Impact-induced K–T boundary mass wasting across the Blake Nose, western North Atlantic. *Geology*, **28**, 319–322.

KROON, D., NORRIS, R. D., KLAUS, A., ODP Leg 171B Scientific Party & 'extreme climate' working group 1999. Variability of extreme Cretaceous–Palaeogene climates: evidence from Blake Nose. *In*: ABRANTES, F. & MIX, A. (eds) *Reconstructing Ocean History: A Window into the Future*. Kluwer–Plenum, New York, 295–319.

MACLEOD, K. G. & HUBER, B. T. 2001. The Maastrichtian record at Blake Nose (western Atlantic) and implications for global palaeoceanographic and biotic changes. *This volume*.

MACLEOD, N. 1993. The Maastrichtian–Danian radiation of microperforate planktonic foraminifera following the K/T mass extinction event. *Marine Micropalaeontology*, **21**, 47–100.

MARTÍNEZ-RUIZ, F., ORTEGA-HUERTAS, M., KROON, D., SMIT, J., PALOMO-DELGADO, I. & ROCCHIA, R. 2001*a*. Geochemistry of the Cretaceous–Tertiary boundary at Blake Nose (ODP Leg 171B). *This volume*.

MARTÍNEZ-RUIZ, F., ORTEGA-HUERTAS, M., PALOMO-DELGADO, I. & SMIT, J. 2001*b*. K–T boundary spherules from Blake Nose (ODP Leg 171B) as a record of the Chicxulub ejecta deposits. *This volume.*

MERCONE, D., THOMSON, J., CROUDACE, I. W., SIANI, G., PATERNE, M. & TROELSTRA, S. 2000. Duration of S1, the most recent sapropel in the eastern Mediterranean Sea, as indicated by accelerator mass spectrometry radiocarbon and geochemical evidence. *Palaeoceanography*, **15**, 336–347.

MILLER, A. R., TCHERNIA, P. & CHARNOCK, H. 1970. *Mediterranean Sea Atlas of Temperature, Salinity, Oxygen Profiles and Data from Cruises of the R. V.* Atlantis *and R. V.* Chain, *Vol. III.* Woods Hole Oceanographic Institution, Woods Hole, MA.

MILLER, A. R., BARRERA, E., OLSSON, R. K., SUGARMAN, P. J. & SAVIN, S. M. 1999. Does ice drive early Maastrichtian eustacy? *Geology*, **27**, 783–786.

MILLER, A. R., FAIRBANKS, R. G. & MOUNTAIN, G. S. 1987. Tertiary oxygen isotope synthesis, sea level history and continental margin erosion. *Palaeoceanography*, **2**, 1–19.

NORRIS, R. D. & RÖHL, U. 1999. Carbon cycling and chronology of climate warming during the Palaeocene/Eocene transition. *Nature*, **401**, 775–778.

NORRIS, R. D. & WILSON, P. A. 1998. Low-latitude sea-surface temperatures for the mid-Cretaceous and the evolution of planktonic foraminifera. *Geology*, **26**, 823–826.

NORRIS, R. D., HUBER, B. T. & SELF TRAIL, J. 1999. Synchroneity of the K–T oceanic mass extinction and meteorite impact: Blake Nose, western North Atlantic. *Geology*, **27**, 419–422.

NORRIS, R. D., KLAUS, A. & KROON, D. 2000. Mid-Eocene deep water, the Late Palaeocene Thermal Maximum and continental slope mass wasting during the Cretaceous–Palaeogene impact. *This volume.*

NORRIS, R. D., KROON, D., KLAUS, A. *et al.* (eds) 1998. *Proceedings of the Ocean Drilling Program, Initial Reports*, **171B**, Ocean Drilling Program, College Station, TX.

OGG, J. G., RÖHL, U. & GEIB, T. 1999. Astronomical tuning of Aptian–Albian boundary interval: Oceanic Anoxic Event OAE 1b through lower Albian magnetic polarity subchron M-2r. *EOS Transactions, American Geophysical Union*, **80**, F491–492.

OLSSON, R. K., HEMLEBEN, C., BERGGREN, W. A. & LIU, C. 1992. Wall texture classification of planktonic foraminifera genera in the lower Danian. *Journal of Foraminiferal Research*, **22**, 195–213.

PLETSCH, T. 2001. Palaeoenvironmental implications of palygorskite clays in Eocene deep-water sediments from the western central Atlantic. *This volume.*

RÖHL, U., OGG, J. G., GEIB, T. L. & WEFER, G. 2001. Astronomical calibration of the Danian time scale. *This volume.*

SANFILIPPO, A. & BLOME, C. D. 2001. Biostratigraphic implications of mid-latitute Palaeocene–Eocene radiolarians fauna from Hole 1051A, ODP Leg 171B, Blake Nose, western North Atlantic. *This volume.*

SELF-TRAIL, J. M. 2001. Biostratigraphic subdivision and correlation of upper Maastrichtian sediments from the Atlantic Coastal Plain and Blake Nose, Western Atlantic. *This volume.*

SHACKLETON, N. J. 1986. Palaeogene stable isotope events. *Palaoegeography, Palaeoclimatology, Palaeoecology*, **57**, 91–102.

SHACKLETON, N. J. & HALL, M. A. 1984. Oxygen and carbon isotope data from Leg 74 sediments. *In*: MOORE, T. C., Jr, RABINOWITZ, P. D. *et al.* (eds) *Initial Reports Deep Sea Drilling Project*, **74**. US Government Printing Office, Washington, DC, 613–619.

SHIPLEY, T. H., BUFFLER, R. T. & WATKINS, J. S. 1978. Seismic stratigraphy and geologic history of Blake Plateau and adjacent western Atlantic continental margin. *AAPG Bulletin*, **62**, 792–812.

SLOAN, L. C. & HUBER, M. 2001. North Atlantic climate variability in early Palaeogene time: a climate modelling sensitivity study. *This volume.*

SMIT, J. 1982. Extinction and evolution of planktonic foraminifera after a major impact at the Cretaceous/Tertiary boundary. *In*: SLIVER, T. & SCHULTZ, P. H. (eds) *Geological implications of impacts of large asteroids and comets on the Earth.* Geological Society of America, Special Publications, **190**, 329–352.

SMIT, J. 1999. The global stratigraphy of the Cretaceous–Tertiary boundary impact ejecta. *Annual Review of Earth and Planetary Science*, **27**, 75–113.

STOLL, H. M. & SCHRAG, D. P. 2000. High-resolution stable isotope records from the Upper Cretaceous rocks of Italy and Spain: glacial episodes in a greenhouse planet? *Geological Society of America Bulletin*, **112**, 308–319.

STOTT, L. D. & KENNETT, J. P. 1989. New constraints on early Tertiary palaeoproductivity from carbon isotopes in foraminifera, *Nature*, **342**, 526–529.

STOTT, L. D. & KENNETT, J. P. 1990. The palaeoceanographic and palaeoclimatic signature of the Cretaceous/Palaeogene Boundary in the Antarctic: stable isotopic results from ODP Leg 113. *In*: BARKER, P. F., KENNETT, J. P. *et al.* (eds) *Proceedings of the Ocean Drilling Program, Scientific Results*, **113**. Ocean Drilling Program, College Station, TX, 829–848.

THOMAS, E. & ZACHOS, J. C. 1999. Isotopic, palaeontologic, and other evidence for multiple transient thermal maxima in the Palaeocene and Eocene. *EOS Transactions, American Geophysical Union*, **80**(46), F487.

TUCHOLKE, B. E. & VOGT, P. R. 1979. Western North Atlantic: sedimentary evolution and aspects of tectonic history. *In*: TUCHOLKE, B. E., VOGT, P. R. *et al.* (eds) *Initial Reports of the Deep Sea Drilling Project*, **43**. US Government Printing Office, Washington, DC, 791–825.

VAN MOURIK, C. A., BRINKHUIS, H. & WILLIAMS, G. L. 2001. Mid- to late Eocene organic-walled diniflagellate cysts from ODP Leg 171B, offshore Florida. *This volume.*

WADE, B. S., KROON, D. & NORRIS, R. D. 2001. Orbitally forced climate change in late mid-Eocene time at Blake Nose (Leg 171B): evidence from stable isotopes in foraminifera. *This volume.*

WILSON, P. W., NORRIS, R. D. & ERBACHER, J. 1999. Tropical surface temperature records and black shale deposition in the mid-Cretaceous western Atlantic (Blake Nose and Demerara Rise). *EOS Transactions, American Geophysical Union*, **80**(46), F488.

WING, S. 1998. Late Palaeocene–Early Eocene floral and climatic change in the Bighorn Basin, Wyoming. *In*: AUBRY, M.-P., LUCAS, S. & BERGGREN, W. A. (eds) *Late Palaeocene–Early Eocene Climatic and Biotic Events in the Marine and Terrestrial Records.* Columbia University Press, New York, 380–400.

ZACHOS, J. C. & ARTHUR, M. A. 1986. Palaeoceanography of the Cretaceous/Tertiary boundary event: inferences from stable isotopic and other data. *Palaeoceanography*, **1**, 5–26.

ZACHOS, J. C. & DICKENS, G. 1999. An assessment of the biogeochemical feedback response to the climatic and chemical perturbations of the LPTM. *Geologiska Föreningens i Stockholm Förhandlingar*, **122**, 188–189.

ZACHOS, J. C., ARTHUR, M. A. & DEAN, W. E. 1989. Geochemical evidence for suppression of pelagic marine productivity at the Cretaceous/Tertiary boundary, *Nature*, **337**, 61–64.

ZACHOS, J. C., STOTT, L. D. & LOHMANN, K. C. 1994. Evolution of early Cenozoic marine temperatures. *Palaeoceanography*, **9**, 353–387.

Mid-Eocene deep water, the Late Palaeocene Thermal Maximum and continental slope mass wasting during the Cretaceous–Palaeogene impact

R. D. NORRIS[1], A. KLAUS[2] & D. KROON[3]

[1]*MS-23, Woods Hole Oceanographic Institution, Woods Hole, MA 02543-1541, USA*

[2]*Ocean Drilling Program, 1000 Discovery Drive, Texas A&M University, College Station, TX 77845-9547, USA*

[3]*Department of Geology and Geophysics, University of Edinburgh, West Mains Road, Edinburgh EH9 3JW, UK*

Abstract: A series of widespread Maastrichtian and Palaeogene reflectors in the western North Atlantic have been interpreted to record episodes of vigorous bottom-water circulation produced by periodic flooding of the deep North Atlantic basins with southern source waters. In general, the ages of these reflectors have been poorly known with estimated ages spanning several million years. New seismic and core data from Ocean Drilling Program Leg 171B tightly constrain the ages of several of the most prominent reflectors and demonstrate that several of them are associated with geologically short-lived events associated with major palaeoclimatic, palaeoceanographic and evolutionary transitions. On Blake Nose in the western North Atlantic, Reflector A^c formed shortly after the close of the early Eocene warm period between 48 and 49 Ma. The reflector corresponds to an abrupt inception of vigorous deep-water circulation that winnowed foraminiferal sands at 2000–2500 m water depth and caused mass wasting into the deep basins of the Bermuda Rise. Reflector A^c is correlative with a sequence of unconformities present in nearly every part of the global ocean from the shallow shelf to the deep sea, suggesting that this time interval is associated with a global change in ocean circulation, including a major sea-level lowstand. The reflector and unconformities are roughly equivalent in age to glacial tillites on the Antarctic Peninsula, suggesting a link to an early phase of southern hemisphere glaciation. Another widespread reflector, A^b, has a late Palaeocene to earliest Eocene age on the Bermuda Rise. On Blake Nose, the equivalents of Reflector A^b consist of a stack of three closely spaced hiatuses ranging from early late Palaeocene (58.5–60.5 Ma) to latest Palaeocene (*c.* 55.5 Ma) age. The youngest of these hiatuses is associated with the carbon isotope excursion at the Late Palaeocene Thermal Maximum (LPTM), when there was a major reorganization of deep-water circulation and dramatic, transient warming of high latitudes. Bottom currents appear to have prevented the widespread deposition of sediments at water depths shallower than *c.* 2200 m from the LPTM until early mid-Eocene time. Erosion on Blake Nose was produced by a strengthened, southward-flowing deep western boundary current at the same time that a southern source watermass produced extensive erosion on the Bermuda Rise. We suggest that the increased flow of the deep western boundary current reflects a stronger outflow of warm intermediate waters shallower than 2000 m from Tethys. The combination of warmer intermediate waters and erosion along the margin may have helped to trigger slope failure of gas hydrate reservoirs around the North Atlantic margin and set the LPTM–greenhouse feedback system in motion. Reflector A* is correlative with highly deformed Maastrichtian sediments on Blake Nose and Maastrichtian chalk interbedded with red claystone on Bermuda Rise. Seismic and coring evidence from Blake Nose shows that the K–P boundary slumping was associated with the magnitude *c.* 11–13 Richter Scale earthquake generated by the Chicxulub impact event. The chalk sequence on the Bermuda Rise appears to represent the distal turbidites produced by slumping of the margin. Correlation of the chalk beds with Reflector A* shows that the mass wasting deposits are found over nearly the entire western North Atlantic basin. Apparently, much of the eastern seaboard of North America must have catastrophically failed during the K–P impact event, creating one of the largest submarine landslides on the face of the Earth.

From: KROON, D., NORRIS, R. D. & KLAUS, A. (eds). 2001 *Western North Atlantic Palaeogene and Cretaceous Palaeoceanography*. Geological Society, London, Special Publications, **183**, 23–48.

Ever since the development of seismic profiling for correlation of deep-sea deposits, it has been recognized that there are widespread, smooth reflectors in the deep North Atlantic basins. Ewing & Ewing (1963) designated one of the more prominent horizons Horizon 'A' and suggested that it represents a sequence of turbidites. Subsequently, various refinements of Atlantic acoustic stratigraphy were made, including recognition of a reflector that apparently represented sediments filling in lows over ocean crust (Horizon β), and a number of subdivisions of the 'A' reflector (Fig. 1; Ewing 1965; Ewing & Hollister 1972; Tucholke 1979; Mountain & Miller 1992). With the advent of deep-sea drilling and better seismic systems, it has been possible to provide general dates for many of the stronger reflectors and map their distributions (Fig. 2; Tucholke 1979, 1981; Pak & Miller 1992). For example, Mountain and Miller (1992) showed that Reflector A^u is of early Oligocene age (*c.* 32–33.7 Ma), Horizon A^b dates between the middle part of magnetochron C24r and C26r (×54.5–59 Ma), and Reflector A* is of late Maastrichtian age (*c.* 67–65 Ma) where the numerical ages have been adjusted to the time scale of Berggren *et al.* (1995). Hence, these reflectors date tantalizingly close to a suite of major palaeoceanographic events or 'critical boundaries', including the onset of major Antarctic glaciation near the Eocene–Oligocene boundary, the Late Palaeocene Thermal Maximum, and the Cretaceous–Palaeogene boundary (K–P boundary) bolide impact and mass extinction.

Ocean Drilling Program (ODP) Leg 171B drilled a transect of sites on the edge of the Blake Plateau to recover a record of Palaeogene and Cretaceous palaeoceanography in the western North Atlantic (Fig. 3). Cores, logs and reprocessed seismic lines from the Blake Plateau provide precise dates for the continental margin equivalents of several major reflectors present in the adjacent deep western North Atlantic and demonstrate that a number of these reflectors do correlate very closely with global critical boundaries. The Blake Plateau is an ideal place for such work, as most of the plateau surface is covered with manganese–phosphorite nodules and pavements that have preserved underlying sediments, often of great age, from substantial erosion (Dillon *et al.* 1985; Dillon & Popenoe 1988; Norris *et al.* 1998). The eastern edge of the Blake Plateau falls away to depths of more than 5000 m at the Blake Escarpment on slopes of >45°. A prominent exception is at the Blake Nose, a NE–SW-trending salient in the escarpment where a gentle ramp extends from about 1000 m depth to more than 2700 m before dropping off into the abyss (Fig. 4).

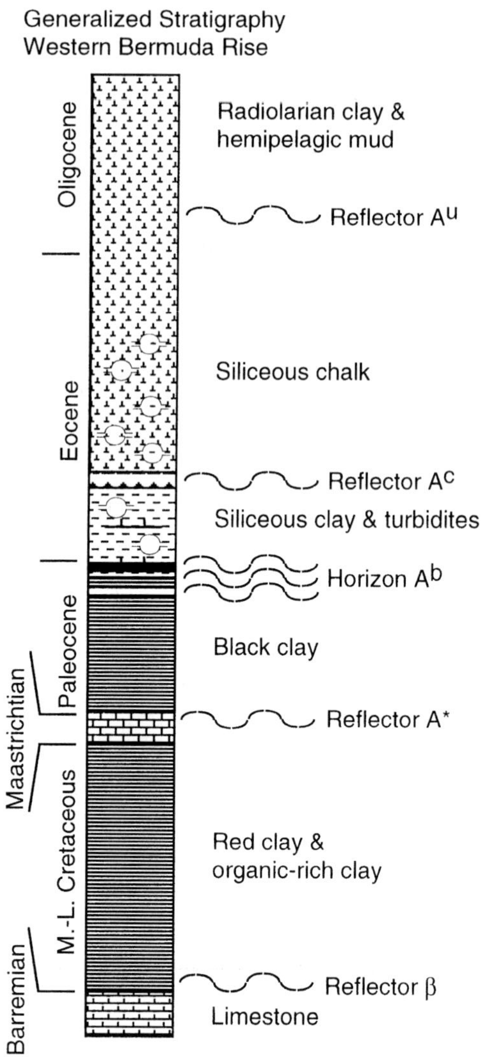

Fig. 1. Generalized stratigraphy of western North Atlantic Palaeogene and Cretaceous acoustic reflectors, Bermuda Rise (after Tucholke 1979; Mountain & Miller 1992).

The Blake Nose can be used to establish a detailed record of the western North Atlantic region, as it preserves deposits laid down at a wide variety of depths under different water-masses and sedimentation regimes. The armouring of the bottom by phosphate–manganese sand and nodules also has preserved an exquisite sequence of Eocene and older strata as ooze and soft chalk that can be used to produce a high-fidelity palaeoceanographic history of the area.

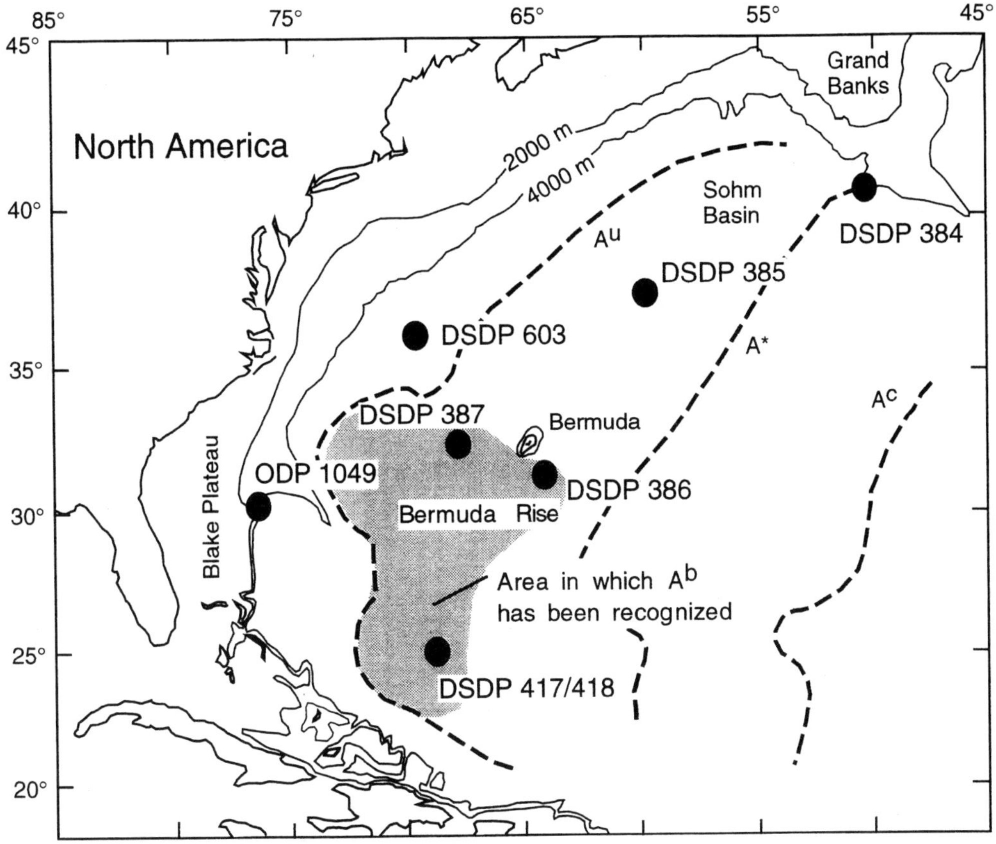

Fig. 2. Location of ODP and DSDP sites in the western North Atlantic that have been used to define the ages and palaeoceanographic context of major acoustic reflectors. Also shown is the distribution of the eastern limits of major Palaeogene and Cretaceous reflectors (after Tucholke 1979; Tucholke & Mountain 1979). The distribution of Reflector A^b (from a survey by Mountain & Miller 1992) is almost certainly larger than shown here, as it is present on Blake Nose (at Site 1049) and on the New Jersey Coastal Plain.

ODP Leg 171B drilled five sites on Blake Nose to document the depositional systems, stratigraphic development, and oceanographic controls on sedimentation in the Blake Plateau Basin (Figs 3 and 4; Norris *et al.* 1998). In this paper, we present the lithofacies and geological history of the region deduced from these cores, associated seismic records and well logs.

Previous work on the Blake Nose

The Blake Plateau has been crisscrossed by a dense grid of single-channel seismic reflection (SCS) and multichannel seismic reflection (MCS) profiles (Dillon & Popenoe 1988; Sheridan *et al.* 1988; EEZ-Scan 87 Scientific Staff 1991). These include surveys of the R.V. *Gloria Farnella* during the survey of the US Exclusive Economic Zone, the TD MCS lines shot by Teledyne under contract for the US Geological Survey (USGS), and cruises of the R.V. *Eastwood*, in the late 1970s and 1980s. The area just north of Blake Nose has been the focus of deep seismic refraction experiments to analyse the structure of the volcanic rifted margin. A recent summary of USGS seismic data has been published by Hutchinson *et al.* (1995).

Before ODP Leg 171B, the Blake Plateau and Blake Nose were the focus of a number of coring programmes. Some of the earliest are the Joint Oceanographic Institutions' Deep Earth Sampling Program (JOIDES) holes 1–6 that established a relatively detailed picture of the Cenozoic stratigraphy of the Blake Plateau Basin (Charm *et al.* 1969). One Atlantic Slope Project (ASP) site, ASP-3, was drilled by a consortium of

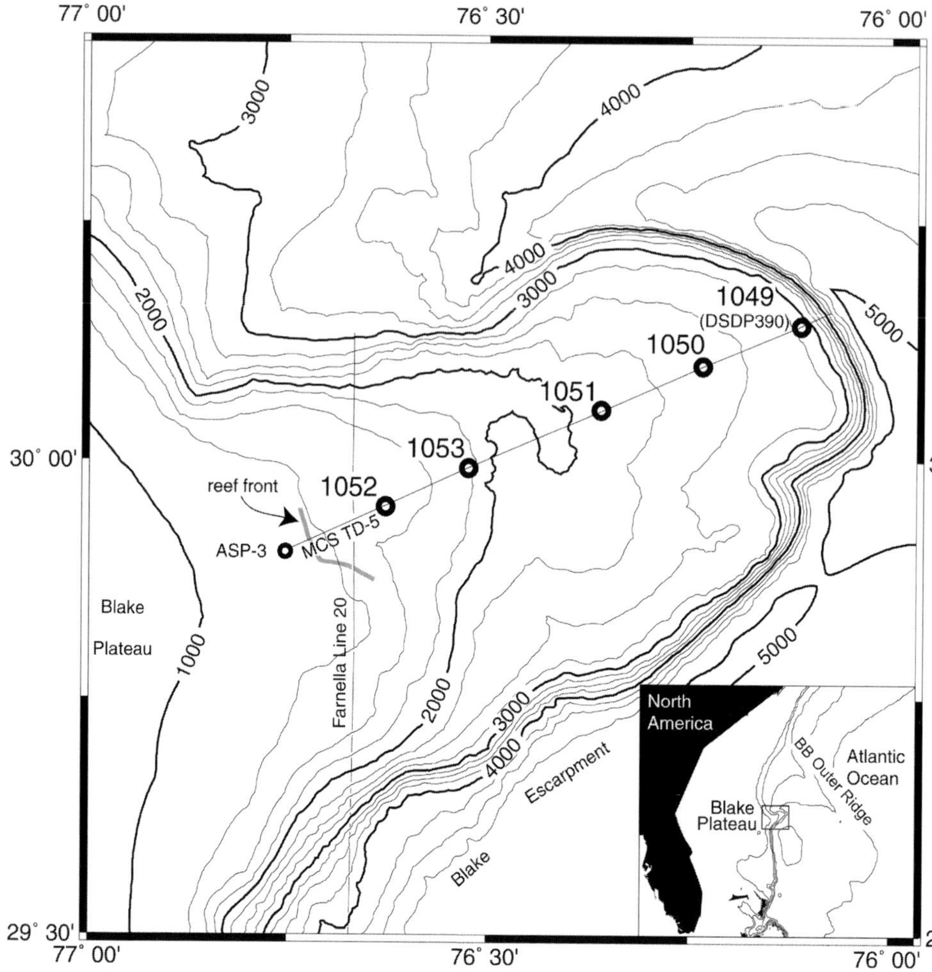

Fig. 3. Bathymetry of Blake Nose showing ODP Leg 171B drill locations and location of MCS seismic profile TD-5. ASP-3 and DSDP 390 are sites previously drilled on Blake Nose. Insert: location map of Blake Plateau, Blake Nose, study area (box), and Blake–Bahama Outer Ridge.

oil companies on the head of the Blake Nose (Fig. 3). Drilling by the Deep Sea Drilling Project (DSDP, Leg 44; Benson *et al.* 1978) provided a glimpse of the Palaeogene and Mesozoic history of the area. More recently, the ODP (Leg 171B; Norris *et al.* 1998) drilled on Blake Nose using hydraulic piston coring technology and managed to recover middle Cretaceous to Eocene strata to test earlier interpretations of the seismic records from the area. Additional information on the Cretaceous stratigraphy has come from the deep submergence research submarine *Alvin* and dredge hauls that have collected rocks from the Blake Escarpment (Mullins *et al.* 1982; Dillon & Popenoe 1988; EEZ-Scan 87 Scientific Staff 1991).

Seismic methods and description

The MCS line TD-5 (shot by Teledyne for the USGS) bisects the Blake Nose and is crossed by a series of R.V. *Gloria Farnella* SCS lines (EEZ-Scan 87 Scientific Staff 1991). The major reflectors and their interpretation of MCS line TD-5 have been published by Dillon *et al.* (1985), Dillon & Popenoe (1988) and Hutchinson *et al.* (1995). Those workers used results for DSDP Site 390 to interpret the updip structure. Unfortunately, DSDP Site 390 was drilled in a highly condensed sequence near the northeastern end of Blake Nose and it is not straightforward to trace the major reflectors updip of this site.

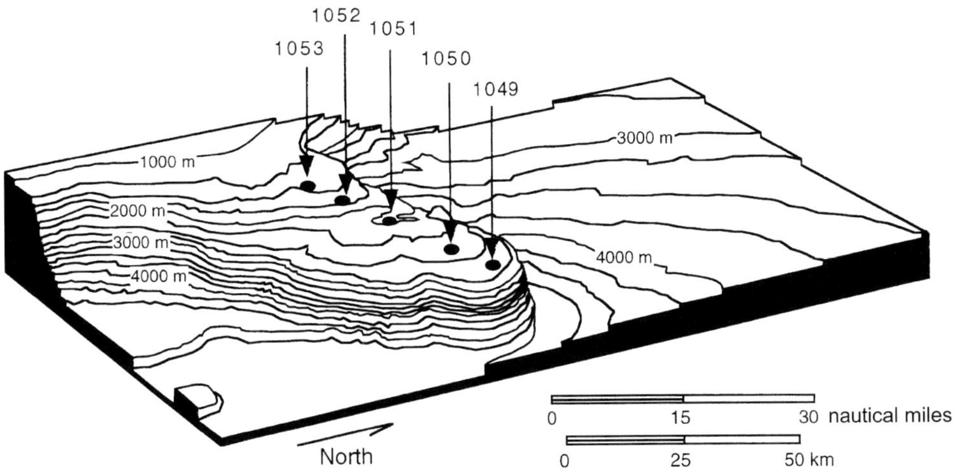

Fig. 4. Three-dimensional view of Blake Nose (looking NW toward the Carolina coast) illustrating the ramp-like profile of Blake Nose and the steep slope of the Blake Escarpment.

Previous versions of the TD-5 MCS profile were processed to enhance deeper structures by using filters that removed all data above 30 Hz (Hutchinson *et al.* 1995). Stratigraphic interpretations were substantially affected, as this filtering severely degraded the resolution of the profile. We completely reprocessed the TD-5 MCS profile from the original field data through migration to obtain the highest resolution record possible for the upper kilometre of the section (Fig. 5). This included retaining frequencies up to 100 Hz. Our reprocessing was successful in bringing out detailed seismic character and stratigraphic relationships that were not evident in the original version (Figs 5 and 6).

Synthetic seismograms were used to correlate drilling results with the seismic reflection data (Figs 7–9). The synthetic seismograms were calculated using velocity and density data from downhole log and shipboard laboratory measurements (Norris *et al.* 1998). Reflection coefficients were calculated every 0.5 m, which was the same spacing as for the downhole log data. Shipboard laboratory measurements of velocity and density obtained from core at all sites were used to fill in the upper 100–200 m below sea floor where no downhole log data exist. At one site (1051), no sonic logs were available, so only laboratory velocity and density measurements were used (Fig. 8). The laboratory data were interpolated and resampled at the same 0.5 m interval to fill in gaps in core recovery and other gaps in analyses. For the sites where we merged the laboratory data in the upper part of the hole with the log data, we shifted the laboratory values so that the laboratory curves matched the logging data. We constructed a source wavelet from the original seismic data by stacking the sea-floor reflection from at least five traces near each site where there were no sub-sea-floor reflections and the sea-floor topography was minimal. This wavelet was then convolved with the reflection coefficient series to produce the synthetic seismogram.

Palaeobathymetry

Shallow-water limestone of Barremian age was recovered at DSDP Site 390 on the seaward toe of Blake Nose and contains a foraminiferal fauna that suggests depths of deposition of <50 mbsl (metres below sea level) (Benson *et al.* 1978). Water depths increased substantially by late Albian time. Aptian–Albian pelagic oozes at the eastern end of the Blake Nose were probably deposited at >500 m water depth based on ratios of planktonic and benthic foraminifers (Benson *et al.* 1978). Indeed, if we assume that the reef was at sea level during Albian–Aptian time and that the depth gradient in the fore-reef deposits has not changed since that time, then the present difference in depth between the reef top and the seaward edge of the Blake Nose (*c.* 1500 m) probably reflects the water depth at which the oozes were deposited on Blake Nose during the Aptian–Albian interval. Upper Cretaceous and younger sediments were probably deposited somewhat below their present range of water depths (1100–2700 m; Benson *et al.* 1978) given the generally higher eustatic sea levels of those times compared with the present sea level. However, the relative depth difference

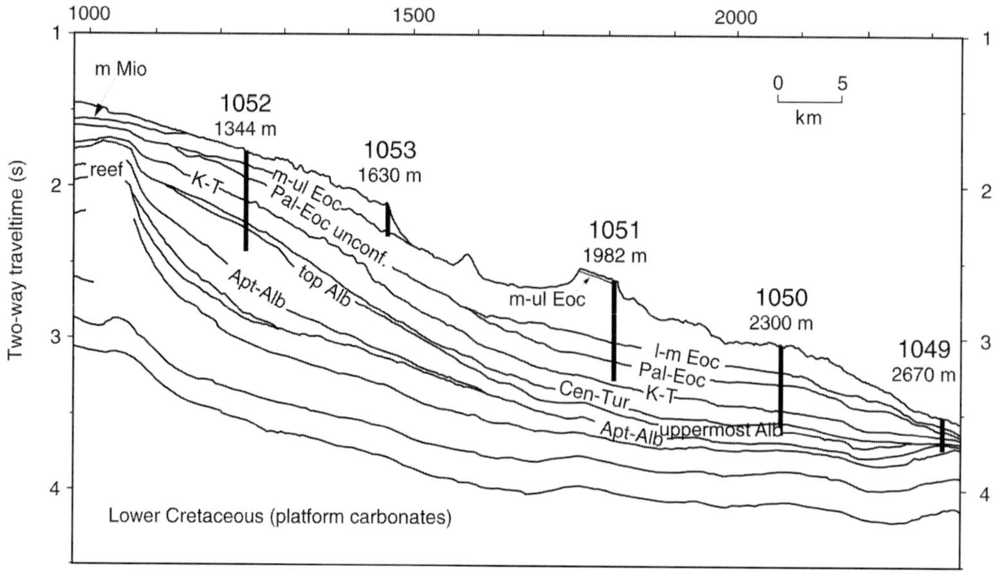

Fig. 6. Schematic interpretation of MCS line TD-5 showing Leg 171B drill sites.

between sites should have been nearly the same throughout deposition of the Cenozoic and Upper Cretaceous sections. Consequentially, we have unusually accurate estimates of palaeowater depth for all the sediments recovered by Leg 171B once adjustments for eustatic sea level and sediment compaction are taken into consideration.

Lithostratigraphy and seismic stratigraphy of Blake Nose

The sedimentary record at Blake Nose consists of Eocene carbonate ooze and chalk that overlie Palaeocene claystones as well as Maastrichtian and possibly upper Campanian chalk (Fig. 10). In turn, Campanian strata rest unconformably upon Albian to Cenomanian claystone and clayey chalk that appear to form a conformable sequence of clinoforms (Fig. 6). A short condensed section of Coniacian–Turonian nannofossil chalks, hardgrounds and debris beds is found between Campanian and Cenomanian rocks on the deeper part of Blake Nose. Aptian claystones are interbedded with Barremian periplatform debris, which shows that the periplatform material is reworked from older rocks. The entire middle Cretaceous and younger sequence rests on a Lower Cretaceous, and probably Jurassic, carbonate platform that is more than 5 km thick in the region of Blake Nose (Figs 5 and 6; Shipley *et al.* 1978; Dillon *et al.* 1985; Dillon & Popenoe 1988).

Seismic records show the presence of buried reef build-ups at the landward end of the Blake Nose (Figs 5 and 6). Fore-reef deposits and pelagic oozes, built seaward of the reef front, rest on relatively flat-lying Barremian shallow-water carbonates and serve largely to define the present bathymetric gradient along the Blake Nose (Benson *et al.* 1978; Dillon *et al.* 1985; Dillon & Popenoe 1988). SCS lines collected by the *Glomar Challenger* over DSDP Site 390 and our reprocessed version of MCS line TD-5 lines show that more than 800 m of strata are present between a series of clinoforms that overlap the reef complex and the sea bed (Fig. 6). ODP Leg 171B demonstrated that most of the clinoform sequence consists of Albian–Cenomanian strata and that a highly condensed sequence of Santonian–Campanian rocks is present in places between the lower Cenomanian and Maastrichtian sequences (Fig. 6; Norris *et al.* 1998). The Maastrichtian section is overlapped by a set of roughly parallel, continuous reflectors interpreted as being of Palaeocene and Eocene age that become discontinuous updip. Most of the Eocene section is incorporated in a major clinoform complex that reaches its greatest thickness downdip of the Cretaceous clinoforms.

A comparison of sedimentation rates among the five sites drilled during Leg 171B (Fig. 11) shows correlative unconformities that date to late mid-Eocene time (*c.* 37 Ma), the early Eocene–mid-Eocene transition (*c.* 49 Ma) and Campanian–Cenomanian timie (*c.* 76–91 Ma).

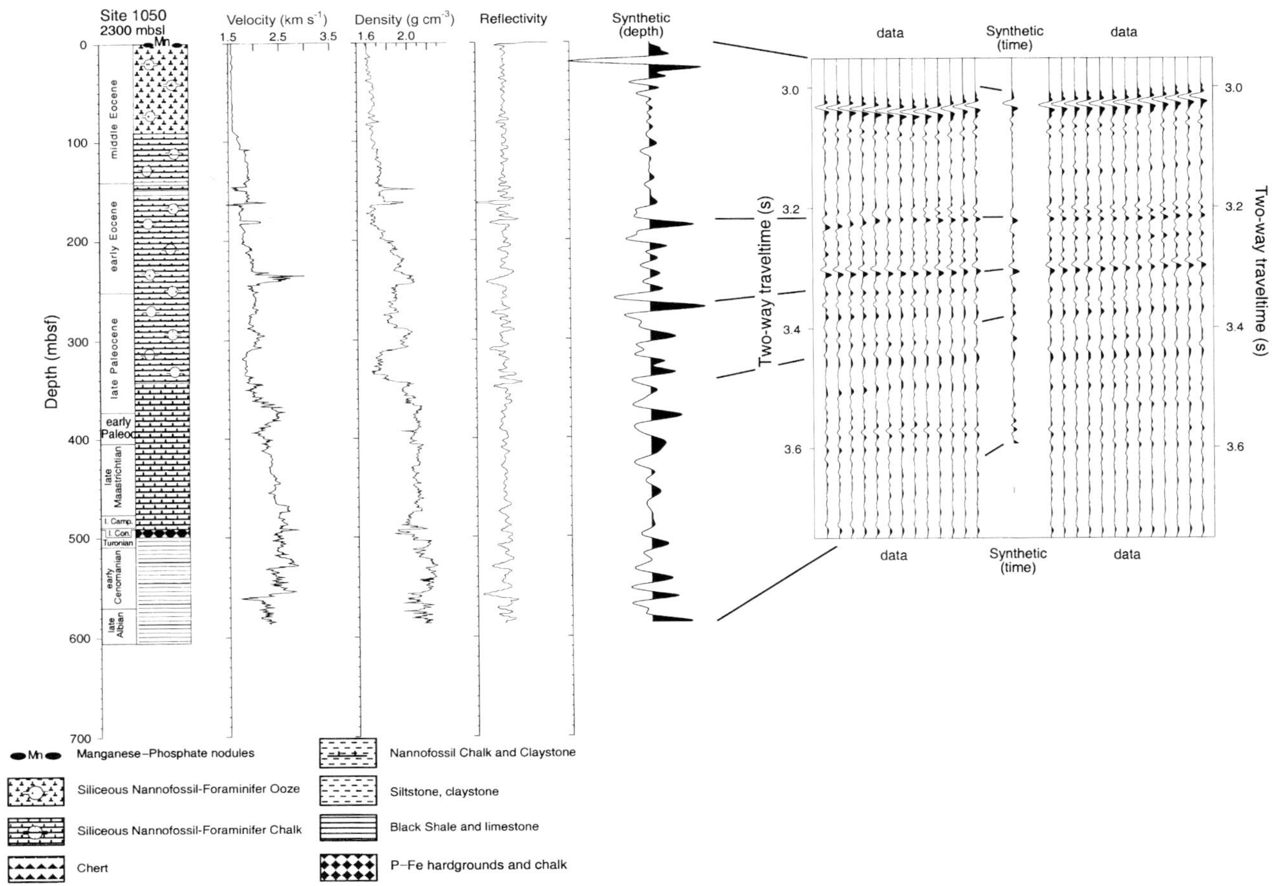

Fig. 7. Correlation of ages and lithologies with seismic reflection data at Site 1050. Downhole log data (velocity, density) were used to generate the synthetic seismogram except for above 109 m below sea floor (mbsf), where shipboard laboratory analyses were used. The synthetic data are displayed along with the part of the seismic data at the location of the drill site. The seismic data are shown in Fig. 5 (foldout). The correlations between depth below sea floor and seismic travel time are shown by the continuous lines.

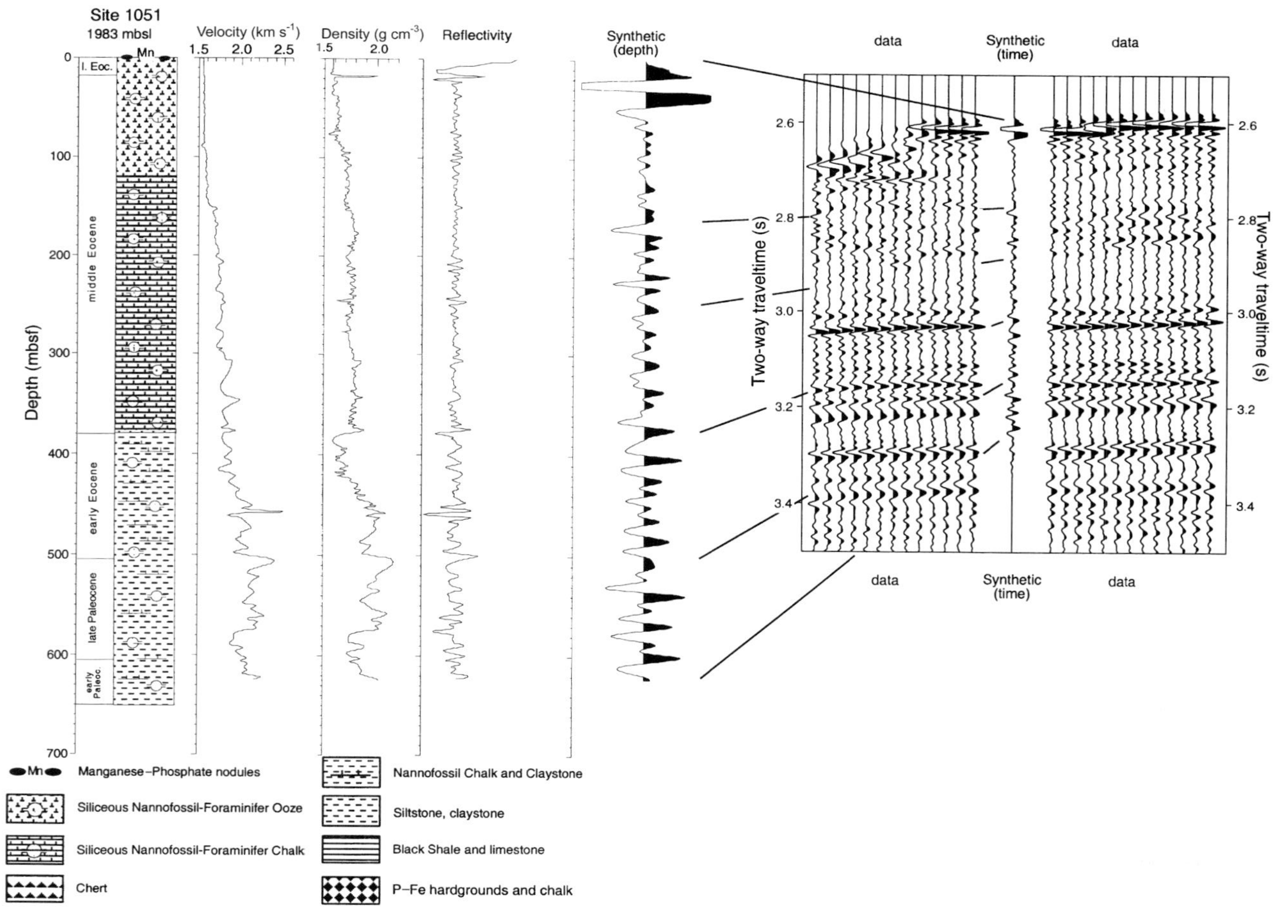

Fig. 8. Correlation of ages and lithologies with seismic reflection data at Site 1051. Shipboard laboratory measurements of velocity and density were used to generate the synthetic seismogram as no downhole velocity logs were obtained at this site. The synthetic data are displayed along with the part of the seismic data at the location of the drill site. The seismic data are shown in Fig. 5 (foldout). The correlations between depth below sea floor and seismic travel time are shown by the continuous lines.

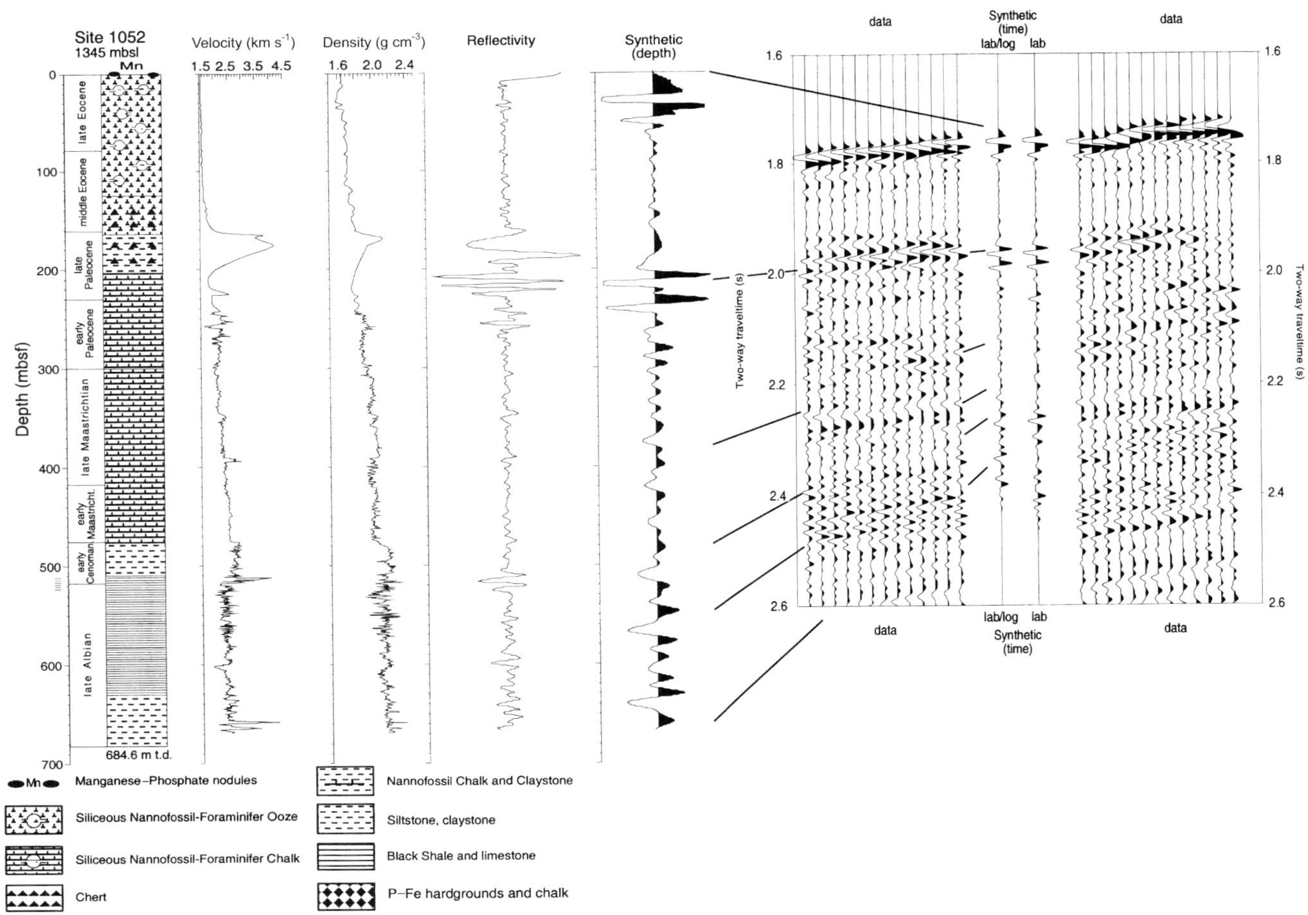

Fig. 9. Correlation of ages and lithologies with seismic reflection data at Site 1052. Downhole log data (velocity, density) were used to generate the synthetic seismogram except for above 240 mbsf, where shipboard laboratory analyses were used. The synthetic data are displayed along with the part of the seismic data at the location of the drill site. The seismic data are shown in Fig. 5 (foldout). The correlations between depth below sea floor and seismic travel time are shown by the continuous lines.

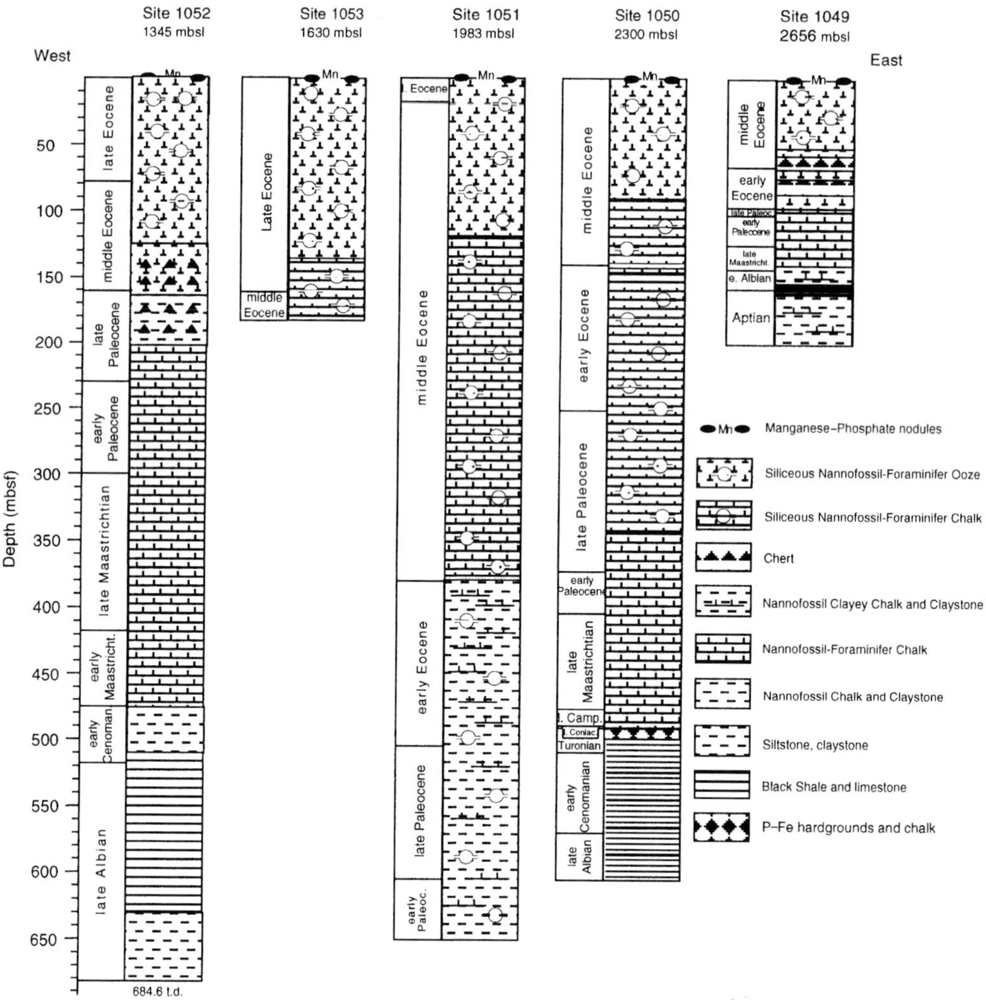

Fig. 10. Simplified lithostratigraphy of ODP Leg 171B sites (after Norris *et al.* 1998).

The uppermost Palaeocene section also contains a suite of unconformities with ages between 55.5 Ma and *c.* 60 Ma. Sedimentation rates were generally high during Palaeogene time, reaching rates of *c.* 5.6 cm ka^{-1} at Site 1052 (Norris *et al.* 1998). High rates of accumulation also characterize the upper Albian sequence. Both intervals of relatively high sedimentation rates occur during the main growth phases of large sediment sequences (Fig. 6).

Eocene sedimentation on Blake Nose

The middle to upper Eocene sequence is exposed on the sea floor across much of Blake Nose. The unconsolidated oozes are protected from erosion by a layer of manganese sand and nodules, which is up to about 3 m thick. The manganiferous sand contains abundant Pleistocene–Recent planktonic foraminifera as well as scattered bivalves and gooseneck barnacle plates (Norris *et al.* 1998). In the shallower parts of the Blake Nose, mixed assemblages of Oligocene to middle Miocene foraminifers are present in the manganiferous sand and an 80 m thick section of middle Miocene phosphatic marls has been reported from the ASP 3 well on the western end of Blake Nose (Poag, pers. comm.).

The Eocene sequence consists largely of green siliceous nannofossil ooze and chalk. Vitric ash beds are present through much of the middle and upper Eocene sequence. The thicker ash beds, up to 3–4 cm thick, serve as excellent correlation markers. Elevated silica concentrations in the

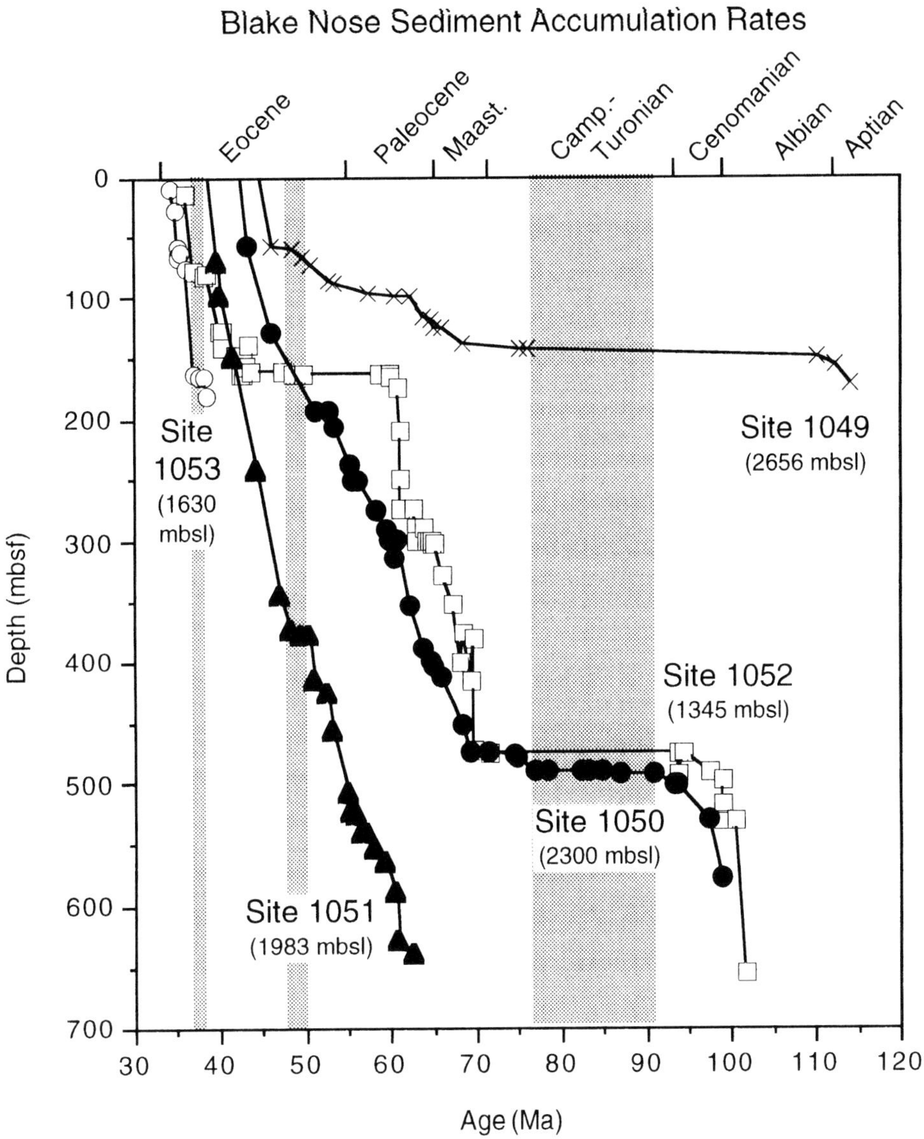

Fig. 11. Age–depth plot for Leg 171B sites. Vertically shaded bars drawn through major hiatuses to illustrate the minimum duration of unconformities on Blake Nose. The highest sedimentation rates are near the middle of Blake Nose in Palaeogene and latest Cretaceous time corresponding to the centre of the Palaeogene sedimentary sequence.

pore waters are consistent with significant alteration of biogenic and volcaniclastic siliceous sediments, particularly in the lower Eocene and Palaeocene sequence (Norris *et al.* 1998). Excellent preservation of radiolarians around ash layers, especially at Site 1051, may indicate that the volcaniclastic deposits are the more important of these two silica sources in the Blake Nose area (Norris *et al.* 1998). Apparently, the Blake Plateau was downwind of a major, long-lived volcanic centre for most of Palaeocene and Eocene time.

A major unconformity in the Palaeocene–Eocene sequence formed between 48 and 49 Ma near the early Eocene–mid-Eocene transition and corresponds to a set of reflectors that can be traced across the entire Blake Nose. The hiatus is represented by laminated to cross-stratified,

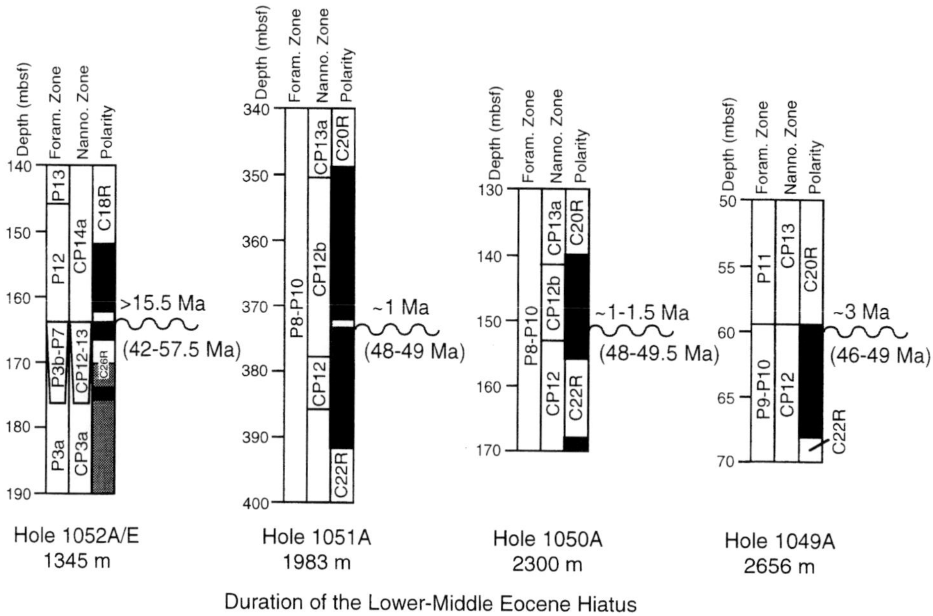

Fig. 12. Stratigraphy and duration of the lower–middle Eocene hiatus on Blake Nose based on planktonic foraminifer and calcareous nannofossil biozones and magnetostratigraphy. Estimated duration of the hiatus given to the right of each column along with the approximate beginning and end points of the hiatus. (Note that the hiatus is both longer and typically younger at the updip and downdip ends of the Blake Nose and is minimized near the centre of the drilling transect.)

silicified foraminifer grainstones, and chert interbedded with dark green claystones that may represent altered volcanic ash. The unconformity is confined to Chronozone C21r at Site 1051 (380 mbsf (metres below sea floor); 3.05 s, Fig 8) suggesting a hiatus between *c.* 48 Ma and 49 Ma (Fig. 12). In the extreme updip and downdip parts of Blake Nose, all of the lower Eocene sequence has been removed by this erosional event, which also cut into the upper Palaeocene sediments (e.g. Site 1052 *c.* 160 mbsf; 1.95 s (Fig. 9) and Site 1049 (Fig. 12)). The hiatus cuts out the top of the Palaeocene units to the lower middle Eocene sequence on the upper part of Blake Nose, where foraminiferal packstones and nannofossil claystones contain highly mixed assemblages within a stratigraphic interval only about 5 m thick.

Below the lower Eocene–middle Eocene unconformity, lower Eocene sediments lap onto an upper Palaeocene–lower Eocene surface between shot points 1600 and 1720 on MCS line TD-5 (Fig. 4). The hiatus is either absent or very short near the centre of the clinoform stack where the Palaeocene–Eocene transition is biostratigraphically complete (Sites 1050 and 1051). The onlapping nature of the reflectors suggests that there was an erosional event dated close to the Palaeocene–Eocene boundary. Sedimentation shifted from a gentle drape to more localized deposition in the middle of Blake Nose, with the thickest deposits accumulating near Site 1051.

The Late Palaeocene Thermal Maximum

Leg 171B recovered an apparently complete, or nearly complete, upper Palaeocene carbonate sequence that should help resolve many of the issues concerning the biochronology and geochemistry of this period. The Palaeocene–Eocene (P–E) transition includes a record of the carbon isotope event and benthic foraminiferal extinction during the Late Palaeocene Thermal Maximum (LPTM; Katz *et al.* 1999; Norris & Röhl, 1999). The LPTM is recognized at Site 1051 by a *c.* 2‰ decrease in the $\delta^{13}C$ of bulk carbonate, a *c.* 3‰ decrease in the $\delta^{13}C$ of planktonic foraminifera, and the presence of the 'excursion fauna' of planktonic foraminifera, including species such as *Acarininia africana* and *Morozovella allisonensis*, which are closely associated with the LPTM in the Pacific and Tethys (Norris & Röhl, 1999). Katz *et al.* (1999) have shown that the LPTM at Site 1051 is also associated with a *c.* 3‰ decrease in benthic foraminifer $\delta^{13}C$ and the last occurrence of several species of benthic foraminifera, as well as a distinctive chalk breccia.

At both Sites 1050 and 1051, siliceous nannofossil chalks are slightly better cemented near the LPTM and show an increase in seismic velocity and sediment density that produces a couplet of three reflectors: two strong reflectors with a weaker reflector between them. The uppermost reflector in this set occurs at *c*. 512 mbsf in Site 1051 (3.02 s, Fig. 8) and at *c*. 230 mbsf in Site 1050 (3.27 s, Fig. 7) and corresponds to the LPTM (*c*. 55.5 Ma), whereas the lowermost reflector of the three is correlated with a gradational increase in clay content near the CP3–CP4 (=NP4–5) zonal boundary at *c*. 59.7 Ma. The middle reflector correlates with an unconformity that has a early late Palaeocene age (Biozone CP6 = NP7–8; *c*.56.9–57.5 Ma). The upper Palaeocene sequence is biostratigraphically complete at Site 1051 save for an erosion surface in lower Chronozone C25r where Calcareous Nannofossil Biozone CP6 is apparently absent.

The LPTM itself occurs in a sequence of greenish grey siliceous nannofossil chalk that displays distinct cyclicity on a 23–25 cm scale in sediment colour, magnetic susceptibility, and resistivity, among other things. The regular cyclicity is broken by a 20 cm thick layer of angular chalk clasts with a chalk matrix. The chalk clasts are composed of lithologies very much like the underlying sediment and contain rare specimens of the upper Palaeocene planktonic foraminifer, *Globanomalina pseudomenardii*, the marker species for foraminiferal Zone P4 (*c*. 55.9–59.2 Ma). The sediment directly above the chalk clast horizon is faintly laminated over an interval about 30 cm thick and passes upward into dark green–grey, siliceous nannofossil chalk. The carbon isotope anomaly and benthic foraminifer extinction level marking the start of the LPTM occur immediately above the chalk clast horizon. Hence the LPTM and the chalk clast bed are both coeval with the angular unconformity marked by an acoustic reflector.

Cretaceous–Palaeogene (K–P) boundary on Blake Nose

A biostratigraphically complete K–P boundary interval was recovered at Site 1049 (Fig. 13), and partial K–P boundary sections were recovered in Hole 1052E and Hole 1050C. The lowest bed associated with the boundary at Site 1049 is a faintly laminated layer consisting largely of green spherules that range in size from 1 to 3 mm. The spherules are accompanied by clasts of limestone, dolomite and chalk, and bits of chert, mica books and schist, and have been interpreted as impact ejecta (Norris *et al.* 1997, 1999; Smit, 1999). The spherulitic layer is capped by a 3 mm orange limonitic layer that contains flat goethite concretions. The limonitic layer is overlain by 3–7 cm of dark, burrow mottled clay which contains abundant quartz, limestone chips and large Cretaceous planktonic foraminifera as well as occasional specimens of *Parvularugoglobigerina eugubina*, a planktonic foraminifer characteristic of early Cenozoic time (Norris *et al.* 1997, 1999). The final bed in the sequence is a 5–15 cm thick, white foraminiferal–nannofossil ooze that contains a Biozone Pα foraminiferal assemblage and fossils typical of calcareous nannofossil Biozone CP1a.

Maastrichtian sections at Sites 1049, 1050, and 1052 are disturbed by slumping (Fig. 14). The Maastrichtian ooze at Site 1049 is 16.8 m thick and is pervasively deformed. Bedding is inclined at an average of *c*. 15° and exhibits extensive microfaulting and pull-apart structures. Relatively little of the Maastrichtian section has been lost, as the sequence is biostratigraphically complete and in the correct stratigraphic order. Maastrichtian sediments are *c*. 89 m thick at Site 1050, of which the lower 30 m display spectacular recumbent folds in both the core and FMS logs. In contrast, it is the upper part of the Maastrichtian section that is most intensely deformed at Site 1052, where the entire upper Cretaceous section is *c*. 170 m thick. Even in the relatively undeformed section of the Maastrichtian sequence at Site 1052 bedding is tilted at *c*. 20° from horizontal and the chalk contains high-angle faults with slickensided surfaces. Deformation typical of the Maastrichtian section is largely absent from the overlying Palaeogene sequence.

The K–P boundary forms a strong reflector at Site 1050 (405 mbsf; 3.46 s, Fig. 7) that is easily traced updip to about shotpoint 1600 on MCS line TD-5 (Fig. 5). At Site 1052 (Fig. 9) and below Site 1053, the K–P reflector is harder to trace but appears to follow an undulating surface. Reflectors within the Maastrichtian section are also difficult to trace laterally and often dip at much higher angles than reflectors in the underlying Albian section or the overlying Palaeocene section (Fig. 5).

The Maastrichtian sequence unconformably overlies white ooze that contains nannofossil and planktonic foraminifera characteristic of the upper Campanian sequence at Site 1049. A somewhat thicker sequence of upper Campanian strata are present at Site 1050 and they overlie a highly condensed section of

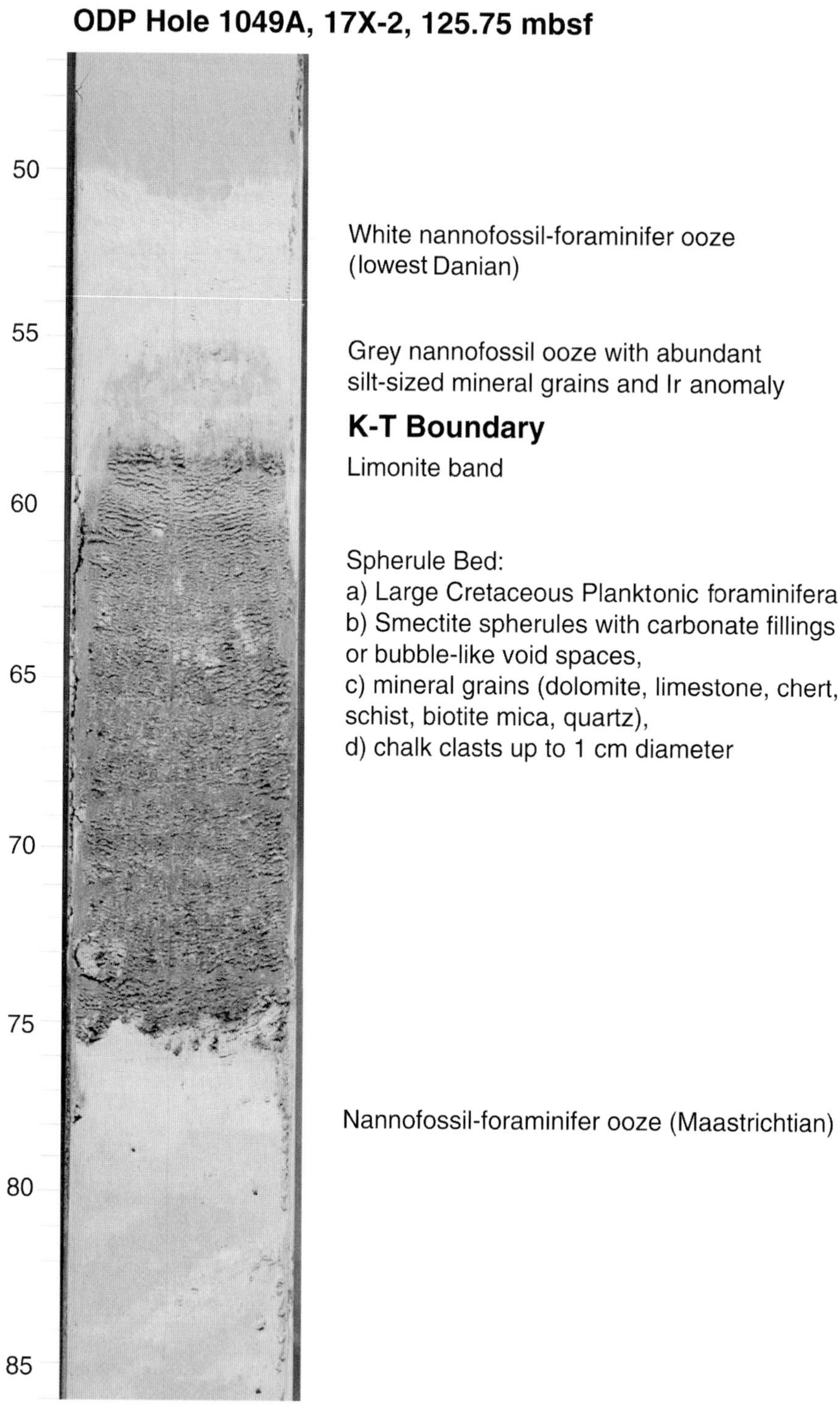

Fig. 13. K–P boundary bed in Hole 1049A. An iridium anomaly has been reported from the dark grey bed immediately overlying the spherule layer in Hole 1049B by Smit *et al.* (1997).

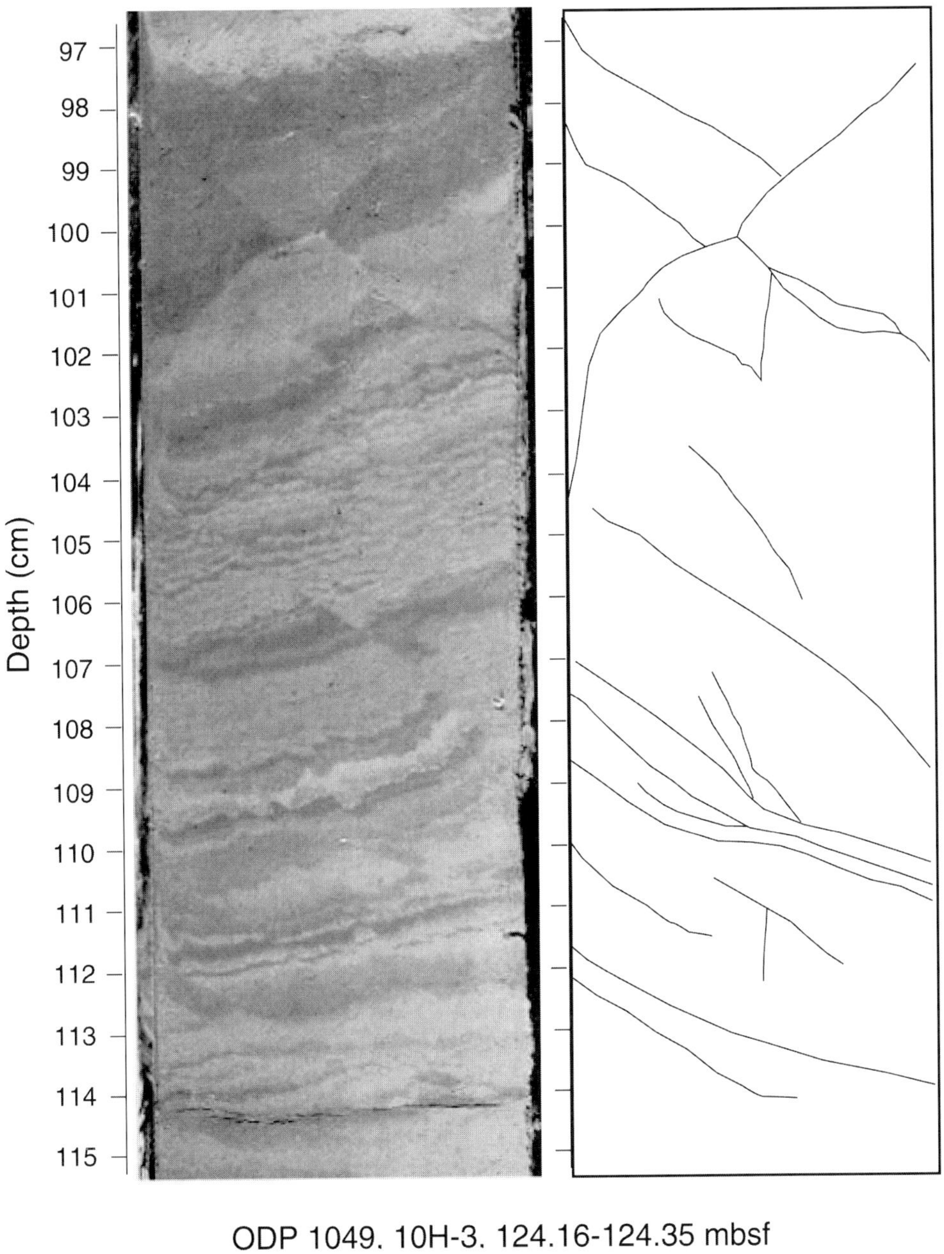

Fig. 14. Deformed Maastrichtian ooze about 12 m below the spherule bed in Hole 1049C.

Coniacian–Turonian hardgrounds that is only *c.* 9 m thick. Updip at Site 1052, Campanian nannofossils are mixed into the lower Maastrichtian chalk, which rests directly upon Cenomanian units. Evidently, upper Campanian–Turonian sediments were deposited on Blake Nose but were largely eroded before deposition of the Maastrichtian sequence.

Stratigraphy of the Palaeogene and upper Cretaceous deposits on Bermuda Rise

The eastern and southern Bermuda Rise was near or below the carbonate compensation depth (CCD) for much of late Cretaceous and Palaeogene time, and consequently preserves a record of pelagic clay, siliceous chalk and porcellanites

(Fig. 1). Drill sites close to the continental margin of eastern North America generally have very incomplete records of Palaeogene sedimentation owing to erosion by the deep western boundary current during Oligocene and Neogene time. In addition, many of the DSDP sites drilled on the Bermuda Rise had very incomplete core recovery (e.g. DSDP Sites 4–10, 28). The most complete records are DSDP Sites 385–387 on the central and northern Bermuda Rise and DSDP Sites 417 and 418 on the southern Bermuda Rise (Fig. 2). The DSDP sites show that the Oligocene sequence consists primarily of pelagic claystone, the Eocene section is dominated by radiolarian claystone and siliceous turbidites, the Palaeocene units by red and black claystone and the upper Cretaceous sequence by red and green claystone (Fig. 1).

Until now, the main acoustic reflectors on the Bermuda Rise have been most precisely tied to the sediment record at DSDP Sites 386 and 387. For example, Reflector A^c has been correlated with a sequence of Eocene chert. Unfortunately, Reflector A^c does not have the same age at every site, suggesting that the reflector is partly a product of diagenesis and chert development. At DSDP Sites 386 and 387, Reflector A^c correlates with the top of a succession of chert and dates to calcareous nannofossil Zone CP13b (=NP15b; *c.* 45–46 Ma), whereas at DSDP Sites 384 and 385, the reflector correlates with lower Eocene sediments (*c.* CP11–CP12b = NP13–NP14b; *c.* 48–51 Ma). The apparent mismatch in age of Reflector A^c may not be as bad as it first appears; the reflector is tied to the uppermost part of a thick sequence of chert that accumulated from early Eocene to mid-Eocene time at DSDP Sites 386 and 387, whereas the chert horizons are restricted to early Eocene time at the other sites. Hence, it appears that the chert horizons began forming in late early Eocene time, but cherts continued to be deposited well into mid-Eocene time in some locations.

The most prominent acoustic horizon in the upper Palaeocene sequence is Reflector A^b. Mountain & Miller (1992) identified the reflector in seismic lines shot across Sites 386, 387 and 417–418 on the western and southern Bermuda Rise. At Sites 417 and 418, Reflector A^b correlates with an angular unconformity between middle Eocene claystone and upper Cretaceous claystone. It is possible that this unconformity concatenates several acoustic horizons, as reflectors A^c, A^b, and A* all have ages compatible with the age range of the unconformity. The age of Reflector A^b is better constrained at DSDP Site 387, where its youngest possible age is within calcareous nannoplankton Zone CP9 (=NP10–11; 52.85–55.0 Ma) and its oldest age is no older than CP3 (=NP4; *c.* 59.7–62.2 Ma). Unfortunately, there is a 10 m coring gap at DSDP Site 387 at the level of the reflector that prevents a precise age determination. Reflector A^b cannot be traced directly to DSDP Site 386 owing to intervening buried topography, but Mountain & Miller (1992) noted that there is an apparent age gap between the top of a sequence of Cretaceous chalk in Core 35 and upper Palaeocene claystone (Zone CP7 (=NP8, 56.2–57.3 Ma)) near the top of this core. Therefore, the unconformity associated with Reflector A^b is of early Late Palaeocene age according to those workers.

We agree that erosion associated with Reflector A^b could be as old as *c.* 60 Ma, but argue that the data from the Bermuda Rise site also allow erosion to have occurred as late as the LPTM (*c.* 55.5 Ma). The age of the Palaeocene claystone at DSDP Site 386 is constrained by recognition of a series of early Eocene and late Palaeocene biozones in core 34, the oldest of which belongs to CP8b (the *Campylosphaera eodela* Biozone = NP9b; 55–55.5 Ma: Okada & Thierstein, 1979). Mountain & Miller (1992) incorrectly attributed the base of core 34 to the slightly older *Discoaster multiradiatus* Biozone, CP8a (=NP9, 55.5–56.2 Ma); their incorrect dating would prevent Reflector A^b from being potentially correlative with the LPTM. There is a *c.* 20 m coring gap between cores 34 and 35. The top of core 35 is assigned to calcareous nannofossil Biozone CP7 (=NP8; *c.* 56.2–57.3 Ma). Hence, although there may be an unconformity within Core 35, there could easily be other late Palaeocene erosion surfaces within the coring gap between Cores 34 and 35. Accordingly, we think the most secure age range for Reflector A^b is the upper bound provided by the age assignment of Core 34 at DSDP Site 386 (*c.* 55–55.5 Ma) and the lower bound provided by DSDP Site 387 (59.7–62.2 Ma). The age range for the hiatus and acoustic reflector on Bermuda Rise are very similar to those for the packet of upper Palaeocene reflectors on Blake Nose, which range in age from 55.5 to 59.7 Ma.

Horizon A* was identified by Ewing & Hollister (1972) near DSDP Site 105 off Cape Hatteras but it is most reflective just south of Bermuda. Drilling at DSDP Sites 386 and 387 demonstrated that Horizon A* correlates with a sequence of Maastrichtian chalk and limestone (Tucholke 1979) that has been named the Crescent Peaks Member of the Plantagenet Formation (Jansa *et al.* 1979). The reflector becomes indistinct east and west of the Bermuda Rise and also dies out in the Magnetic Quiet

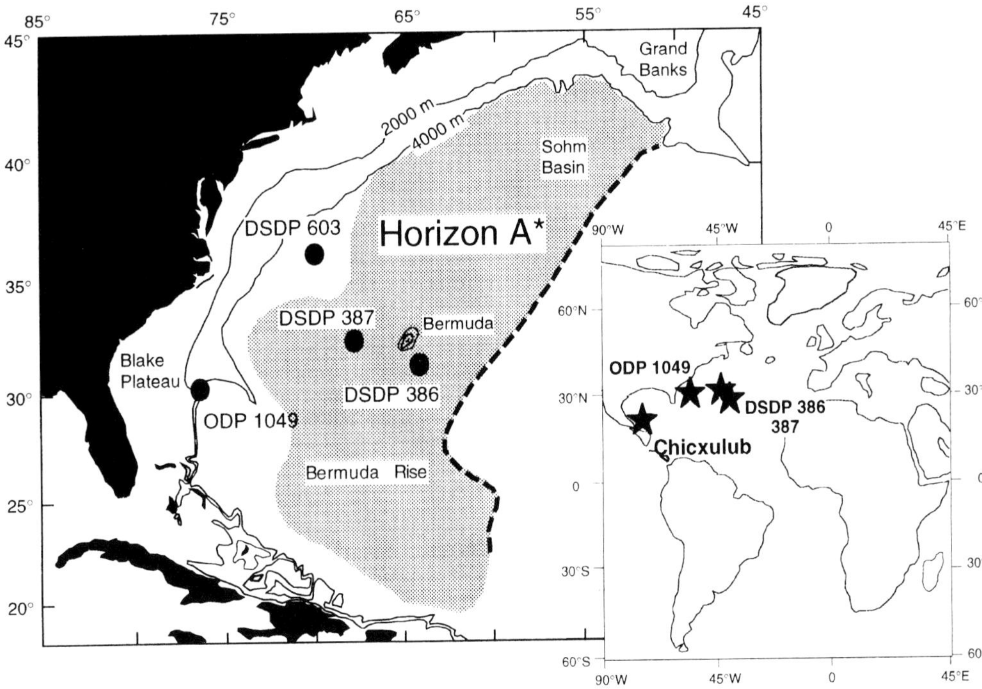

Fig. 15. Distribution of Reflector A* in relationship to DSDP sites that recovered mass wasting deposits at the K–P boundary.

Zone east of Bermuda (Fig. 15; Tucholke & Mountain 1979). The western disappearance of Horizon A* is partly due to erosional stripping beneath Horizon A^u close to the continental margin, whereas its eastern pinchout may reflect either a loss of acoustic character in the carbonate sediments near the ridge crest or the genuine disappearance of the carbonate lithofacies.

At DSDP Site 386, Maastrichtian carbonates were recovered in Core 35, where there are two layers of laminated to cross-laminated, light grey to white limestone that total about 2.3 m thick (Tucholke & Vogt 1979*a*). In contrast, at DSDP Site 387, the carbonates are a much thicker sequence of chalks recovered in Cores 27 and 28 (Tucholke & Vogt 1979*a*). Core 27 recovered 6.25 m of chalk below a sharp contact with grey to black claystone (of Palaeocene age) whereas Core 28 recovered only 65 cm of chalk and failed to preserve the contact with the underlying reddish brown claystone found in Core 29. Hence, the chalk sequence must be somewhere between 6.9 and 26.6 m thick at DSDP Site 387 if we allow for the incomplete core recovery in Cores 28 and 29.

The chalk at both DSDP Sites 386 and 387 consists entirely of very fine-grained micritic carbonate and nannofossils with rare, mostly juvenile, planktonic foraminifera dominated by typically small taxa such as *Guembelitria cretacea* and heterohelicids (Tucholke & Vogt 1979*a*). Indeed, McNulty (1979) reported that the foraminifera were so small that only sparse assemblages were retained on a 63 μm screen. Examination of a few chalk samples from DSDP Sites 386 and 387 by one of us (R.D.N.) confirms McNulty's observations that the sediments at both sites contain a dwarfed planktonic foraminiferal assemblage including minute representatives of *Heterohelix*, *Hedbergella*, *Guembelitria*, *Globigerinelloides* and *Globotruncanella* as well as rare juveniles of globotruncanids. The tiny size of the foraminifera is unusual, as upper Cretaceous foraminifer assemblages are usually dominated by globotruncanids, rugoglobigerinids and serially coiled species that typically reach sizes of several hundred microns in diameter.

The foraminifer species composition suggests a Maastrichtian age for the chalk sequence. The chalk at Site 387 contains the calcareous nannofossil marker, *Micula murus*, for late Maastrichtian time. The upper chalk bed at DSDP Site 386 also contains *M. murus*. In contrast, the lower chalk bed at that site reportedly does not contain *M. murus* but does contain the mid-Maastrichtian marker *Lithraphidites quadratus* as well as rare specimens of mid-Cretaceous nannoconids (Okada & Thierstein 1979). Both the chalk beds at DSDP Site 386 are otherwise similar in lithology and foraminifer assemblages, and both display reversed magnetic polarity (Keating & Helsley 1979). Therefore the chalk and Reflector A* are of late Maastrichtian age at the youngest, and of mid-Maasthrichtian age at the oldest.

Discussion: Palaeogene and upper Cretaceous seismic reflectors and their palaeoceanographic significance

Horizon A^u: cessation of siliceous sedimentation on Blake Nose

A major unconformity is associated with the upper Eocene and Oligocene section along most of the deep eastern margin of North America (Fig. 1). The unconformity correlates with seismic Horizon A^u, which locally crosscuts older reflectors of Eocene to early Cretaceous age. Downcutting reached lower Cretaceous rocks (of Barremian age and older) beneath the Blake–Bahama Outer Ridge and northwest of the Bahama platform as well as near the mouth of the Laurentian Channel (Tucholke & Mountain 1979) (Fig. 2). An erosional bench is present in places along the foot of the Blake Escarpment (Paull & Dillon 1980; EEZ-Scan 87 Scientific Staff 1991). The bench at the foot of the escarpment on the northeastern tip of the Blake Nose is *c.* 10–15 km wide (Paull & Dillon 1980). The top of the erosional bench is equivalent to the A^u reflector and is overlapped by Miocene turbidites. The eastern limit of Horizon A^u has been mapped by Mountain & Tucholke (1985), who showed that it is confined to a broad belt between the continental slope and the Bermuda Rise that is about 1100 miles wide off the Blake Escarpment but only *c.* 550 miles wide off the Laurentian channel (Fig. 2). The distribution of Horizon A^u shows that it must have been cut by the deep western boundary current.

Mountain & Miller (1992) obtained lowermost Oligocene sediment corresponding to planktonic foraminifer Biozone P18 from a piston core taken just below Horizon A^u. Hence, the erosional event must be younger than the Eocene–Oligocene boundary. Tucholke & Mountain (1986) estimated the age of the reflector at *c.* 28 Ma. This age suggests that the A^u reflector is younger than the onset of major Antarctic glaciation. Tucholke & Mountain suggested that the unconformity was cut by northern source waters overflowing the subsiding Greenland–Scotland Ridge. They speculated that northern component waters were strengthened by the global development of strong thermohaline deep-water production associated with the gradual thermal isolation and glaciation of Antarctica.

It is probably no coincidence that the youngest sediments on Blake Nose are of latest Eocene age. The Gulf Stream assumed its present course in Oligocene time and cut into the surface of the Florida Straits and the Blake Plateau (Dillon & Popenoe 1988). A sea-level highstand in late Oligocene time shifted sedimentation from the shelf to the coastal plain, starving the outer shelf and slope landward of the Blake Escarpment. Erosion along the base of the Blake Escarpment presumably would have been intense in late Eocene and Oligocene time because the large mass of Neogene sediments in the Blake–Bahama Outer Ridge had just begun to accumulate (Ewing & Hollister 1972; Sheridan *et al.* 1978; Shipley *et al.* 1978; Markl & Bryan 1983) but probably was not large enough to deflect the deep western boundary current away from the Escarpment as it partly does today (Heezen *et al.* 1966).

Erosion of the base of the Blake Escarpment occurred during Oligocene time as well. The northern side of Blake Nose is a relatively gentle ramp. Gently dipping Cretaceous rocks that make up most of the platform are truncated and overlapped by a veneer of Palaeogene strata. In contrast, the southeastern side of Blake Nose features well-preserved slump blocks that are not overlapped by younger pelagic sediments (EEZ-Scan 87 Scientific Staff 1991). Strata on top of Blake Nose terminate abruptly at the southeastern escarpment without obvious thinning or pinch-outs. The absence of sediment ponds on top of the slumps suggests that the slope failure is recent and suggests that the Blake Escarpment is being maintained by undercutting at its base. An erosional bench at the foot of the Blake Escarpment also suggests that the escarpment has been cut back as much as 15 km since Cretaceous time (Paull & Dillon 1980). Amos *et al.* (1971) showed that corrosive Antarctic Bottom Water flows into the Blake–Bahama Basin at the foot of the Blake Escarpment and forms an eddy on the southern flank of the Blake

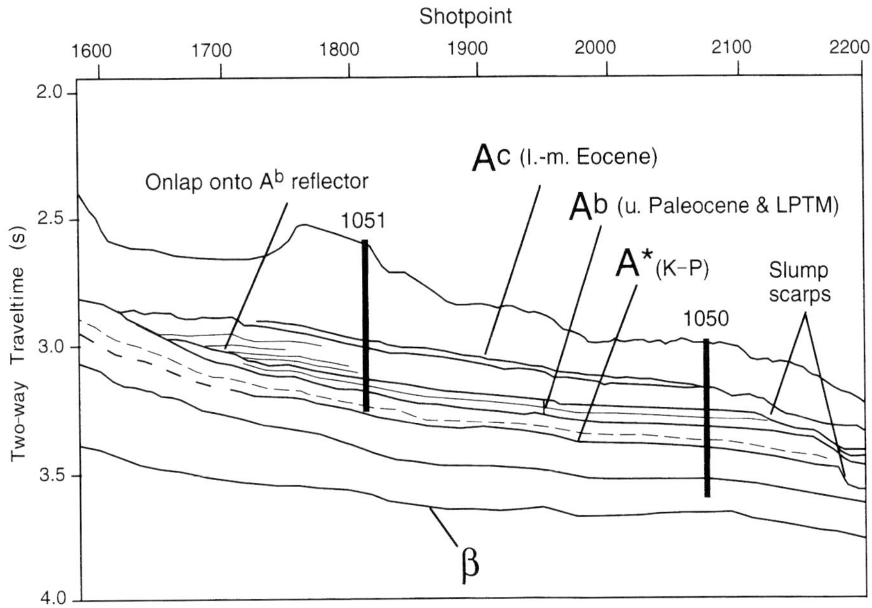

Fig. 16. Sketch of a segment of MCS line TD-5 (shotpoints 1600–2200) showing interpreted correlation with major reflectors in the western North Atlantic. Reflectors that onlap onto Reflector A^b do continue downdip of Site 1050, but are shown only near their termination against the Reflector A^b. Reflector A^b consists of three discrete reflectors. The lowest one correlates with a hiatus in Chronozone C26r, the middle one with a hiatus in Chronozone C25r, and the upper one with the Late Palaeocene Thermal Maximum in early Chronozone C24r. Reflector β approximates the contact between Barremian–Aptian limestone and overlying Aptian–Albian nannofossil claystone.

Nose. Currents flowing southward along the Escarpment have velocities of *c.* 17 cm s^{-1} below 2800 m (Amos *et al.* 1971) and probably help maintain the steep slopes in this area.

Horizon A^c: early mid-Eocene submarine erosion

Aubry (1995) has shown that an uppermost lower Eocene hiatus occurs throughout the North and South Atlantic. The erosion surface is typically associated with calcareous nannofossil Biozone NP14b and Chronozone C21r at localities where the duration of the unconformity is minimized (Aubry 1995). Tucholke (1981) mapped the distribution of the seismic reflector associated with this unconformity and found that its eastern limit in the western North Atlantic is approximately associated with Magnetic Anomaly 21 (Fig. 2). Browning *et al.* (1997) identified a pair of closely spaced hiatuses in onshore New Jersey ODP boreholes. The upper most of these occurs within Chronozone C21r and is dated between 49.6 and 48.6 Ma whereas the lower event cuts out part of Chronozone C22r (*c.* 50–51 Ma; Browning *et al.* 1997).

On Blake Nose, the presence of well-sorted foraminiferal grainstones suggests that the early to mid-Eocene hiatus (*c.* 48–49 Ma) represents a renewal of a deep western boundary current or a shoaling of such a current that may have been restricted earlier to depths below the top of the Blake Escarpment. The short duration of the hiatus is consistent with Browning's results from the New Jersey margin and shows that the erosion surface developed over a wide area of the eastern seaboard. The presence of siliceous nannofossil chalk above and below the unconformity on Blake Nose suggests that long-term patterns of sedimentation were not changed by the erosional event. Therefore, erosion was probably caused by a brief intensification of bottom currents before current activity slackened and pelagic biogenic sediments began to accumulate again.

The Chronozone C21r hiatus has been correlated with the E5 sequence boundary of Haq *et al.* (1987), which occurs between Chrons 22r and 22n. Browning *et al.* (1997) suggest that the sequence boundary in New Jersey is not associated with a sea-level lowstand although Olsson & Wise (1987) suggested that there was a significant

50–90 m sea-level drop at this time. An oxygen isotope stratigraphy for tropical Pacific ODP Site 865 (Bralower *et al.* 1995) does not show any major inflection of surface water $\delta^{18}O$ during this erosional period, suggesting the unconformity was not produced by glacial eustacy. None the less, the widespread occurrence of this surface or surfaces suggests an essentially a global hiatus reflecting one or more erosional events within a span of *c.* 1 Ma or less. There is tantalizing evidence for a glacial event near the early Eocene–mid-Eocene boundary, as Birkenmajer (1988) reported glacial tills underlying a tuff bed dated at 49 ± 5 Ma from the South Shetland Islands, Antarctica. Hence, it is possible that an increase in southern source water contributed to the erosion of deep-sea sediments. The same oceanographic change could have also produced an increase in upwelling and fertility of the western North Atlantic to contribute to widespread siliceous sedimentation there.

Horizon A^b: submarine erosion associated with the Late Palaeocene Thermal Maximum

Mountain & Miller (1992) have examined the age and geographical distribution of Horizon A^b in great detail (Fig. 2). Their correlation of Reflector A^b with DSDP Sites 386 and 387 on the Bermuda Rise suggests that the reflector corresponds to a hiatus of late Palaeocene to early Eocene age. Those workers concluded that the unconformity was cut by northward flowing bottom waters released into the North Atlantic as a result of the opening of a deep passage to the Southern Ocean.

We suggest that all three of the reflectors in the uppermost Palaeocene sequence on Blake Nose (Fig. 16) constitute the temporal equivalents of 'Reflector' A^b on the Bermuda Rise. The upper Palaeocene erosion surfaces on Blake Nose have, in combination, essentially the same ages as Reflector A^b on Bermuda Rise. The unconformities on Blake Nose and Bermuda Rise also correlate closely with erosion surfaces in the onshore New Jersey boreholes (Liu *et al.* 1997; Miller *et al.* 1997; Pak *et al.* 1997). It must be emphasized that there are at least two late Palaeocene erosion surfaces on Blake Nose, one at 55.5 Ma (the LPTM) and the other at *c.* 56.7–57.5 Ma, so it is possible that only one of these surfaces is the temporal equivalent of Reflector A^b on Bermuda Rise. However, it is also plausible that Reflector A^b represents the concatenation of several discrete erosion surfaces that are distinguishable as separate events on Blake Nose.

At the LPTM, sediments were eroded on the upper portion of Blake Nose and the depocentre shifted from a gentle drape over the whole region to much more localized sedimentation on the lower half of Blake Nose. There cannot have been much erosion on the upper half of Blake Nose, as the youngest part of the Palaeocene sequence below the unconformity at Site 1052 was deposited during Chronozone C26r or possibly C26n (60–57.6 Ma; Norris *et al.* 1998). In addition, part of the erosion of the upper slope of Blake Nose could have occurred during early Mid-Eocene time associated with Horizon A^c. Still, there must have been some mass wasting, as the reflector corresponding to the LPTM appears to truncate an underlying reflector near shot-point 2120 on MCS line TD-5 (Fig. 16). In addition, the chalk clast bed immediately below the LPTM at ODP Site 1051 contains the foraminifer *Globanomalina pseudomenardii* (LAD *c.* 55.9 Ma) indicating that erosion had removed at least several hundred thousand years of sedimentation updip of this site.

Katz *et al.* (1999) have suggested that the chalk clast bed represents a debris flow associated with venting of methane from gas hydrates on the upper part of Blake Nose. Their primary evidence is (a) the local derivation of the chalk clasts, and (b) seismic evidence for chaotic reflections in the vicinity of ODP Site 1052 that they attribute to deformation caused by gas escape. We agree that the chalk clast bed is a debris flow but we find the evidence for gas escape to be equivocal. The cores and well logs from the Palaeocene sequence at ODP Site 1052 do not show tilted bedding or evidence of deformation expected if the hole had penetrated slumps associated with a gas hydrate reservoir. Modern gas hydrate escape structures can penetrate *c.* 400–500 m into underlying sediment to the depth of the bottom-simulating reflector that marks the gas hydrate stability zone (e.g. Dillon *et al.* 1998). Hence, we should expect significant deformation in the Palaeocene and Cretaceous sediments if a deeply buried hydrate reservoir had failed on Blake Nose. Indeed, there is good evidence for slumping in the upper Cretaceous sequence (see below), but that deformation is associated with the K–P boundary, not the LPTM.

It is possible that the gas hydrate stability zone may have been near the Palaeocene sea floor on the upper part of Blake Nose rather than being deeply buried. Gas hydrates are known to outcrop on the modern continental slope where the hydrate stability zone shoals toward the sea floor. Hence, it is possible that slope instability would have been confined to the upper 50–100 m of the sediment column in the Palaeocene sequence. Core recovery was poor in the upper 50 m of the Palaeocene sequence at ODP Site 1052 so it is possible that deformation structures produced by shallow mass wasting were simply not recovered. However, if there was a slump triggered by gas hydrates on Blake Nose, the slump seems likely to have been a relatively small and shallowly seated feature.

We believe that the debris flow at the LPTM is related to a sudden intensification of the deep western boundary current. We agree with Katz *et al.* (1999) that the debris flow could be related to gas escape, perhaps triggered by erosion of overburden by the deep western boundary current or a rise in temperature of this current. Lower Eocene sediments lap onto the LPTM reflector suggesting that sedimentation effectively stopped above *c.* 2200 m palaeo-depth on Blake Nose until mid-Eocene time. The absence of channels cut on the upper part of Blake Nose suggests that bottom currents were not strong enough to cause widespread erosion but were strong enough to prevent sediment from accumulating. Hence, it appears that the LPTM was intimately associated with an increase in intermediate water current velocity and ventilation at the same time that erosion was occurring along the southwestern Bermuda Rise in response to more vigorous lower deep water currents.

Mountain & Miller's reconstruction suggests that the bottom current that formed Reflector A^b looped over the southern Bermuda Rise like modern Antarctic Bottom Water (AABW). Today, the southward-flowing deep western boundary current flows counter to northward flowing Antarctic Bottom Water. When AABW is strong, the deep western boundary current shifts to shallow depths (as upper North Atlantic Intermediate Water) whereas when AABW is weak, most of the flow from the deep North Atlantic is concentrated in North Atlantic Deep Water. We suggest that erosion near the P–E boundary on Blake Nose reflects a shift toward stronger northern component intermediate water that began precisely at the LPTM and lasted throughout early Eocene time. Stronger flow of intermediate water was countered by stronger southern source water flowing into the deep North Atlantic that eroded Palaeocene and early Eocene sediments to form Reflector A^b on the southern Bermuda Rise.

DSDP Sites 385 and 387 recovered black and green–grey Palaeocene clay (Figs 1 and 2) that are locally organic rich (up to 1.3 wt% organic carbon) suggesting that the deep western North Atlantic was very poorly ventilated during early and mid-Palaeocene time before the erosion associated with Reflector A^b (Tucholke & Vogt 1979*b*). The flushing of the western North Atlantic by the southern-source deep water replaced the oxygen-deficient deep waters and ended deposition of organic-rich mudrocks in this area.

What does a strengthening of intermediate water flow have to do with the LPTM? The close timing of the LPTM carbon excursion and the strengthening of intermediate water circulation suggests that a major shift in deep ocean current systems is somehow connected to the global warming and melting of gas hydrates at the LPTM. The source of the intermediate water that sculpted sediments on Blake Nose is unknown but was probably a northern component water conditioned by Tethyan outflow. Many researchers have suggested that Tethyan water filled the deep oceans near the LPTM either as a watermass that conditioned deep-water formation in the North Atlantic like modern Mediterranean Sea outflow (Corfield & Norris 1996, 1998), or as 'Warm Saline Deep Water' (Pak & Miller 1992; Bralower *et al.* 1997). We suggest that erosion on Blake Nose around the LPTM may be evidence of a Tethyan-sourced intermediate watermass. The strengthening of warm intermediate waters could have been the trigger necessary to cause clathrate decomposition and initiation of the carbon isotope anomaly.

Horizon A: slope failure during the K–P boundary impact event*

Tucholke (1979) and Tucholke & Vogt (1979*b*) speculated that the carbonate sequence correlative with Reflector A* was deposited during a drop in the carbonate compensation depth, as the Maastrichtian chalk is sandwiched between Cretaceous and Palaeocene claystones containing little, if any carbonate. In their view, the chalk sequence is a normal pelagic sediment.

On Blake Nose, the age equivalents of the Crescent Peaks Member consist of a drape of slumped Maastrichtian foraminifer–nannofossil ooze and chalk. A very thin sequence of upper Campanian foraminifer ooze is present at Site 1049, but no sediment of corresponding age has been recovered from the shallower parts of the slope (Norris *et al.* 1998). The Maastrichtian sequence contains numerous slumps, including one at the Maastrichtian–Cenomanian contact at Site 1052, so it is possible that Campanian sediments were removed from the area of Site 1052 by downslope transport before

Maastrichtian time. Alternatively, the Campanian sediments may never have been very thick on Blake Nose and deformed sequences in the Maastrichtian and Campanian sequences may represent a single, large slump that failed along a décollement in the soft Campanian deposits and lower Maastrichtian chalk.

At least part of the slumping on Blake Nose must have immediately preceded deposition of the ejecta at Sites 1052 and 1049, as spherule-bearing chalk and ooze rests directly on slumped sediments (Klaus *et al.* 1997, 2000). The variable thickness of the impact ejecta bed (which is absent from Sites 1050 and 1052 and ranges from 7 to 17 cm thick in three holes at Site 1049) may be due either to post-depositional slumping or fallout of ejecta onto the hummocky topography of the slumped Maastrichtian sequence (Martínez-Ruiz *et al.* this volume *a&b*). In either scenario, it seems possible that the slumping may have been produced by the large-magnitude earthquake associated with the Chicxulub impact. Slumping may have been followed by one of more phases of redeposition of ejecta on the unstable slope (Klaus *et al.* 1997, 2000). Despite the slumping, the K–P boundary sections at all Blake Nose sites preserve the earliest Danian biozones as well as the nannofossil markers for the latest Maastrichtian period.

Sediments slumped off Blake Nose and elsewhere along the continental escarpment should have accumulated on the continental rise and the abyssal plains of the western North Atlantic. Unfortunately, Oligocene erosion associated with Reflector A^c has removed Palaeogene and upper Cretaceous sediments along most of the continental rise from Canada to the Blake–Bahama Abyssal Plain. Hence, there is little prospect of finding proximal mass wasting deposits offshore of Blake Nose. Possible K–P boundary turbidites have been recovered at DSDP Site 603 off New Jersey where a layer of cross-bedded spherules is preserved in a layer *c.* 20 cm thick (Klaver *et al.* 1987). The spherules are composed of smectite and are morphologically and geochemically similar to those in the K–P ejecta bed at ODP Site 1049. However, the spherule bed at DSDP Site 603 is interbedded with sandstone turbidites that lack age diagnostic microfossils. Hence, we must look to more distal sites on the Bermuda Rise for evidence of K–P boundary mass wasting.

Palaeontological and sedimentological evidence suggests that the chalk deposits on the Bermuda Rise represent the distal equivalents of slumps on Blake Nose. The absence of a diverse assemblage of large Cretaceous planktonic foraminifera strongly suggests that the carbonates at Sites 386 and 387 are not normal pelagic sediments as hypothesized by Tucholke & Vogt (1979*b*) but instead represent the winnowed, distal equivalents of sediments redeposited from the continental margin. The extremely small size of planktonic foraminifera within the chalk beds at DSDP Sites 386 and 387 is consistent with the size-sorting of grains that should occur within large turbidites originating from the continental margin. Interpretation of the chalk as a mass flow deposit is also supported by its occurrence within red claystone. Either the chalk records a rapid drop in the CCD followed by an equally fast rise, or the chalk itself was deposited suddenly below the CCD. The two chalk beds at DSDP Site 386 both have sharp basal contacts with the underlying claystone, further suggesting that the calcareous sediments were deposited rapidly rather than by a gradual increase in carbonate preservation associated with a shoaling of the CCD.

Interpretation of the chalk beds as K–P boundary turbidites can also explain two otherwise puzzling aspects of the deposits. The shipboard party reported that the chalk beds at DSDP Site 386 are laminated, and cross laminated which is consistent with high-energy deposition for a turbidity current but not with pelagic sedimentation in *c.* 4782 m of water. In addition, both chalk beds at DSDP Site 386 are lithologically similar, contain dwarfed foraminifera, display reversed magnetic polarity and are of about the same thickness, yet one contains upper Maastrichtian calcareous nannofossils and the other contains middle Maastrichtian zone fossils. The two chalk beds could represent separate turbidites, the lower one deposited during mid-Maastrichtian reversed polarity zone C30r and the upper one deposited 2.7 Ma later during Chron C29r. Alternatively, they could both be of K–P boundary age but derived from different parts of the slope. Indeed, the presence of the mid-Cretaceous calcareous nannofossil, *Nannoconus* in the lower chalk bed suggests that the chalk includes a mixture of sediment of different ages, as we should expect for turbidities.

Bralower *et al.* (1998) have documented redeposited sediments at the K–P boundary in the Gulf of Mexico and the Caribbean. These sediments consists of a ‘cocktail’ of sediments of diverse Cretaceous ages as well as lithic fragments and spherules derived from the Chicxulub impact structure. In contrast, the Maastrichtian deposits at Sites 386 and 387 do not include abundant exotic fossils (e.g. McNulty 1979; Okada & Thierstein 1979) although Okada & Thierstein (1979) reported lower Maastrichtian,

Campanian and middle Cretaceous calcareous nannofossils from the lower parts of the carbonate sequence at both sites. We suggest that these pre-upper Maastrichtian fossils are reworked and that the chalk on Bermuda Rise represents the distal equivalents of K–P boundary slumps documented on Blake Nose and presumed to exist elsewhere along the continental slope and rise.

Horizon A* marking the top of the Crescent Peaks Member can be recognized as far as 1000–1300 km east of the continental escarpment of North America (Fig. 15; Tucholke & Mountain 1979; Tucholke & Vogt 1979*b*). The reflector has been recognized in the Sohm Basin north of the New England Seamounts and south of the Grand Banks in Conrad MCS line C21-152, where it is continuous from the continental margin of Nova Scotia as far east as the J-Anomaly Ridge (Ebinger & Tucholke 1988). Furthermore, the Crescent Peaks Member on the Bermuda Rise is age equivalent to the carbonate Wyandot Formation on the Scotian shelf and upper slope (Ebinger & Tucholke 1988). Hence, we suggest that the Chicxulub impact destabilized the entire eastern seaboard of North America, causing one or more massive slumps to spread across nearly the whole of the western North Atlantic basin.

Conclusions

The combination of coring, logging and analysis of seismic reflection profiles from Blake Nose has tightly constrained the age of a series of widespread palaeoceanographic events in Palaeogene and Cretaceous time in the western North Atlantic. We find that sedimentation effectively halted on Blake Nose during late Eocene or earliest Oligocene time, essentially synchronous with erosion that produced Reflector A^u in the deep western North Atlantic. Reflector A^c is associated with chert development on the Bermuda Rise and elsewhere in the deep ocean but correlates partly with a clearly defined and short-lived episode of erosion on Blake Nose and the New Jersey margin between *c.* 48 and 49 Ma. The strong bottom currents that formed the lower middle Eocene erosion surface may be related to one of the first pulses of high-latitude glaciation in Eocene time. The Late Palaeocene Thermal Maximum (*c.* 55.5 Ma) is also associated with a prominent reflector that is approximately age equivalent to Reflector A^b on the Bermuda Rise and erosion surfaces on the New Jersey margin. We suggest that erosion relates to an abrupt increase in the strength of the deep western boundary current over Blake Nose synchronous with a flood of southern-source waters into the deep basins of the southwestern North Atlantic. The increase in flow of the deep western boundary current may be one of the first pieces of sedimentological evidence for outflow of warm, saline Tethyan waters, which may have triggered the failure of gas hydrate reservoirs by heating and physically unroofing them. Finally, sedimentological and seismic evidence for extensive slumping at the K–P boundary leads us to re-evaluate the origins of Maastrichtian chalk beds on Bermuda Rise. The sedimentology and palaeontology of the chalk beds is consistent with their origins as K–P boundary turbidites derived by large-scale mass wasting from the eastern seaboard of North America. The association of the chalk with Reflector A*, which can be traced from the Grand Banks to the southern Bermuda Rise, suggests that mass wasting occurred along the entire margin, creating one of the most extensive mass wasting deposits on Earth.

We thank the crew of the *JOIDES Resolution*, her captain, E. Oonk, drilling supervisor, R. Grout, the shipboard technical staff, and all those in the ODP planning structure and shore staff who made our cruise to the Blake Nose both possible and enjoyable. Our post-cruise work was supported by grants from JOI-USSAC, National Science Foundation and the Ocean Drilling Program. We also thank M. Holerichter for assistance with seismic processing and synthetic seismograms.

References

AMOS, A. F., GORDON, A. L. & SCHNEIDER, E. D. 1971. Water masses and circulation patterns in the region of the Blake–Bahama Outer Ridge. *Deep-Sea Research*, **18**, 145–165.

AUBRY, M.-P. 1995. From chronology to stratigraphy: interpreting the lower and middle Eocene stratigraphic record in the Atlantic Ocean. *In*: BERGGREN, W. A., KENT, D. V., AUBRY, M.-P. & HARDENBOL, J. (eds) *Geochronology, Time Scales and Global Stratigraphic Correlation.* SEPM Special Publications, **54**, 213–274.

BENSON, W. E., SHERIDAN, R. E. *et al.* (eds) 1978. *Initial Reports of the Deep Sea Drilling Project: vol, 44* US Government Printing Office, Washington, DC.

BERGGREN, W. A., KENT, D. V., SWISHER, C. C., III & AUBRY, M.-P. 1995. A revised Cenozoic geochronology and chronostratigraphy. *In*: BERGGREN, W. A., KENT, D. V., AUBRY, M.-P. & HARDENBOL, J. (eds) *Geochronology, Time Scales and Global Stratigraphic Correlations. A Unified Temporal Framework for an Historical Geology.* SEPM, Special Publications, **54**, 129–212.

BIRKENMAJER, K. 1988. Tertiary glacial and interglacial deposits, South Shetland Islands, Antarctica: geochronology versus biostratigraphy (a progress report). *Bulletin of the Polish Academy of Sciences, Earth Sciences*, **36**, 133–144.

BRALOWER, J. J., PAULL, C. K. & LECKIE, R. M. 1998. The Cretaceous/Tertiary boundary cocktail: Chixulub impact triggers margin collapse and extensive sediment gravity flows. *Geology*, **26**, 331–334.

BRALOWER, T. J., THOMAS, D., ZACHOS, J. *et al.* 1997. High resolution records of the late Palaeocene thermal maximum and circum-Caribbean volcanism: is there a causal link? *Geology*, **25**, 963–966.

BRALOWER, T. J., ZACHOS, J. C., THOMAS, E. *et al.* 1995, Late Palaeocene to Eocene palaeoceanography of the equatorial Pacific Ocean: stable isotopes recorded at Ocean Drilling Program Site 865, Allison Guyot: *Palaeoceanography*, **10**, 841–865.

BROWNING, J. V., MILLER, K. G., VAN FOSSEN, M., LIU, C., PAK, D. K., AUBRY, M.-P. & BYBELL, L. M. 1997. Lower to middle Eocene sequences of the New Jersey Coastal Plain and their significance for global climate change. *In*: MILLER, K. G., AUBRY, M.-P. BROWNING, J. V. *et al.* (eds) *Proceedings of the Ocean Drilling Program, Scientific Results*, **150X**. Ocean Drilling Program, College Station, TX, 229–242.

CHARM, W. B., NESTEROFF, W. D. & VALDES, S. 1969. Detailed stratigraphic description of the JOIDES cores on the continental margin off Florida. US Geological Survey Professional Papers, **581-D**, D1–D13.

CORFIELD, R. M. & NORRIS, R. D. 1996. Deep water circulation in the Palaeocene ocean. *In*: KNOX, R. W., CORFIELD, R. M. & DUNAY, R. E. (eds) *Correlation of the Early Palaeogene in Northwest Europe*. Geological Society, London, Special Publications, **101**, 443–456.

CORFIELD, R. M. & NORRIS, R. D. 1998. The oxygen and carbon isotopic context of the Palaeocene–Eocene Epoch boundary. *In*: AUBRY, M.-P., LUCAS, S. & BERGGREN, W. A. (eds) *Late Palaeocene–Early Eocene Climate and Biotic Events in the Marine and Terrestrial Records*. Columbia University Press, New York, 124–137.

DILLON, W. P. & POPENOE, P. 1988, The Blake Plateau Basin and Carolina Trough. *In*: SHERIDAN, R. E. & GROW, J. A. (eds) *The Atlantic Continental Margin. Volume I-2: The Geology of North America*. Geological Society of America, Boulder, CO, 291–328.

DILLON, W. P., DANFORTH, W. W., HUTCHINSON, D. R., DRURY, R. M., TAYLOR, M. H. & BOOTH, J. S. 1998. Evidence for faulting related to dissociation of gas hydrate and release of methane off the southeastern United States. *In*: HENRIET, J.-P. & MIENERT, J. (eds) *Gas Hydrates*. Geological Society of London, Special Publications, **137**, 293–302.

DILLON, W. P., PAULL, C. K. & GILBERT, L. S. 1985. History of the Atlantic continental margin off Florida: the Blake Plateau Basin. *In*: POAG, C. W. (ed.) *Geologic Evolution of the United States Atlantic Margin*. Van Nostrand Reinhold, New York, 189–215.

EBINGER, C. J. & TUCHOLKE, B. E. 1988. Marine geology of Sohm Basin, Canadian Atlantic Margin. *AAPG Bulletin*, **72**, 1450–1468.

EEZ-Scan 87 Scientific Staff, 1991. *Atlas of the U.S. Exclusive Economic Zone, Atlantic Continental Margin*. US Geological Survey Miscellaneous Investigations Series, **I-2054**.

EWING, J. I. & HOLLISTER, C. H. 1972. Regional aspects of deep sea drilling in the western North Atlantic. *In*: HOLLISTER, C. D., EWING, J. I. *et al* (eds) *Initial Reports of the Deep Sea Drilling Project*, **11**, US Government Printing Office, Washington, DC, 951–973.

EWING, M. 1965. The sediments of the Argentine Basin. *Quarterly Journal of the Royal Astronomical Society*, **6**, 10–27.

EWING, M. & EWING, J. 1963. Sediments at proposed LOCO drilling sites: *Journal of Geophysical Research*, **68**, 251–256.

HAQ, B. U., HARDENBOL, J. & VAIL, P. R. 1987. Chronology of fluctuating sea levels since the Triassic. *Science*, **235**, 1156–1167.

HEEZEN, B. C., HOLLISTER, C. D. & RUDDIMAN, W. F. 1966. Shaping the continental rise by deep geostrophic contour currents. *Science*, **152**, 502–508.

HUTCHINSON, D. R., POAG, C. W. & POPENOE, P. 1995. *Geophysical database of the East coast of the United States: southern Atlantic margin—stratigraphy and velocity from multichannel seismic profiles*. US Geological Survey Open File Report, **95-27**.

JANSA, L. F., ENOS, P., TUCHOLKE, B. E., GRADSTEIN, F. M. & SHERIDAN, R. E. 1979. Mesozoic–Cenozoic sedimentary formations of the North American Basin: western North Atlantic. *In*: TALWANI, M., HAY, W. & RYAN, W. B. F. (eds) *Deep Drilling Results in the Atlantic Ocean: Continental Margins and Palaeoenvironment, Maurice Ewing Symposium, Vol. 3*. American Geophysical Union, Washington, DC, 1–57.

KATZ, M. E., PAK, D. K., DICKENS, G. R. & MILLER, K. G. 1999. The source and fate of massive carbon input during the Latest Palaeocene Thermal Maximum: new evidence from the North Atlantic Ocean. *Science*, **286**, 1531–1533.

KEATING, B. H. & HELSLEY, C. E. 1979. Magnetostratigraphy of Cretaceous sediments from DSDP Site 386. *In*: TUCHOLKE, B. E., VOGT, P. R. *et al.* (eds) *Initial Reports of the Deep Sea Drilling Project*, **43**, US Government Printing Office, Washington, DC, 781–784.

KLAUS, A., NORRIS, R. D., KROON, D. & SMIT, J. 2000. Impact-induced mass wasting at the L–T boundary: Blake Nose, western North Atlantic. *Geology*, **28**, 319–322.

KLAUS, A., NORRIS, R. D., SMIT, J., KROON, D. & MARTINEZ-RUIZ, F. 1997. Impact-induced K–T boundary mass wasting across the Blake Nose, w. North Atlantic: evidence from seismic reflection and core data. *Transactions of the American Geophysical Union, Abstracts*, **78**, F371.

KLAVER, G. T., VAN KEMPEN, T. M. G., BIANCHI, F. R. & VAN DER GAAST, S. J. 1987. Green spherules as indicators of the Cretaceous/Tertiary boundary in Deep Sea Drilling Project Hole 603B. *In*: VAN HINTE, J. E., WISE, S. W. Jr. *et al.* (eds) *Initial Reports of the Deep Sea Drilling Project*, **93**, US Government Printing Office, Washington, DC, 1039–1056.

LIU, C., BROWNING, J. V., MILLER, K. G. & OLSSON, R. K. 1997. Palaeocene benthic foraminiferal biofacies and sequence stratigraphy, Island Beach borehole, New Jersey. *In*: MILLER, K. G., AUBRY, M.-P. *et al* (eds) *Proceedings of the Ocean Drilling Program, Scientific Results*, **150X**. Ocean Drilling Program, College Station, TX, 267–275.

MARKL, R. G. & BRYAN, G. M. 1983. Stratigraphic evolution of Blake Outer Ridge. *AAPG Bulletin*, **67**, 666–683.

MARTÍNEZ-RUIZ, F., ORTEGA-HUERTAS, M., KROON, D., SMIT, J., PALOMO-DELGADO, I. & ROCCHIA, R. 2001*a*. Geochemistry of the Cretaceous–Tertiary boundary at Blake Nose (ODP Leg 171B). *This volume*.

MARTÍNEZ-RUIZ, F., ORTEGA-HUERTAS, M., PALOMO-DELGADO, I. & SMIT, J. 2001*b*. K–T boundary spherules from Blake Nose (OPD Leg 171B) as a record of the Chixulub ejecta deposits. *This volume*.

MCNULTY, C. L. 1979. Smaller Cretaceous foraminifers of Leg 43, Deep Sea Drilling Project. *In*: TUCHOLKE, B. E. & VOGT, P. R. (eds) *Initial Reports of the Deep Sea Drilling Project*, **43**. US Government Printing Office, Washington, DC, 487–505.

MILLER, K. G., BROWNING, J., PEKAR, S. F. & SUGARMAN, P. J. 1997. Cenozoic evolution of the New Jersey Coastal Plain: changes in sea level, tectonics, and sediment supply. *In*: MILLER, K. G., AUBRY, M.-P. *et al* (eds) *Proceedings of the Ocean Drilling Program, Scientific Results*, **150X**, Ocean Drilling Program, College Station, TX, 361–373.

MOUNTAIN, G. S. & MILLER, K. G. 1992. Seismic and geologic evidence for early Palaeogene deep-water circulation in the western North Atlantic. *Palaeoceanography*, **7**, 423–439.

MOUNTAIN, G. S. & TUCHOLKE, B. E. 1985. Mesozoic and Cenozoic geology of the U.S. Atlantic continental slope and rise. *In*: POAG, W. C. (ed.) *Geologic Evolution of the United States Atlantic Margin*. Van Nostrand Reinhold, New York, 293–341.

MULLINS, H., KELLER, G., KOFOED, J., LAMBERT, D., STUBBLEFIELD, W. & WARME, J. 1982. Geology of Great Abaco Submarine Canyon (Blake Plateau): observations from the research submersible 'Alvin'. *Marine Geology*, **48**, 239–257.

NORRIS, R. D. & RÖHL, U. 1999. Carbon cycling and chronology of climate warming during the Palaeocene/Eocene transition. *Nature*, **401**, 775–778.

NORRIS, R. D., HUBER, B. T. & SELF-TRAIL, J. 1999. Synchroneity of the K–T oceanic mass extinction and meteorite impact: Blake Nose, western North Atlantic. *Geology*, **27**, 419–422.

NORRIS, R. D., KROON, D., KLAUS, A. *et al.* 1998. Blake Nose Palaeoceanographic Transect, Western North Atlantic. *In*: NORRIS, R. D., KROON, D., KLAUS, A. *et al.* (eds) *Proceedings of the Ocean Drilling Program, Initial Reports*, **171B**. Ocean Drilling Program, College Station, TX, 1–749.

NORRIS, R. D., KROON, D., SMIT, J. & LEG 171B SCIENCE PARTY 1997. Anatomy of the apocalypse: K–T boundary beds from ODP Leg 171B. *Geological Society of America Annual Meeting, Abstracts with Programs*, **29**, A142.

OKADA, H. & THIERSTEIN, H. R. 1979. Calcareous nannoplankton—Leg 43, Deep Sea Drilling Project. *In*: TUCHOLKE, B. E., VOGT, P. R. *et al.* (eds) *Initial Reports of the Deep Sea Drilling Project*, **43**. US Government Printing Office, Washington, DC, 507–573.

OLSSON, R. K. & WISE, S. W. 1987. *Upper Palaeocene to middle Eocene depositional sequences and hiatuses in the New Jersey Atlantic Margin*. Cushman Foundation for Foraminiferal Research, Special Publication, **24**, 99–112.

PAK, D., MILLER, K. G. & BROWNING, J. 1997. Global significance of an isotopic record from the New Jersey Coastal Plain: linkage between the shelf and deep sea in the late Palaeocene to early Eocene. *In*: MILLER, K. G., AUBRY, M.-P. *et al* (eds) *Proceedings of the Ocean Drilling Program, Scientific Results*, **150X**. Ocean Drilling Program, College, Station, TX, 305–315.

PAK, D. K. & MILLER, K. G. 1992. Palaeocene to Eocene benthic foraminiferal isotopes and assemblages: implications for deepwater circulation. *Palaeoceanography*, **7**, 405–422.

PAULL, C. K. & DILLON, W. P. 1980. Erosional origin of the Blake Escarpment: an alternative hypothesis. *Geology*, **8**, 538–542.

SHERIDAN, R. E., ENOS, P., GRADSTEIN, F. & BENSON, W. E. 1978. Mesozoic and Cenozoic sedimentary environments of the western North Atlantic. *In*: BENSON, W. E., SHERIDAN, R. E. *et al.* (eds) *Initial Reports of the Deep Sea Drilling Project*, **44**. US Government Printing Office, Washington, DC, 971–979.

SHERIDAN, R. E., GROW, J. A. & KLITGORD, K. D. 1988. Geophysical data. *In*: SHERIDAN, R. E. & GROW, J. A. (eds) *The Atlantic Continental Margin, Volume I-2: The Geology of North America*. Geological Society of America, Boulder, CO, 177–197.

SHIPLEY, T. H., BUFFLER, R. T. & WATKINS, J. S. 1978. Seismic stratigraphy and geologic history of Blake Plateau and adjacent western Atlantic continental margin. *AAPG Bulletin*, **62**, 792–812.

SMIT, J. 1999. The global stratigraphy of the Cretaceous–Tertiary boundary impact ejecta. *Annual Review of Earth and Planetary Sciences*, **27**, 75–113.

SMIT, J., ROCCHA, R. & ROBIN, E. 1997. Preliminary iridium analysis from a graded spherule layer at the K/T boundary and late Eocene ejecta from ODP Sites 1049, 1052, 1053, Blake Nose, Florida. *Geological Society of America Annual Meeting, Abstracts with Programs*, **29**, A141.

TUCHOLKE, B. & MOUNTAIN, G. S. 1986. Tertiary palaeoceanography of the western North Atlantic. *In*: VOGT, P. & TUCHOLKE, B. (eds) *The Western North Atlantic Region, Volume M: The Geology of North America*. Geological Society of America, Boulder, CO.

TUCHOLKE, B. E. 1979. Relationships between acoustic stratigraphy and lithostratigraphy in the western North Atlantic Basin. *In*: TUCHOLKE, B. E., VOGT, P. R. *et al.* (eds) *Initial Reports of the Deep Sea Drilling Project*, **43**. US Government Printing Office, Washington, DC, 827–846.

TUCHOLKE, B. E. 1981. Geologic significance of seismic reflectors in the deep western North Atlantic Basin. *In*: WARME, J. E., DOUGLAS, R. G. & WINTERER, E. L. (eds) *The Deep Sea Drilling Project: a Decade of Progress*. SEPM Special Publication, **32**, 23–37.

TUCHOLKE, B. E. & MOUNTAIN, G. S. 1979. Seismic stratigraphy, lithostratigraphy and palaeosedimentation patterns in the North Atlantic Basin. *In*: TALWANI, M., HAY, W. & RYAN, W. B. F. (eds) *Deep Drilling Results in the Atlantic Ocean: Continental Margins and Palaeoenvironment, Maurice Ewing Symposium, Vol. 3*. American Geophysical Union, Washington, DC, 58–86.

TUCHOLKE, B. E. & VOGT, P. R. (eds) 1979*a*. *Initial Reports of the Deep Sea Drilling Project: vol, 43*. US Government Printing Office, Washington, DC.

TUCHOLKE, B. E. & VOGT, P. R. 1979*b*. Western North Atlantic: sedimentary evolution and aspects of tectonic history. *In*: TUCHOLKE, B. E., VOGT, P. R. *et al.* (eds) *Initial Reports of the Deep Sea Drilling Project*, **43**. US Government Printing Office, Washington, DC, 791–825.

Deposition of sedimentary organic matter in black shale facies indicated by the geochemistry and petrography of high-resolution samples, Blake Nose, western North Atlantic

CHARLES E. BARKER[1], MARK PAWLEWICZ[2] & EMILY A. COBABE[2,3]

[1]*US Geological Survey, Box 25046, MS 977, Denver, CO 80225, USA (e-mail: barker@usgs.gov)*

[2]*Department of Geosciences, University of Massachusetts, Amherst, MA 01003, USA*

[3]*Present address: Department of Geological Sciences, Montana State University, Bozeman, MT 59717, USA*

Abstract: A transect of three holes drilled across the Blake Nose, western North Atlantic Ocean, retrieved cores of black shale facies related to the Albian Oceanic Anoxic Events (OAE) 1b and 1d. Sedimentary organic matter (SOM) recovered from Ocean Drilling Program Hole 1049A from the eastern end of the transect showed that before black shale facies deposition organic matter preservation was a Type III–IV SOM. Petrography reveals that this SOM is composed mostly of degraded algal debris, amorphous SOM and a minor component of Type III–IV terrestrial SOM, mostly detroinertinite. When black shale facies deposition commenced, the geochemical character of the SOM changed from a relatively oxygen-rich Type III–IV to relatively hydrogen-rich Type II. Petrography, biomarker and organic carbon isotopic data indicate marine and terrestrial SOM sources that do not appear to change during the transition from light-grey calcareous ooze to the black shale facies. Black shale subfacies layers alternate from laminated to homogeneous. Some of the laminated and the poorly laminated to homogeneous layers are organic carbon and hydrogen rich as well, suggesting that at least two SOM depositional processes are influencing the black shale facies. The laminated beds reflect deposition in a low sedimentation rate (6m Ma^{-1}) environment with SOM derived mostly from gravity settling from the overlying water into sometimes dysoxic bottom water. The source of this high hydrogen content SOM is problematic because before black shale deposition, the marine SOM supplied to the site is geochemically a Type III–IV. A clue to the source of the H-rich SOM may be the interlayering of relatively homogeneous ooze layers that have a widely variable SOM content and quality. These relatively thick, sometimes subtly graded, sediment layers are thought to be deposited from a Type II SOM-enriched sediment suspension generated by turbidites or direct turbidite deposition.

This study focuses on three holes that recovered dark Cretaceous age sediments, during the Ocean Drilling Program (ODP) Leg 171B transect across the Blake Nose (Figs 1 and 2). These dark sediments, sometimes laminated, typically sedimentary organic matter (SOM) enriched, were deposited during Early Albian and Late Albian time, and are related to the Oceanic Anoxic Events (OAE) 1b and 1d and, herein, are collectively termed the black shale facies. These sediments were deposited in a continuously widening ocean basin and on the Blake Nose, which were both largely developed by Early Cretaceous time. The Blake Nose is a gentle slope that projects eastward towards the abyssal plain from the continental margin and the Blake Plateau (Figs 2 and 3). Palaeo-water depth during Cretaceous time was from about 100 m toward the west end of the transect to 1500 m to the east, whereas present-day depth spans 1000–2700 m. The Blake Nose, never deeply buried, is covered by Jurassic to Palaeogene age strata. The sedimentary cover remains unmodified by significant burial diagenesis. Thus, the recovered Cretaceous black shale samples vary with palaeo-water depth, age and position relative to terrigenous sediment supply from the North American continent and marine-generated allochthonous sediment.

Because of their importance to the oil industry as source rocks and their often enigmatic origin, information on the location, origin and

From: KROON, D., NORRIS, R. D. & KLAUS, A. (eds). 2001 *Western North Atlantic Palaeogene and Cretaceous Palaeoceanography*. Geological Society, London, Special Publications, **183**, 49–72.

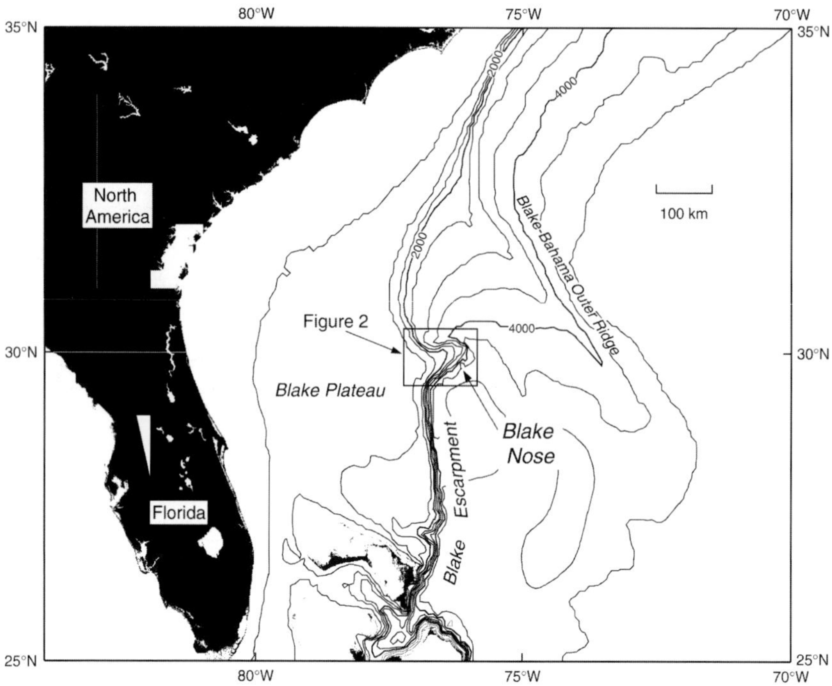

Fig. 1. Selected features in the Blake Nose region, western North Atlantic. Contour interval is 1000 m. Geographical details and drill sites occupied during ODP Leg 171B shown in Fig. 2. Figure modified from Norris *et al.* (1998).

preservation of organic matter in black shale facies is of wide interest. Consequently, there are also many ideas, some conflicting, on the relative importance and degree of interplay of natural factors in their origin. Summaries by Tyson (1987) and Wignall (1994) indicated six common factors important in controlling black shale deposition: (1) sediment texture (in particular grain size); (2) water depth (reworking); (3) primary productivity; (4) rate of sedimentary organic matter (SOM) supply (including both marine and terrestrial sources); (5) rate of total sediment accumulation (site isolation); (6) preservation of the accumulated sediment and SOM (usually attributed to bottom-water oxygenation, or chances of burial). In particular, the geological setting of the Blake Nose makes it possible to isolate and study the impact of sediment supply, water depth, and SOM sources and their preservation on the character of the black shale facies recovered during Leg 171B.

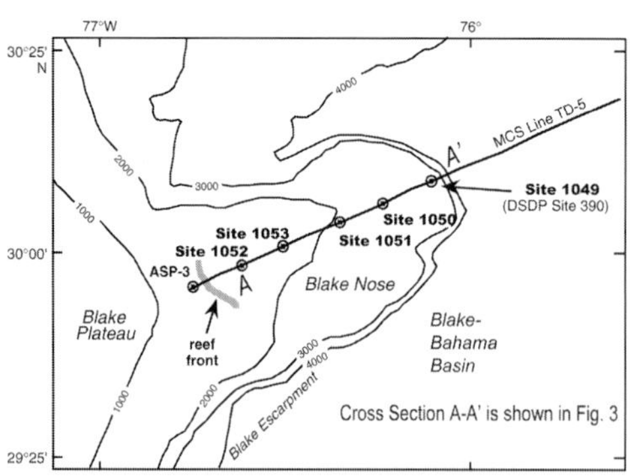

Fig. 2. Drill sites occupied during ODP Leg 171B, location of the seismic reflection profile and selected features in the Blake Nose region, western North Atlantic. Contour interval is 1000 m. Interpretation of the seismic line shown in Fig. 3. Figure modified from Norris *et al.* (1998).

Geological setting

Summarizing Norris *et al.* (1998), the Blake Nose rests upon a basement complex largely composed of intrusive and volcanic rocks that penetrated a continental crust thinned during extension and rifting associated with the opening of the Atlantic Ocean in late Triassic to early Jurassic time. Subsequent to rifting, up to 10 km of Jurassic to mid-Cretaceous carbonate platform sediments were deposited on this basement complex. By Barremian–Aptian time the reef

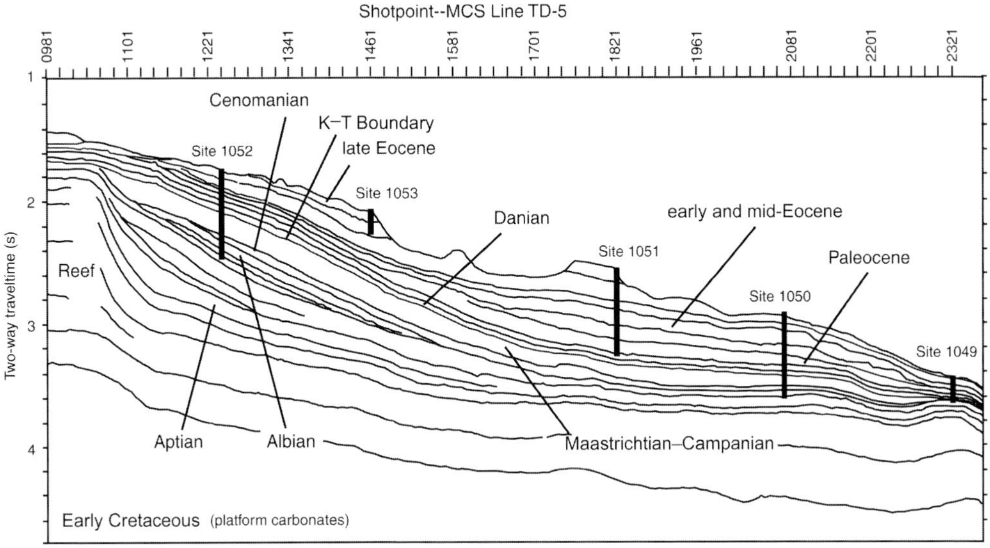

Fig. 3. Interpretation of major reflectors and their ages along seismic reflection profile TD-5 by Norris *et al.* (1998). Also shown are the location of holes drilled during ODP Leg 171B. Location of line shown in Fig. 2. For depth reference Hole 1052E reached a total depth of 685 mbsf.

tract initially formed near the Early Cretaceous margin had stepped back some 40–50 km to the head of the Blake Nose as the Atlantic Ocean continued to open. Clinoforms composed of Aptian–Early Albian age carbonate debris extend eastward down the Blake Nose from the reef tract and partly overlap the crest of the reef itself. As the reef was covered by sediment in late Aptian time, its growth ceased thereby terminating the supply of carbonate debris to the head of the Blake Nose. Possibly related to this event is the deposition of a condensed section of greenish grey to reddish brown clays near the toe of the Blake Nose at Site 1049. The crest of the Aptian reef to the west does not appear to have been substantially eroded and is taken as indicating near sea level in mid-Cretaceous time. Given that the depth difference from head to toe down the Blake Nose is about 1500 m, the deepest nannofossil clays must have been deposited near this depth.

The lowermost Albian sequence at Site 1049 contains a 46 cm thick black shale facies correlative with the black shale facies related to OAE 1b (Fig. 4). The laminated black shale subfacies units consist of dark, finely bedded, nannofossil claystone with pyrite. The laminated layers are interbedded with homogeneous or more thickly laminated subfacies units that represent oozes composed of silt-sized dolomite rhombs nannofossil chalk and clay. The more homogeneous subfacies units show little or no evidence of bioturbation and sometimes show subtle graded bedding. The rate of sediment accumulation was 6 m Ma^{-1} during deposition of the black shale facies at Site 1049 (Norris *et al.* 1998). The features of the condensed Albian section found at Site 1049 suggest deposition under generally sediment-starved conditions.

The Upper Albian clinoform (Fig. 3) was partially cored at Sites 1050 and 1052. The black shale facies recovered consists of a light to dark greenish grey clay-rich, laminated mudstone that contains silt-sized organic debris, carbonate shell debris, quartz, feldspar and other detrital grains. The mudstone beds typically alternate with an SOM-poor, light grey marl also containing silt-sized organic debris, quartz, feldspar and other detrital grains. The repetition of sediment packages over tens of metres suggests cyclic deposition, with each cycle 0.5–1.5 m thick. Infrequently, the mudstones contain thin beds of mixed siliciclastic and carbonate grainstone that may show cross-bedding. Bioturbation is common in the non-laminated portions of the black shale facies at both sites but rare in the laminated portions. In some dark to light layering transitions bioturbation increases upward from the laminated dark portions to the lighter marls. Hole 1050C black shale facies often show chaotic to convolute bedding, syn-sedimentary fracturing, micro-faulting and other features consistent with soft sediment deformation that imply pervasive slumping.

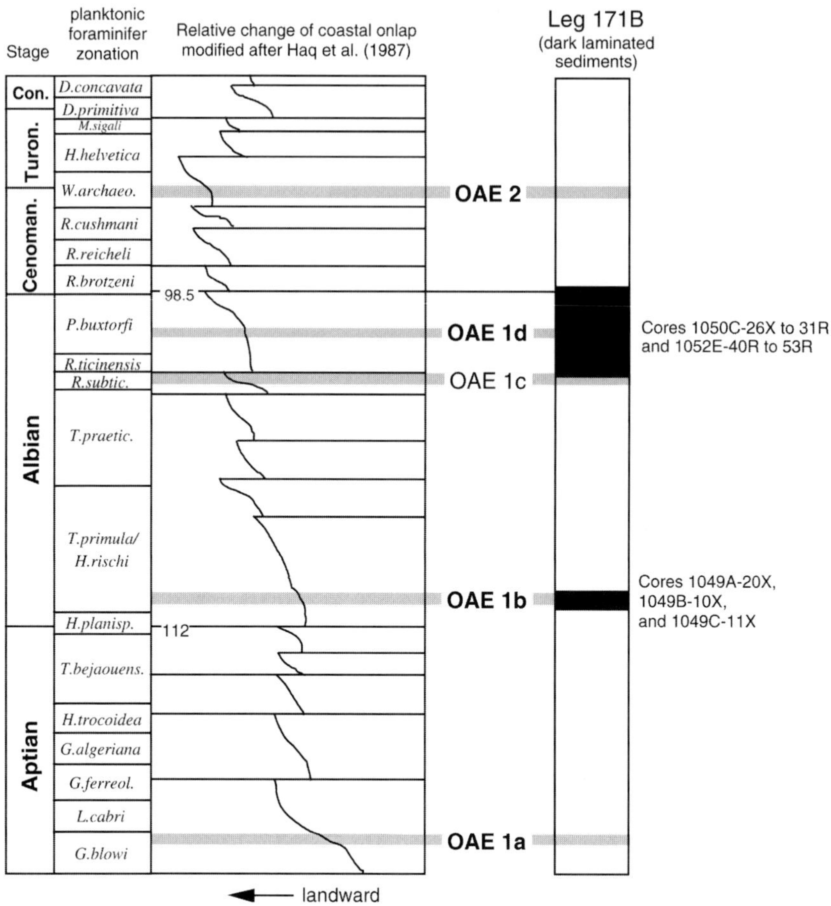

Fig. 4. The relationship of the black shale facies recovered in holes drilled during ODP Leg 171B to the onlap curve and Ocean Anoxic Events correlated by Haq *et al.* (1987). From Norris *et al.* (1998).

Hole 1052E contains some sediment packages that show occasionally cross-bedded claystone with calcareous microfossil ooze that grades upward to fine dark laminated claystone. These dark greenish grey, variably laminated mudstones have slightly elevated SOM contents shown by the average total organic carbon (TOC) values of 0.6 wt % in Hole 1050C and 0.7 wt % in Hole 1052E. The low overall TOC contents may be related to sediment dilution as a result of a rate of sediment accumulation of 17 m Ma^{-1} at Site 1050C and 43 m Ma^{-1} at 1052E. In both holes, this intermittently SOM-enriched laminated section is chronologically related to OAE 1d, which was deposited during a oceanic lowstand (Fig. 4).

Towards the base of Hole 1052E, the Albian sediments are variably bioturbated, dark greenish grey, sandy siltstones thought to have been deposited in middle to outer shelf environments at near storm wave base. Grain size decreases and lamination increases upward within the upper clinoform, suggesting that water depth increased as deposition proceeded. Decompaction of the clay-rich strata in the Late Albian clinoforms as well as the presence of marine rocks on a then-submerged Blake plateau suggest that these sediments were deposited at depths of about 100–200 m (Norris *et al.* 1998); much shallower than the 1500 m indicated for Sites 1049 and 1050.

The early Albian and Aptian age strata encountered from 144–158 mbsf (metres below sea floor) at Site 1049 are mostly composed of non-carbonate materials ranging from 14 to 56 % and averaging 34 wt % $CaCO_3$ that appear rich in clay minerals (Norris *et al.* 1998). Within this interval, an early Albian-age black shale, occurring just below 140 mbsf in Hole 1049B, is locally rich in sapropelic kerogen. Only a small piece of this black shale was recovered during drilling at Site 390 and, consequently its significance was not recognized at that time (Erdman & Schorno 1978). The black shale at Site 1049 consists of medium-grey clay-rich mudstone interbedded with sparse to locally abundant, homogeneous to subtly graded layers of cream dolomite laminae, nannofossil ooze and translucent brownish black sapropelic layers (Fig. 5). In Holes 1049A and 1049C, a 46 cm thickness of black shale facies was found. In Hole 1049B, however, this sample interval was compressed during coring to about 20% of its original thickness and only 9 cm was recovered. The samples for Hole 1049B were not used in this study but the initial results of their analysis were reported by Norris *et al.* (1998). In all three holes, the coring process caused drag folding and sometimes thinning and homogenization of the beds towards the contact with the core barrel liner (Fig. 5).

The Lower Albian black shale (OAE 1b) at Site 1049 was selected for high-resolution (sub-centimetre scale) sampling because of the apparent high content of sapropel in a compact 46 cm interval compared with tens of metres of Upper Albian black shale recovered from Holes 1052 and 1050. The black shale facies plus a few centimetres of the bounding chalk beds were divided into 76 layers primarily using each layer's unique degree of lamination and colour as a guide (Fig. 5; also see Appendix).

Burial history

The Blake Nose structural development and subsidence was largely complete in latest Jurassic to Early Cretaceous time before the early Albian black shale strata were deposited (Norris *et al.* 1998). The area is now a positive topographic feature and the early Cretaceous history of continuing pelagic sedimentation suggests that it was also a positive feature during deposition of the black shale in early Albian time. Therefore, continental and shelf-derived sediment seems to have largely bypassed this positive area, allowing the SOM-rich black shale facies to accumulate. The minor amounts of terrestrial SOM and mineral debris are attributed to redeposition of terrestrial and shelf sediment, rafting and wind transport.

In contrast, the Blake–Bahama Basin just east of the Blake Nose on the abyssal plain appears to have been a negative feature since Jurassic to Early Cretaceous time, and consequently seems to have received more terrestrial and shelf-derived debris, which has diluted the marine SOM that was simultaneously deposited there (Dow 1978; Kendrick 1978; Herbin *et al.* 1983; Katz 1983).

As discussed in detail by Norris *et al.* (1998), deposition of pelagic sediment continued into Palaeocene time on the Blake Nose but burial of the Cretaceous rocks never exceeded a few hundred metres. Given a bottom-water temperature of near 0 °C across the Blake Nose and a low geothermal gradient typical of the area from mid-Cretaceous time (a post rift phase passive margin) to the present (Sclater & Wixon 1986), these sediments have likely not been heated above 20 °C. These sediments are immature as indicated by a Rock-Eval T_{max} of about 405 °C (see Table 3, below; Rock-Eval data interpreted using the terminology of Merrill (1991)).

Site 1049

By its very nature, the origin of amorphous SOM is not easily characterized or interpreted by reflected light or transmitted light microscopy, both of which are dependent upon observing internal plant-derived structure to identify the SOM. In recent studies, bulk chemical techniques, such as Rock-Eval, have been found to be essential for interpreting the source of amorphous SOM (reviewed by Tyson (1995)). Further, in recent studies, it has proven useful to compare the degree of lamination, and other evidence of bioturbation and metabolic activity, with chemical characteristics to interpret depositional setting. In this study we use the petrology, primarily colour and lamination of the subfacies units, as a way to differentiate them. Consequently, we report the petrographic, petrological and geochemical results together and follow these data with a combined interpretation. This study used four major categories of geochemical techniques: ultimate (C–H–N–S elemental), proximate (Rock-Eval), stable isotopes and biomarker analyses described in the Appendix.

Additional details on geochemical and petrographic techniques applied in this paper including interpretation of Rock-Eval, petrographic differentiation of vitrinite, bitumen v. vitrinite and data have been discussed in the Explanatory Notes (organic geochemistry) and the site reports of ODP Leg 171B (Norris *et al.* 1998) as well as by Emeis & Kvenvolden (1986). It should be noted that for chemical, mineralogical and techno-logical reasons, only samples with over 0.5 wt % TOC are interpreted for Rock-Eval analyses (Peters 1986; Bordenave *et al.* 1993). The geochemical criteria for interpretation of Rock-Eval data in terms of source rock potential are listed by Merrill (1991).

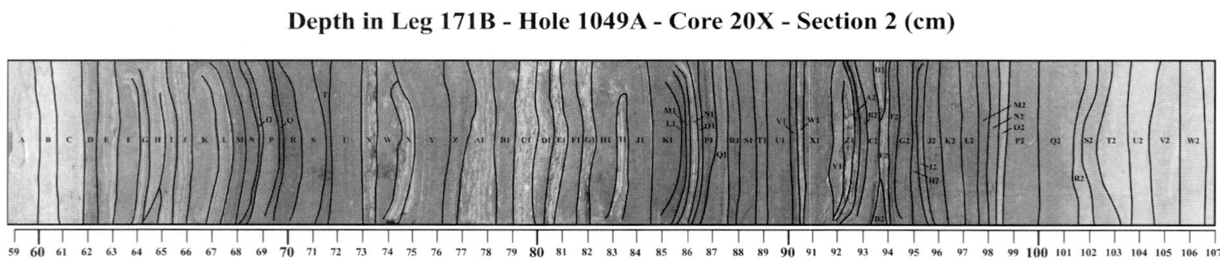

Fig. 5. Photograph of the black shale facies interval, Hole 1049A, ODP Leg 171B, shown with a digital copy of the overlay used to identify and sample the subfacies units. Because the photograph was shot under fluorescent lighting and not filter corrected, the colour shown in this reproduction is more greenish than the core looks to the eye. (Note how coring process appears to have deformed some layers.)

The black shale facies recovered from Hole 1049A can be poor to very rich in SOM as shown by a TOC ranging from below 0.5 wt % (the lower analytical accuracy limit of Rock-Eval) to 12.3 wt % and averaging 4.6 wt % over a *c.* 46 cm thick interval (Fig. 6). Hydrogen indices (HI) of the SOM range from 99 to 790 with a mean of 492 mg hydrocarbon (HC) g^{-1} carbon (Fig. 7). Oxygen content of the SOM was low to moderate with oxygen indices (OI) values ranging from 33 to 394 with a mean of 87 mg CO_2 g^{-1} carbon (Fig. 7). The H/C index (the ratio of Rock-Eval parameters S_2 to S_3) computed from these analyses ranged from 0.35 to 22 with a mean of 7.8 mg HC mg^{-1} CO_2, indicating that the SOM is oil prone. Finally, another computed value, the genetic potential of the SOM to generate hydrocarbons (the sum of Rock-Eval parameters S_1 and S_2) is very good to excellent, ranging from 6.7 to 93 and averaging 26 mg HC g^{-1} sediment. The SOM is immature, as shown by an average T_{max} of about 405 °C (see Table 3, below). If heated into the oil window, these data indicate that most of the black shale facies would constitute a rich, oil-prone source rock.

Geochemical typing of SOM in the black shale facies of Site 1049 (Fig. 8) indicates that the black shale layer contains mostly a Type II SOM. Microscopy indicates that overall the amorphous SOM is composed of mostly granular opaque to dark brown granular bituminite-like material to a translucent brown lamellar to globular material of probably algal affinity. The amorphous SOM when less degraded can be brightly fluorescent. Rock-Eval analyses and organic petrography of these samples suggest that the sample contains SOM derived from detrital alginite fragments and bituminite typically derived from cyanobacteria (Tyson 1987). Microscopy shows the structured SOM to be mostly dark brown to opaque vascular plant debris. At this site, smear slides and prepared palaeontological and petrographic mounts show that traces of humic (terrestrial source) debris occur as a background constituent throughout the Cretaceous strata and within the black shale beds. Most of this material is detroinertinite, diagnosed by its relatively high reflectance and often showing the shard-like shape characteristic of pyrofusinite or open ovoid-like sclerotinite. Maceral analysis (see Appendix) of selected samples confirms that structured organic matter generally constitutes a subordinate proportion of the SOM (Fig. 9). The proportion of preserved fluorescent SOM varies widely in the black shale facies and this variation is positively correlated with the HI. A well-defined trend of increasing carbon and hydrogen content is indicated by the strong correlation of TOC content with the Rock-Eval parameter S_2 (related to hydrogen content by HI S_2/TOC) (Fig. 8). This correlation is typical for marine sediments in the northern Atlantic Ocean (Summerhayes 1987) and is commonly interpreted as an indication of increasing preservation yielding an increasingly carbon- and hydrogen-rich SOM. Preservation is also indicated by an increasing TOC content corresponding to a decreasing OI at Site 1049 (Fig. 10). These data imply that changes in the degree of preservation are probably caused by oxygen-induced degradation that produces two end members: a relatively preserved geochemical Type II SOM during black shale deposition and an altered geochemical Type III–IV SOM before and after black shale facies was deposited (Fig. 7). Within the black shale facies variation in degrees of preservation lead to petrographically mixed samples between these two end members with widely varying concentration in the sediment. This variation in preservation seemingly increases in intensity over time as shown by the two increasing upward trends in the carbon and hydrogen contents (Figs 7 and 8).

Biomarkers

For this work, the distribution of *n*-alkanes within the 1049A black shale was evaluated with regards to overall abundance, biogenic sources and their relationship to other geochemical proxies. Alkanes have been used to provide information with regard to the biogenic sources found in sediment. The most common distinction made in the evaluation of *n*-alkanes is based on the chain length of the compounds, used to determine the relative contributions of algae and higher plants (e.g. Ishiwatari *et al.* 1980; Cranwell *et al.* 1987). Shorter-chain alkanes ($<C_{25}$ with a median value of C_{19}) are generally regarded as having been produced by algae found in the water column (Gelpi *et al.* 1970; Tissot & Welte 1984), whereas longer-chain alkanes ($>C_{23}$ with a median value of C_{29}) are indicative of alkanes generated in the epicuticular leaf waxes of higher plants (Tissot & Welte 1984). These leaf waxes also contain 'odd-over-even' carbon preference, the result of the biosynthetic pathway that generates the compounds. The result is that higher-plant alkanes are generally dominated by C_{27}, C_{29} and C_{31} alkanes (Brassell *et al.* 1978; Tissot and Welte 1984).

In the six biomarker samples evaluated from just below, within and just above the OAE 1b black shale interval from Hole 1049A, the distribution of the alkanes is relatively uniform, ranging from C_{16} to C_{38} (Fig. 11). Only two of the

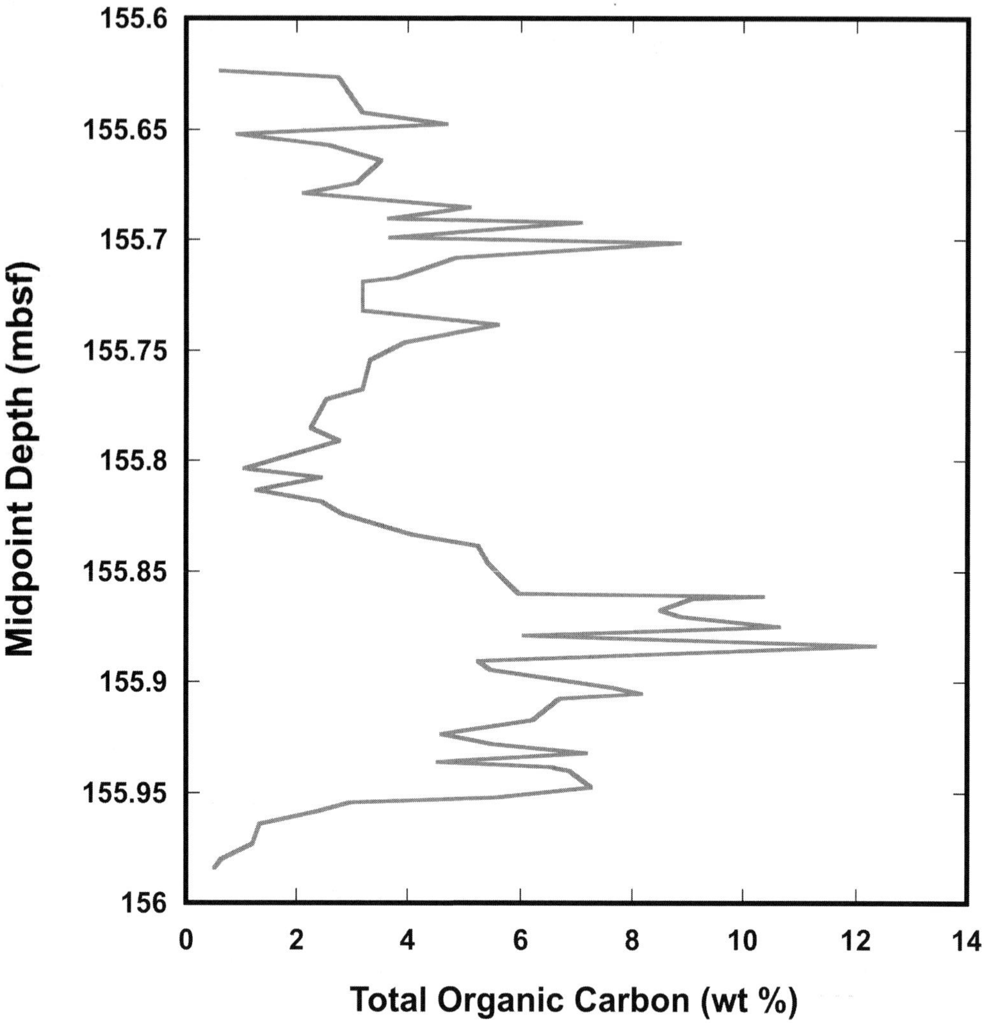

Fig. 6. Total organic carbon (TOC) results for the black shale interval and bounding beds taken from Hole 1049A. Only those samples with a TOC content >0.5 wt % are shown.

samples (II and VI) do not contain the C_{36}, C_{37} or C_{38} alkanes. The sediments have an abundance of shorter-chain alkanes (C_{16}–C_{22}), although each sample has a slightly different composition, often dominated by the C_{17} and either C_{20} or C_{22} alkanes. The preponderance of the shorter-chain alkanes (typically 80–200 ng g^{-1} of sediment per compound) suggests that there is a considerable algal component in the lipid fractions.

All six samples contain long-chain alkanes (C_{23}–C_{38}) with the characteristic odd-over-even distribution found in higher plants. In almost every sample, the concentration of the long-chain alkanes is much smaller than that of the shorter-chain compounds (typically no more than 10–30 ng g^{-1} of sediment per compound), suggesting a much smaller contribution of higher-plant material in the lipid components of these sediments. The exception to this pattern is sample II. The concentrations of the alkanes in the sediment roughly correspond to the overall amount of organic carbon in the sediments. In sediments that have low TOC values (<72.5 %) (biomarker samples I, III, and VI, Appendix), the maximal concentrations of specific alkanes are found in the range of 30–80 ng g^{-1} of sediment. In sediments with higher TOC (4–10 %) (biomarker samples II, VI and V; see Appendix), individual alkanes are found in concentrations as high as 100–270 ng g^{-1} of sediment.

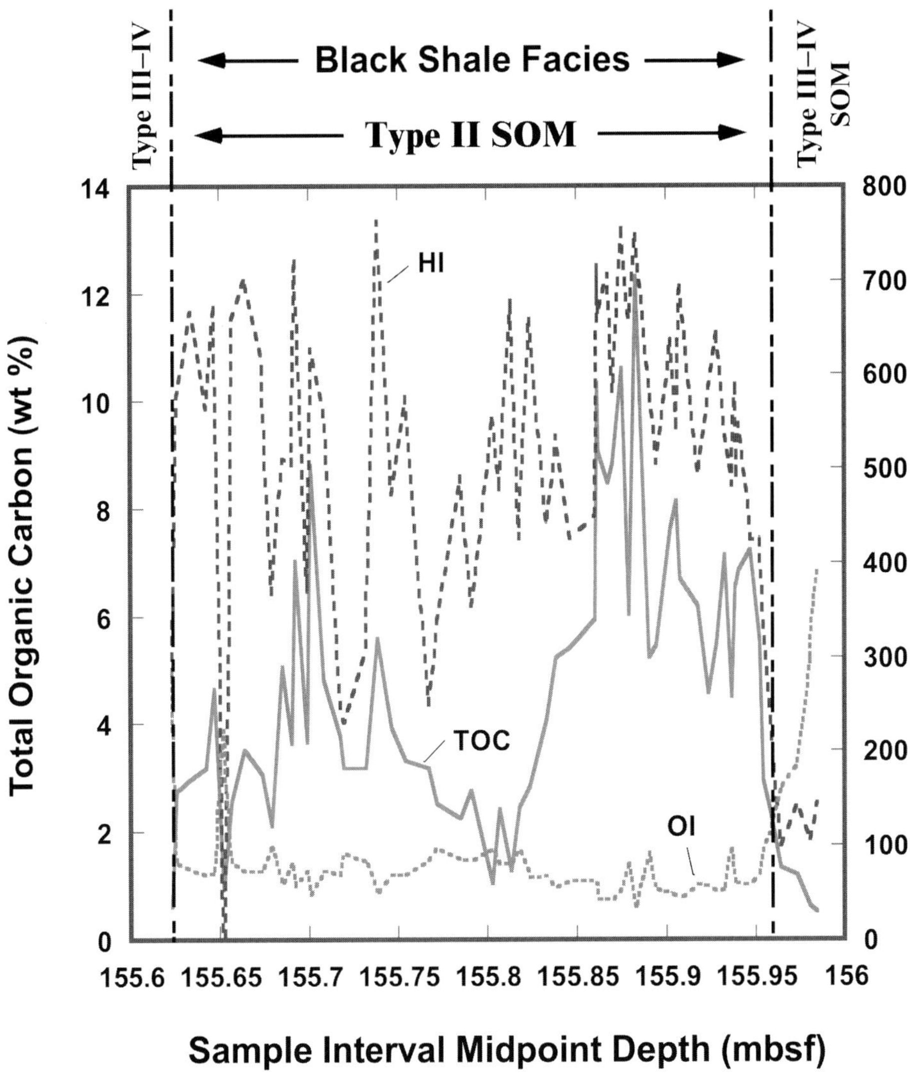

Fig. 7. Hydrogen index (HI) divided by 10, oxygen index (OI) divided by 10 and TOC results for the black shale interval and bounding beds taken from Hole 1049A. The parameters are divided by 10 to allow convenient scaling for comparison. Only those samples with a TOC content >0.5 wt % are shown.

These results suggest that there is a minimal qualitative difference in algae and higher-plant sources in the sediments, during and after the OAE. The amounts of the compounds increase, but the distribution is consistent through the interval. Thus, the biomarker analysis is consistent with the organic petrography.

Sedimentology

The observation that increasing TOC correlates with increased preservation of SOM (i.e. relatively high HI and low OI SOM) can be used to interpret the TOC content of the facies subunits plotted as a function of depth (Fig. 6). The plot of the TOC profile shows two cycles of upward-increasing TOC distinguished by low TOC intervals at the base, midsection and top of the black shale facies. The upward-increasing TOC profile suggests that establishment of a dysoxic-bottom water environment leads to the deposition of black shale, the intensity of the preservation process increases with time within each cycle and then abruptly decreases. This

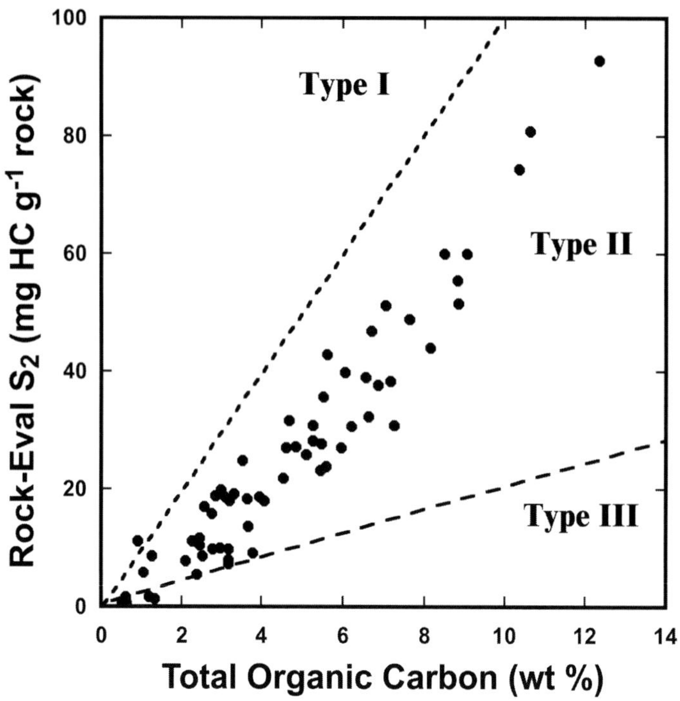

Fig. 8. Sedimentary organic matter typing based on TOC and the Rock-Eval pyrolysis parameter S_2 data for Hole 1049A, based on the method of Langford & Blanc-Valleron (1990). Only those samples with a TOC content >0.5 wt % are shown.

process also seems related to degree of lamination within subfacies units that, considering laminated beds alone, shows an overall upward-increasing number of laminations per layer (Fig. 12) throughout the black shale facies.

Figure 12 also shows that the highly laminated layers alternate with more homogeneous layers (plies per layer near zero). The HI is reliably high when lamination is high, suggesting the greatest SOM preservation is likely to be associated with laminated layers. The more homogeneous subfacies units have a variable but often high HI and TOC (Figs 13 and 14) as well. This relationship suggests that the preservation of SOM is accomplished by a combination of processes. The alternation of laminated beds with poorly laminated beds (i.e. low or no lamination) (Fig. 2) suggests that the deposition of laminated layers was interrupted, with thicker layers being laid down in a solitary, relatively rapid, event. Although possibly related to increased productivity or perhaps destruction of lamination by bioturbation, we envisage a dysoxic bottom water where a low sediment deposition rate and a low degree of bioturbation lead to highly laminated subfacies units. The more homogeneous beds with the variable but sometimes high TOC and HI that interleave with the laminated layers (Fig. 12) suggest that increased deposition rates can be composed of varying amounts of SOM along with mineral matter. As discussed below, we attribute the two processes to a background of gravity settling from overlying water punctuated by a rapid suspension settling of sediment generated by dispersing of turbidite flows. The turbidites would introduce the clay-rich sediment that typifies the black shale facies as shown by the marked decrease in carbonate in black shale interval from 155.6 to 156.1 mbsf in Hole 1049A (Fig. 15).

Organic matter elemental and carbon isotopic composition

Marine sourced SOM has $\delta^{13}C$ values ranging from −20‰ to −25‰ and terrestrial SOM has $\delta^{13}C$ values ranging from −23‰ to −33‰, with a typical value for land plants about −30‰ (Wignall 1994). Given the overlap in ranges between the two SOM sources carbon isotope values are useful, but not conclusive, evidence of

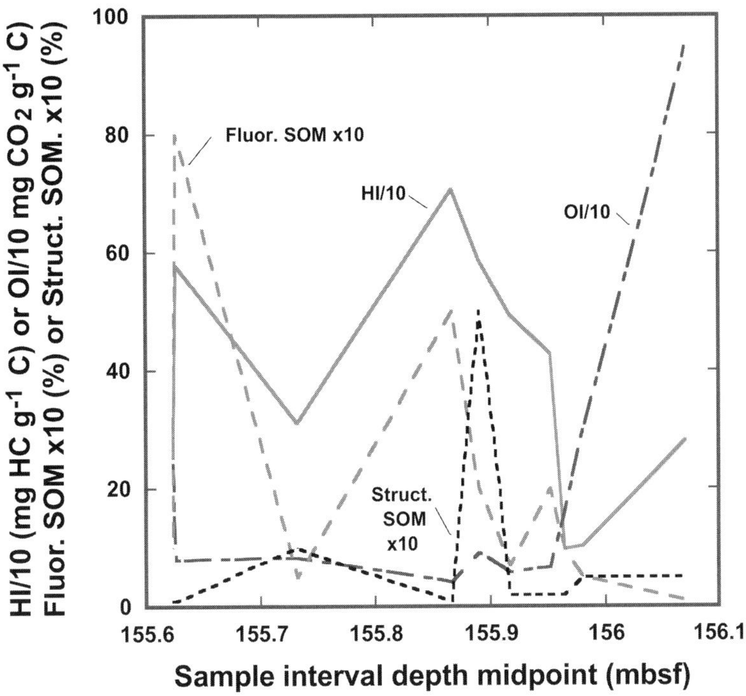

Fig. 9. Maceral analysis of selected subfacies units compared with the HI and OI, Hole 1049A. The percentage of fluorescent (fluor.) and structured (struct.) SOM multiplied by 10 are compared with the HI and OI divided by 10 to allow convenient scaling and comparison.

origin. Interpretation of $\delta^{13}C_{org}$ values is also clouded by SOM preservation and thermal diagenesis altering isotopic composition (Tyson 1995). Thermal diagenesis has a negligible influence on these sediments but preservation varies widely. The results from $\delta^{13}C$ analysis indicate a range between −28.28 and 29.07‰ from core samples from Hole 1049C that approximately duplicate the biomarker samples assembled from the subfacies units in Hole 1049A (see Appendix Table A). These values fall in an intermediate range between the $\delta^{13}C$ marine and terrestrial SOM that are consistent with mostly marine with subordinate terrestrial SOM petrography of these samples. What is surprising is the less than 1‰ difference in these samples that span a range of carbon contents from 0.4 to 13.3 wt % and wide range of preservation indicated by vatiation of the H/C ratio from near 0.98 to 13.2. Inexplicably, no influence of enhanced preservation on the isotopic values is observed.

Interpretation of C/N ratios is problematic in that the natural ranges for marine and terrestrial SOM show a wide overlap (Tyson 1995). Superimposed on these diffuse and overlapping compositional fields are early diagenetic preservation and burial diagenetic effects that can strongly modify C/N ratios. The typically high C/N ratios (Table 1) show that the carbon-enriched zones contain a moderately to strongly nitrogen-depleted SOM that may reflect stronger terrestrial input in the amorphous SOM component than recognized by petrography. A high C/N ratio could also be a reflection of a mixture of degraded, reduced-nitrogen content marine SOM with a minor component of typically high C/N ratio terrestrial SOM.

The S/C ratios are related to degree of oxygenation of the benthic environment (Arthur *et al.* 1984) because the early diagenetic incorporation of organic sulphur appears to be related to the intensity of sulphate reduction (Tyson 1995). Anoxic conditions lead to sulphate reduction and a high sulphur content in the sediments, as indicated by S/C weight ratios of greater than a 0.15 atomic ratio. Anoxic to anaerobic conditions are strongly indicated for biomarker sample IV. Anoxic to anaerobic conditions are consistent with this sample's typically laminated bedding and relatively high

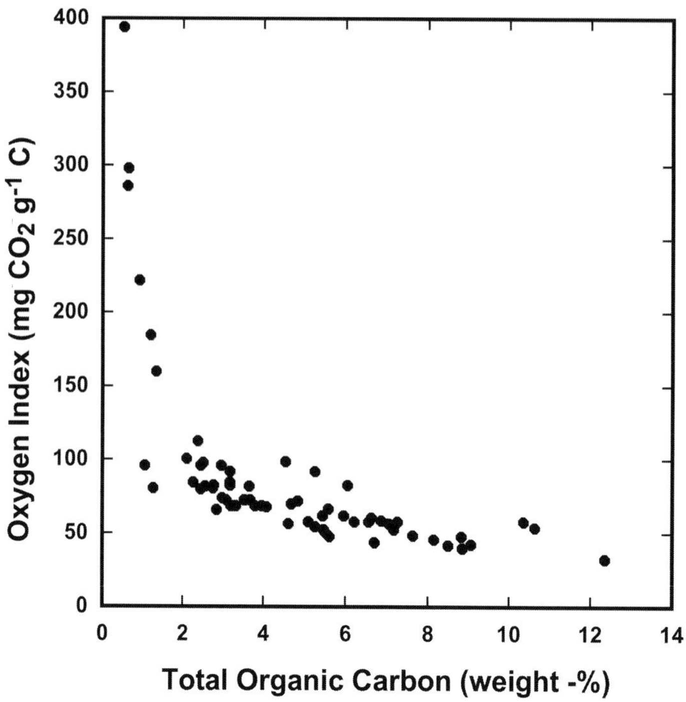

Fig. 10. Comparison of TOC and OI data for Hole 1049A. Only those samples with a TOC content >0.5 wt % are shown.

Table 1. *Results of C–H–N–S element analysis showing selected element contents (in wt %), selected atomic ratios and organic carbon (C_{org}) isotopic composition*

Biomarker sample	Combined subfacies units	TOC range (%)	H/C	C/N	S/C	$\delta^{13}C$ [midrange value] (duplicate analyses)
I	A, B, C	<1	0.98	23.54	0.031	[−28.28] (−28.23; −28.32)
II	M–R	4–6	0.69	20.05	0.021	[−28.59] (−28.56; −28.61)
III	Z–G1	<2.5	1.86	51.29	0.058	[−28.92] (−28.96; −28.87)
IV	L1–Q1	5–10	13.16	29.82	0.41	[−29.01] (−29.06; −28.94)
V	Z1–G2	4–7	2.32	7.12	0.072	[−28.75] (−28.78; −28.72)
VI	N2–W2	<1	1.04	37.51	0.032	[−29.07] (−29.02; −29.12)

These analyses are based on decarbonated samples from Hole 1049C selected to parallel the biomarker sample suite in Hole 1049A (Table A1 in Appendix). $\delta^{13}C_{org}$ values are reported with respect to PDB and the value shown in parentheses is the mean of the duplicate analyses listed below.

TOC and hydrogen contents, which are probably caused by preservation from oxidation.

Site 1052

On the western end of Leg 171B transect is Hole 1052E, which penetrated the Upper Albian black shale facies related to OAE 1d, towards the head of the Blake Nose (Figs 2 and 3). Sediments recovered in Hole 1052E were deposited closer to the shelf break and at a shallower palaeo-water depth than those at Sites 1049 and 1050 to the east. The black shale facies accumulated at the highest sedimentation rate at Site 1052 (Table 2). The black shale facies recovered from Hole 1052E is lean in SOM, as shown by a TOC ranging from less than 0.5 to 1.06 wt % and averaging 0.7 wt %. over a *c.* 175 m thick interval (Fig. 16). HI values of the SOM are also low, ranging from 40 to 192 with a mean of 83 mg HC g^{-1} carbon. Oxygen content of the SOM was moderate to high with OI values ranging from 129 to 272 with a mean of 194 mg CO_2 g^{-1} carbon (Fig. 16). The H/C index

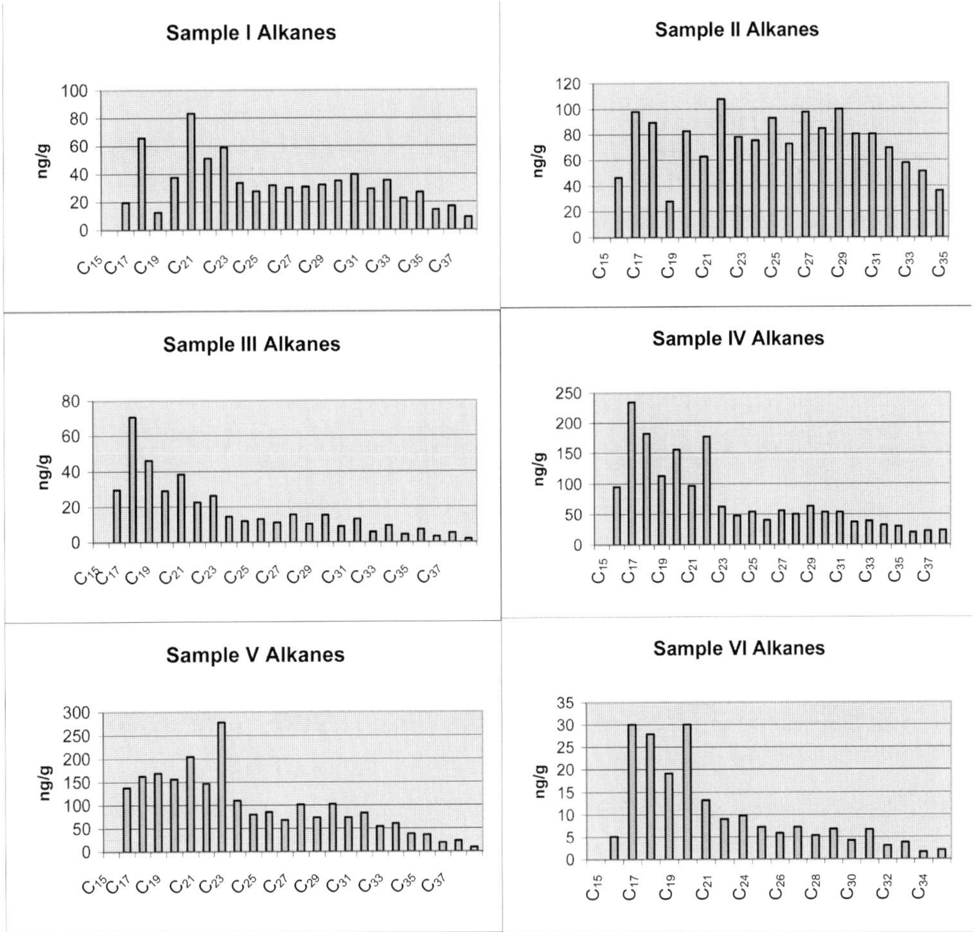

Fig. 11. Biomarker analyses for composite samples I–VI (see Appendix; in Hole 1049A). (See text for discussion.)

Table 2. *Sedimentation rate during black shale facies deposition compared with average of selected Rock-Eval results*

Leg 171B Hole	Present distance to shelf break (km)	Palaeo-water depth during black shale facies deposition (m)	Sedimentation rate (m Ma^{-1})	Mean TOC (wt %)	Mean HI (mg HC g^{-1} C)	Mean OI (mg CO_2 g^{-1} C)	Mean S_2/S_3
1052E	20	100–200	43	0.6	90	208	0.42
1050C	50	~1500	17	0.7	89	266	0.38
1049A	60	1500	6	4.6	492	87	7.8

Sedimentation rate data from Norris *et al.* (1998). The shelf break is taken as the 1000 m bathymetry contour shown in Fig. 2. S_2/S_3 is a measure of the H/C content and hence the preservation of the SOM; in general, the higher the S_2/S_3, the greater the hydrocarbon generative capacity.

computed from these analyses ranged from 0.23 to 0.92 with a mean of 0.42 mg HC mg^{-1} CO_2, indicating that the SOM is gas prone. The potential of the SOM to generate hydrocarbons is poor, indicated by the sum of Rock-Eval parameters S_1 and S_2 which ranges from 0 to 1.6 and averages 0.47 mg HC g^{-1} sediment (Merrill 1991). The SOM is immature, as shown by an average T_{max} of about 370 °C for 30 samples with TOC |280.5 wt %. Because these sediments contain hydrocarbon-bearing SOM they have some innate capacity to generate hydrocarbons

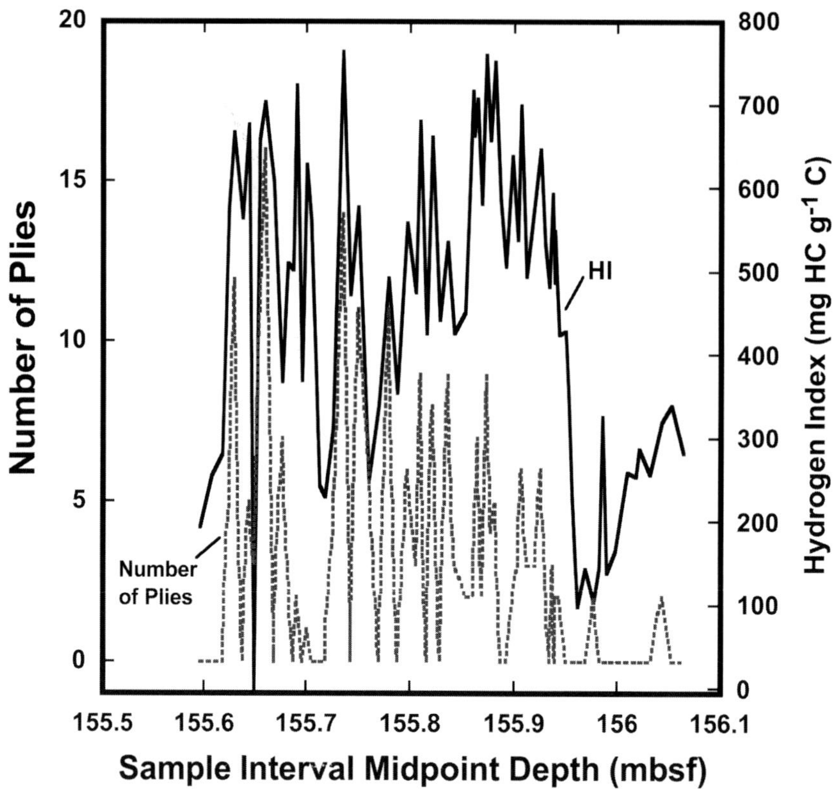

Fig. 12. Comparison of plies (laminae) per subfacies unit (or layer) and HI with depth in Hole 1049A.

but are unlikely to be able to expel the hydrocarbon (Bordenave *et al.* 1993) and therefore are not viable source rocks.

Geochemical typing of SOM in the black shale facies at Site 1052E indicates that the facies contains a Type III–IV SOM (Fig. 17), the most oxidized, least reactive form of SOM. Overall, microscopy indicates that SOM at 1052E is composed of sparse darkened to opaque SOM. Smear slides and prepared palaeontological mounts show traces of humic (terrestrial source) debris occurring sporadically throughout the Cretaceous strata and in the black shale beds. Most of this material is detroinertinite, diagnosed by its relatively high reflectance and typical shard-like shape characteristic of pyrofusinite.

We attribute the source of the poorly preserved SOM in 1052E to its position at the head of Blake Nose, where there was a greater input of terrestrial SOM and sediment, which produces a more dilute, less hydrogen-rich Type III SOM to start with. HI and OI do show an overall increase upward in Hole 1052E (Fig. 16). This increasing-upward trend may be related to an increase in palaeo-water depth causing an enhanced preservation by reducing oxygen levels and decreasing reworking of the SOM. Regardless, far less SOM is preserved compared with Site 1049 and the poor preservation results in a lean generally Type IV SOM-bearing sediment. Overall, the sediments show a geochemical and petrographic character attributable to dilution of SOM by other sediments and poor preservation possibly related to sediment recycling, oxidation and degradation during transport across the shelf.

Site 1050

At Site 1050, the black shale facies was recovered only from Hole 1050C, which lies nearer the toe of the Blake Nose and also nearer Site 1049 (Figs 2 and 3). Consequently, sediment deposition at Site 1050 occurred at palaeo-water depth similar to that of Site 1049 at nearly 1500 m, and sediments were deposited only slightly more rapidly at 17 m Ma^{-1}. Yet the SOM in Hole 1050C shows geochemical and petrographic affinities (Figs 18 and 19) like those at Hole 1052E. The SOM is Type III–IV in low

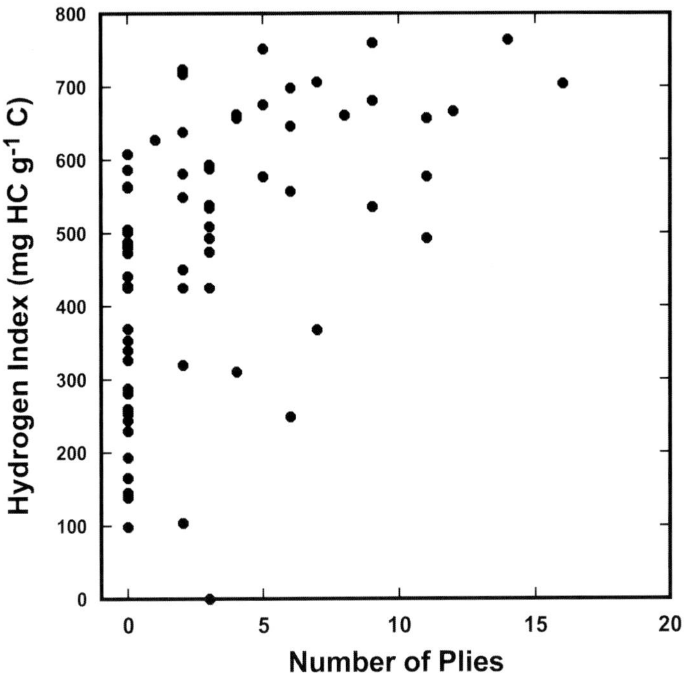

Fig. 13. Comparison of the HI with plies (laminae) per subfacies unit (or layer) in Hole 1049A.

quantities as indicated by a TOC ranging from less than 0.5 to 1.4 wt % and averaging 0.8 wt % over a *c.* 45 m thick interval (Fig. 18). The HI values of the SOM are also low, ranging from 47 to 148 with a mean of 83 mg HC g^{-1} carbon (Fig. 18). Oxygen content of the SOM was moderate to high, with OI values ranging from 119 to 307 with a mean of 216 mg CO_2 g^{-1} carbon (Fig. 18). The H/C index computed from these analyses ranged from 0.15 to 0.66 with a mean of 0.40 mg HC mg^{-1} CO_2, indicating that the SOM is gas prone. The genetic potential of the SOM to generate hydrocarbons (the sum of Rock-Eval parameters S_1 and S_2) ranges from 0.30 to 1.32, averaging 0.76 mg HC g^{-1} sediment. The SOM is immature, as shown by an average T_{max} of 373 °C for 22 samples that have TOC >0.5 wt %. All of these data ranges and averages are similar to those found in Hole 1052E. Overall, the sediments in Hole 1050C contain a hydrocarbon-bearing SOM that may have some innate hydrocarbon-generating capacity. Despite this, they are unlikely to expel the hydrocarbon (Bordenave *et al.* 1993) and therefore are not viable source rocks.

We attribute the source of the SOM in 1050C, in spite of its distal position on the Blake Nose, to deposition of sufficient terrestrial and shelf-derived sediment bearing a less hydrogen-rich Type III SOM and poor preservation after deposition. Massive slumping of the sediment is common in Hole 1050C (Norris *et al.* 1998) and sediment rollover would re-expose the buried sediment to further bioturbation and metabolic oxidation of the hydrogen and carbon in the SOM. The overall low TOC content, the low HI and high OI SOM signature, along with evidence of bioturbation, indicate a relatively oxygenated benthic environment. Thus, the marine Type II SOM is introduced into an depositional environment that did not preserve it and allowed its degradation to a geochemical Type III–IV SOM. Alternatively, the marine Type II SOM contribution is of such low volume (low productivity) that it is geochemically unrecognizable when diluted in the abundant mineral matter that accumulated along with the SOM.

Discussion

To reiterate, the factors important in SOM-rich black shale deposition are: (1) sediment texture (sealing effects by finer grain sizes); (2) water depth (deeper water tends to be dysoxic water, and increases isolation from sediment dilution); (3) primary SOM productivity (local generation); (4) rate of allochthonous SOM supply; (5) rate of total sediment accumulation (dilution of SOM in

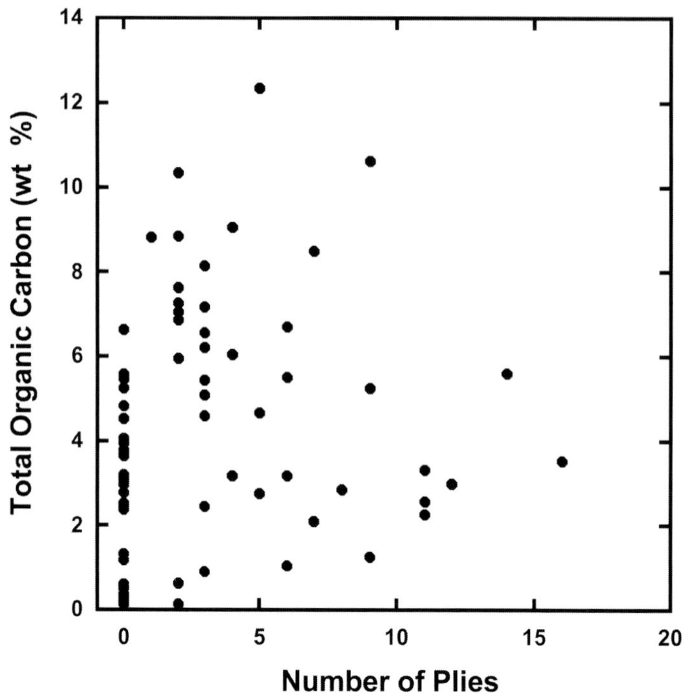

Fig. 14. Comparison of the TOC with plies (laminae) per subfacies unit (or layer) in Hole 1049A.

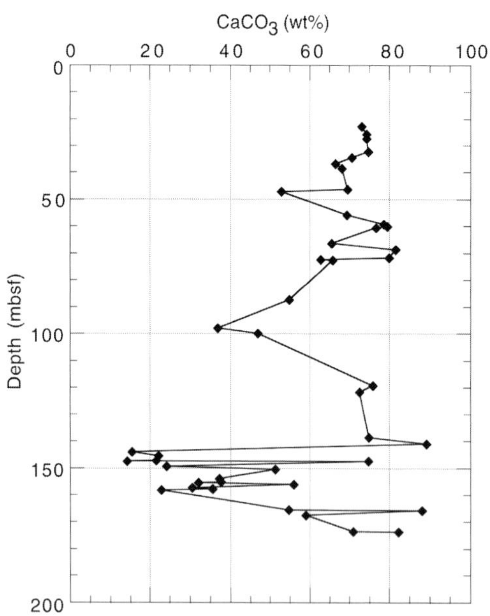

Fig. 15. Comparison of calcium carbonate ($CaCO_3$) content with depth in Hole 1049A. From Norris *et al.* (1998).

mineral sediment); (6) preservation of the accumulated sediment (usually related to bottom-water oxygenation or burial). This study can offer only a limited perspective on the effects of grain size, as most of these beds are texturally similar and fine to very fine grained. In particular, the well-known geological setting of the Blake Nose makes it possible to isolate and study the impact of sediment supply and rate of burial (i.e. increasing isolation from mineral sediment) on the black shale facies recovered during Leg 171B. In addition, the analytical techniques that we applied allow us to comment only on SOM preservation and its possible sources in the context of potential regional sediment provenance.

SOM preservation

At a water depth of 4000 m, <1 % of the organic matter produced in surface layers reaches the ocean floor (Arthur *et al.* 1984). Type II SOM is rapidly consumed by the benthic biomass and much of this SOM reaching the ocean floor is an inert, poorly preserved, geochemical Type III–IV. Consequently, most ocean-floor sediment contains about 0.3 wt % TOC (McIver 1975) of

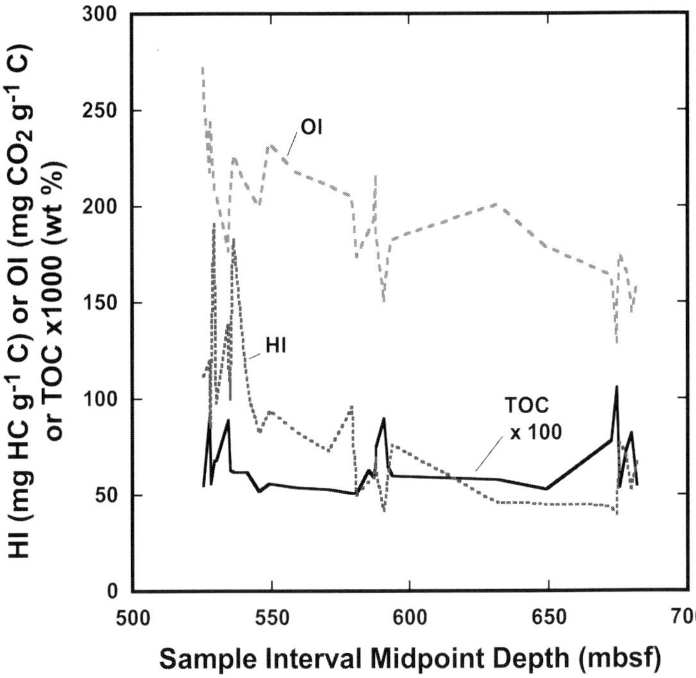

Fig. 16. Comparison of HI, OI and TOC times 100 with depth in Hole 1052E. TOC is multiplied by 100 to allow for convenient scaling and comparison. Only those samples with a TOC content >0.5 wt % are shown.

mostly inert SOM. Although the Blake Nose Albian sediments were deposited at 1500 m or shallower they reflect this process and generally contain <0.5 wt % TOC (Norris *et al.* 1998). Therefore, in the black shale facies there is a marked enrichment in TOC and hydrogen content that may be in part due to preservation of marine SOM. In modern ocean basins, dysoxic conditions in the water column reduce the benthic biomass and metabolic oxidation of the SOM. In the black shale facies, increasingly dysoxic conditions are shown by increasing TOC corresponding to a decreasing oxygen content in the SOM (Fig. 10). As stated above, we interpret this trend as showing an increasing preservation of the SOM; that is, decreasing metabolic oxidation of carbon and hydrogen in SOM that results in a trend of increasing TOC content is highly correlated with increasing H content (Fig. 8). It is not due to a change in SOM sources. Petrography shows that this SOM, before black shale facies deposition, is composed mostly of marine amorphous SOM that has been altered to a geochemical Type III and is mixed with a minor component of Type III–IV terrestrial SOM, mostly detroinertinite. At this time organic matter preservation was moderate, with the sediment showing up to 1 wt % TOC.

However, when black shale facies deposition commenced, the geochemical character of the SOM changed from a relatively oxygen-rich Type III to relatively hydrogen-rich Type II, fluorescent SOM (Fig. 9). Petrographic observations consistent with biomarker and geochemical data indicate a mixture of marine SOM with minor terrestrial SOM that is independent of lithology. Thus, the SOM sources do not appear to change during the transition from light grey calcareous ooze to the black shale facies and back. In the black shale layers, TOC increases to a mean of 4.6 wt % as the HI increases to a mean of 492 mg HC g^{-1} C whereas the OI decreases markedly (Fig. 7). Again, we attribute this trend to a reduced oxygen environment (Fig. 10) leading to increased preservation of TOC and hydrogen both in the water column and on the ocean bottom as indicated by the SOM-enriched layers being non-bioturbated and mostly highly laminated.

Preservation of SOM on the Blake Nose is not enhanced by rapid burial or water depth. The moderate to high sedimentation rates during black shale facies deposition at Sites 1050 and 1052, relative to Site 1049, lead to low concentrations of poor quality SOM (Table 2). The water depth at Sites 1050 and 1049, at about

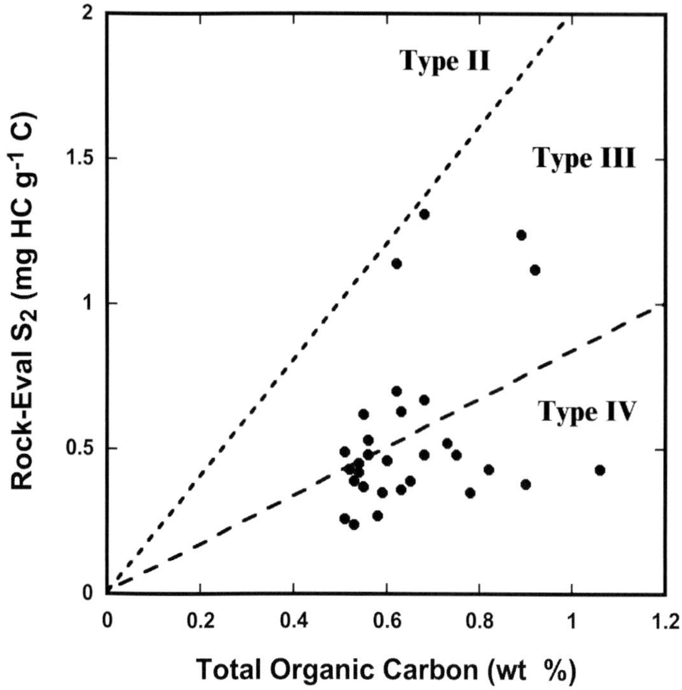

Fig. 17. Sedimentary organic matter typing based on TOC and the Rock-Eval pyrolysis parameter S_2 data for Hole 1052E, based on the method of Langford & Blanc-Valleron (1990). Only those samples with a TOC content >0.5 wt % are shown.

1500 m is far deeper than that at Site 1052 (200 m), so this factor does not seem to be a controlling element. Apparent distance from the palaeo-shelf break (Table 2) is also not a strong control because redeposition processes can move large amounts of sediment laterally across the Blake Nose as indicated by the widespread occurrence of turbidites and slumping features. Other sedimentological evidence for the processes of SOM preservation or deposition is that during black shale deposition, carbonate content is strongly decreased as non-carbonate matter increases (mostly SOM, silica, clay minerals, etc.) (Fig 15). Because mid-oceanic waters typically show low productivity and marine SOM is mostly generated in the overlying water and deposited by settling, the typical abyssal sediment is a calcareous to siliceous ooze containing about 0.3 % TOC (McIver 1975). If the bottom water is anoxic, as is indicated during black shale facies deposition here, decomposition of organic matter in the water column and sediment pore waters may lower the pH to well below carbonate saturation and produce dissolution of any carbonate settling in the water column (see discussions on both sides of this idea by Einsele *et al.* (1991)). Thus, some sapropels may result from elimination of the commonly available carbonate from the system leaving an insoluble residue that is SOM rich. However, because calcareous nannofossils and early diagenetic dolomite rhombs are commonly well preserved in the black shale facies (Norris *et al.* 1998), carbonate dissolution is not indicated. We attribute the concentration of SOM to the isolation of Site 1049 from sedimentation allowing the marine SOM to accumulate without much dilution. It also appears that the rate of SOM supply was higher than expected in the mid-ocean. Because this is usually an area of low productivity and good chances of degradation of the SOM during vertical settling, other sources of H-rich SOM are indicated.

The source of a marine H-rich SOM from simple gravity settling from the overlying water is problematic because, before black shale deposition, the marine SOM supplied to the site is geochemically a Type III–IV, so simply increasing supply or enhancing preservation at best can only provide a Type III SOM. A clue is that these beds are clay rich and also show interlayering of poorly laminated to homogeneous layers that have a widely variable SOM content and quality. These poorly laminated, sometimes subtly

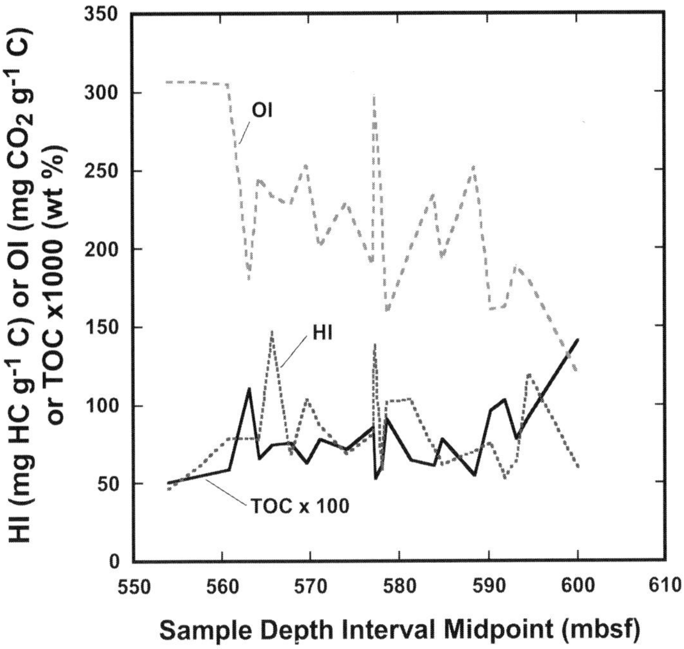

Fig. 18. Comparison of the HI, OI and TOC times 100 with depth in Hole 1050C. TOC is multiplied by 100 to allow for convenient scaling and comparison. Only those samples with a TOC content >0.5 wt % are shown.

graded, layers are thought to be deposited from a sediment suspension related to turbidite passage from the shelf edge to the west around the palaeo-topographically high Blake Nose down to the Blake–Bahama Basin in the east. During OAE 1b and 1d there was a sea-level lowstand and shelf sedimentation moved eastward as the sea regressed. Albian black shale facies related to OAE 1d in Holes 1050C and 1052E reflect an increasing degree of terrestrial SOM input and mineral sediment dilution because of the proximity of the shelf edge to its terrestrial and shelf sediment supply. Similar shelf edge conditions probably existed west of Site 1049 during the early Albian lowstand as well. As regression of the sea moved the high-productivity coastal waters towards the shelf edge this provided an increased sediment supply and slope conditions for sediment redeposition as turbidites. As the turbidites moved from the shelf edge to the ocean floor they would tend to flow off the palaeo-high Blake Nose, forming a low sedimentation near its toe at Site 1049 (Fig. 1). Other turbidites originating from the Blake Plateau and Blake escarpment would flow out subparallel to the Blake Nose and potentially generate large amounts of suspended sediment near Site 1049. The hydrogen-rich algal and amorphous SOM, which can have a density near that of water (Tyson 1995), would be strongly fractionated into this suspension and inert SOM about 1.3 g cm^{-3} would be either settle more rapidly or tend to remain in the turbidite flow. This concentration of H-rich SOM in the dispersing suspended sediment cloud would settle relatively rapidly, producing the homogeneous layers on the laminated layers. The subtly graded layers may present a density fractionation from the suspended sediment cloud or the toes of turbidites themselves. This model asserts that the key to depositing a high-quality black shale facies is some mechanism, such as sediment bypass, for reducing mineral sedimentation while an allochthonous source of H-rich SOM is simultaneously supplied from the overlying ocean or sediment redeposition. The influence of redeposited shelf- and terrestrial-derived sediments on the black shale facies explains how an SOM-rich and clay-rich black shale facies is deposited in a low-productivity, isolated deep-ocean setting.

Regional petroleum prospects

The black shale facies recovered from Site 1049A, which averages 4.6 wt % TOC, has a mean HI of 492 mg HC g^{-1} C and a genetic potential ($S_1 + S_2$) of nearly 26 mg HC g^{-1} rock (Table 3), and is a significant oil source rock

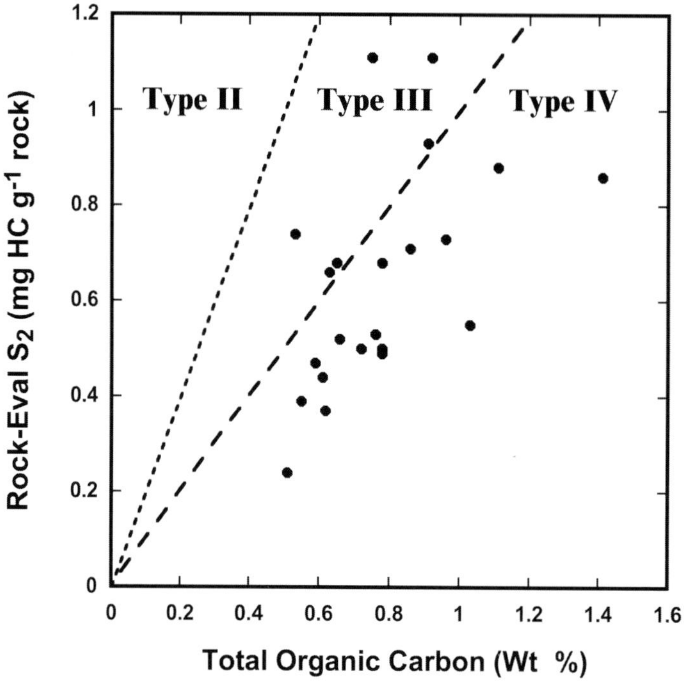

Fig. 19. Sedimentary organic matter typing based on TOC and the Rock-Eval pyrolysis parameter S_2 data for Hole 1050C, based on the method of Langford & Blanc-Valleron (1990). Only those samples with a TOC content >0.5 wt % are shown.

Table 3. *Summary of Rock-Eval data for black shale facies sub-units C to M2 (shown in Fig. 5) in Hole 1049A*

	T_{max} (°C)	S_1 (mg HC g^{-1} rock)	S_2 (mg HC g^{-1} rock)	S_1+S_2 (mg HC g^{-1} rock)	S_3 (mg CO_2 g^{-1} rock)	PI	S_2/S_3	S_4	TOC (wt %)	HI (mg HC g^{-1} TOC)	OI (mg CO_2 g^{-1} TOC)
Mean	405	0.02	25.93	25.95	3.04	0.00	7.83	2.16	4.63	492	87
Maximum	507	0.16	92.90	93.06	6.02	0.01	22.33	7.75	12.34	764	394
Minimum	387	0.00	6.7	6.7	1.01	0.00	0.35	0.05	0.51	99	33
Range	120	0.16	92.23	92.39	5.01	0.01	21.98	7.70	11.83	764	361

These results represent a core interval that is about 40 cm thick.

(Merrill 1991) that is capable of expulsing appreciable quantities of oil (Bordenave *et al.* 1993). In the western North Atlantic, beds rich in oil-prone Type II kerogen like those found at Site 1049 have been penetrated only in the deep basin (Dillon *et al.* 1986; De Graciansky *et al* 1987; Summerhayes 1987). The seismic survey (Fig. 3), enhanced by the stratigraphic age information measured on cores gathered during Leg 171B, indicates that this source rock interval may extend far to the west of Site 1049. It may well approach the reef facies at the edge of the continental shelf (Fig. 3), where it could potentially charge reef facies reservoir rocks, if the black shale reaches thermal maturity there.

The lower Albian black shale is not thermally mature at Site 1049A as shown by a production index (PI) of near zero and a T_{max} of 405 °C (Table 3), and based on the sparse data available it does not appear thermally mature anywhere across the region. Type II source rocks reach thermal maturation for oil at about 50–150 °C (Tissot & Welte 1984). Thermal maturation measurements at the nearby Site 391 in the Blake–Bahama Basin indicate a depth of 1250 mbsf to reach 0.4 % mean vitrinite reflectance measured on non-suppressed, hydrogen-poor terrestrial SOM-rich samples (Cardoso *et al.* 1978). This vitrinite reflectance indicates a maximum palaeotemperature of about 60 °C

(Barker & Pawlewicz 1994). The sediments appear to have reached an early-mature stage at about 1250 m at Site 391. The deepest stratigraphic occurrence of black shale (the early Albian age OAE 1B) in the Blake Nose is found near the base of Hole 1049A. The OAE 1B black shale interval when correlated across the Blake Nose seismic profile appears to be buried to about 1.5 times the total depth of Hole 1052E (Fig. 3) or about 1000 m deep, that is, above the depth required for thermal maturation. Further, because the geothermal gradients are typically lower towards the continental margin than in the deeper portions of the continental margin and abyssal plain drilled at Site 391 (Sclater & Wixon 1986) the black shale interval related to OAE 1b is likely to be less thermally mature than at an equivalent depth in the Blake–Bahama Basin. The lack of thermal maturation, however, may be mitigated by the sulphur-rich character of some of the black shale facies subunits. The sulphur-rich SOM, as seen in biomarker sample IV, may generate hydrocarbons at relatively low thermal maturities because some S–C bonds are weaker than C–C bonds (Tyson 1995).

Conclusions

(1) In general, the better quality (moderate to high HI and TOC) SOM-rich layers within black shale facies across the Blake Nose appear to have been deposited under sediment- and oxygen-starved conditions that led to highly laminated beds.

(2) The variable TOC but typically high to very high HI SOM in highly laminated beds is interpreted as occurring by mid-ocean marine gravity settling style of sedimentation on an isolated sediment-starved palaeo-high influenced by anaerobic to anoxic waters. These laminated beds alternate with more homogeneous beds that may be TOC and hydrogen enriched, which are attributed to dispersing sediment suspensions related to turbidites. Lower-density Type II SOM would be markedly concentrated in a turbidite sediment suspension over Type III–IV SOM and mineral sediment.

(3) The influence of redeposited shelf- and terrestrial-derived sediments on the black shale facies could explain the SOM-rich and clay-rich nature of the black shale facies in a low-productivity, deep-ocean setting generally isolated from sedimentation.

(4) In the distal portion of the Blake Nose, deposition and preservation of a regionally significant oil source rock occurred in the lower Albian black shale facies. However, this source rock is not known to be thermally mature anywhere in the region.

(5) During the late Albian the, sedimentation closer to the shelf break on Blake Nose was more rapid and the SOM was less preserved and more diluted by mineral sedimentation.

The high-resolution sediment samples from the 1049 core were the result of a co-operative effort with L. J. Clarke and F. Martínez-Ruiz. L. Clarke provided the digitized photograph to which we applied the layer map. The authors also acknowledge the palaeontological, sedimentological and geochemical data generated by the ODP Leg 171B Shipboard Scientific Party, which were used extensively in our geological history and discussion portion of this paper. The Scientific Party and personnel at the ODP Bremen core repository also helped sample and ship the Site 1050 and 1052 cores. This paper greatly benefited from all of these efforts and would never have come to publication without their aid. We also thank T. Daws of the USGS Denver for the Rock-Eval analyses; he amazingly still completed the analyses while under great pressure in the last week before he retired. We also thank M. Pribil of the USGS, Denver, for running the C–H–N–S elemental analyses, and M. Emmons of Mountain Mass Spectroscopy (Evergreen, CO) for the stable isotope analyses.

Appendix: Methods

High-resolution sampling

The black shale and related strata were divided into facies sub-units based on lithology, lamination, distinctiveness from its neighbouring layers and colour (Fig. 5). Colour is especially important in black shale facies studies because it is usually related to the TOC content of the layer. In these layers, it is visually related to the content of brownish black, translucent-appearing, sapropel-rich layers. High-resolution sampling was started by sawing the core in half. Then transparent semi-rigid plastic sheet was fixed to lie over the core half and the layers to be sampled were drawn on the overlay. This overlay was taped to one side of the plastic core holder such that it could be rotated in to identify the layer to be sampled or out of the way, so that the cut line scribed onto the surface of the sediment. This technique is important in controlling mixing between subfacies layers and for precisely cutting the samples out. Sample integrity is thought to be very good as a result of this procedure. After delineation, a Boker brand ceramic knife was used to cut the sample away from the adjoining layer. (Note that the use of brand names is for descriptive purposes only and does not constitute endorsement by the US Geological Survey.) After cutting out the subunit, the homogenized disturbed sediment at the contact with the core liner was trimmed off. The disturbed sediment was not analysed further because it may contain sediment slurry derived during coring or from strata outside the sample interval. After cutting and trimming each layer was placed in a vial and sealed, and the vials were labelled in sequence from the bottom of the core to the top. Preparation of these samples for geochemical and petrographic analysis involved simply crushing and grinding the sample using an agate mortar and pestle. The mortar and pestle were cleaned in distilled water before grinding each sample. Samples for geochemical analysis were either submitted as whole-rock splits from these powdered samples or as decarbonated powders after digestion in 6N HCl.

Maceral analysis

Point counting of amorphous SOM is also limited because of the difficulty in distinguishing it from mineral matter and other debris found in sediment. Therefore, traditional point counting of the amorphous SOM composition was not carried out. Rather, we used a percentage estimate based on the charts of Compton (1962, appendix 3) when the sample was under white light and then UV light excitation in reflected light microscopy mode. This method has been found to be similar in accuracy to the point counting method for an experienced microscopist (Bustin 1991).

Because of the intrinsic (structureless) character of amorphous SOM, only simple petrographic observations are usually possible using white light or UV light excitation in reflected or transmitted mode: (1) amorphous v. detrinite SOM content; (2) fluorescent v. non-fluorescent content; and (3) shape or form of amorphous SOM (Whelan & Thompson-Rizer 1993). If structured detrinitic debris are present, other information may be gleaned from it, such as palynology of spores and pollen or the origin of woody debris (often charcoal in deep marine deposits).

Layer lamination and thickness

Counting of the number of laminations in each subfacies layer involved scaling a transparent copy of the plastic overlay to the scale of a core photograph, like that shown in Fig. 5. The number of laminae in each subfacies layer was counted using an Edmund brand transparent base 9 × magnifier placed on a photograph and overlay composite. The thickness of each layer near the centre of the core was also noted. Layer thickness is considered suspect because the deformation of layers that occurred during coring may have significantly altered the *in situ* thickness.

Biomarker analysis

Samples from the Lower Albian black shale facies from Hole 1049A were, individually, too small for biomarker analysis and were combined into larger samples (Table A1). The decisions regarding which samples should be combined were made on the basis of TOC and Rock-Eval patterns in the black shale facies interval (Figs 6 and 7).

Table A1. *Rock-Eval criteria for the combination of samples for biomarker analysis*

Biomarker sample	Combined subfacies units	TOC (%)	OI
I	A, B, C	<1	High
II	M–R	4–6	Low
III	Z–G1	<2.5	Low
IV	L1–Q1	5–10	Low
V	Z1–G2	4–7	Low
VI	N2–W2	<1	High

Samples were powdered, weighed and extracted in chloroform–methanol (2 : 1 v/v, 3 × 30 min, ultrasonication). After filtration, the lipid extracts were decanted to vials, evaporated to dryness under a stream of dry nitrogen, and then refrigerated. Lipids were separated on a silica gel flash column. Neutral lipids were eluted from the column with hexane. Samples were doped with C_{12} alkane (25 ng μl^{-1}) as an internal standard. High-temperature gas chromatography (GC) was carried out on a Hewlett–Packard 5890 Series II gas chromatograph fitted with a DB-5 fused-silica capillary column (0.25 mm film thickness ×30 m × 0.32 mm i.d.). After an isothermal hold at 50 °C (2 min), the oven temperature was raised to 160 °C at 5 °C min^{-1}, then raised to 290 °C at 4 °C min^{-1}, followed by a 25 min isotherm. GC–MS (mass spectrometry) analyses were performed using a Hewlett–Packard 5972 MSD (electron energy 70 eV, filament current, 0.35 mA, source temperature 200 °C) interfaced with a Hewlett–Packard 6890 gas chromatograph. The column type and operating conditions are the same as those described above. Helium was used as the carrier gas.

Organic matter elemental ratios and carbon isotopic composition

Elemental analyses and organic carbon isotopic compositions were determined on samples decarbonated in dilute HCl. After no further reaction was observed when fresh acid was added, the samples were repeatedly rinsed in distilled water until the wash water had a neutral pH. C–H–N–S elemental analyses used a CE Instruments Model EA 1110 machine. Mountain Mass Spectroscopy (Evergreen, CO) performed the stable isotope analyses.

References

ARTHUR, M. A., DEAN, W. E. & STOW, D. A. V. 1984. Models for deposition of Mesozoic–Cenozoic fine grained organic-carbon-rich sediment in the deep sea. *In*: STOW, D. A. V. & PIPER, D. J. W. (eds) *Fine-Grained Sediments*. Geological Society, London, Special Publications, **15**, 527–560.

BARKER, C. E. & PAWLEWICZ, M. J. 1994. Calculation of vitrinite reflectance from thermal histories and peak temperature—a comparison of methods. *In*: Mukhopadhyay, P. K. & DOW, W. G. (eds)

Reevaluation of Vitrinite Reflectance as a Maturity Parameter. American Chemical Society Symposium Series, **570**, 216–229.

BORDENAVE, M. L., ESPITALIE, J., LEPLAT, P., OUDIN, J. L. & VANDENBROUCKE, M. 1993. Screening techniques for source rock evaluation. *In*: BORDENAVE, M. L. (ed.) *Applied Petroleum Geochemistry*. Technip, Paris, 217–279.

BRASSELL, S. C., EGLINTON, G., MAXWELL, J. R. & PHILP, R. P. 1978. Natural background of alkanes in the aquatic environment. *In*: HUTZINGER, O., VAN LELYVELD, L. H. & ZOETEMAN, B. C. J. (eds) *Aquatic Pollutants: Transformation and Biological Effects*. Pergamon, Oxford, 69–86.

BUSTIN, R. M. 1991. Quantifying macerals: some statistical and practical considerations. *International Journal of Coal Geology*, **17**, 213–238.

CARDOSO, J. N., WARDROPER, A. M. K., WATTS, C. D. *et al.* 1978. Preliminary organic geochemical analyses, Site 391, Leg 44 of the Deep Sea Drilling Project. *In*: BENSON, W. E. & SHERIDAN, R. E. (eds) *Initial Reports of the Deep Sea Drilling Project*, **44**. US Government Printing Office, Washington, DC, 617–624.

COMPTON, R. R. 1962. *Manual of Field Geology*. Wiley, New York.

CRANWELL, P. A., EGLINTON, G. & ROBINSON, N. 1987. Lipids of aquatic organisms as potential contributors to lacustrine sediments—II. *Organic Geochemistry*, **11**, 513–527.

DE GRACIANSKY, P. C., BROSSE, E., DEROO, G. *et al.* 1987. Organic-rich sediments and palaeoenvironmental reconstructions of the Cretaceous North America. *In*: BROOKS, J. & FLEET, A. J. (eds) *Marine Petroleum Source Rocks*. Geological Society, London, Special Publications, **26**, 317–344.

DILLON, W. P., MANHEIM, F. T., JANSA, L. F., PALMASON, G., TUCHOLKE, B. E., LANDRUM, R. S. 1986. Resource potential of the western North Atlantic Basin. *In*: VOGT, P. R. & TUCHOLKE, B. E. (eds) *The Geology of North America, Volume M, The Western North Atlantic Region*. Geological Society of America, Boulder, CO, 661–676.

DOW, W. G. 1978. Geochemical analysis of samples from Holes 391A and 391C, Leg 44, Blake–Bahama Basin. *In*: BENSON, W. E. & SHERIDAN, R. E. (eds) *Initial Reports of the Deep Sea Drilling Project*, **44**. US Government Printing Office, Washington, DC, 625–634.

EINSELE, G., RICKEN, W. & SEILACHER, A. (eds) 1991. *Cycles and Events in Stratigraphy*. Springer, Berlin.

EMEIS, K. C. & KVENVOLDEN, K. A. 1986. Shipboard organic geochemistry on JOIDES Resolution Ocean Drilling Program Technical Note, **7**.

ERDMAN, J. G. & SCHORNO, K. S. 1978. Geochemistry of carbon: Deep Sea Drilling Project Leg 44. *In*: BENSON, W. E. & SHERIDAN, R. E. (eds) *Initial Reports of the Deep Sea Drilling Project*, **44**. US Government Printing Office, Washington, DC, 605–615.

GELPI, E., SCHNIEDER, H., MANN, J. & ORO, J. 1970. Hydrocarbons of geochemical significance in microscopic algae. *Phytochemistry*, **9**, 603–612.

HAQ, B. U., HARDENBOL, J. & VAIL, P. R. 1987. *The New Chronostratigraphic Basis of Cenozoic and Mesozoic Sea Level Cycles*. Special Publication, Cushman Foundation Foraminifera Research, **24**, 713.

HERBIN, J. P., DEROO, G., ROUCACHÉ, J. 1983. Organic geochemistry in the Mesozoic and Cenozoic formations of Site 534, Leg 76, Blake Bahama Basin, and comparison with Site 391, Leg 44. *In*: SHERIDAN, R. E. & GRADSTEIN, F. M. *et al.* (eds) *Initial Reports of the Deep Sea Drilling Project*, **76**. US Government Printing Office, Washington, DC, 481–493.

ISHIWATARI, R., OGURA, K. & HORIE, S. 1980. Organic geochemistry of a lacustrine sediment (Lake Haruna, Japan). *Chemical Geology*, **29**, 261–280.

KATZ, B. J. 1983. Organic geochemical character in some Deep Sea Drilling Project cores from Legs 76 and 44. *In*: SHERIDAN, R. E., GRADSTEIN, F. M. *et al.* (eds) *Initial Reports of the Deep Sea Drilling Project*, **44**. US Government Printing Office, Washington, DC, **76**. 463–468.

KENDRICK, J. W., HOOD, A. & CASTANO, J. R. 1978. Petroleum-generating potential of sediments from Leg 44, Deep Sea Drilling Project. *In*: BENSON, W. E. & SHERIDAN, R. E. (eds) *Initial Reports of the Deep Sea Drilling Project*, **44**. US Government Printing Office, Washington, DC, 599–603.

LANGFORD, F. F. & BLANC-VALLERON, M. M. 1990. Interpreting Rock-Eval pyrolysis data using graphs of pyrolizable hydrocarbons vs total organic carbon. *AAPG Bulletin*, **70**, 799–804.

MCIVER, R. 1975. Hydrocarbon occurrences from JOIDES Deep Sea Drilling Project. *Proceedings of the Ninth World Petroleum Congress*, Tokyo, 2, 269–280.

MERRILL, R. K. 1991, Preface to the Volume. *In*: MERRILL, R. K. (ed.) *Source and Migration Processes and Evaluation Techniques*. AAPG Treatise of Petroleum Geology, xvii.

NORRIS, R. D., KROON, D., KLAUS, A. *et al.* 1998. *Proceedings of the Ocean Drilling Program, Initial Reports*, **171B**. Ocean Drilling Program, College Station, TX.

PETERS, K. E. 1986. Guidelines for evaluating petroleum source rock using programmed pyrolysis. *AAPG Bulletin*, **70**, 318–329.

SCLATER, J. G. & WIXON, L. 1986. The relationship between depth and age and heat flow and age in the western North Atlantic. *In*: VOGT, P. R. & TUCHOLKE, B. E. (eds) *The Geology of North America, Volume M, The Western North Atlantic Region*. Geological Society of America, Boulder, CO 257–270.

SUMMERHAYES, C. P. 1987. Organic-rich Cretaceous sediments from the North Atlantic. *In*: BROOKS, J. & FLEET, A. J. (eds) *Marine Petroleum Source Rocks*. Geological Society, London, Special Publications, **26**, 301–316.

SUMMERHAYES, C. P. & MASRAN, T. C. 1983. Organic facies of Cretaceous and Jurassic sediments from Deep Sea Drilling Project Site 534 in the Blake Bahama Basin, Western North Atlantic. *In*: SHERIDAN, R. E. & GRADSTEIN, F. M. *et al.* *Initial Reports of the Deep Sea Drilling Project*, **76**. US Government Printing Office, Washington, DC, 469–480.

TISSOT, B. P. & WELTE, D. H. 1984. *Petroleum Formation and Occurrence*. Springer, New York.

TYSON, R. V. 1987. The genesis and palynofacies characteristics of marine petroleum source rocks. *In*: BROOKS, J. & FLEET, A. J. (eds) *Marine Petroleum Source Rocks*. Geological Society, London, Special Publications, **26**, 47–67.

TYSON, R. V. 1995. *Sedimentary Organic Matter*. Chapman and Hall, London.

WHELAN, J. K. & THOMPSON-RIZER, C. L. 1993. Chemical methods for assessing kerogen and protokerogen types and maturity. *In*: ENGEL, M. H. & MACKO, S. A. (eds) *Organic Geochemistry*. Plenum, New York, 289–354.

WIGNALL, P. B. 1994. *Black Shales*. Clarendon Press, Oxford.

No extinctions during Oceanic Anoxic Event 1b: the Aptian–Albian benthic foraminiferal record of ODP Leg 171

ANN HOLBOURN & WOLFGANG KUHNT

Institut für Geowissenschaften, Christian Albrechts Universität, Olshausenstr. 40, 24118 Kiel, Germany (e-mail: ah@gpi.uni-kiel.de)

Abstract: Outstandingly well-preserved benthic foraminiferal successions from upper Aptian–lower Albian sediments at Site 1049 (Leg 171, Blake Nose escarpment, western North Atlantic) provide a detailed record of the faunal turnover across Oceanic Anoxic Event 1b (OAE 1b). Changes in abundance, diversity and species composition reflect strong fluctuations in carbon flux and bottom-water oxygenation. Before the onset of black shale sedimentation, the originally diverse assemblages are replaced by low-diversity associations, dominated by species inferred to be opportunistic phytodetritus feeders and thriving on an enhanced carbon flux to the sea floor. The 46 cm thick laminated black shale horizon corresponding to OAE 1b is virtually devoid of benthic foraminifers or contains highly impoverished assemblages, suggesting that intense eutrophication and/or strong stratification triggered near anoxia at the sea floor during black shale deposition. Above the black shale, reoccurrence of the pre-black shale fauna points to relatively rapid bottom-water reoxygenation. The benthic foraminiferal record of Leg 171 provides clear evidence that no major extinctions occurred across OAE 1b, as most of the species occurring below the black shale reappear above it. In contrast to other Cretaceous anoxic events, OAE 1b may have been more limited in duration or in geographical and water-depth extent, allowing recolonization from adjacent, more hospitable areas, once local conditions improved at the sea floor. Prolific radiation within the suborder Rotaliina and diversification of Textulariina with calcareous cement appear to have started in the Aptian time before OAE 1b, and continued into early Albian time to give rise to many of the modern lineages.

High atmospheric CO_2 concentrations, warm climates, elevated sea levels and changes in the focus of deep-water formation triggered major fluctuations in oceanic circulation and surface water productivity during mid-Cretaceous time (Brass *et al.* 1982; Barron & Peterson 1990; Frakes *et al.* 1992; Barron *et al.* 1993, 1995; Hay 1995). These changes induced several global episodes of severe dysoxia or anoxia in mid-Cretaceous oceans, which may have sparked off widespread extinctions in the marine realm. However, most previous studies of the impact of anoxic events on benthic foraminifer communities have concentrated on sediments from the Cenomanian–Turonian boundary (Eicher & Worstell 1970; Koutsoukos *et al.* 1990; Tronchetti & Groshenny 1991; Kaiho & Hasegawa 1994; Coccioni *et al.* 1995; Kuhnt & Wiedmann 1995; Peryt & Lamolda 1996). In fact, relatively little is known about the timing of extinctions and evolutionary events in relation to anoxic episodes in Aptian and Albian time, as the stratigraphic ranges of many species within that interval are not well constrained because of the scarcity of stratigraphically complete, well-preserved sedimentary successions. Additionally, planktonic markers and calcareous benthic species are often missing in deep-water sections, as the calcite compensation depth (CCD) in Aptian–Albian oceans was generally elevated. Previous studies of Aptian–Albian benthic foraminifers have focused mainly on assemblages from epicontinental seas (Ten Dam 1950; Bartenstein & Brand 1951; Bartenstein & Bolli 1973, 1986; Magniez-Jannin 1975; Neagu 1975; Bartenstein 1978, 1987; Haig 1980, 1982; Tronchetti 1981; Bartenstein & Kovatcheva 1982; Haig & Lynch 1993; Prokoph *et al.* 1999) or from sub-CCD basins (Geroch 1966; Geroch & Nowak 1984; Weidich 1990). However, only relatively few records of benthic foraminiferal assemblages exist from deep-water settings containing the full range of planktonic and benthic foraminifers (Guerin 1981; Riegraf & Luterbacher 1989; Holbourn & Kaminski 1997).

The recovery of outstandingly well-preserved benthic foraminiferal successions in upper Aptian–lower Albian sediments from Site 1049

From: KROON, D., NORRIS, R. D. & KLAUS, A. (eds). 2001 *Western North Atlantic Palaeogene and Cretaceous Palaeoceanography*. Geological Society, London, Special Publications, **183**, 73–92.
0305-8719/01/$15.00

provides the opportunity to investigate faunal changes during a period of acute global environmental change, and to decipher the significance of Ocean Anoxic Event (OAE) 1b for the evolution of mid-Cretaceous benthic foraminifers. The main objectives of this work are: (1) to document the distribution of late Aptian–early Albian benthic foraminifers at Site 1049, in comparison with published records from Alpine–Carpathian basins, North Atlantic marginal basins and the Indian Ocean; (2) to evaluate how faunal changes relate to sedimentary and palaeoenvironmental changes; (3) to determine the timing of any faunal turnover and assess whether OAE 1b had a major impact on the evolution of mid-Cretaceous benthic foraminifers; (4) to speculate about the factors influencing evolutionary change during Aptian and early Albian time.

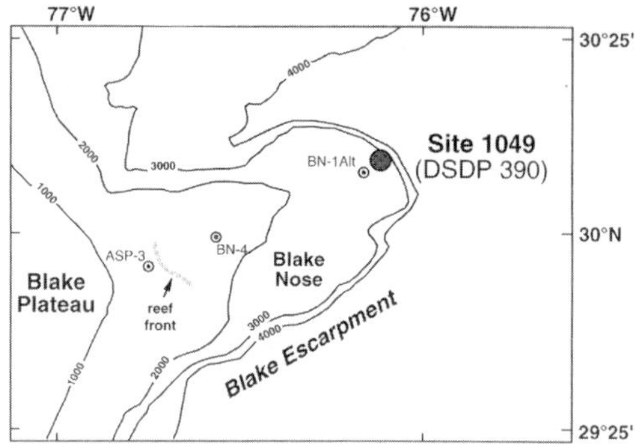

Fig. 1. Location of Site 1049 on the Blake Nose escarpment, western North Atlantic.

Material and methods

Ocean Drilling Program (ODP) Leg 171B recovered a thick succession of upper Aptian to lower Albian sediments at Holes 171B-1049A, -B and -C on the Blake Nose escarpment in the western North Atlantic (Fig. 1). This succession includes a 46 cm thick black shale, representing the local expression of the early Albian oceanic anoxic event (OAE 1b). These mid-Cretaceous sediments, which are overlain by *c*. 150 m of Upper Cretaceous to Eocene sediments and probably did not undergo deeper burial (Norris *et al.* 1998), contain extremely well-preserved benthic foraminiferal assemblages. Benthic foraminiferal tests appear to have suffered little recrystallization or dissolution during diagenesis and generally do not have secondary calcite infillings.

A total of 49 samples from Holes 171B-1049A and -C were examined. Samples from Section 171B-1049C-12X-3, which comprises a 46 cm thick laminated black shale corresponding to OAE 1b, were integrated with 28 samples from Hole 171B-1049A, dated from planktonic foraminifers as Aptian to early Albian by Bellier *et al.* (2000). We visited the ODP Core Repository in Bremen to select additional samples and to make detailed lithological descriptions of cores from Holes 1049A and 1049C.

Samples were dried, weighed and initially treated with a mixture of H_2O_2 (15%) and chalk to avoid carbonate dissolution by sulphuric acid from oxidation of disseminated pyrite. Samples were then wet-sieved through a 63 μm mesh. Residues were dry sieved into 63–100 or 125 μm, 100 or 125–250 μm and 250–630 μm fractions. Samples were picked from splits of each fraction and total numbers were recalculated accordingly. We aimed to count approximately an equal number of specimens from each fraction. Thus in the large size fraction, we were usually able to count all specimens, whereas in the small size fraction often only a small split could be counted because of the extremely high abundance of tests. Splitting was not necessary for samples within the black shale, where foraminiferal density was extremely low. Foraminiferal density is given in specimens per 10 g dried sediment. The revised dataset of Erbacher *et al.* (1999) was used together with new counts of benthic foraminifers from Hole 1049A to produce the final dataset. The dataset of Erbacher *et al.* (1999) from Hole 1049C was updated taxonomically and counts from the 63–100 μm fraction of their original samples were added to ensure consistency of the final dataset. A taxonomic list is given in the Appendix. Detailed taxonomy of the Aptian–Albian benthic foraminifers from North Atlantic Deep Sea Drilling Program (DSDP) and ODP sites and coastal sections is in preparation.

We used Fisher's α as diversity index, as it is easier to use than any other method for rarefaction or abundifaction, yet is as reliable as other methods (Hayek & Buzas 1997). Fisher's α values were calculated from n (number of individuals) and S (number of species) using a program written by Weinholz and Altenbach revised for Mac-Systems in FORTRAN 77 by Pflaumann. The resulting α values compare well with values given in appendix 4 of Hayek & Buzas (1997).

We used correspondence analysis to discriminate faunal assemblages and to identify major trends of faunal change. Correspondence analysis was carried out using the software package ECOLOGIX written in 1982 by Rioux (Montpellier University). ECOLOGIX was run on a CRAYVAX computer at the University of Kiel. Quantitative distribution charts of benthic foraminifers were combined into a coded matrix for a total of 49 samples with frequency counts of 66 taxa. Rare taxa with fewer than four occurrences were not included into the correspondence analysis. Original counts of benthic foraminifers were converted into frequency counts for the correspondence analysis as follows:

Number of foraminifers	Frequency
0	0
1–3	1
4–10	2
10–100	3
100–1000	4
>1000	5

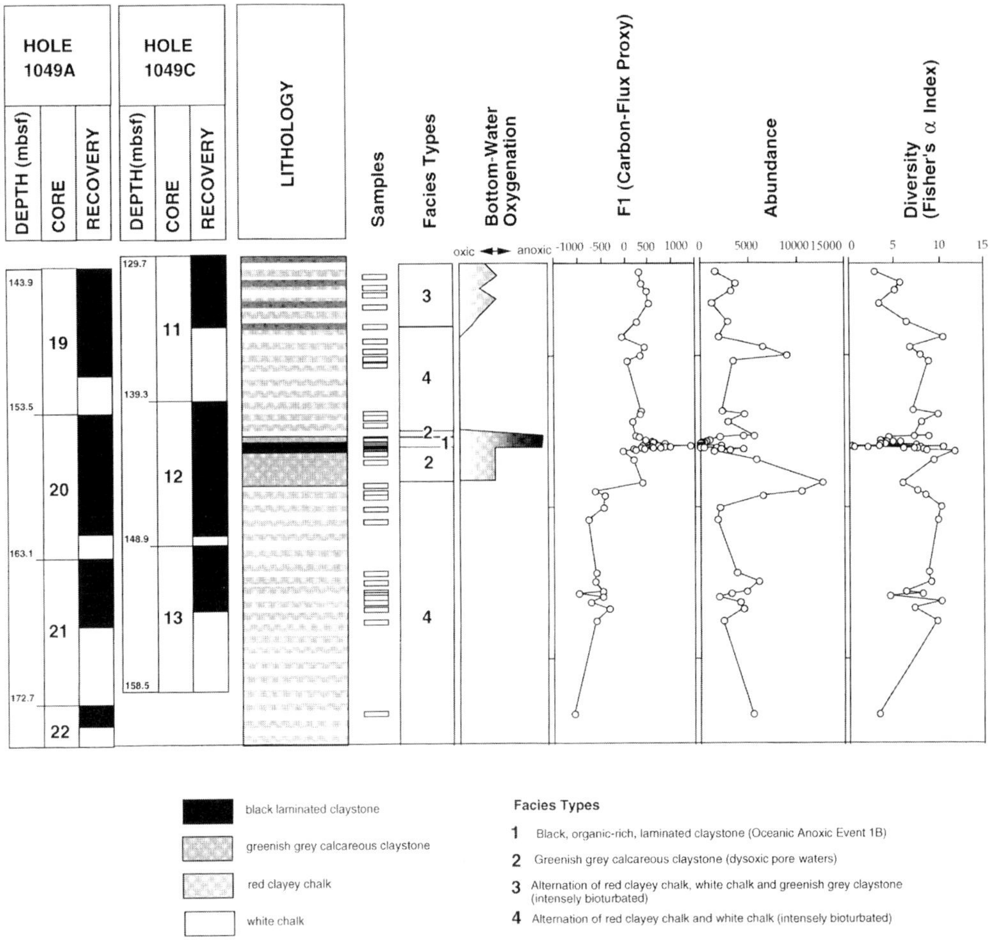

Fig. 2. Benthic foraminiferal distribution data for Holes 1049A and 1049C: correspondence analysis Factor 1 values, number of specimens per 10 g of sediment, number of species, Fisher's α index values. Lithology interpolated in non-recovered portions of cores.

A more detailed description of the application of correspondence analysis to micropalaeontological datasets has been given by Saint-Marc & Berggren (1988), Kuhnt & Moullade (1991) and Kuhnt *et al.* (1999).

Results

Abundance and diversity patterns

Preservation of benthic foraminifers is generally good to excellent, except in a few samples where tests appear to have suffered some degree of recrystallization (near the base of the studied interval) or dissolution (in the uppermost samples). Abundance and diversity within the assemblages show significant variations, particularly below and above the laminated black shale horizon (Fig. 2). Long-term variations in abundance and diversity clearly relate to marked lithological changes, whereas short-term fluctuations do not coincide with distinct lithological boundaries. Benthic foraminiferal data are shown in Table 1 and Fig. 2. Original counts of benthic foraminifers and full results of correspondence analysis are available at http://www.odsn.de/odsn/data/lit/hk_gssp_00.

Long-term variations. High abundance and low diversity characterize Sample 1049A-22X-2, 20–25 cm at the base of Hole 1049A. This sample of undetermined Aptian age (Bellier

Table 1. *Benthic foraminiferal data for Holes 1049A and 1049C (number of >63 μm specimens per 10 g sediment)1*

	Ammolagena clavata	Ammopshaeroidina spp.	Arenobulimina chapmani	Astacolus calliopsis	Astacolus parallellus	Astacolus sp.	Berthelina intermedia	Bulbobaculites humei	Charltonia australis	Clavulinoides gaultinus	Conorboides sp.	Conorotalites aptiensis	Ellipsoidella cuneata	Falsoguadryinella moesiana	Frondicularia hastata	Fursenkoina viscida
1049A-19X-1, 62–64 cm	.	.	2	1	.	.	.	.	.	.	.	.	.	1	.	.
1049A-19X-1, 135.5–137.5 cm	.	.	.	13	.	13	.	.	13	.	.	.	.	.	.	.
1049A-19X-2, 35–39 cm	.	.	.	14	.	.	12	.	6	9	.	.	.	.	.	.
1049A-19X-2, 116–118 cm	.	.	.	.	.	7	1	.	.	.	.	.	.	.	.	.
1049A-19X-3, 93–95 cm	.	.	.	12	.	.	10	.	.	.	.	.	.	.	.	.
1049A-19X-4, 40–44 cm	.	.	.	12	10	.	186	.	.	24	.	.	.	109	.	16
1049A-19X-4, 107–109 cm	.	.	0	52	.	.	72	.	7	19	.	.	.	.	.	.
1049A-19X-5, 8.5–10.5 cm	.	.	1	17	6	.	176	.	188	.	.	.	.	.	.	.
1049A-19X-5, 49–53 cm	.	.	166	99	25	.	116	.	.	.	.	.	.	.	.	.
1049A-20X-1, 15–17 cm	.	.	41	31	.	3	87	.	3	.	.	.	.	.	.	7
1049A-20X-1, 42–44 cm	.	.	.	29	.	10	104	.	.	13	.	3	10	.	.	10
1049A-20X-1, 90–92 cm	.	.	1	8	1	.	136	.	.	.	.	.	.	.	.	.
1049C-12X-3, 02–04 cm	.	.	3	40	.	16	38	.	33	.	.	107	.	.	.	.
1049A-20X-2, 36–40 cm	.	.	7	38	19	13	144	.	.	7	.	32	.	.	.	.
1049C-12X-3, 15–15 cm	.	.	.	20	4	.	36	.	.	.	.	.	.	.	.	.
1049C-12X-3, 34–36 cm	.	.	.	.	.	.	.	.	.	.	.	.	.	.	.	.
1049C-12X-3, 39–41 cm	.	.	.	1	.	1	.	.	.	.	.	5	169	.	.	13
1049C-12X-3, 43–44 cm	.	.	.	5	.	.	2	.	.	.	.	5	21	.	.	7
1049C-12X-3, 44–45 cm	.	.	.	.	.	.	3	.	.	.	.	.	5	.	.	.
1049C-12X-3, 52–54 cm	.	.	.	.	.	.	.	.	.	.	.	.	3	.	.	.
1049C-12X-3, 55–57 cm	.	.	.	.	.	.	.	.	.	.	.	.	.	.	.	.
1049C-12X-3, 65–66 cm	.	.	.	3	.	.	7	.	.	.	.	.	.	.	.	3
1049C-12X-3, 67–69 cm	.	.	.	.	.	5	1	.	.	.	.	.	191	.	.	628
1049C-12X-3, 70–71 cm	.	.	.	.	.	.	.	.	.	.	.	.	.	.	.	.
1049C-12X-3, 74–76 cm	.	.	.	.	.	.	.	.	.	.	2	.	.	.	.	2
1049C-12X-3, 77–78 cm	.	.	.	.	.	.	.	.	.	.	.	.	.	.	.	6
1-49C-12X-3, 78–79 cm	.	.	.	.	.	.	.	.	.	.	.	.	.	.	.	41
1049C-12X-3, 79–80 cm	.	.	.	.	.	.	.	.	.	.	.	.	.	.	.	21
1049C-12X-3, 81–83 cm	.	.	.	.	.	6	47	.	.	6	.	.	2	.	.	163
1049C-12X-3, 84–86 cm	.	.	.	6	.	20	133	.	.	54	.	.	.	.	.	339
1049C-12X-3, 87–89 cm	.	2	.	10	.	\5	203	.	.	12	.	.	.	22	.	.
1049C-12X-3, 89–91 cm	.	2	2	21	.	16	233	.	.	51	.	.	.	.	.	42
1049C-12X-3, 98–100 cm	.	11	4	29	.	11	198	.	.	18	.	.	.	.	4	38
1049C-12X-3, 103–105 cm	.	2	1	6	.	10	110	.	.	22	.	.	.	16	1	8
1049-20X-3, 40.5, 44 cm	.	.	.	5	.	18	467	.	.	95	.	.	9	1	.	171
1049A-20X-4, 45–50 cm	.	.	.	.	.	1	1935	.	.	45	.	.	.	.	.	227
1049A-20X-4, 97–100 cm	.	.	.	125	.	.	3099	.	.	.	.	.	.	220	.	.
1049A-20X-4, 125–127 cm	.	.	.	14	1	.	956	.	.	.	.	.	.	.	.	.
1049A-20X-5, 54–60 cm	5	.	.	19	.	.	130	.	.	.	.	.	53	248	.	11
1049A-20X-6, 37–43	.	.	.	1	.	.	111	0	.	.	1	.	.	6	.	.
1049A-21X-1, 124–126 cm	.	.	.	3	0	.	242	.	.	.	3	2	.	99	.	50
1049A-21X-2, 35–40 cm	.	.	.	6	.	.	924	.	.	.	.	.	.	186	.	.
1049A-21X-2, 100–105 cm	.	.	.	15	0	30	414	.	.	.	.	.	.	7	.	.
1049A-21X-2, 117–119 cm	.	.	.	15	0	.	208	.	.	.	.	.	.	19	.	.
1049A-21X-2, 133–135 cm	.	.	.	2	.	.	221	3	.	.	.	.	.	.	.	.
1049A-21X-3, 18–20 cm	.	.	.	15	.	1	277	.	.	.	.	27	.	845	.	.
1049A-21X-3, 60.5–62.5 cm	.	.	.	11	.	24	481	9	.	.	89	.	16	.	.	11
1049A-21CC, 33–38 cm	.	.	.	11	.	.	344	.	.	.	22	11	50	.	6	.
1049A-22X-2, 20–25 cm	.	.	.	.	.	35	2649	.	.	.	.	.	.	.	.	.

	Gaudryina dividens	Gaudryina gradata	Gaudryina jendrejakovae	Gaudryina spp.	Globulina lacrima	Globulina prisca	Globulina spp.	Glomospira charoides	Glomospirella gaultina	Gyropidinoides infracretaceus	Hemirobulina bullata	Laevidentalina communis	Laevidentalina debilis	Laevidentalina nana	Laevidentalina oligostegia	Laevidentalina soluta
1049A-19X-1, 62–64 cm	1	.	.	.	.	.	.	.	.	.	.	.	.	.	.	.
1049A-19X-1, 135.5–137.5 cm	2	1	.	.	.	.	.	13	.	133	13	.	0	.	.	.
1049A-19X-2, 35–39 cm	12	.	.	.	.	.	.	76	.	43	.	.	.	3	.	3
1049A-19X-2, 116–118 cm	0	.	.	.	.	.	.	.	.	14	.	.	0	.	.	.
1049A-19X-3, 93–95 cm	18	.	.	.	.	15	.	3	.	34	13	.	.	1	.	.
1049A-19X-4, 40–44 cm	3	4	.	.	14	59	.	25	.	25	2	.	2	18	.	6
1049A-19X-4, 107–109 cm	33	.	.	.	.	.	.	.	.	62	3	3	.	21	.	3
1049A-19X-5, 8.5–10.5 cm	57	.	.	.	28	22	.	.	.	144	1	.	.	6	.	6
1049A-19X-5, 49–53 cm	141	17	398	.	.	50	.	8	.	257	8	66	.	75	17	8
1049A-20X-1, 15–17 cm	11	.	97	.	.	7	.	3	.	228	7	1	.	3	.	.
1049A-20X-1, 42–44 cm	1	.	0	.	3	21	10	.	.	139	3	6	3	13	.	0
1049A-20X-1, 90–92 cm	10	0	.	.	1	11	.	.	.	47	3	3	.	.	.	15
1049C-12X-3, 02–04 cm	.	.	.	.	.	5	.	6	.	043	2	2	2	14	24	10
1049A-20X-2, 36–40 cm	.	0	.	.	32	.	5	.	572	9	.	16	13	33	2	169
1049C-12X-3, 15–15 cm	.	.	.	.	.	.	.	2	.	153	.	.	1	.	2	.
1049C-12X-3, 34–36 cm	.	.	.	.	.	.	.	.	.	10	.	.	.	.	.	2
1049C-12X-3, 39–41 cm	.	.	.	.	.	1	.	.	.	5	.	.	.	.	2	.
1049C-12X-3, 43–44 cm	.	.	.	.	.	.	.	.	7	.	.	.	.	.	2	.
1049C-12X-3, 44–45 cm	.	.	.	.	.	.	.	.	.	14	.	3	.	.	.	.
1049C-12X-3, 52–54 cm	.	.	.	.	.	.	.	.	.	8	.	.	.	.	.	.
1049C-12X-3, 55–57 cm	.	.	.	.	.	.	.	.	.	.	.	.	.	.	.	.
1049C-12X-3, 65–66 cm	.	.	.	.	.	.	.	.	.	3	.	.	.	7	.	.
1049C-12X-3, 67–69 cm	.	.	.	.	.	6	.	.	.	.	1	1	.	.	.	.
1049C-12X-3, 70–71 cm	.	.	.	.	.	.	.	.	.	.	.	.	.	.	.	.
1049C-12X-3, 74–76 cm	.	.	.	.	.	.	.	.	.	.	.	.	.	.	.	.
1049C-12X-3, 77–78 cm	.	.	.	.	.	.	.	.	.	.	.	.	.	4	.	.
1-49C-12X-3, 78–79 cm	.	.	.	.	.	.	.	.	.	2	.	.	.	.	.	.
1049C-12X-3, 79–80 cm	.	.	.	.	.	.	.	.	.	3	.	.	.	.	.	.
1049C-12X-3, 81–83 cm	.	.	.	.	.	4	.	.	.	39	2	.	.	.	.	2
1049C-12X-3, 84–86 cm	.	.	.	.	.	8	.	8	.	73	.	11	.	6	.	.
1049C-12X-3, 87–89 cm	2	.	2	.	.	7	.	225	5	.	5	.	2	7	22	2
1049C-12X-3, 89–91 cm	.	.	16	4	16	.	.	.	1044	4	12	4	19	27	7	65
1049C-12X-3, 98–100 cm	.	4	13	36	4	7	.	.	.	763	13	9	.	13	9	9
1049C-12X-3, 103–105 cm	5	.	4	27	3	17	.	.	.	219	2	5	4	5	9	4
1049-20X-3, 40.5, 44 cm	.	19	.	.	9	45	4	.	.	91	0	13	.	22	27	1
1049A-20X-4, 45–50 cm	57	.	5	.	.	.	.	.	.	234	.	.	.	.	.	14
1049A-20X-4, 97–100 cm	990	0	.	.	.	.	.	.	1663	.	.	.	0	.	.	219
1049A-20X-4, 125–127 cm	161	3	.	.	.	.	.	56	.	430	.	.	.	112	.	.
1049A-20X-5, 54–60 cm	168	37	.	.	3	8	.	13	3	144	8	16	5	3	8	.
1049A-20X-6, 37–43	22	2	.	.	6	.	0	.	.	358	1	1	.	3	.	.
1049A-21X-1, 124–126 cm	5	0	.	.	.	1	.	.	.	304	0	4	.	50	50	.
1049A-21X-2, 35–40 cm	203	55	.	.	17	34	132	17	.	1055	.	34	.	17	34	17
1049A-21X-2, 100–105 cm	.	0	.	.	.	.	.	68	.	93	.	22	.	.	29	0
1049A-21X-2, 117–119 cm	19	3	.	.	.	.	.	0	.	526	1	6	52	129	6	.
1049A-21X-2, 133–135 cm	.	.	.	.	.	2	.	2	.	236	.	30	.	.	.	.
1049A-21X-3, 18–20 cm	.	59	.	.	.	26	.	.	.	446	7	.	.	26	1	0
1049A-21X-3, 60.5–62.5 cm	.	0	.	.	.	25	3	15	.	574	.	.	.	1	12	.
1049A-21CC, 33–38 cm	0	.	.	.	28	.	44	6	17	.	.	94	13	89	1	6
1049A-22X-2, 20–25 cm	.	.	.	.	.	.	.	.	.	102	.	1	.	34	.	.

	Laevidentalina spp.	Lenticulina muensteri	Lenticulina oligostegia	Lenticulina ouachensis	Lenticulina saxocretacea	Lenticulina spp.	Lenticulina subangulata	Lenticulina subgaultina	Lingulina loryi	Lingulina sp. 1	Lingulogavelinella albiensis	Lingulonodosaria nodosaria	Marginulinopsis jonesi	Marginulina pyramidilis	Marginulina spp.
1049A-19X-1, 62–64 cm	2	.	3	.	.	11	5	.	.	.	.	.	.	.	.
1049A-19X-1, 135.5–137.5 cm	33	.	39	.	1	34	.	1	13	.	.	.	.	.	.
1049A-19X-2, 35–39 cm	134	3	15	.	2	9	.	.	.	.	70	.	.	.	.
1049A-19X-2, 116–118 cm	20	0	2	.	.	27	9	.	.	.	.	.	.	.	.
1049A-19X-3, 93–95 cm	13	3	15	.	.	19	.	.	37	.	.	.	.	.	3
1049A-19X-4, 40–44 cm	44	.	23	.	.	17	.	.	39	.	.	1	.	.	.
1049A-19X-4, 107–109 cm	244	52	45	.	1	75	.	.	.	.	8	.	.	.	.
1049A-19X-5, 8.5–10.5 cm	145	22	110	.	1	78	.	.	22	.	.	22	.	.	.
1049A-19X-5, 49–53 cm	133	.	.	.	8	8	.	.	41	.	.	8	.	.	.
1049A-20X-1, 15–17 cm	17	3	21	.	.	31	.	.	21	7	.	.	.	.	.
1049A-20X-1, 42–44 cm	108	0	68	.	0	41	.	.	21	10	.	73	.	.	0
1049A-20X-1, 90–92 cm	33	.	11	.	0	26	.	.	42	54	.	.	.	.	0
1049C-12X-3, 02–04 cm	34	6	27	.	6	2	.	.	26	19	.	.	.	.	.
1049A-20X-2, 36–40 cm	9	34	.	13	184	13	7	70	7	.	5	.	2	.	.
1049C-12X-3, 15–15 cm	34	13	10	.	.	42	.	.	28	.	.	.	.	.	.
1049C-12X-3, 34–36 cm	.	2	14	.	5	26	.	.	5	.	.	.	.	.	.
1049C-12X-3, 39–41 cm	2	2	11	.	.	12	.	.	3	.	.	.	.	.	.
1049C-12X-3, 43–44 cm	5	.	.	9	.	.	2	.	.	.	.	.	.	.	.
1049C-12X-3, 44–45 cm	.	3	3	.	.	.	.	.	.	.	.	.	.	.	.
1049C-12X-3, 52–54 cm	.	.	.	.	.	.	.	.	.	.	.	.	.	.	.
1049C-12X-3, 55–57 cm	1	.	.	.	.	1	.	.	.	.	.	.	.	.	.
1049C-12X-3, 65–66 cm	3	.	.	.	.	.	.	.	.	.	.	.	.	.	.
1049C-12X-3, 67–69 cm	15	.	.	.	.	2	.	.	.	.	.	.	.	.	.
1049C-12X-3, 70–71 cm	.	.	.	.	.	.	.	.	.	.	.	.	.	.	.
1049C-12X-3, 74–76 cm	.	.	.	.	.	.	.	.	.	.	.	.	.	.	.
1049C-12X-3, 77–78 cm	.	.	.	.	.	2	.	.	.	.	.	.	.	.	.
1-49C-12X-3, 78–79 cm	.	.	.	.	.	.	.	.	.	.	.	.	.	.	.
1049C-12X-3, 79–80 cm	.	.	.	.	.	3	.	.	.	.	.	.	.	.	.
1049C-12X-3, 81–83 cm	2	2	.	.	.	6	.	.	6	.	.	4	.	.	.
1049C-12X-3, 84–86 cm	54	3	3	.	.	37	.	.	48	6	.	8	.	.	.
1049C-12X-3, 87–89 cm	.	.	.	29	.	.	22	27	.	10	.	.	.	.	.
1049C-12X-3, 89–91 cm	4	25	.	14	1344	.	.	19	21	.	5	.	.	9	.
1049C-12X-3, 98–100 cm	45	4	27	.	11	25	.	.	27	18	.	.	.	.	7
1049C-12X-3, 103–105 cm	63	1	11	.	8	22	.	.	10	.	.	1	.	.	.
1049-20X-3, 40.5, 44 cm	90	19	72	.	36	126	.	.	50	40	.	4	.	.	.
1049A-20X-4, 45–50 cm	1	14	.	.	14	289	.	.	1	.	.	.	.	.	.
1049A-20X-4, 97–100 cm	.	32	.	0	1	.	.	63	.	.	.	.	63	.	.
1049A-20X-4, 125–127 cm	29	.	28	.	4	60	.	1	28	.	.	.	.	.	.
1049A-20X-5, 54–60 cm	27	3	.	.	43	64	.	.	16	.	.	5	.	13	.
1049A-20X-6, 37–43	15	.	38	.	4	5	.	.	54	.	.	.	.	.	.
1049A-21X-1, 124–126 cm	2	.	4	.	3	11	.	.	2	.	.	5	.	.	.
1049A-21X-2, 35–40 cm	55	.	21	.	.	125	.	.	17	17	.	.	.	.	.
1049A-21X-2, 100–105 cm	.	.	45	.	.	45	.	.	.	15	.	.	.	.	.
1049A-21X-2, 117–119 cm	208	.	26	.	0	60	.	.	.	0	.	.	.	.	.
1049A-21X-2, 133–135 cm	0	7	36	.	.	58	.	2	.	.	.	.	.	.	.
1049A-21X-3, 18–20 cm	138	2	47	.	4	108	1	13	20	.	.	13	7	.	.
1049A-21X-3, 60.5–62.5 cm	53	3	24	.	14	93	24	.	.	3	.	.	.	.	11
1049A-21CC, 33–38 cm	266	.	7	.	.	.	.	6	.	12	.	.	.	.	.
1049A-22X-2, 20–25 cm	34	.	.	.	.	188	.	.	.	.	.	.	1	.	.

Table 1. *Continued*

Sample	Marginulinopsis spp.	Neobulimina minima	Oolina globosa	Oolina sp.	Oolina sulcata	Osangularia schloenbachi	Palmula sp.	Patellina subcretacea	Planularia crepidularis	Plectorecurvoides alternans	Pleurostomella reussi	Praebulimina elata	Praebulimina nannina	Praebulimina sp. 1	Protomarssonella oxycona	Protomarssonella trochus
1049A-19X-1, 62–64 cm	.	.	.	.	.	295	.	.	.	.	219	.	55	.	.	.
1049A-19X-1, 135.5–137.5 cm	.	.	.	.	.	2014	.	.	.	.	210	203	51	.	.	.
1049A-19X-2, 35–39 cm	.	.	.	.	.	1230	.	.	.	49	213	695	110	.	.	.
1049A-19X-2, 116–118 cm	.	.	.	.	.	201	.	.	.	.	20	72	7	633	.	.
1049A-19X-3, 93–95 cm	.	.	12	25	.	1200	.	.	.	.	195	372	397	.	.	.
1049A-19X-4, 40–44 cm	.	.	2	10	.	453	.	.	.	12	78	156	96	.	.	.
1049A-19X-4, 107–109 cm	.	.	.	.	.	3021	.	.	.	.	459	767	55	.	.	4
1049A-19X-5, 8.5–10.5 cm	.	.	1	50	.	4922	.	.	.	0	330	880	138	.	.	.
1049A-19X-5, 49–53 cm	.	.	.	25	.	1020	.	.	.	91	240	.	17	.	.	50
1049A-20X-1, 15–17 cm	.	.	.	4	.	771	.	.	.	22	90	3346	83	.	.	.
1049A-20X-1, 42–44 cm	.	.	10	42	.	1905	.	.	.	6	197	1205	21	.	.	.
1049A-20X-1,90–92 cm	.	.	.	44	.	755	.	.	.	4	77	1154	32	.	.	.
1049C-12X-3,02–04 cm	.	.	14	13	.	2286	.	.	.	5	227	1044	320	.	.	.
1049A-20X-2, 36–40 cm	.	.	14	27	.	1746	.	.	.	2	182	437	121	.	.	.
1049C-12X-3, 13–15 cm	.	.	.	16	.	1234	.	.	.	4	71	90	21	.	.	.
1049C-12X-3, 34–36 cm	.	.	.	5	.	604	.	.	.	.	60	110	10	.	.	.
1049C-12X-3, 39–41 cm	.	.	.	.	.	220	.	.	.	.	126	8	20	.	.	.
1049C-12X-3, 43–44 cm	.	.	.	.	.	180	.	.	.	.	65	2	14	.	.	.
1049C-12X-3, 44–45 cm	.	.	.	.	.	27	.	.	.	.	11	.	3	.	.	.
1049C-12X-3, 52–54 cm	.	.	.	.	.	3	.	.	.	.	1	.	.	.	.	.
1049C-12X-3, 55–57 cm	.	.	.	.	.	11	.	.	.	.	1	.	1	.	.	.
1049C-12X-3, 65–66 cm	.	.	.	3	.	40	3	.	.	.	13	7	7	.	.	.
1049C-12X-3, 67–69 cm	.	1	.	.	.	585	1	.	.	.	621	15	49	.	.	.
1049C-12X-3, 70–71 cm	.	.	.	.	.	18	.	.	.	.	.	.	.	.	.	.
1049C-12X-3, 74–76 cm	.	.	.	.	.	6	.	.	.	.	2	.	.	.	.	.
1049C-12X-3, 77–78 cm	.	.	.	.	.	2	.	.	.	.	2	2	2	.	.	.
1049C-12X-3, 78–79 cm	.	.	.	.	.	.	.	.	.	.	.	.	.	.	.	.
1049C-12X-3, 79–80 cm	.	.	.	.	.	9	.	.	.	.	.	.	.	.	.	.
1049C-12X-3, 81–83 cm	.	6	.	2	.	35	.	.	.	.	26	18	.	.	.	.
1049C-12X-3, 84–86 cm	3	79	3	6	.	424	.	.	.	3	90	96	.	.	.	.
1049C-12X-3, 87–89 cm	.	34	15	5	.	719	.	.	.	.	148	136	.	.	.	.
1049C-12X-3, 89–91 cm	.	16	2	2	.	1424	.	.	.	.	177	564	5	.	.	.
1049C-12X-3, 98–100 cm	.	.	13	4	.	936	.	.	.	.	126	209	2	.	.	.
1049C-12X-3, 103–105 cm	1	.	.	9	6	297	1	.	.	.	75	192	1	.	.	.
1049A-20X-3, 40.5, 44 cm	.	.	.	4	.	526	.	.	.	1	162	1869	81	.	.	.
1049A-20X, 45–50 cm	.	.	.	113	.	4667	.	.	.	.	921	794	340	.	.	.
1049A-20X-4,97–100 cm	.	.	.	63	.	657	.	407	.	.	63	751	375	.	.	32
1049A-20X04, 125–127 cm	1	.	.	.	.	644	.	14	.	1	98	1569	70	.	.	.
1049A-20X-5, 54–60 cm	5	.	.	.	.	130	.	3	.	.	19	.	24	.	.	.
1049A-20X-6, 37–43 cm	.	.	.	3	.	46	.	.	.	46	.	.	.	.	3	.
1049A-21X-1, 124–126 cm	.	.	1	2	.	.	.	.	.	.	1	.	440	.	.	.
1049A-21X-2, 35–40 cm	.	.	.	17	.	.	68	.	.	.	660	85	.	.	27	70
1049A-21X-2, 100–105 cm	.	.	.	87	.	.	.	.	.	.	7	1393	36	.	.	7
1049A-21X-2, 117–119 cm	.	.	.	.	.	.	.	52	.	.	.	52	.	.	.	7
1049A-21X-2, 133–135 cm	.	.	.	0	.	.	.	28	.	.	.	140	28	.	.	.
1049A-21X-3, 18–20 cm	.	.	7	.	.	.	13	.	.	7	39	.	.	.	23	13
1049A-21X-3, 60.5–62.5 cm	.	.	1	15	.	.	.	13	.	.	.	863	.	.	.	.
1049A-21CC, 33–38 cm	.	.	22	44	.	.	.	55	.	.	.	6	.	.	.	7
1049A-22X-2, 20–25 cm	.	.	.	1	.	.	.	205	1	.	.	.	.	.	.	.

Sample	Pseudonodosaria humilis	Pseudopatellina sp.	Psilocitharella kochii	Psilocitharella recta	Pyramidulina sceptrum	Pyramidulina spp.	Pyramidulina zippei	Pyrulina cylindroides	Pyrulinoides acuminata	Quadromorphina allomorphinoides	Ramulina aculeata	Ramulina tetrahedralis	Rhizammina spp.	Saracenaria spinosa	Saracenaria spp.	Spirillina minima
1049A-19X-1, 62–64 cm	.	.	.	.	.	.	.	.	.	.	.	.	.	.	.	27
1049A-19X-1, 135.5–137.5 cm	.	.	.	.	.	.	.	13	.	38	25	.	.	.	7	.
1049A-19X-2, 35–39 cm	.	.	.	.	.	.	.	.	.	.	26	.	.	.	.	.
1049A-19X-2, 116–118 cm	.	.	.	.	.	.	.	.	.	.	.	.	.	.	.	.
1049A-19X-3, 93–95 cm	3	.	3	.	.	.	.	.	.	149	.	.	.	.	9	.
1049A-19X-4, 40–44 cm	24	31	.	.	.	.	.	4	45	47	.	.	.	.	4	.
1049A-19X-4, 107–109 cm	.	.	.	3	0	.	.	14	.	27	205	14	.	.	.	.
1049A-19X-5, 8.5–10.5 cm	22	.	.	.	.	.	6	22	.	132	0	.	.	.	6	.
1049A-19X-5, 49–53 cm	17	.	.	.	.	.	8	.	8	.	33	17	.	.	25	17
1049A-20X-1, 15–17 cm	.	.	.	3	.	.	.	10	3	45	.	.	.	.	1	.
1049A-20X-1, 42–44 cm	4	.	.	.	.	.	.	.	31	52	34	.	.	.	.	.
1049A-20X-1,90–92 cm	1	.	.	1	.	11	.	.	.	11	42	36	.	.	.	0
1049C-12X-3,02–04 cm	.	.	.	3	.	13	.	3	6	19	5	2	.	.	19	.
1049A-20X-2, 36–40 cm	4	.	.	.	.	.	.	.	2	58	7	.	.	.	.	.
1049C-12X-3, 13–15 cm	.	.	.	.	.	1	.	1	.	2	2	.	.	.	.	.
1049C-12X-3, 34–36 cm	.	.	.	.	.	.	.	.	.	.	2	.	.	.	.	.
1049C-12X-3, 39–41 cm	3	.	.	5	.	.	.	.	.	.	3	.	.	.	.	.
1049C-12X-3, 43–44 cm	.	.	.	.	.	.	.	.	.	.	.	.	.	.	.	.
1049C-12X-3, 44–45 cm	3	.	.	.	.	.	.	.	.	.	.	.	.	.	.	.
1049C-12X-3, 52–54 cm	.	.	.	.	.	.	.	.	.	.	.	.	.	.	.	.
1049C-12X-3, 55–57 cm	.	.	.	.	.	.	.	.	.	.	.	.	.	.	.	.
1049C-12X-3, 65–66 cm	.	.	.	.	.	.	.	.	.	.	.	.	.	.	.	.
1049C-12X-3, 67–69 cm	.	.	.	.	.	.	.	.	.	1	.	.	.	.	1	.
1049C-12X-3, 70–71 cm	.	.	.	.	.	.	.	.	.	.	.	.	.	.	.	.
1049C-12X-3, 74–76 cm	.	.	.	.	.	.	.	.	2	.	.	.	.	.	.	.
1049C-12X-3, 77–78 cm	.	.	.	.	.	.	.	.	.	.	.	.	.	.	.	.
1049C-12X-3, 78–79 cm	.	.	.	.	.	.	.	.	.	.	.	.	.	.	.	.
1049C-12X-3, 79–80 cm	.	.	.	.	.	.	.	.	.	.	.	.	.	.	.	.
1049C-12X-3, 81–83 cm	.	.	.	.	.	.	.	.	.	.	.	.	.	.	2	.
1049C-12X-3, 84–86 cm	6	.	.	3	.	6	3	6	.	.	.	.	.	.	14	.
1049C-12X-3, 87–89 cm	24	.	.	.	5	.	.	.	.	.	2	.	.	.	12	.
1049C-12X-3, 89–91 cm	4	.	.	12	.	5	4	2	9	53	25	.	.	.	25	2
1049C-12X-3, 98–100 cm	4	.	.	2	.	4	.	29	29	.	.	11	.	.	18	.
1049C-12X-3, 103–105 cm	5	27	.	5	.	4	1	18	3	32	1	1	.	.	3	59
1049A-20X-3, 40.5, 44 cm	0	.	.	.	.	9	9	9	18	63	27	.	.	.	4	63
1049A-20X, 45–50 cm	.	.	.	1	14	.	.	.	.	.	.	.	.	.	1	113
1049A-20X-4,97–100 cm	63	125	.	.	.	.	.	63	.	125	0	.	.	.	.	971
1049A-20X04, 125–127 cm	28	.	.	.	.	.	.	.	126	16	.	.	.	.	1067	154
1049A-20X-5, 54–60 cm	.	27	.	5	3	.	.	.	.	8	11	.	.	.	24	240
1049A-20X-6, 37–43 cm	.	.	.	3	.	.	23	23	.	.	.	.	6	692	40	4
1049A-21X-1, 124–126 cm	0	.	.	.	.	.	.	.	.	.	.	.	.	.	49	1839
1049A-21X-2, 35–40 cm	.	.	.	.	.	17	51	19	68	.	.	.	.	40	1423	85
1049A-21X-2, 100–105 cm	58	.	.	.	.	.	.	.	.	.	.	.	.	.	.	1729
1049A-21X-2, 117–119 cm	.	.	.	.	.	0	.	.	.	.	.	.	.	.	1	942
1049A-21X-2, 133–135 cm	.	.	.	.	.	.	.	.	.	.	.	.	.	.	62	1043
1049A-21X-3, 18–20 cm	124	7	.	.	13	.	.	13	13	59	20	.	.	.	39	1079
1049A-21X-3, 60.5–62.5 cm	9	159	.	.	.	.	.	.	.	.	11	.	2	0	5	1454
1049A-21CC, 33–38 cm	.	.	.	.	.	.	.	.	.	.	.	.	.	6	962	.
1049A-22X-2, 20–25 cm	1	.	204	.	.	.	.	.	.	.	.	.	.	.	.	1941

Sample	Spiroloculina sp.	Spiroplectinate annectens	Spiroplectinata complanata	Spiroplectinata spp.	Textularia champania	Tristix acutangula	Tristix excavata	Turrispirillina subconica	Vaginulinopsis excentricica	Vaginulinopsis harpa	Vaginulinopsis spp.	Valvulineria gracillima	Valvulineria spp.	No.of species	No.of specimens
1049A-19X-1, 62–64 cm	.	.	.	.	.	.	.	.	.	.	.	865	.	13	1486
1049A-19X-1, 135.5–137.5 cm	.	.	.	.	.	.	.	.	.	7	.	636	.	25	3525
1049A-19X-2, 35–39 cm	.	.	.	.	.	.	.	.	.	12	.	369	49	24	3165
1049A-19X-2, 116–118 cm	.	.	.	.	.	.	.	.	.	.	.	150	7	17	1170
1049A-19X-3, 93–95 cm	.	.	.	.	.	.	.	.	.	.	.	217	.	25	2780
1049A-19X-4, 40–44 cm	.	.	.	4	.	.	16	.	.	.	.	209	16	39	1908
1049A-19X-4, 107–109 cm	.	.	.	.	.	.	.	.	0	14	.	996	123	32	6406
1049A-19X-5, 8.5–10.5 cm	.	.	.	.	.	.	6	.	.	0	.	1430	44	36	4037
1049A-19X-5, 49–53 cm	.	.	.	.	.	.	.	.	.	.	.	149	.	34	3366
1049A-20X-1, 15–17 cm	.	.	.	.	.	.	3	.	.	.	.	184	.	33	2206
1049A-20X-1, 42–44 cm	.	.	.	.	.	.	.	.	.	.	.	233	31	41	4478
1049A-20X-1,90–92 cm	11	.	.	.	.	.	.	.	.	2	.	135	.	35	2719
1049C-12X-3,02–04 cm	.	26	.	.	.	.	.	.	3	2	.	165	27	42	5588
1049A-20X-2, 36–40 cm	.	.	.	5	.	.	.	.	2	12	.	290	23	43	4387
1049C-12X-3, 13–15 cm	.	.	.	.	.	.	.	.	.	.	11	153	9	25	1962
1049C-12X-3, 34–36 cm	.	.	.	.	.	.	.	.	2	.	.	55	.	15	912
1049C-12X-3, 39–41 cm	.	.	.	.	.	.	.	.	.	1	.	15	.	22	631
1049C-12X-3, 43–44 cm	.	.	.	.	.	.	.	.	.	.	.	2	.	15	329
1049C-12X-3, 44–45 cm	.	.	.	.	.	.	.	.	.	.	.	.	.	10	73
1049C-12X-3, 52–54 cm	.	.	.	.	.	.	.	.	.	.	.	3	.	6	19
1049C-12X-3, 55–57 cm	.	.	.	.	.	.	.	.	.	.	.	4	.	6	20
1049C-12X-3, 65–66 cm	.	.	.	.	.	.	.	.	.	.	.	.	.	12	100
1049C-12X-3, 67–69 cm	.	.	.	.	.	.	.	.	.	.	.	.	.	17	2126
1049C-12X-3, 70–71 cm	.	.	.	.	.	.	.	.	.	.	.	.	.	1	18
1049C-12X-3, 74–76 cm	.	.	.	.	.	.	.	.	.	.	.	.	.	5	13
1049C-12X-3, 77–78 cm	.	.	.	.	.	.	.	.	.	.	.	.	.	7	21
1049C-12X-3, 78–79 cm	.	.	.	.	.	.	.	.	.	.	.	.	.		43
1049C-12X-3, 79–80 cm	.	.	.	.	.	.	.	.	.	.	.	.	.		37
1049C-12X-3, 81–83 cm	.	.	.	.	.	.	.	.	.	.	.	4	.	21	384
1049C-12X-3, 84–86 cm	.	.	.	.	.	.	.	.	.	.	.	42	.	32	1599
1049C-12X-3, 87–89 cm	.	.	.	.	.	.	.	.	.	.	.	356	5	32	2092
1049C-12X-3, 89–91 cm	.	.	16	.	.	.	.	.	.	.	5	233	21	47	4330
1049C-12X-3, 98–100 cm	.	.	.	.	.	.	.	.	245	16	43	3015			
1049C-12X-3, 103–105 cm	.	.	.	.	.	.	2	12	.	.	.	4	20	52	1390
1049A-20X-3, 40.5, 44 cm	.	.	.	.	.	.	.	.	.	0	.	1389	94	43	5766
1049A-20X, 45–50 cm	.	.	.	.	.	.	.	.	.	30	.	2835	.	24	12667
1049A-20X-4,97–100 cm	.	.	0	.	.	.	.	219	0	.	125	.	.	30	10514
1049A-20X04, 125–127 cm	3	7	.	.	.	.	350	.	.	.	420	.	32	6452	
1049A-20X-5, 54–60 cm	8	133	104	.	.	.	.	21	8	.	3	189	.	44	2018
1049A-20X-6, 37–43 cm	20	.	.	.	.	182	.	.	.	9	.	34	1729		
1049A-21X-1, 124–126 cm	49	.	6	.	.	.	.	445	2	.	.	15	.	33	3678
1049A-21X-2, 35–40 cm	114	68	.	.	2	.	118	.	.	.	271	.	37	6036	
1049A-21X-2, 100–105 cm	.	.	15	.	.	.	.	239	.	.	.	370	.	24	4727
1049A-21X-2, 117–119 cm	6	64	57	.	.	6	.	621	1	.	.	.	.	30	3091
1049A-21X-2, 133–135 cm	.	.	.	.	.	.	.	3	.	0	.	.	.	20	1903
1049A-21X-3, 18–20 cm	39	43	269	26	.	.	.	59	7	1	.	.	.	44	4008
1049A-21X-3, 60.5–62.5 cm	.	.	.	3	.	.	.	142	.	8	3	242	.	38	4436
1049A-21CC, 33–38 cm	12	11	.	.	.	.	11	.	0	6	67	.	34	2265	
1049A-22X-2, 20–25 cm	.	.	.	.	.	.	.	70	.	.	.	.	.	16	5479

et al. 2000) contains assemblages that are strongly dominated by patellinids, spirillinids and gavelinellids. Above this sample, diversity remains relatively high in Samples 1049A-21XCC, 33–38 cm to 1049A-20X-4, 97–100 cm, except for Samples 1049A-20X-21X-2, 133–135 cm, and 1049A-21X-2, 100–105 cm, which contain fewer nodosariids. First Factor values (F1) are consistently low and negative within this interval.

A marked change in assemblage composition coincides with a sharp lithological boundary in Interval 1049A-20X-4, 57–58 cm, marking the change from alternating white and red clayey chalk to greenish grey calcareous claystone (Fig. 3). In Samples 1049A-20X-4, 45–50 cm, to 1049C-12X-3, 81–83 cm, just below the black shale, abundance decreases sharply overall, as the number of specimens falls from 12 667 to 384 per 10 g of sediment. A concurrent decline in diversity is observed at the top of this interval. F1 values are mostly positive and significantly higher than in the underlying interval. Within the laminated black shale (Samples 1049C-12X-3, 79–80 cm to 44–45 cm), diversity and abundance are generally extremely low, except in Sample 1049C-12X-3, 67-69 cm, which contains over 2000 specimens per 10 g sediment and 17 species. The highest positive F1 values are recorded within the black shale.

Within the greenish grey claystone above the laminated black shale, abundance initially rises gradually (Samples 12X-3, 43–44 cm to 13–15 cm), as the number of specimens climbs from 329 to 1962 per 10 g of sediment, whereas diversity remains low. Many species present in samples below the black shale reappear in samples above it. However, a sharp increase in diversity and abundance occurs in Sample 1049A-20X-2, 36–40 cm, which contains 43 species and 4387 specimens per 10 g of sediment. Diversity remains high in the overlying interval consisting of alternating white and red clayey chalk (Samples 1049A-20X-2, 36–40 cm, to 1049A-19X-4, 40–44 cm), although fluctuations in abundance are significant. F1 values within this interval are positive and comparable with those immediately below the black shale.

By contrast, the uppermost samples (Samples 1049A-19X-3, 93–95 cm, to 1049A-19X-1, 62–64 cm) are characterized by low diversity and relatively low to moderate abundance. A coincident lithological change is observed in this interval, which consists of alternating red clayey chalk, white chalk and greenish grey claystone. First factor values (F1) are positive and slightly higher than in the preceding interval.

Short-term variations. Superimposed over the long-term variations are pronounced short-term fluctuations in abundance and diversity, both below and above the black shale (Fig. 2). For instance, below the black shale in Sections 1049A-21CC to 1049A-20X-4, the number of tests fluctuates between 1729 and 12 667 per 10 g of sediment, and the number of species varies between 20 and 44. Fluctuations in abundance within this interval mainly reflect changes in the numbers of *Gyroidinoides infracretaceus*, *Praebulimina elata* and *Berthelina intermedia*. Above the black shale, in Sections 1049A-20X-1 to 1049A-19X-1, the number of tests fluctuates between 1170 and 9037 per 10 g of sediment and the number of species varies between 13 and 41. These short-term fluctuations do not correlate with marked lithological changes.

Benthic foraminiferal associations

Correspondence analysis allows distinct patterns in species distribution to be recognized in our samples. A late Aptian–early Albian association, characterized by *Spirillina minima*, *Patellina subcretacea*, *Turrispirillina subconica*, *Spiroplectinata annectens*, *Spiroplectinata complanata*, *Falsogaudryinella moesiana*, *Protomarssonella trochus*, *Berthelina intermedia*, *Gyroidinoides infracretaceus* and *Spiroloculina* sp., can be identified at the base of Hole 1049A (Samples 1049A-22X-2, 20–25 cm, to 1049A-20X-4, 97–100 cm). These basal samples and their diagnostic species display high negative F1 values, and form distinct clusters on the left side of the F1 axis in Fig. 4.

Samples within the laminated black shale are distinguished by the association of *Fursenkoina viscida*, *Ellipsoidella cuneata*, *Neobulimina minima*, *Osangularia schloenbachi* and *Pleurostomella reussi*. The black shale samples and their distinctive species exhibit high F1 and F2 values, and plot in the right side of Fig. 4. *Osangularia schloenbachi* and *Pleurostomella reussi*, also abundant in samples above and below the black shales, plot closer to the central cluster of species near the intersection of the F1 and F2 axes.

Species in the central cluster, close to the intersection of the F1 and F2 axes, occur both in the samples below the black shale (Samples 1049A-20X-4, 45-50 cm, to 1049C-12X-3, 81–83 cm) and above the black shale (Samples 1049C-12X-3, 43–44 cm, to 1049A-19X-1, 62–64 cm). Correspondence analysis shows that the benthic foraminiferal association remains relatively similar in composition within these two intervals, although fluctuations in diversity

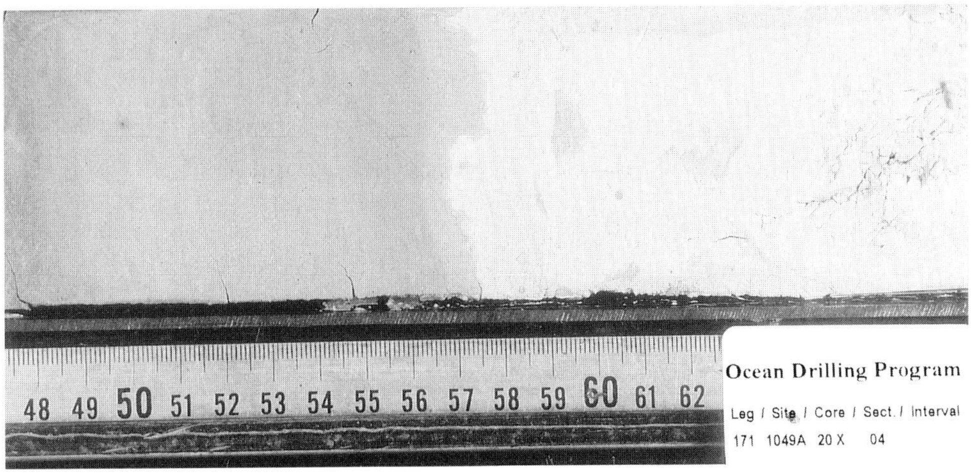

Fig. 3. Sharp lithological boundary in Interval 1049A-20X-4, 57–58 cm, marking the change from alternating white and red clayey chalk to greenish grey calcareous claystone.

and abundance are evident. Characteristic taxa are *Berthelina intermedia*, *Gyroidinoides infracretaceus*, *Osangularia schloenbachi*, *Pleurostomella reussi*, *Praebulimina elata*, *Praebulimina nannina*, *Clavulinoides gaultinus* and *Valvulineria gracillima*, whereas patellinids, spirillinids and *Spiroplectinata* spp. are virtually absent.

Stratigraphic distribution of benthic foraminiferal species

Figure 5 shows that most of the species that disappear during deposition of the black shale reoccur in sediments above it. Only the following eight species have their last occurrence (LO) below the black shale: *Pyramidulina sceptrum*, *Spiroplectinata complanata*, *Patellina subcretacea*, *Marginulinopsis jonesi*, *Tristix acutangula*, *Saracenaria spinosa*, *Planularia crepidularis* and *Lenticulina ouachensis*. Only one species, *Neobulimina minima*, disappears within the black shale. Most LOs, in fact, occur well above the black shale, within the uppermost studied samples. However, only few of these LOs represent rare species.

First occurrences (FOs) take place in a stepwise fashion, both below and above the black shale (Fig. 5). However, most FOs are recorded in the four samples at the base of Hole 1049A, well below the black shale (Samples 1049A-22X-2, 20–25 cm, to 1049A-21X-3, 18–20 cm). Sixteen species have their FOs in the overlying samples that precede the black shale (Samples 1049A-21X-2, 133–135 cm, to 1049C-12X-3, 103–105 cm). No new species appears within the black shale. The following five species have their FO above the black shale: *Charltonina australis*, *Lingulogavelinella albiensis*, *Protomarssonella oxycona*, *Textularia chapmani* and *Praebulimina* sp. 1.

Palaeoenvironment

General trends

General trends in faunal distribution are deduced from abundance, diversity and main trends in assemblage composition reflected by the F1 value of the correspondence analysis (Fig. 2). The F1 value of modern benthic foraminiferal assemblages generally shows a significant correlation with carbon flux rates in the modern oceans (Kuhnt *et al.* 1999; Wollenburg & Kuhnt 2000; Weinelt *et al.* 2001). A similar relationship of F1 values, palaeoproductivity and carbon flux has been already suggested for Late Cretaceous deep-sea benthic foraminiferal assemblages of North Atlantic DSDP–ODP Sites (Kuhnt & Moullade 1991). Overall, availability of food rather than oxygen appears to be the main factor controlling benthic foraminiferal distribution also within our Aptian–Albian assemblages except during OAE 1b.

Oligotrophic to mestotrophic conditions during late Aptian and earliest Albian time are indicated by relatively diverse assemblages with negative F1 values at the base of Hole 1049. An outer shelf or upper slope depositional environment is suggested by the high numbers of spirillinids and, patellinids within the

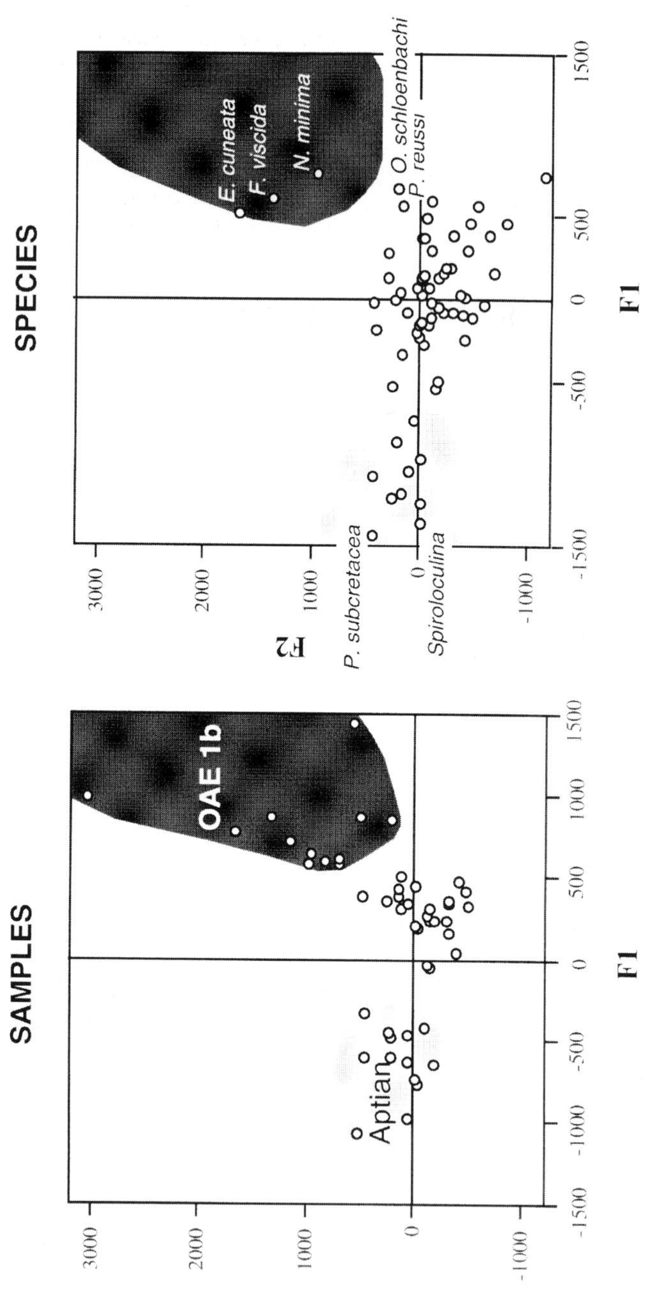

Fig. 4. Results of correspondence analysis: position of species and samples on correspondence analysis Factors 1 and 2.

assemblages. The marked change in assemblage composition above the sharp lithological boundary in Interval 1049A-20X-4, 57–58 cm (Fig. 3) may reflect a rapid increase in water depth as typical shelf–upper bathyal taxa (patellinids, spirillinids and *Spiroplectinata* spp.) virtually disappear above this boundary. The depositional environment from this interval may have been a mid-slope oxygen minimum zone. Positive F1 values and high numbers of *Gyroidinoides infracretaceus*, *Berthelina intermedia*, *Osangularia schloenbachi*, *Pleurostomella reussi*, *Praebulimina elata* and *Valvulineria gracillima* suggest that assemblages became increasingly dominated by opportunistic phytodetritus feeders thriving on an enhanced carbon flux to the sea floor and tolerating some degree of oxygen depletion. Depletion of pore-water oxygenation is also suggested by the transition from white–reddish chalk and claystone to greenish grey claystone in Interval 1049A-20X-4, 57–58 cm. The dramatic increase in F1 values and the concurrent drop in diversity and abundance in samples immediately below the black shale (Samples 1049C-12X-3, 89–91 cm, to 1049C-12X-3, 81–83 cm) suggest intense eutrophication before the onset of black shale sedimentation.

Extremely impoverished assemblages within the black shale indicate virtually anoxic conditions at the sea floor, perhaps induced by strong stratification in the overlying water column. Detailed documentation of the benthic foraminiferal successions within the black shale and palaeoenvironmental interpretations have been presented by Erbacher *et al.* (1999) and Holbourn *et al.* (2001). Increasing abundance, decreasing F1 values and the gradual return of the pre-black shale fauna in the 30 cm interval above the black shale signals the progressive recovery of the ecosystem. Improved ventilation is also indicated by the light sediment colour and the disappearance of laminations. Relatively high diversity in Samples 1049A-20X-2, 36–40 cm, to 1049A-19X-4, 40–44 cm, indicates rapid reoxygenation at the sea floor. However, the F1 values remain significantly higher than in the late Aptian assemblages, which may be an indication of generally higher carbon fluxes in early Albian time. Renewed eutrophication is indicated by a decrease in diversity, increasing F1 values and the strong dominance of taxa inferred to be opportunistic phytodetritus feeders in the uppermost lower Albian samples (Samples 1049A-19X-3, 93–95 cm, to 1049A-19X-1, 62–64 cm). Depleted pore-water oxygenation is suggested by the dominance of greenish grey nannofossil chalks and claystones within this upper interval.

Rapid fluctuations

Sharp fluctuations in abundance and in diversity also characterize the intervals both below and above the black shale. As preservation is relatively good in samples with lower diversity and abundance, such fluctuations are unlikely to reflect preservational bias. Eicher & Diner (1991) proposed that strong fluctuations in productivity in mid-Cretaceous oceans were the main cause of limestone–marlstone rhythms in epicontinental seas. These productivity cycles were driven by contrasting circulation modes, as the focus of deep-water formation alternated between high and low latitudes during mid-Cretaceous time. The fluctuations in benthic foraminiferal abundance and diversity at Site 1049 suggest an unstable water column, characterized by rapid overturns and by frequent changes in carbon flux and/or oxygenation during late Aptian and early Albian time, both before and after deposition of the black shale.

Timing of faunal turnover

Five of the eight LOs that occur below the black shale do not have stratigraphic significance (Fig. 6). Indeed, the stratigraphic ranges of *Pyramidulina sceptrum*, *Spiroplectinata complanata*, *Patellina subcretacea*, *Marginulinopsis jonesi* and *Tristix acutangula* are known to extend further into late Albian or Late Cretaceous time. Only the LOs of *Saracenaria spinosa*, *Planularia crepidularis* and *Lenticulina ouachensis* may be considered to represent real extinction events, as they coincide with the reported LOs of these species in the literature. *Saracenaria spinosa*, a characteristic Aptian taxon, disappears in latest Aptian–earliest Albian (Bartenstein 1987; King *et al.* 1989). *Planularia crepidularis*, a typical late Valanginian to Barremian taxon becomes extinct in late Aptian time (Bartenstein & Bettenstaedt 1962; Meyn & Vespermann 1994). *Lenticulina ouachensis*, common in Berriasian to Barremian sediments, disappears before the end of Aptian time (Sigal 1952; Moullade 1966; Riegraf & Luterbacher 1989; Weidich 1990). LOs recorded within and above the black shale are either the LOs of long-ranging species with sporadic occurrences or are facies related. Significantly most LOs occur in the five uppermost studied samples, where a marked drop in diversity is recorded.

In contrast to LOs, a higher number of FOs are of stratigraphic significance: 22 FOs correspond to, or even precede, FOs reported in the literature (Table 2); other FOs represent rare

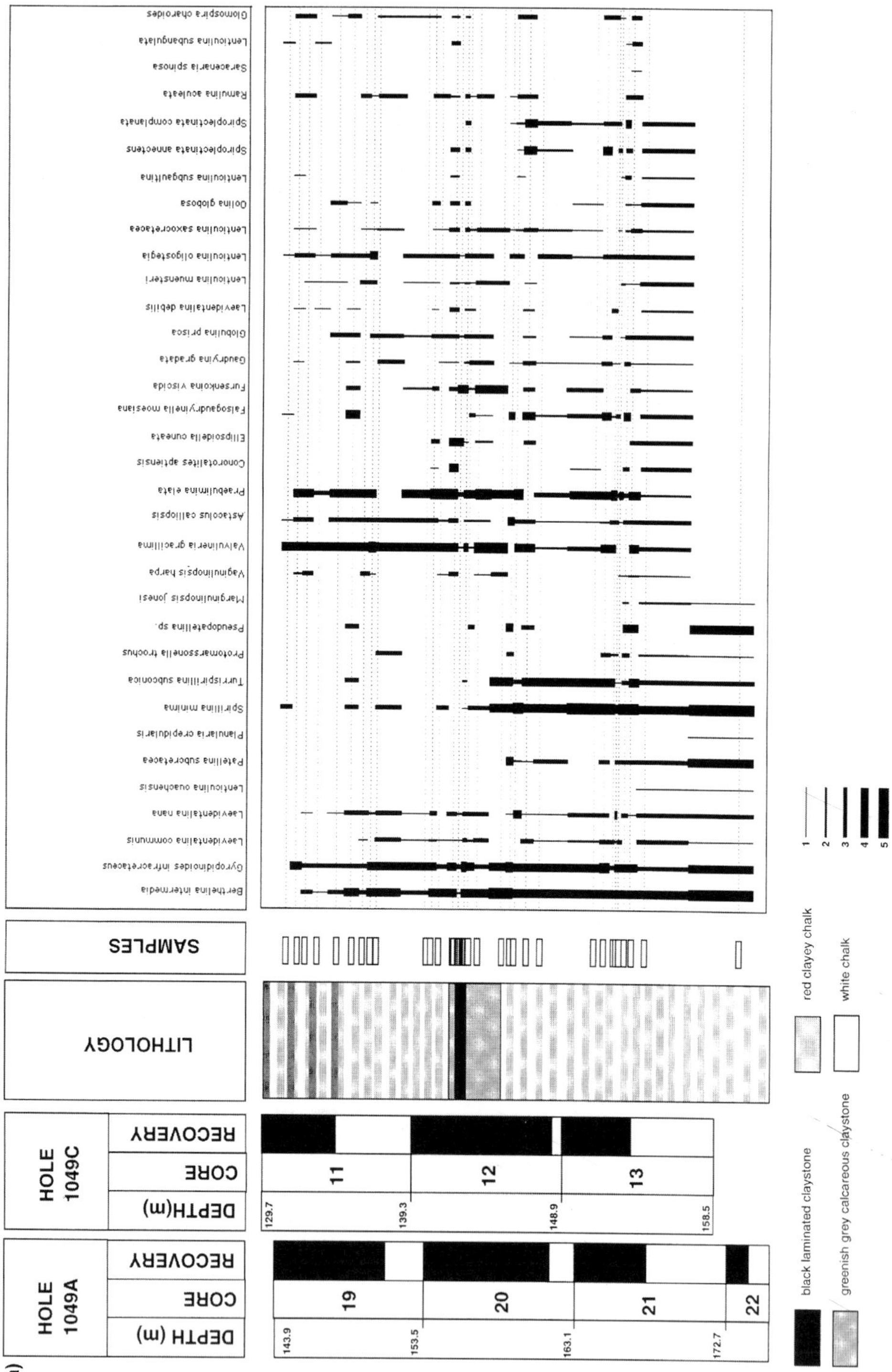

Fig. 5. Stratigraphic distribution of Aptian–Albian benthic foraminifers at Site 1049. Planktonic zonation from Bellier *et al.* (2000). Lithology interpolated in non-recovered portions of cores.

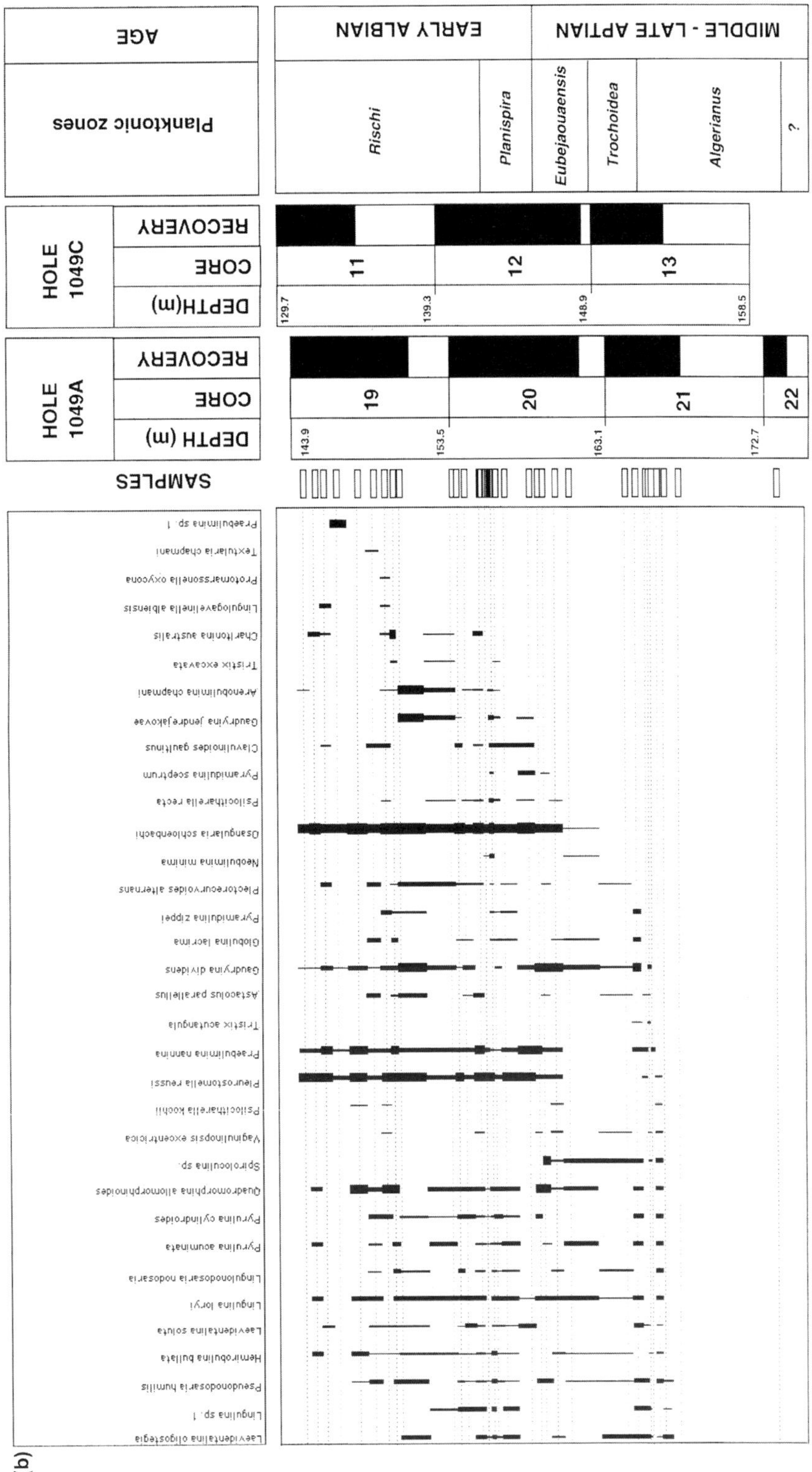

Fig. 5. Stratigraphic distribution of Aptian–Albian benthic foraminifers at Site 1049. Planktonic zonation from Bellier *et al.* (2000). Lithology interpolated in non-recovered portions of cores.

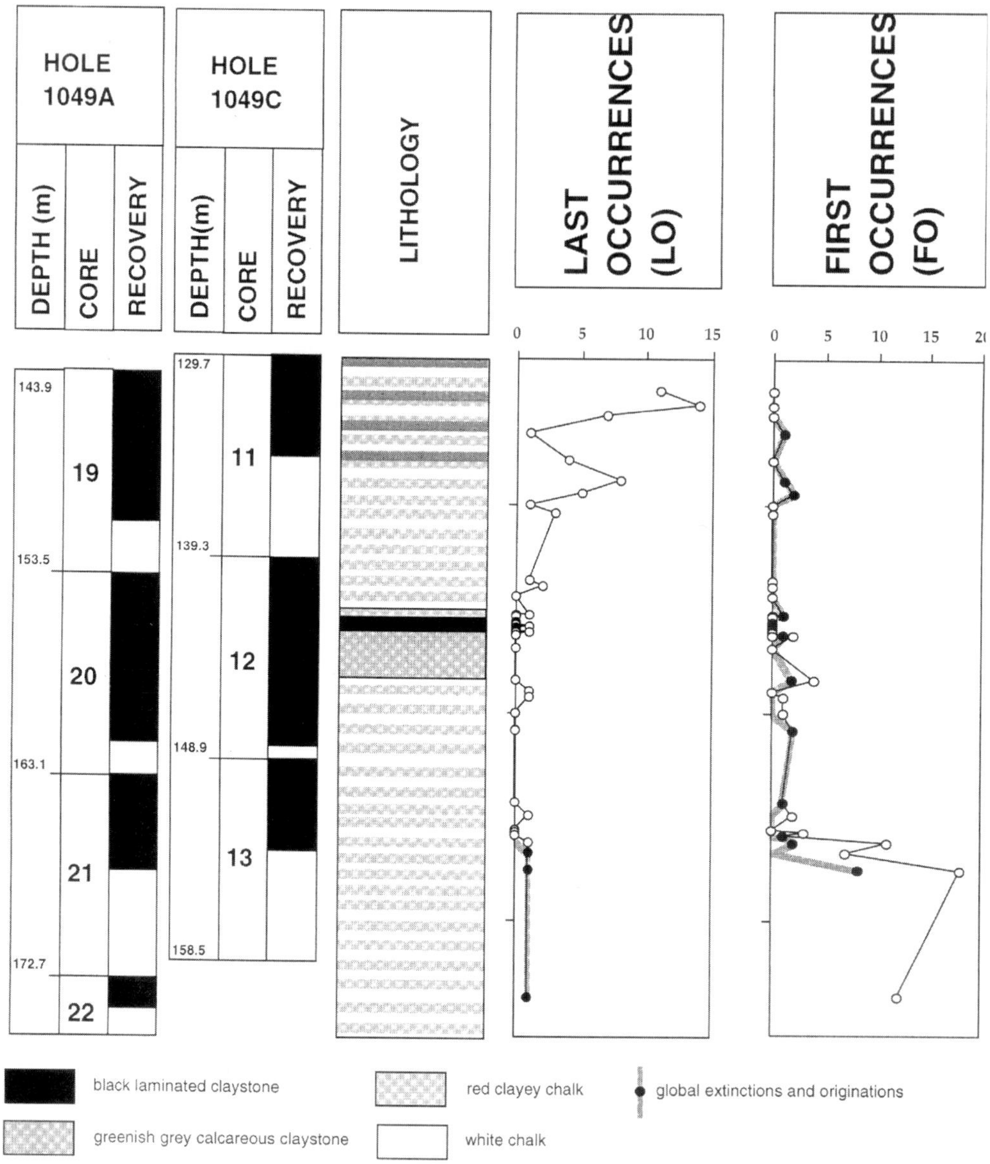

Fig. 6. LOs and FOs in late Aptian–early Albian benthic foraminiferal assemblages from Site 1049. Lithology interpolated in non-recovered portions of cores.

species or are facies related. However, only five of the 22 stratigraphically significant FOs occur above the black shale (Fig. 6), indicating that most speciations occurred in late Aptian and early Albian time before to the onset of black shale sedimentation. Significantly, the new species that appear in late Aptian and early Albian time are either representatives of the suborder Rotaliina or are members of the suborder Textulariina with calcareous cement (with the exception of *Plectorecurvoides alternans*). These two phylogenetic lines appear to have been extremely proliferous in late Aptian and early Albian time, when the typical Early Cretaceous fauna, characterized by nodosariids and non-calcareous agglutinates, declined and precursors of many modern lineages first appeared. We observed that the

Table 2. *Stratigraphically significant FOs at Site 1049; planktonic zones from Bellier* et al. *(2000)*

	FO at Site 1049			
Calcareous species	Sample	Zone	Reported FO	References
Charltonina australis	1049C-12X-3, 02–04 cm	Rischi	Late Aptian	Holbourn & Kaminski (1997)
Ellipsoidella cuneata	1049A-21XCC, 33–38 cm	Algerianus	Albian	Loeblich & Tappan (1946)
Fursenkoina viscida	1049A-21XCC, 33–38 cm	Algerianus	Mid-Albian	Khan (1950)
Lingulogavelinella albiensis	1049A-19X-4, 107–109 cm	Rischi	Early Albian	Malapris (1965)
Neobulimina minima	1049A-20X-6, 37–43 cm	Eubejaouensis	Albian	Tappan (1940)
Osangularia schloenbachi	1049A-20X-6, 37–43 cm	Eubejaouensis	Late Aptian–early Albian	Crittenden (1983)
Praebulimina elata	1049A-21XCC, 33–38 cm	Algerianus	Early Albian	Magniez-Jannin (1975)
Praebulimina nannina	1049A-21X-2, 133–135 cm	Trochoidea–Algerianus	Early Albian	Holbourn & Kaminski (1997)
Pleurostomella reussi	1049A-21X-3, 18–20 cm	Trochoidea–Algerianus	Early Albian	Riegraf & Luterbacher (1989)
Praebulimina sp. 1	1049A-19X-2, 116–118 cm	Rischi	Early Turonian	Holbourn & Kuhnt (1998)
Quadrinorphina allomorphinoides	1049A-21X-3, 18–20 cm	Algerianus	early Albian	Holbourn & Kaminski (1997)
Valvulineria gracillima	1049A-21XCC, 33–38 cm	Algerianus	Albian	Ten Dam (1947)
Arenobulimina chapmani	1049C-12X-3, 103–105 cm	Rischi	Mid-Albian	Magniez-Jannin (1975)
Clavulinoides gaultinus	1049A-20X-4, 45–50 cm	Planispira	Early Albian	Riegraf & Luterbacher (1989)
Falsogaudryinella moesiana	1049A-21XCC, 33–38 cm	Algerianus	Late Aptian	Kaminski *et al.* (1995)
Gaudrina gradata	1049A-21XCC, 33–38 cm	Algerianus	Late Aptian	Weidich (1990)
Gaudryina jendrejakovae	1049A-20X-4, 45–50 cm	Planispira	Late Albian	Weidich (1990)
Plectorecurvoides alternans	1049A-21X-1, 124–126 cm	Trochoidea	Late early Albian	Geroch & Nowak (1984)
Protomarssonella oxycona	1049A-19X-4, 107–109 cm	Rischi	Mid-Albian	Magniez-Jannin (1975)
Spiroplectina annectens	1049A-21XCC, 33–38 cm	Algerianus	Late Aptian? or early Albian	Weidich (1990)
Spiroplectinata complanata	1049A-21XCC, 33–38 cm	Algerianus	late Aptian? or early Albian	Weidich (1990)
Textularia chapmani	1049A-19X-4, 40–44 cm	Rischi	late Albian	Magniez-Jannin (1975)

genera *Berthelina*, *Osangularia*, *Lingulogavelinella*, *Gyroidinoides*, *Gaudryina* and *Spiroplectinata* exhibit a high degree of morphological variability in our samples. Transitional forms seem to transgress taxonomic boundaries, suggesting that these genera had not yet stabilized and were still evolving rapidly in late Aptian and early Albian time.

Our data, therefore, show that no major faunal turnover occurred at Blake Nose across OAE 1b, as most of the species present below the black shale reappear above it (Fig. 6). Only three stratigraphically significant extinctions occurred in late Aptian–earliest Albian time, well before the onset of black shale deposition. No extinction occurred either during or after the black shale event. However, 17 new species first appear in sediments below OAE 1b and five new species in sediments above OAE 1b, suggesting that a significant faunal change took place during late Aptian to early Albian time.

Factors influencing evolutionary change during the late Aptian–early Albian

Mid-Cretaceous time was characterized by several major episodes of severe dysoxia or even anoxia. An extensive episode of global dysoxia occurred in early Aptian time, which is widely known as the Selli event or OAE 1a (Bralower *et al.* 1993, 1994). The impact of this more prolonged and intense episode of dysoxia on benthic foraminiferal evolution is, however, not well documented, mainly because of the lack of stratigraphically complete, fossil-bearing land sections and deep-sea cores. It might be speculated that this event had devastating repercussions for global benthic foraminiferal communities, triggering numerous extinctions and eventually sparking off a faunal renewal to fill vacant niches, once conditions improved at the sea floor. However, the composition of late Aptian assemblages at Site 1049 demonstrates that many lineages of nodosariids with stratigraphic ranges extending earlier into Early Cretaceous time survived this event. This suggests that extinctions were limited in extent during OAE 1a, and did not wipe out the majority of the Early Cretaceous fauna.

However, changing oceanographic conditions during Aptian time and Albian probably imposed different selection pressures on the existing Early Cretaceous fauna and favoured the selection of lineages demonstrating better adaptation to new oxygenation and carbon flux regimes. The benthic foraminiferal record of Site 1049 suggests that speciations occurred throughout late Aptian and early Albian time, during a period of relatively rapid environmental change, indicated by rapid fluctuations in benthic foraminiferal abundance and diversity. Significantly, members of the suborder Rotaliina and representatives of the suborder Textulariina with calcareous cement appear to have shown greater potential for evolution than the more conservative Early Cretaceous fauna, and were probably the main groups to exploit newly created niches in changing Aptian and Albian oceans. The late Aptian–early Albian faunal change eventually led to the decline of the typical Early Cretaceous fauna and heralded the development of many modern lineages.

Conclusion

The benthic foraminiferal record of Site 1049 suggests that OAE 1b was not a catalyst for evolutionary change in early Albian time. No distinct faunal turnover occurs across OAE 1b, as most of the species present below the event reoccur above it. However, a major faunal turnover appears to have been initiated in late Aptian–earliest Albian time during a period of relatively rapid environmental change, which culminated in several acute and widespread anoxic episodes. Most of the new taxa that evolved during late Aptian and earliest Albian time belong to the suborder Rotaliina or are members of the suborder Textulariina with calcareous cement. These two main phylogenetic lines continued to evolve in early Albian time, after OAE 1b, eventually giving rise to many of the modern lineages. In contrast to other Cretaceous anoxic events, OAE 1b appears to have been more limited in duration or in geographical and water depth extent, allowing the survival of benthic foraminifers on more hospitable sea floors. Once local conditions improved at Blake Nose, migration and recolonization from adjacent areas could take place at a relatively rapid pace.

We are especially grateful to O. Hermelin and J. Tyszka for critical reviews and to J. Erbacher for fruitful discussions and for providing samples from Hole 1049C. We extend our thanks to the staff of the SEM Laboratory and Photographic Unit of the Institut für Geowissenschaften at the Christian Albrechts University in Kiel, and to the staff of the ODP Core Repository in Bremen. We gratefully acknowledge the Deutsche Forschungsgemeinschaft for financial support within the German ODP-Schwerpunkt (grants KU 649/7 and KU649/8).

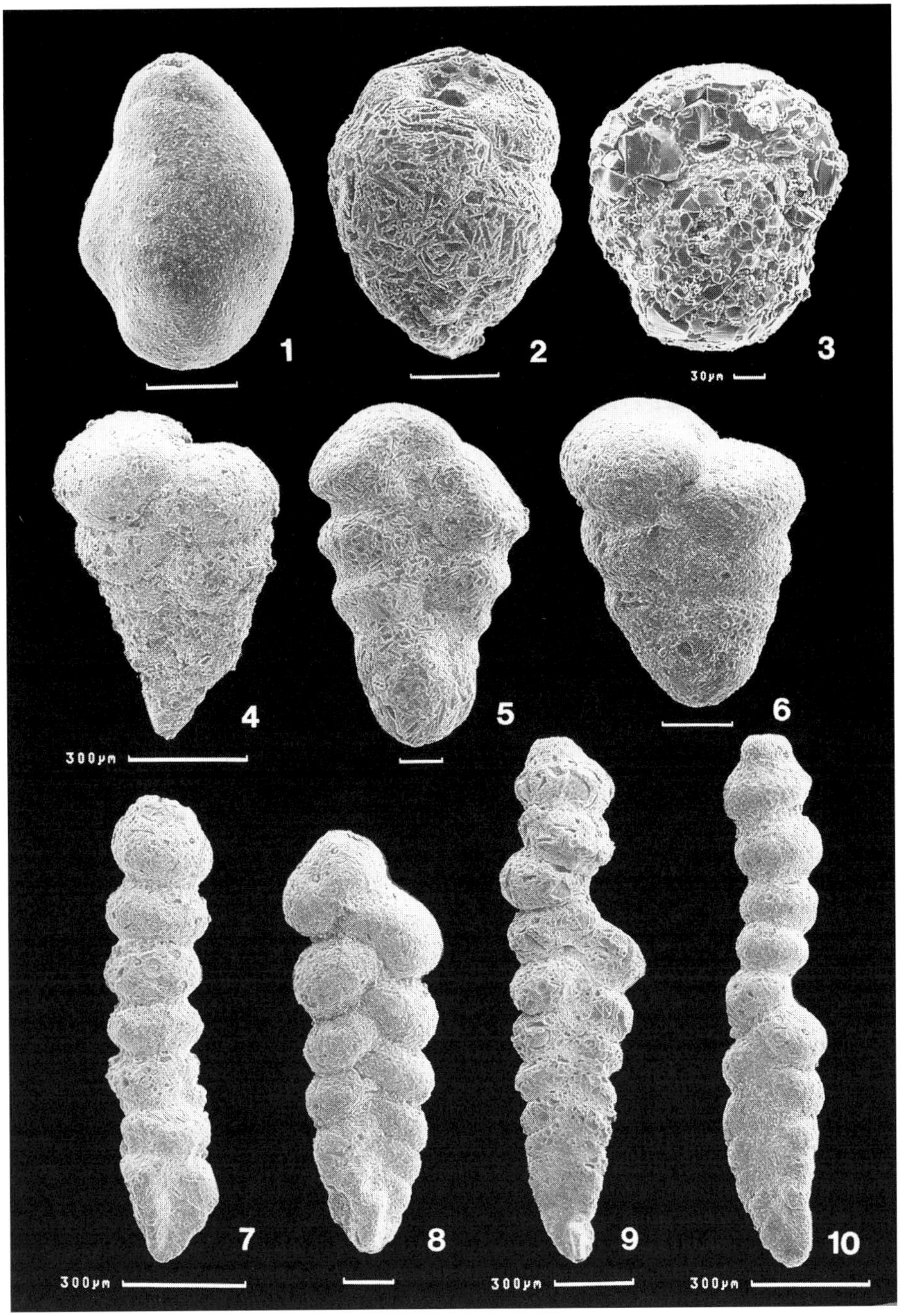

Fig. 7. Unless otherwise indicated scale bar represents 100 μm. (1) *Falsogaudryinella moesiana*, Sample 1049A, 21X-1, 124–126 cm. (2) *Arenobulimina chapmani*, Sample 1049A, 20X-1, 90–92 cm. (3) *Plectorecurvoides alternans*, Sample 1049A, 20X-1, 90–92 cm. (4) *Textularia chapmani*, Sample 1049A, 19X-4, 40–44 cm. (5) *Gaudryina jendrejakovae, Sample 1049A, 20X-1, 15–17 cm. (6) Gaudryina gradata*, Sample 1049A, 21X-2, 117–119 cm. (7) *Clavulinoides gaultinus*, Sample 1049A, 20X-3, 40.5–44 cm. (8) *Spiroplectinata complanata*, Sample 1049A, 21X-2, 117–119 cm. (9) *Spiroplectinata complanata* (with well-developed uniserial portion), Sample 11049A, 21X-2, 117–119 cm. (10) *Spiroplectinata annectens*, Sample 1049A, 21X-2, 117–119 cm.

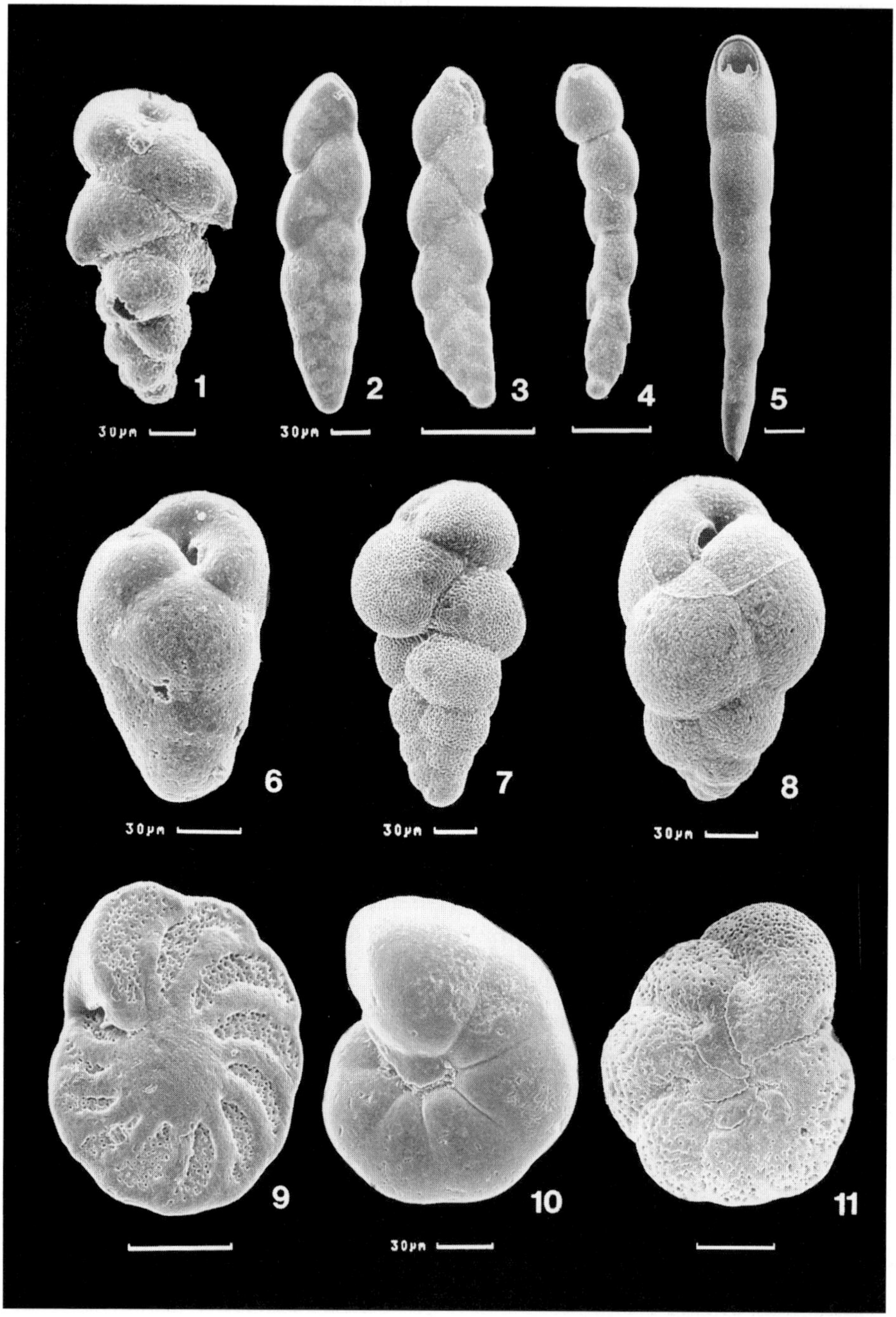

Fig. 8. Unless otherwise indicated scale bar represents 100 µm. (1) *Praebulimina* sp. 1 (scale bar represents 30 µm), Sample 1049A, 19X-2, 116–118 cm. (2) *Fursenkoina viscida*, Sample 1049C, 12X-3, 81–83 cm. (3) *Fursenkoina viscida*, Sample 1049C, 12X-3, 67–69 cm. (4) *Ellipsoidella cuneata* Sample 1049C, 12X-3, 67–69 cm. (5) *Pleurostomella reussi*, Sample 1049A, 20X-3, 40.5–44 cm. (6) *Praebulimina elata*, Sample 1049C, 12X-3, 89–91 cm. (7) *Neobulimina minima*, Sample 1049C, 12X-3, 67–69 cm. (8) *Praebulimina nannina*, Sample 1049C, 12X-3, 67–69 cm. (9) *Osangularia schloenbachi*, Sample 1049C, 12X-3, 39–41 cm. (10) *Valvulineria gracillima*, Sample 1049C, 12X-3, 87–89 cm. (11) *Lingulogavelinella albiensis*, Sample 1049A, 19X-4, 107–109 cm.

Appendix

Taxonomy

Species with stratigraphically important FOs are illustrated in Figs 7 and 8.

Ammolagena clavata (Jones & Parker 1860)
Arenobulimina chapmani Cushman 1936
Astacolus calliopsis (Reuss 1863)
Astacolus parallellus (Reuss 1863)
Berthelina intermedia (Berthelin 1880)
Bulbobaculites humei (Nauss 1947)
Charltonina australis Scheibnerová 1978
Clavulinoides gaultinus (Morozowa 1948)
Conorboides sp.
Conorotalites aptiensis (Bettenstaedt 1952)
Ellipsoidella cuneata (Loeblich & Tappan 1946)
Falsogaudryinella moesiana Neagu 1965
Frondicularia hastata Roemer 1842
Fursenkoina viscida (Khan 1950)
Gaudryina dividens Grabert 1959
Gaudryina gradata (Berthelin 1880)
Gaudryina jendrejakovae Weidich 1990
Globulina lacrima (Reuss 1845)
Globulina prisca (Reuss 1863)
Glomospira charoides (Jones & Parker 1860)
Glomospirella gaultina (Berthelin 1880)
Gyroidinoides infracretaceus Morozova 1948
Hemirobulina bullata (Reuss 1860)
Laevidentalina communis (d'Orbigny 1826)
Laevidentalina debilis (Berthelin 1880)
Laevidentalina nana (Reuss 1863)
Laevidentalina oligostegia (Reuss 1845)
Laevidentalina soluta (Reuss 1851)
Lenticulina muensteri (Roemer 1839)
Lenticulina oligostegia (Reuss 1860)
Lenticulina ouachensis Sigal 1952
Lenticulina saxocretacea Bartenstein 1954
Lenticulina subangulata (Reuss 1863)
Lenticulina subgaultina Bartenstein 1962
Lingulina loryi (Berthelin 1880)
Lingulina sp. 1
Lingulogavelinella albiensis Malapris 1965
Lingulonodosaria nodosaria (Reuss 1863)
Marginulinopsis jonesi (Reuss 1863)
Neobulimina minima Tappan 1940
Oolina globosa (Montagu 1803)
Oolina sulcata (Walker & Jacob 1798)
Osangularia schloenbachi (Reuss 1863)
Palmula sp. *Patellina subcretacea* Cushman & Alexander 1930
Planularia crepidularis Roemer 1842
Plectorecurvoides alternans Noth 1952
Pleurostomella reussi Berthelin 1880
Praebulimina elata Magniez-Jannin 1975 (see note below)
Praebulimina nannina (Tappan 1940) (see note below)
Praebulimina sp. 1
Protomarssonella oxycona (Reuss 1860)
Protomarssonella trochus (d'Orbigny 1840)
Pseudonodosaria humilis (Roemer 1841)
Pseudopatellina sp.
Psilocitharella recta (Reuss 1863)
Pyramidulina sceptrum (Reuss, 1863)
Pyramidulina zippei (Reuss 1845)
Pyrulina cylindroides (Roemer 1838)
Pyrulinoides acuminata (d'Orbigny 1840)
Quadrimorphina allomorphinoides (Reuss 1860)
Ramulina aculeata (d'Orbigny 1840)
Ramulina tetrahedralis Ludbrook 1966
Saracenaria spinosa (Eichenberg 1935)
Spirillina minima Schacko 1897
Spiroplectinata annectens (Parker & Jones 1863)
Spiroplectinata complanata (Reuss 1860)
Textularia chapmani Lalicker 1935
Tristix acutangula (Reuss 1863)
Tristix excavata (Reuss 1863)
Turrispirillina subconica Tappan 1943
Vaginulinopsis excentricica (Cornuel 1848)
Vaginulinopsis harpa (Reuss 1860)
Valvulineria gracillima Ten Dam 1947

According to the International Code of Zoological Nomenclature, *Praebulimina elata* should be treated as a junior synonym of *Praebulimina nannina*, as type material of *P. nannina* includes morphotypes corresponding to *P. elata*. Here, we use *P. elata* for morphotypes that have less inflated chambers and often a more elongate outline than *P. nannina*.

References

BARRON, E. J. & PETERSON, W. H. 1990. Mid-cretaceous ocean circulation: results from model sensitivity studies. *Palaeoceanography*, **5**, 319–338.

BARRON, E. J., PETERSON, W. H., W. H., POLLARD, D. & THOMPSON, S. 1993. Past climate and the role of ocean heat transport: model simulation for the Cretaceous. *Palaeoceanography*, **8**, 785–798.

BARRON, E. J., FAWCETT, P. J. & PETERSON, W. H. 1995. A 'simulation' of mid-Cretaceous climate. *Paleoceanography*, **10**, 953–962.

BARTENSTEIN, H. 1978. Phylogenetic sequences of Lower Cretaceous benthic foraminifera and their use in biostratigraphy. *Geologie en Mijnbouw*, **57**(1), 19–24.

BARTENSTEIN, H. 1987. Micropalaeontological synopsis of the Lower Cretaceous in Trinidad, West Indies. Remarks on the Aptian/Albian boundary. *Newsletters on Stratigraphy*, **17**, 143–152.

BARTENSTEIN, H. & BETTENSTAEDT, F. 1962. Marine Unterkreide (Boreal and Tethys). *In*: SIMON, W. & BARTENSTEIN, H. (eds) *Leitfossilen der Mikropaläontologie*. Borntraeger, Berlin, 225–298.

BARTENSTEIN, H. & BOLLI, H. M. 1973. Die Foraminiferen der Unterkreide von Trinidad, W.I. Dritter Teil: Maridaleformation (Co-Typolokalität). *Eclogae Geologicae Helvetiae*, **66**(2), 389–418.

BARTENSTEIN, H. & BOLLI, H. M. 1986. The Foraminifera in the Lower Cretaceous of Trinidad, W. I. Part 5: Maridale Formation, upper part; *Hedbergella rohri* zone. *Eclogae Geologicae Helvetiae*, **79**(3), 945–999.

BARTENSTEIN, H. & BRAND, E. 1951. Mikropaläontologie Untersuchungen zur Stratigraphie des nordwestdeutschen Valendis. *Abhandlungen der Senckenbergischen Naturforschenden Gesellschaft*, **485**, 239–336.

BARTENSTEIN, H. & KOVATCHEVA, T. 1982. A comparison of Aptian Foraminifera in Bulgaria and North West Germany. *Eclogae Geologicae Helvetiae*, **75**(3), 621–667.

BELLIER, J. P., MOULLADE, M. & HUBER, B. 2000. Mid-Cretaceous planktonic foraminifers from Blake Nose: revised biostratigraphic framework. *In*: NORRIS, R. D., KROON, D., KLAUS, A. *et al.* (eds) *Proceedings of the Ocean Drilling Program, Scientific Results*, **171B**. Available from world wide web: http// www-odp.tamu.edu/publications/171B_SR/ chap_03.htm

BRALOWER, T. J., ARTHUR, M. A., LECKIE, R. M., SLITER, W. V., ALLARD, D. J. & SCHLANGER, S. O. 1994. Timing and palaeoceanography of oceanic dysoxia/anoxia in the Late Barremian to Early Aptian (Early Cretaceous). *Palaios*, **9**, 335–369.

BRALOWER, T. J., SLITER, W. V., ARTHUR, M. A., LECKIE, R. M., ALLARD, D. J. & SCHLANGER, S. O. 1993. Dysoxic/anoxic episodes in the Aptian–Albian (Early Cretaceous). *In*: PRINGLE, M. S., SAGER, W. W., SLITER, W. V. & STEIN, S. (eds). *The Mesozoic Pacific: Geology, Tectonics and Volcanism*. Geophysical Monograph, American Geophysical Union, **77**, 5–37.

BRASS, G. W., SOUTHAM, J. R. & PETERSON, W. H. 1982. Warm saline bottom-water in the ancient ocean. *Nature*, **296**: 620–623.

COCCIONI, R., GALEOTTI, S. & GRAVILI, M. 1995. Latest Albian–earliest Turonian deep-water agglutinated foraminifera in the Bottacione section (Gubbio, Italy)—biostratigraphic and palaeoecologic implications. *Revista Española de Paleontologia* (no. homenaje al Dr Guillermo Colom), 135–152.

CRITTENDEN, S. 1983. *Osangularia schloenbachi* (Reuss, 1863): an index foraminiferid species from the Middle Albian to the Late Aptian of the southern North Sea. *Neues Jahrbuch für Geologie und Paläontologie, Abhandlungen*, **167**, 40–64.

EICHER, D. L. & DINER, S. R. 1991. Environmental factors controlling Cretaceous limestone–marlstone rhythms. *In*: EINSELE, G., RICKEN, W. & SEILACHER, A. (eds) *Cycles and Events in Stratigraphiy*. Springer, Berlin, 79-93.

EICHER, D. L. & WORSTELL, P. 1970. Cenomanian and Turonian Foraminifera from the Great Plains, United States. *Micropalaeontology*, **16**, 269–324.

ERBACHER, J., HEMLEBEN, C., HUBER, B. & MARKEY, M. 1999. Correlating environmental changes during early Albian Oceanic Anoxic Event 1B using benthic foraminiferal palaeoecology. *Marine Micropalaeontology*, **38**, 7–28.

FRAKES, L. A., FRANCIS, J. E. & SYKTUS, J. I. 1992. *Climate Models of the Phanerozoic*. Cambridge University Press, Cambridge.

GEROCH, S. 1966. Lower Cretaceous small foraminifera of the Silesian series, Polish Carpathians. *Rocznik Polskiego Towarzystwa Geologicznego*, **36**(4), 414–480.

GEROCH, S. & NOWAK, W. 1984. Proposal of zonation for the late Tithonian-late Eocene, based upon arenaceous foraminifera from the outer Carpathians, Poland. *In*: OERTLI, H. J. (ed.) *Benthos '83, 2nd International Symposium on Benthic Foraminifera, Pau, April 1983*, Elf Aquitaine, Esso REP and Total CFP, Pau and Bordeaux, 225–239.

GUERIN, D. 1981. *Utilisation des foraminifères planctiques et benthiques dans l'étude des paléo-environnements océaniques au Crétacé moyen: application au materiel des forages DSDP de l'Atlantique Nord et Sud. Comparaison avec la Téthys*. Thèse Doctorat 3ème Cycle, Université de Nice.

HAIG, D. W. 1980. Early Cretaceous textulariine foraminiferids from Queensland. *Palaeontographica*, **A170**, 87–138.

HAIG, D. W. 1982. Early Cretaceous Milioline and Rotaliine benthic foraminiferids from Queensland. *Palaeontographica*, **A177**, 1–88.

HAIG, D. W. & LYNCH, D. A. 1993. A late early Albian marine transgressive pulse over north-eastern Australia, precursor to epeiric basin anoxia: foraminiferal evidence. *Marine Micropalaeontology*, **22**, 311–362.

HAY, W. W. 1995. Paleoceanography of marine organic carbon-rich sediments. *In*: HUC, A. Y. & SCHNEIDERMANN, N. (eds) *Paleogeography, Paleoclimate and Source Rocks*. AAPG Studies in Geology, **40**, 21–59.

HAYEK, L. C. & BUZAS, M. A. 1997. *Surveying Natural Populations*. Columbia University Press, New York.

HOLBOURN, A. E. L. & KAMINSKI, M. A. 1997. *Lower Cretaceous Deep-water Benthic Foraminifera of the Indian Ocean*. Grzybowski Foundation Special Publications. **4**.

HOLBOURN, A. E. L., KUHNT, W. 1998. Turonian–Santonian benthic foraminiferal assemblages from Site 959D (Côte d'Ivoire–Ghana Transform Margin, Equatorial Atlantic): indication of a Late Cretaceous oxygen minimum zone. *In*: *Proceedings of the Ocean Drilling Program, Scientific Results*, **159**. Ocean Drilling Program, College Station, TX, 375–387.

HOLBOURN, A. E. L., KUHNT, W., ERBACHER, J. 2001. Benthic foraminifers from lower Albian black shales (Site 1049, ODP Leg 171): evidence for a non 'uniformitarian' record. *Journal of Foraminiferal Research*, in press.

KAIHO, K. & HASEWAGA, T. 1994. End-Cenomanian benthic foraminiferal extinctions and oceanic dysoxic events in the northwestern Pacific Ocean. *Palaeogeography, Palaeoclimatology, Palaeoecology*, **111**, 29–43.

KAMINSKI, M. A., NEAGU, T. & PLATON, E. 1995. A revision of the Lower Cretaceous foraminiferal genus *Falsogaudryinella* from northwest Europe and Romania, and its relationship to *Uvigerinammina*. *In*: KAMINSKI, M. A. *et al.* (eds) *Proceedings of the Fourth International Workshop on Agglutinated Foraminifera.* Grzybowski Foundation Special Publications, **3**, 145–157.

KHAN, M. H. 1950. On some new Foraminifera from the Lower Gault of Southern England. *Journal of the Royal Microscopical Society, Series 3*, **70**, 268–279.

KING, C., BAILEY, H. W., BURTON, C. A. & KING, A. D. 1989. Cretaceous of the North Sea. *In*: JENKINS, D. G. & MURRAY, J. W. (eds) *Stratigraphical Atlas of Fossil Foraminifera*, 2nd edn. Ellis Horwood, London, 372–417.

KOUTSOUKOS, E. A. M., LEARY, P. M. & HART, M. B. 1990. Latest Cenomanian–earliest Turonian low oxygen tolerant benthonic foraminifera: a case study from the Sergipe Basin (NE Brazil) and the western Anglo-Paris Basin (southern England). *Palaeogeography, Palaeoclimatology, Palaeoecology*, **77**, 145–177.

KUHNT, W. & MOULLADE, M. 1991. Quantitative analysis of Upper Cretaceous abyssal agglutinated foraminiferal distribution in the North Atlantic—palaeoceanographic implications. *Revue de Micropaléontologie*, **34**, 313–349.

KUHNT, W. & WIEDMANN, J. 1995. Cenomanian–Turonian source rocks: palaeobiogeographic and palaeoenvironmental aspects. *In*: HUC, A. Y. & SCHNEIDERMANN, N. (eds) *Paleogeography, Paleoclimate and Source Rocks*. AAPG Studies in Geology, **40**, 213–232.

KUHNT, W., HESS, S. & JIAN, Z. 1999. Quantitative composition of benthic foraminiferal assemblages as a proxy indicator for organic carbon flux rates in the South China Sea. *Marine Geology*, **156**, 123–157.

LOEBLICH, A. J. & TAPPAN, H. 1946. New Washita Foraminifera, *Journal of Paleontology*, **20**, 238–258.

MAGNIEZ-JANNIN, F. 1975. *Les Foraminifères de l'Albien de l'Aube: Paléontologie, Stratigraphie, Écologie*. Cahiers de Paléontologie, Centre National de la Recherche Scientifique.

MALAPRIS, M. 1965. Les Gavelinellidae et formes affinées du gisement Albien de Courcelles (Aube). *Revue de Micropaléontologie*, **8**, 131–150.

MEYN, H. & VESPERMANN, J. 1994. Taxonomische Revision von Foraminiferen der Unterkreide SE-Niedersachsens nach ROEMER (1839, 1841, 1842), KOCH (1851) und REUSS (1863). *Senckenbergiana lethaea*, **74**(1–2), 49–272.

MOULLADE, M. 1966. Étude stratigraphique et micropaléontologique du Crétacé Inférieur de la 'Fosse Vovontienne', *Documents des Laboratoires de Géologie de la Faculté des Sciences de Lyon*, **15**(1–2), 1–369.

NEAGU, T. 1975. Monographie de la Faune des Foraminifères éocrétacés du couloir de Dîmbovicioara, de Codlea et des Monts Persani (Couches de Carhaga). *Memorii Institutul de Geologie si Geofizica, Bucaresti*, **25**, 1–141.

NORRIS, R. D., KROON, D., KLAUS, A. *et al.* (eds) 1998. *Proceedings of the Ocean Drilling Program, Initial Reports*, **171B**, Ocean Drilling Program, College Station, TX.

PERYT, D. & LAMOLDA, M. 1996. Benthonic foraminiferal mass extinction and survival assemblages from the Cenomanian–Turonian Boundary Event in the Menoyo section, northern Spain. *In*: HART, M. B. (ed.) *Biotic Recovery from Mass Extinction Events*. Geological Society, London, Special Publications, **102**, 245–258.

PROKOPH, A., SZAREK, R., KLOSOWSKA, B. & KUHNT, W. 1999. Late Albian benthic foraminiferal biofacies and palaeogeography of Northeast Germany. *Neues Jahrbuch für Geologie und Paläontologie, Abhandlungen*, **212**, 289–334.

RIEGRAF, W. & LUTERBACHER, H. 1989. Benthonische Forminiferen aus der Unterkreide des 'Deep Sea Drilling Project' (Leg 1–79). *Geologische Rundschau*, **78**(3), 1063–1120.

SAINT-MARC, P. & BERGGREN, W. A. 1988. A quantitative analysis of Palaeocene benthic foraminiferal asemblages in Central Tunisia. *Journal of Foraminiferal Research*, **18**(2), 97–113.

SIGAL, J. 1952. Aperçu stratigraphique sur la micropaléontologie du Crétacé. *Alger 19ème Congrès International Géologique Monographies Régionales, Serie 1*, **26**, 1–47.

TAPPAN, H. 1940. Foraminifera from the Grayson Formation of northern Texas. *Journal of Paleontology*, **17**, 93–126.

TEN DAM, A. 1947. Espèces nouvelles ou peu connues de l'Albien des Pays-Bas. *Geologie en Mijnbouw*, **8**, 25–29.

TEN DAM, A. 1950. Les Foraminifères de l'Albien des Pays-Bas. *Mémoires de la Société Géologique de France, Nouvelle Série*, **63**, 1–67.

TRONCHETTI, G. 1981. *Les Foraminifères Crétacés de Provence (Aptien–Santonien). Thèse Doctorat d'État, Université de Provence.* Travaux du Laboratoire de Géologie, Historique et de Paléontologie, **12**, Marseille.

TRONCHETTI, G. & GROSHENNY, D. 1991. Les assemblages de foraminifères benthiques au passage Cénomanien–Turonien à Vergons, SE France. *Geobios*, **24**, 13–31.

WEIDICH, K. F. 1990. Die kalkalpine Unterkreide und ihre Foraminiferenfauna. *Zitteliana*, **17**, Abhandlungen der bayerischen Staatssammmlung für Paläontologie und historische Geologie, 1–312.

WEINELT, M., KUHNT, W., SARNTHEIN, M. *et al.* 2000. Paleoceanographic proxies in the Northern North Atlantic. *In*: SCHÄFER, P., SCHLÜTER, M., SCHRÖDER-RITZRAU, W. & THIEDE, J. (eds). *The Northern North Atlantic: a Changing Environment*. Springer, Berlin, 319–352.

WOLLENBURG, J. & KUHNT, W. 2000. The response of benthic foraminifers to carbon flux and primary production in the Arctic Ocean. *Marine Micropalaeontology*, 189–231.

Biostratigraphic subdivision and correlation of upper Maastrichtian sediments from the Atlantic Coastal Plain and Blake Nose, western Atlantic

JEAN M. SELF-TRAIL

Department of Geosciences, University of Nebraska, Lincoln, NE 68508-0340, USA (e-mail: jstrail@usgs.gov)

Abstract: Detailed biostratigraphic analyses of nine cores from the Atlantic Coastal Plain and two cores from the Blake Nose, western Atlantic Ocean, provide the basis for subdivision and correlation of upper Maastrichtian sediments along a shallow- to deep-water transect. The calcareous nannofossil record from these sites shows distinct differences between the middle to outer neritic Coastal Plain sediments and the lower to upper bathyal Blake Nose sediments. *Micula murus*, a reliable marker species for low- to mid-latitude sites, is shown herein to respond to differing palaeoenvironmental conditions of nearshore v. open-ocean sites. Its usefulness as a biostratigraphic marker for neritic sediments is called into question. The last appearance datum of *Ceratolithoides kamptneri* is documented as a reliable biozone marker for latest Maastrichtian time (within CC26b) in this region. The evolutionary radiation and resulting biostratigraphic utility of species of *Ceratolithoides*, *Lithraphidites* and *Micula* is discussed in detail, and their first and last occurrences are tied to magnetostratigraphic chrons where possible. *Ceratolithoides amplector*, *Ceratolithoides indiensis* and *Ceratolithoides pricei* are shown to be useful, biostratigraphically, in sediments deposited under bathyal conditions. Several species of *Lithraphidites* (*Lithraphidites*? *charactozorro*, *Lithraphidites kennethii* and *Lithraphidites grossopectinatus*) can be used to further subdivide upper Maastrichtian sediments at both neritic and bathyal localities. The first and last occurrence of *Micula praemurus* in Zones CC25a and CC26a, respectively, are shown to be useful biostratigraphic datum points.

Calcareous nannofossil data from Ocean Drilling Program (ODP) Leg 171B, Blake Nose, western Atlantic Ocean, and from the Atlantic Coastal Plain of South Carolina, USA form the basis for an improved biostratigraphy of the upper Maastrichtian succession at mid- to low-latitude sites. The two ODP sites examined for calcareous nannofossil content (Holes 1052E and 1050C; Fig. 1) are representative of middle bathyal and lower bathyal depths, respectively. Coreholes drilled on the South Carolina Coastal Plain by the US Geological Survey and the South Carolina Department of Natural Resources (C-15, Cannon Park, Clubhouse Crossroads, St George, Santee Coastal Reserve, Myrtle Beach, Brittons Neck and Lake City; Fig. 1) are typically representative of middle to outer neritic water depths.

Routine examination of calcareous nannofossil slides from the study area revealed a noticeable difference in the nannofossil assemblage composition between the shallower- and deeper-water sites. Several key taxa used to biostratigraphically subdivide the Upper Cretaceous succession of low latitudes (e.g. *Micula murus*, *Ceratolithoides* spp.) were found to consistently demonstrate different biostratigraphic ranges in open-ocean v. shelf settings. Although diachroneity of a few key Late Cretaceous nannofossil marker species across latitude has been documented previously (Worsley & Martini 1970; Pospichal & Wise 1990), this study is the first to document probable diachroneity of Upper Cretaceous calcareous nannofossil species across a shallow- to deep-water transect.

Recent drilling by the US Geological Survey has provided a significant increase in the number of coreholes drilled through Coastal Plain sediments. Compilation of these data and comparison with nearby bathyal sequences generates a more detailed biostratigraphy for the region, as well as illuminating some differences between nannofossil assemblages from open-ocean and nearshore settings (Fig. 2). The majority of global Cretaceous biozonation schemes have been based on deep-ocean material sampled

From: KROON, D., NORRIS, R. D. & KLAUS, A. (eds). 2001 *Western North Atlantic Palaeogene and Cretaceous Palaeoceanography*. Geological Society, London, Special Publications, **183**, 93–110.

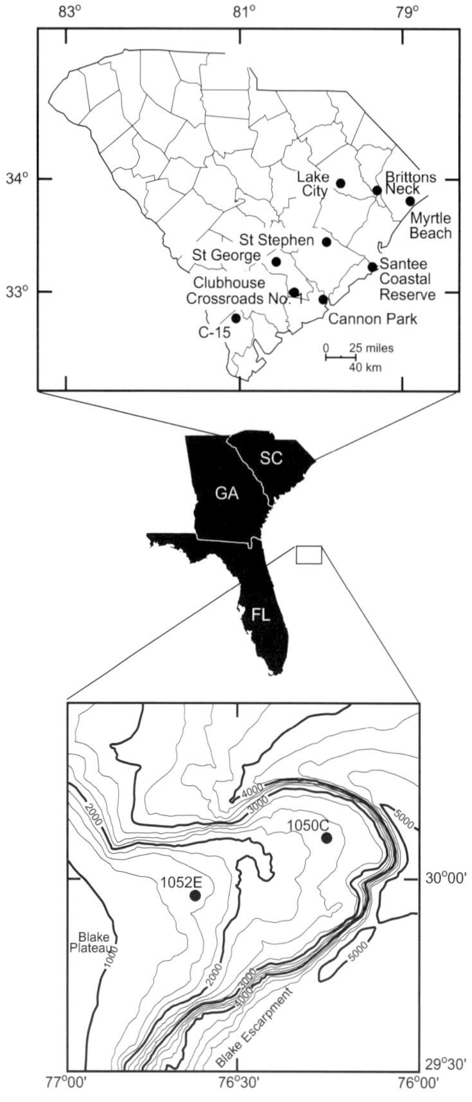

Fig. 1. Map showing the location of coreholes drilled by the US Geological Survey and the South Carolina Department of Natural Resources in the Atlantic Coastal Plain, and by the Ocean Drilling Program, Leg 171B, on the Blake Nose, western Atlantic Ocean. ●, corehole locations. Depth contours in metres.

from ODP or Deep Sea Drilling Program (DSDP) sites (Thierstein 1976; Roth 1978; Watkins *et al.* 1996; Burnett 1998). Only recently has a combination of onshore and ODP studies begun to clarify the role that provincialism plays in determining the nannofossil biostratigraphy of a region (e.g. Burnett 1998). Data from two distinct but adjacent regions were used in this study: the Blake Nose, western Atlantic Ocean, and the Atlantic Coastal Plain of South Carolina, USA. Data were compiled in occurrence charts and can be obtained from the Society Library of the British Library at Boston Spa, Wetherby, West Yorkshire LS23 7BQ, UK, as Supplementary Publication No. SUP-18154 (66 pages).

Blake Nose

Two ODP sites (Hole 1050C and Hole 1052E) that penetrated upper Maastrichtian sediments of the Blake Nose (Leg 171B) were examined for calcareous nannofossil content. Drilling at Hole 1050C penetrated 71 m of Maastrichtian sediments (Norris *et al.* 1998). At 1050C, the basal upper Maastrichtian sequence consists of nannofossil claystones and chalks that are rich in foraminifers and are moderately to heavily bioturbated. The basal section is overlain conformably by a nannofossil claystone that is lithologically similar but contains less clay than the basal interval. Slumping occurs sporadically at the base of the upper Maastrichtian units.

Drilling at Hole 1052E penetrated over 114 m of upper Maastrichtian sediments, which consist of clayey nannofossil chalks that become more enriched in clay upwards. Subtle metre-scale variations between lighter, carbonate-rich sediments and darker, carbonate-poor sediments are present throughout this section (Norris *et al.* 1998). Bioturbation is moderate throughout. A slumped interval at the base of the upper Maastrichtian units marks the disconformable boundary between the upper Maastrichtian and the lower Maastrichtian sequences.

Location information, palaeoenvironmental designation and the age of the sediments overlying and underlying the Maastrichtian section at Holes 1050C and 1052E are outlined in Table 1. The benthic foraminifera fauna from these two sites is indicative of deposition in the lower to upper bathyal region of the pelagic realm (Norris *et al.* 1998).

Atlantic Coastal Plain

Nine boreholes from the Atlantic Coastal Plain that penetrated Upper Cretaceous sediments have been examined for calcareous nannofossil content: C-15 (Gillisonville), DOR-37 (Clubhouse Crossroads), DOR-211 (St. George), CHN-800 (Cannon Park), CHN-803 (Santee Coastal Reserve), BRK-644 (St. Stephens), HOR-1165 (Myrtle Beach), MRN-78 (Brittons Neck) and FLO-274 (Lake City). Truncation of Maastrichtian sediments as a result of erosion is present in all of the updip cores. Therefore,

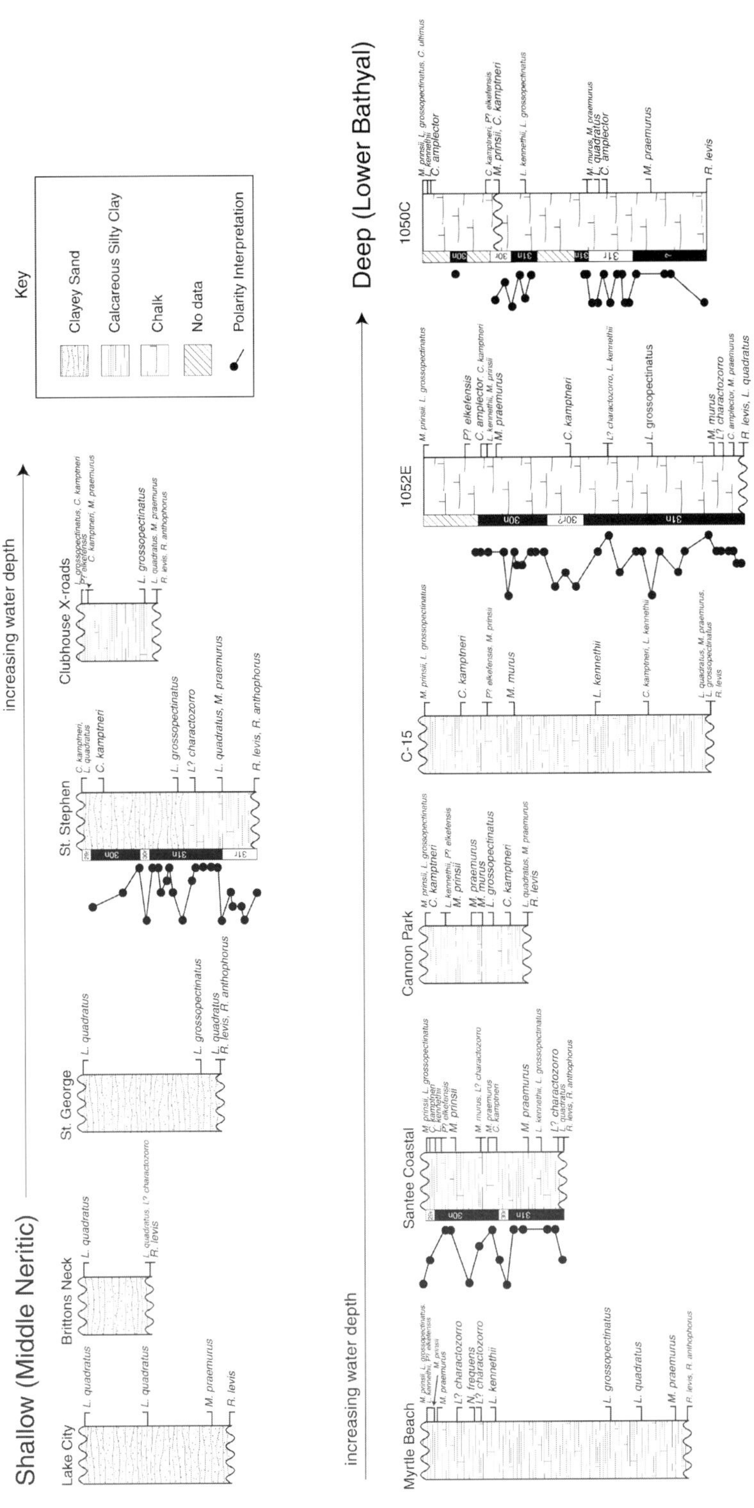

Fig. 2. Changes in lithology, correlation of calcareous nannofossil datum levels and magnetic polarity interpretations along a shallow- to deep-water transect. Magnetic polarity interpretations for Holes 1050C and 1052E from Ogg & Bardot (2000) and for the Santee Coastal Reserve core from Edwards *et al.* (1999). Magnetic polarity interpretation for the St. Stephen core is from Bardot (pers. comm.).

Table 1. *Location information, age of overlying and underlying sediments, and palaeoenvironmental designation for coreholes used in this study*

Corehole	Latitude and longitude	Palaeoenvironment	Lower Maastrichtian sediments present	Overlying sediments	Underlying sediments
1050C	30°05.9953′N, 76°14.0997′W	Lower bathyal	Yes	L. Palaeocene	L. Maastrichtian
1052E	29°57.0794′N, 76°37.6094′W	Middle bathyal	Yes	L. Palaeocene	L. Maastrichtian
C-15	32°37′04″N, 80°59′45″W	Middle–outer neritic	No	L. Palaeocene	U. Campanian
Clubhouse Crossroads	32°53′25″N, 80°21′41″W	Middle–outer neritic	No	L. Palaeocene	U. Campanian
St. George	33°09′25″N, 80°31′18″W	Middle–outer neritic	No	L. Palaeocene	U. Campanian
Cannon Park	32°46′55″N, 79°56′41″W	Middle–outer neritic	No	L. Palaeocene	U. Campanian
Santee Coastal Reserve	33°09′21″N, 79°21′50″W	Middle–outer neritic	No	L. Palaeocene	U. Campanian
St. Stephen	33°24′15″N, 79°56′04″W	Middle–outer neritic	No	L. Palaeocene	U. Campanian
Brittons Neck	33°51′43″N, 79°19′50″W	Middle–outer neritic	No	Pleistocene	U. Campanian
Myrtle Beach	33°43′44″N, 78°54′14″W	Middle–outer neritic	No	Pleistocene	U. Campanian
Lake City	33°51′20″N, 78°46′02″W	Middle–outer neritic	No	Pleistocene	U. Campanian

downdip cores typically have nearly complete upper Maastrichtian sequences (Zones CC25b to CC26b) whereas updip cores typically contain only Zones CC25a and/or CC25b (Table 1).

Sediments drilled in the South Carolina Coastal Plain can be broken down into two broad regions: updip and downdip. Updip sediments have had the uppermost Maastrichtian sediments stripped away by erosion. The lithology of the updip sequence typically consists of fine-grained, bioturbated, clayey sands interbedded with massive, slightly sandy, carbonate-rich clays. Small amounts of glauconite are typically present in both the sands and clays. Calcareous microfossils are scattered throughout. The upper Maastrichtian sequence in the updip section varies in thickness from 18 m at Brittons Neck to 43 m at the Lake City corehole. Sediments become more fine grained downdip, consisting predominantly of dark grey, calcareous silty clay with fine laminations and interspersed semi-lithified horizons. Thickness varies from 37 m at St. George to 19 m at Clubhouse Crossroads. Detailed calcareous nannofossil biostratigraphy was provided by Hattner & Wise (1980) for the Clubhouse Crossroads core, but was re-examined herein. Self-Trail & Gohn (1996) and Self-Trail & Bybell (1997) provided the calcareous nannofossil biostratigraphy for the St. George and C-15 cores, respectively.

Downdip boreholes located along the coast of South Carolina include Myrtle Beach, Santee Coastal Reserve, Cannon Park and C-15. Detailed calcareous nannofossil biostratigraphy for the Cannon Park core was provided by Bybell *et al.* (1998). Self-Trail & Bybell (1997) and Edwards *et al.* (1999) provided the calcareous nannofossil biostratigraphy for the C-15 and Santee Coastal Reserve cores, respectively. The upper Maastrichtian sediments from these localities consist predominantly of homogeneous to massively bioturbated, finely micaceous, calcareous silty clay of the Peedee Formation. Cemented sandstones and nodules up to a metre in thickness occur throughout the section. This unit is bounded disconformably at its base by a phosphatic lag deposit and disconformably at its top by the overlying Danian Rhems Formation. Thickness of the Peedee Formation in the downdip is variable, ranging from 70 m of upper Maastrichtian sediments at C-15 to no greater than 33 m at Santee Coastal Reserve. This is due, in part, to the Cape Fear arch, which is affecting regional sediment distribution in eastern South Carolina. Late Cenozoic uplift and erosion across the arch removed the Tertiary sediments, which were subsequently covered by a Pleistocene cap of fluvial sand. Upper Cretaceous sediments crop out along the flank of the arch (Owens & Gohn 1985; Soller & Mills 1991).

Although bioturbation is a factor in all cores examined for this study, comparison of the ranges of key calcareous nannofossil taxa with geophysical logs suggests that reworking of key species into overlying zones is rare except where a major disconformity is present (Self-Trail & Gohn 1996; Self-Trail & Bybell 1997; Self-Trail unpubl. data). Samples examined from the basal Peedee Formation often contain reworked specimens of *Reinhardtites anthophorus* and *Reinhardtites levis*, and occasional reworked specimens of *Quadrum trifidum*. However, detailed examination of the background assemblage, coupled with the rare abundance and sporadic occurrence of these specimens, is usually enough to determine the reworked nature of these species (Supplemental Publication No. SUP-18154).

Biostratigraphic correlation

Maastrichtian calcareous nannofloras of the western Atlantic Ocean were characterized by a general period of decline in speciation and diversity following the Campanian acme. Increased rates of speciation and diversity occurred after a lull (Zone CC25a/UC19) in early late Maastrichtian time, particularly in the *Ceratolithoides*, Polycyclolithaceae and Microrhabdulaceae (see discussion below). Although it has been documented that late early to early late Maastrichtian time was characterized by low-diversity assemblages (Burnett 1998), the overall trend throughout Maastrichtian time was one of increasing diversity, radiation and extinction. The three taxonomic groups listed above give rise to at least 16 Maastrichtian taxa that are potentially useful as biostratigraphic markers in the northwest Atlantic and possibly globally. Correlation of these events with the biozones of Sissingh (1977), Perch-Nielsen (1985) and Burnett (1998) and the magnetochrons of Gradstein *et al.* (1995), are illustrated in Fig. 3.

Although the stratigraphic ranges of several of the species in question are still uncertain, at least three are sufficiently well documented to be used to correlate accurately from shallow-water to deep-water environments at intermediate latitudes. Several species of *Ceratolithoides* are potential biohorizon markers in sediments deposited in uppermost to upper lower bathyal conditions. These include the first common occurrence (FCO) of *Ceratolithoides amplector*, the last appearance datum (LAD) of *Ceratolithoides pricei*, the first appearance datum (FAD) and LAD of *Ceratolithoides kamptneri*, and the FAD of *Ceratolithoides ultimus*.

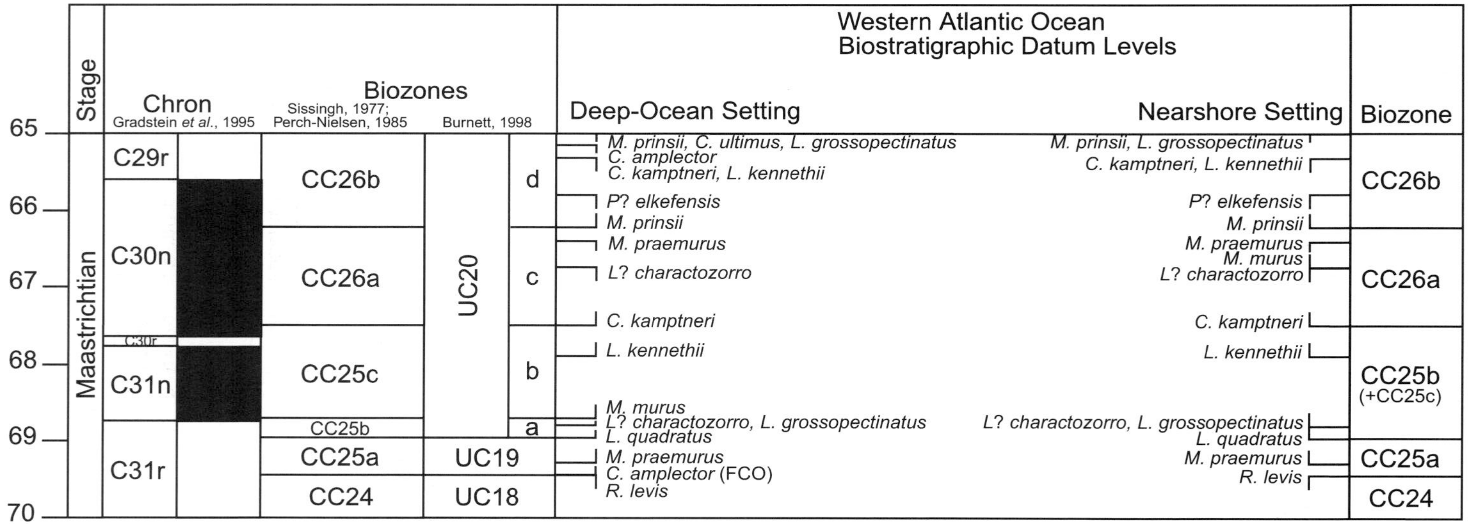

Fig. 3. Potentially useful intermediate-latitude Maastrichtian nannofossil events identified in this study, correlated with the calcareous nannofossil zonations of Sissingh (1977), Perch-Nielsen (1985) and Burnett (1998). Correlation of established nannofossil events with the chronostratigraphic scale of Gradstein *et al.* (1995) was based on Erba *et al.* (1995).

It is important to note that the distribution of biozone markers used in this study was, in some cases, strongly influenced by palaeoenvironment. Thus, although the late Maastrichtian biozonation often relies heavily on cosmopolitan taxa, there are several instances where the biostratigraphic zonation of sediments deposited in a bathyal environment will differ greatly from those deposited in a neritic environment.

Magnetostratigraphic events

Correlation of calcareous nannofossil events with magnetostratigraphic chrons is based primarily on fig. 5 of Norris *et al.* (1998) and secondarily on the zonal scheme of Roth (1978). Roth (1978) placed the FAD of *Micula murus* in C31n and the FAD of *Lithraphidites quadratus* in C31r for sediments examined from DSDP Site 390 on the Blake Plateau, and the current study is in concurrence with his findings. However, it is important to note that placement of several key datum points with regard to the palaeomagnetic chrons differs between this study and Bralower & Siesser (1992) and Bralower *et al.* (1995). Bralower and coworkers would place the FAD of *Ceratolithoides kamptneri* well into C30n, based on its first occurrence above the first occurrence of *M. murus*, which they also place in C30n. However, as illustrated by Fig. 3 and discussed below, the FAD of *M. murus* was controlled by palaeodepth and therefore its correlation with the polarity time scale is dependent on environmental factors at the time of deposition. Higher placement of the FAD of *L. quadratus* by Bralower & Siesser (1992) and Bralower *et al.* (1995) at the base of C30n may, in part, be due to differences in species concepts. Differentiation between *L. quadratus* and *Lithraphidites praequadratus* by the author was based on the criteria set forth by Roth (1978), and may have resulted in an apparent lower first appearance for *L. quadratus* than recorded by Bralower *et al.* (1995).

Ceratolithoides

Burnett (1997) recently documented and described a diverse assemblage of *Ceratolithoides* taxa from a series of ODP and DSDP sites in the Indian Ocean. These sites occupied open-ocean, low to intermediate palaeolatitudes during Campanian and Maastrichtian time. The biostratigraphic utility of several of Burnett's new species is discussed herein. The viability of using both the FAD and LAD of *Ceratolithoides kamptneri* as biostratigraphic events for latest Maastrichtian time is also discussed.

Ceratolithoides kamptneri is a small- to medium-sized, horseshoe-shaped ceratolith that was present in frequent to common abundances in both nearshore and deep-water environments during latest Maastrichtian time (Fig. 4). It is known from the Indian and Atlantic Oceans and from mid- to low-latitude nearshore sediments. Unlike other *Ceratolithoides*, which have a well-defined cone and base, *C. kamptneri* lacks a cone and consists entirely of two long, slender horns. Perch-Nielsen (1985) used the first appearance of this species as a proxy zonal marker for the base of Zone CC26, in lieu of the first appearance of *Nephrolithus frequens*, a species whose FAD is known to be diachronous (Worsley & Martini 1970; Pospichal & Wise 1990).

The first occurrence of *C. kamptneri* appears to be a recognizable event in low- to mid-latitude sites in the Atlantic and Indian Ocean realms (Self-Trail & Bybell 1997; Burnett 1998) and is therefore a useful biostratigraphic marker. Unlike with *N. frequens*, the FAD of this species appears to be synchronous within its known latitudinal and biostratigraphic range. *Ceratolithoides kamptneri* is present in six downdip coreholes from the South Carolina Coastal Plain. In two of these cores, Santee Coastal Reserve and St. Stephens, its first appearance can be tied to palaeomagnetic chrons. In the Santee Coastal Reserve core, the FAD of *C. kamptneri* is coincident with the C30r–C30n boundary (Edwards *et al.* 1999), and in the Saint Stephens core it first occurs just above the C30r–C30n boundary (Bardot, pers. comm.). This species first appears before *N. frequens*, which occurs only sporadically in sediments from this palaeolatitude, including the South Carolina sections. Although *C. kamptneri* is not often common, this species is consistently present throughout its range in Coastal Plain sediments. It seems to have thrived in the shallow, continental shelf waters in the area that is now the South Carolina Coastal Plain, in palaeodepths of 200 m or less.

Ceratolithoides kamptneri is also present at Holes 1050C and 1052E on the Blake Nose. The sediment from these boreholes was deposited in upper and lower bathyal conditions, respectively, and represents water depths far greater than any recorded from Maastrichtian sections in South Carolina. Assemblages recorded from these sites are indicative of a latest Maastrichtian age, and include *Lithraphidites kennethii*, *Micula prinsii*, *Micula murus* and *Pseudomicula quadrata*. *Nephrolithus frequens* is not present in either of these cores. Correlation with palaeomagnetic data from ODP Leg 171B shows that the first appearance of *C. kamptneri* is at the C30r?–C31n boundary in 1052E. This is lower than recorded

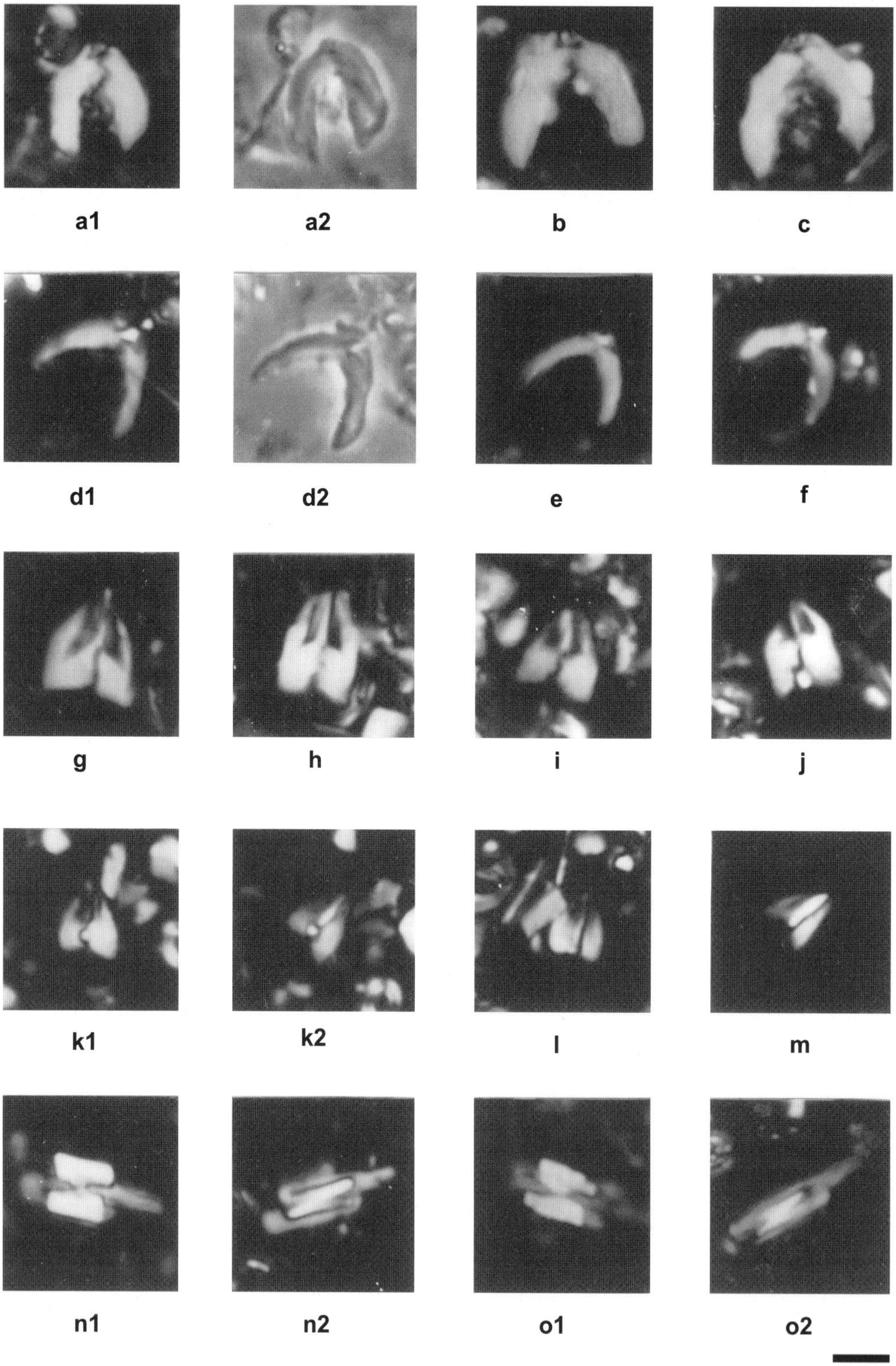
a1
a2
b
c
d1
d2
e
f
g
h
i
j
k1
k2
l
m
n1
n2
o1
o2

from the Santee Coastal Reserve and St Stephens cores, and could be due to either bioturbation or an unreliable palaeomagnetic signal. The first appearance of *C. kamptneri* at Hole 1050C is correlative with a disconformity and is therefore an unreliable event datum.

Comparison of biostratigraphic data from South Carolina and the Blake Nose has shown that the LAD of *Ceratolithoides kamptneri* also is useful as a biostratigraphic datum. In complete upper Maastrichtian stratigraphic sections (i.e. 1052E and 1050C) or nearly complete sections of that age (i.e. Santee Coastal Reserve), *C. kamptneri* disappears well below the Cretaceous–Tertiary (K–T) boundary. At Holes 1050C and 1052E, *C. kamptneri* last occurs *c.* 15 m below the K–T boundary. Comparison of the biostratigraphic data with the palaeomagnetic data at Hole 1050C places the LAD of *C. kamptneri* just above the C29r?–C30n? boundary, whereas at Hole 1052E, it last occurs in C30n. In both cores, the abundance of this species wanes near the top of its range. Preservation remains the same in both cores throughout the range of *C. kamptneri*, and is therefore not a contributing factor to the disappearance of this species. Six cores from the Atlantic Coastal Plain have sediment containing *C. kamptneri*. Of those six, four record the LAD of *C. kamptneri* below, and not equivalent to, the K–T boundary. Unlike in the Blake Nose cores, the K–T boundary event in South Carolina Coastal Plain sediments is always missing, and upper Maastrichtian sediments are truncated disconformably either by the lower Danian Rhems Formation or by Pleistocene sands (Table 1). *Ceratolithoides kamptneri* disappears 0.5 m below the K–T boundary in the Santee Coastal Reserve core, 10 m below in the C-15 core, and 1.5 m below the base of the Pleistocene section in Myrtle Beach. However, its location in the C-15 core may be an artefact of the sampling interval: sidewall cores at this locality were taken approximately every 4 m. Its LAD is recorded at 112.8 m depth in the Santee Coastal Reserve core, just above the C29r–C30n boundary, which occurs at 118.3 m. Again, the abundance of this species decreases near the top of its range. The range of *C. kamptneri* extends to the disconformable top of the Maastrichtian sequence at St Stephens, Clubhouse Crossroads and Cannon Park, indicating that the uppermost Maastrichtian sequence is missing at these three localities.

Because of the paucity of complete upper Maastrichtian sections in other regions, it has been difficult to determine whether the LAD of *C. kamptneri* is indeed below the K–T boundary. However, several previous reports confirm the findings of this study: El Mehaghag (1996) documented the LAD of *C. kamptneri* in Libya at *c.* 3 m below the top of the Maastrichtian sequence; Moshkovitz & Habib (1993) recorded the LAD at 4.9 m below the K–T boundary in the deeper-water marls and carbonates of the Braggs core, Gulf Coastal Plain; Eshet *et al.* (1992) documented the LAD at 2.5 m below the K–T boundary and above the FAD of *Micula prinsii* in the Hor Hahar, Israel section; Rio *et al.* (1983) record the presence of *C. kamptneri* in their Monte Rotondo section, along with *N. frequens* and *M. prinsii*, with *C. kamptneri* disappearing below the top of the section. Correlation of the LAD of *C. kamptneri* with the geochronological time-scale of Gradstein *et al.* (1995) and the palaeomagnetic data of Bardot (pers. comm.) and Edwards *et al.* (1999) indicates that the LAD of *C. kamptneri* occurs near the base of C29r.

Several other species of *Ceratolithoides* were found to be biostratigraphically useful in this study; however, their utility as world-wide markers still needs to be ascertained. *Ceratolithoides indiensis*, *C. amplector*, *C. pricei* and *C. ultimus* were first described by Burnett (1997)

Fig. 4. (**a1**) *Ceratolithoides amplector*, crossed polars (XP), Hole 1052E, 22R-4, 47–50 cm, Subzone CC26a, Blake Nose. (**a2**) Same specimen, phase contrast (PC). (**b**) *Ceratolithoides amplector*, XP, Hole 1052E, 20R-4, 77–80 cm, Subzone CC26b, Blake Nose. (**c**) *Ceratolithoides amplector*, XP, Hole 1052E, 21R-5, 13–14 cm, Subzone CC26a, Blake Nose. (**d1**) *Ceratolithoides kamptneri*, XP, Hole 1052E, 22R-4, 47–50 cm, Subzone CC26a, Blake Nose. (**d2**) Same specimen, PC. (**e**) *Ceratolithoides kamptneri*, XP, Santee Coastal reserve (SCR) core, 119.0 m, Subzone CC25b, South Carolina. (**f**) *Ceratolithoides kamptneri*, XP, Hole 1052E, 22R-4, 47–50 cm, Subzone CC26a, Blake Nose. (**g**) *Ceratolithoides indiensis*, XP, Hole 1050C, 18R-4, 67–68 cm, Zone CC24, Blake Nose. (**h**) *Ceratolithoides indiensis*, XP, Hole 1050C, 15R-6, 34–35 cm, Subzone CC25c, Blake Nose. (**i**) *Ceratolithoides indiensis*, XP, Hole 1052E, 24R-4, 13–14 cm, Subzone CC25c, Blake Nose. (**j**) *Ceratolithoides indiensis*, XP, Hole 1052E, 22R-4, 47–50 cm, Subzone CC26a, Blake Nose. (**k1**) *Ceratolithoides ultimus*, XP, Hole 1050C, 11R-CC, Subzone CC26b, Blake Nose. (**k2**) Same specimen, 45°. (**l**) *Ceratolithoides ultimus*, XP, C-15 borehole, 341.1 m, Subzone CC26a, South Carolina. (**m**) *Ceratolithoides ultimus*, 45°, XP, Hole 1050C, 11R-CC, Subzone CC26b, Blake Nose. (**n1**) *Lithraphidites? charactozorro*, XP, focused down, SCR core, 138.0 m, Subzone CC25b, South Carolina. (**n2**) Same specimen, focused up. (**o1**) *Lithraphidites? charactozorro*, XP, focused down, SCR core, 139.0 m, Subzone CC25b, South Carolina. (**o2**) Same specimen, focused up. Scale bar represents 5 μm.

from the Indian Ocean. These four species were found in this study to occur with regularity in the open-ocean setting of Holes 1050C and 1052E, and sporadically or not at all in the shelf setting of the Atlantic Coastal Plain cores. This discrepancy may imply that these species are palaeoenvironmental indicators.

Of the four species mentioned above, only *Ceratolithoides amplector* occurs with any regularity in both nearshore and open-ocean settings (Fig. 4). Burnett (1997) described this species as a medium-sized ceratolith with a blocky horseshoe-shaped outline and horns forming an 90° angle. This species can be confused with overgrown forms of *C. kamptneri*. Burnett (1997) recorded the presence of *C. amplector* from Holes 217, 761B and 765C in the Indian Ocean, and indicated that this species ranges from the Zone CC24–25 boundary up into Zone CC26. This range is supported here by data from the South Carolina Coastal Plain and the Blake Nose. *Ceratolithoides amplector* occurs consistently throughout Subzone CC25a and up into CC26b at Holes 1050C and 1052E, where its LAD is slightly above the LAD of *C. kamptneri*. Although *C. amplector* occurs somewhat sporadically above its first appearance, near the CC24–CC25a boundary, this datum may prove to be biostratigraphically useful upon further assessment of more complete sections. In open-ocean settings, *C. amplector* usually occurs commonly or frequently. In the shelfal marine sediments of the St. George, Lake City, Cannon Park and C-15 cores, it is sporadic in Subzone CC25b and Zone CC26 and is not useful as a biostratigraphic marker.

Ceratolithoides indiensis (Fig. 4) was described by Burnett (1997) as a blocky, subconical form possessing a bright cone and an obtuse inter-horn angle, and thus far has been documented only from Site 217 in the Indian Ocean and Holes 1050C and 1052E in the western South Atlantic. Burnett (1997) listed the known range of this species as lower Campanian to upper Maastrichtian units (Zones CC20–CC25c). *Ceratolithoides indiensis* was found to occur consistently from the upper Campanian to upper Maastrichtian Zone CC25c, and sporadically from Subzones CC26a and CC26b in cores 1050C and 1052E. This species is present only in the deeper-water cores of the Blake Nose and was not documented from any Atlantic Coastal Plain sites. The FAD and LAD of this species does not appear to be useful biostratigraphically; however, the presence of this species may be indicative of open-ocean environments.

Ceratolithoides pricei, a large blocky form with a square outline and horns that form an extremely obtuse angle, was described from numerous Indian Ocean ODP and DSDP sites by Burnett (1997), who listed its known range as upper Maastrichtian Subzone CC25b to Zone CC26. It is herein documented in the western Atlantic from Holes 1050C and 1052E from Zones CC24 and CC25. The first occurrence of *C. pricei* does not appear to be useful biostratigraphically; however, it has its LAD at or near the top of Subzone CC25c in both cores. Further documentation of the range of this species may prove it to be a useful biostratigraphic marker in sediments deposited in deep-ocean settings. This species was not documented from coeval sediments in South Carolina.

Ceratolithoides ultimus (Fig. 4) is a very small, blocky ceratolith, having a parallel-sided cone that extends down the length of the nannolith and is highly birefringent. Burnett (1997) recorded its range as upper Maastrichtian Subzones CC25c–CC26, from Holes 217, 758A and 761B in the Indian Ocean (Fig. 3). *Ceratolithoides ultimus* was recorded in Zone CC26 from Holes 1050C and 1052E and from borehole C-15 of this study. The short range of this species makes it a useful biostratigraphic marker for further subdividing the latest Maastrichtian period. The FAD of *C. ultimus* lies above the first occurrence of *C. kamptneri* and below the LAD of *M. praemurus*, below the C30n–C29r boundary. It is unclear where the FAD of this species is relative to the last occurrence of *L*? *charactozorro*. Further documentation of the range of these two species is needed to resolve this question. This species ranges to the uppermost Maastrichtian units along with *Micula murus* and *Micula prinsii*.

Lithraphidites

One of the more biostratigraphically useful genera to undergo radiation during late Maastrichtian time was *Lithraphidites*. This genus first evolved in Berriasian time, represented by *Lithraphidites carniolensis*, a cosmopolitan species that became extinct at the Cretaceous–Tertiary boundary. A number of species from this genus have been used as biostratigraphic markers (e.g. *Lithraphidites quadratus*, *Lithraphidites acutus*); however, it is during late Maastrichtian time that this genus undergoes rapid evolution and radiation, producing several species useful for detailed biostratigraphy.

Lithraphidites? *charactozorro* (Fig. 4) is an unusual species first described by Self-Trail & Pospichal (in Self-Trail (1999)) from South Carolina and El Kef, Tunisia, sections. It has a relatively limited range. This species was found to occur in four cores from the Atlantic Coastal Plain (Santee Coastal Reserve, Myrtle Beach, Clubhouse Crossroads and Brittons Neck) and from Holes 1050C and 1052E. The precise range of *L*? *charactozorro* is currently under investigation. *Lithraphidites*? *charactozorro* has its first appearance in Subzone CC25b, above the FAD of *Lithraphidites quadratus* and just below the first occurrence of *Lithraphidites grossopectinatus*. Correlation of the FAD of *L*? *charactozorro* with the palaeomagnetic stratigraphy of the Santee Coastal Reserve and 1050E cores shows that it is near the base of C31n in both cores. The last occurrence datum of *L*? *charactozorro* is a useful biostratigraphic datum for upper Maastrichtian sediments of this region. This species last occurs in Subzone CC26a, above the first occurrence of *Ceratolithoides kamptneri* and below the first occurrence of *Micula prinsii*. In coastal plain sediments, it disappears at approximately the same time as the first occurrence of *Micula murus* (see below). Although this species is a useful biostratigraphic marker, it is often only rare to frequent in its occurrence and can be difficult to discern when broken. However, palaeoenvironmental conditions did not seem to affect the biostratigraphic range of this species, as it is present at both shallow-water and deep-water sites. An aberrant form of *L*? *charactozorro* occurs in Zone CC24 at Hole 1052E, 48 m below the FAD of *L*? *charactozorro sensu stricto*. The form differs from *L*? *charactozorro s.s.* in having shorter blades and a less well-defined set of bars, and an indistinct look when viewed with the light microscope; also specimens are typically smaller. This aberrant form seems to be restricted to Zone CC24.

Lithraphidites grossopectinatus (Fig. 5) has its first occurrence in Subzone CC25b, above the FAD of *Lithraphidites quadratus* and below the FAD of *Lithraphidites kennethii*. The range of this species is sporadic near its base, becoming more consistent above. *Lithraphidites grossopectinatus* becomes extinct at the K–T boundary. It is unclear where the first occurrence of *L. grossopectinatus* falls in relation to the FAD of *L*? *charactozorro*, but both occur near the base of CC25b. Material from sections of similar age, containing both taxa, will need to be investigated before an exact sequence of events can be determined. To date, very few researchers have recorded the existence of *L. grossopectinatus*. This is probably due in part to the small size of this species, coupled with its non-global distribution and its propensity to appear like a ragged *L. praequadratus* with the light microscope, which has led workers to believe the two species are synonymous. However, Risatti (1973) recorded the common presence of *L. grossopectinatus* from the *L. quadratus* Zone (CC25b) in the Prairie Bluff Chalk of Mississippi, corroborating a Zone CC25 first occurrence for this species.

Lithraphidites kennethii (Fig. 5) is documented from seven coreholes in the study area. This species is rare to frequent during its first occurrence in Subzones CC25b–CC25c, but commonly occurs in Subzones CC26a–CC26b. The exact placement of the first appearance of this species is problematic. *Lithraphidites kennethii* first appears in Subzone CC25c at Holes 1050C and 1052E, above the first occurrence of *Micula murus* and below the first occurrence of *C. kamptneri*. However, it does not appear until after the first occurrence of *C. kamptneri* in the shallow-water cores, with two notable exceptions: *Lithraphidites kennethii* first appears sporadically in Subzone CC25b in both the Santee Coastal Reserve and Myrtle Beach cores. Palaeomagnetic data from the Santee Coastal Reserve core suggest that although the basal Subzone CC25c marker species is not present (*Micula murus*), the sediments are of Subzone CC25c age (Edwards *et al.* 1999; see *Micula* discussion, below). It is unclear whether the LAD of *L. kennethii* pre-dates the K–T boundary; however, it seems likely that this species disappeared around the same time as *C. kamptneri*. *Lithraphidites kennethii* is recorded from the uppermost samples of only two cores in the study area (Myrtle Beach and Clubhouse Crossroads), both of which are known to have truncated upper Maastrichtian sections (Hattner & Wise 1980; Self-Trail, unpubl. data). Pospichal (pers. comm.) has corroborated a pre-boundary last appearance for this species.

Micula

Detailed studies of the genus *Micula* over the recent decades has provided a wealth of information about the possible evolutionary trends and palaeoenvironmental preferences of individual species (Bukry 1973; Roth & Bowdler 1979; Jafar 1994). The identification of species can be difficult at times because of the gradual evolution of one species into another. However, despite difficulty in distinguishing intermediate forms, this evolutionary lineage can be very useful for subdividing the upper Maastrichtian units of low- to mid-latitude sections.

The species of most interest to this study is *Micula murus*, which is the subzonal marker for the bases of Subzones CC25c (Perch-Nielsen 1985) and UC20b (Burnett 1998). It has long been recognized that this species is useful as a biostratigraphic marker only in samples from low

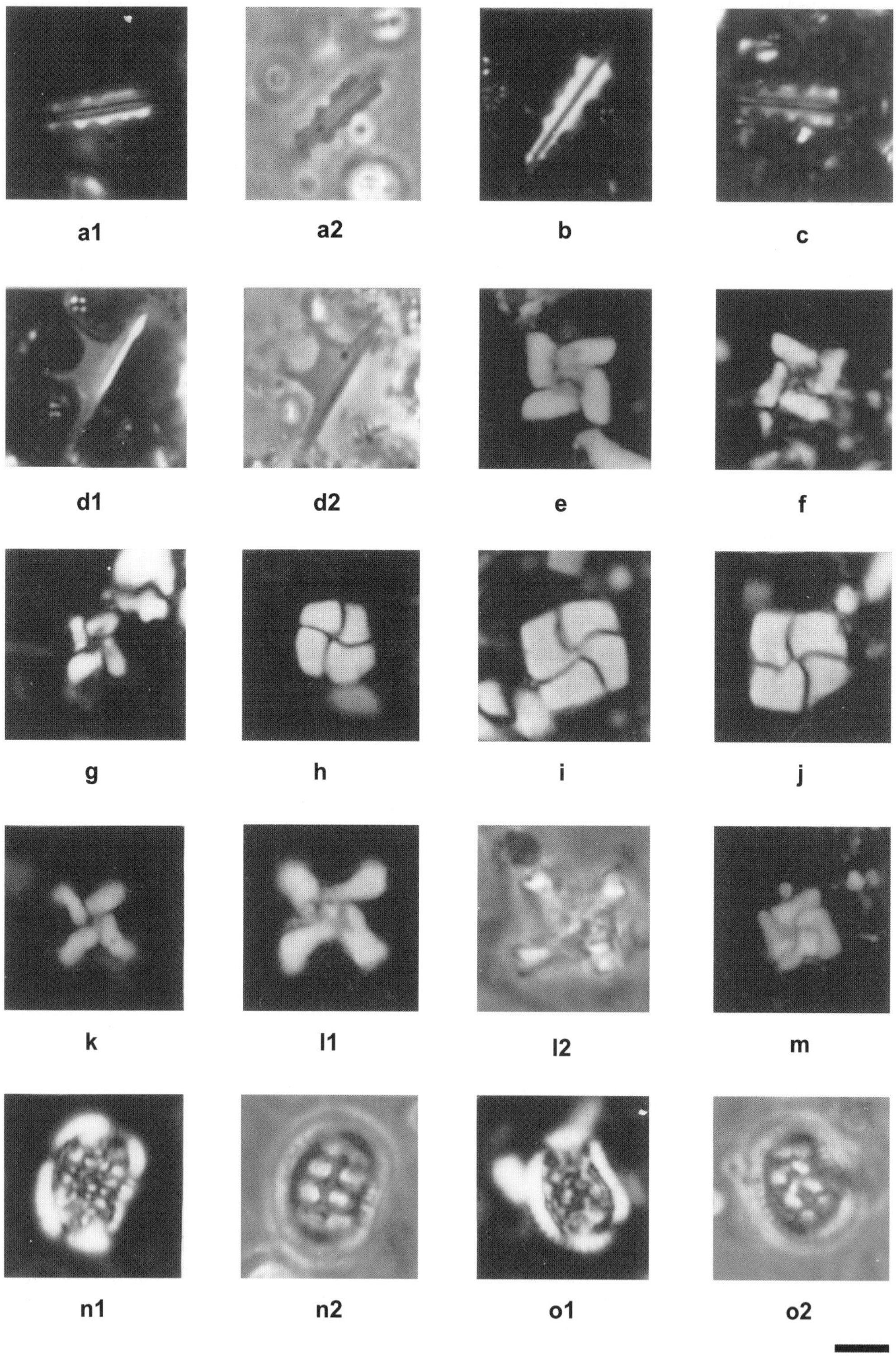
a1
a2
b
c
d1
d2
e
f
g
h
i
j
k
l1
l2
m
n1
n2
o1
o2

to intermediate palaeolatitudes (Worsley & Martini 1970; Burnett 1998); however, this study now confirms that the biostratigraphic range of *M. murus* was also controlled by proximity to shore. *Micula murus* is present in rare abundance in four cores from South Carolina (C-15, Cannon Park, Santee Coastal Reserve and Myrtle Beach) which consist of sediments that accumulated in outer- to middle-neritic water depths (Fig. 6). Because of the sporadic occurrence of *M. murus* in these cores, extra time was spent examining slides from the nearshore sequences to ensure accuracy. The first occurrence of *M. murus* in all four cores lies above the FAD of *Ceratolithoides kamptneri* and below or coincident with the FAD of *Micula prinsii*. This sequence of events was also documented by Jiang & Gartner (1986), Moshkovitz & Habib (1993) and Jafar (1994), who all reported the first occurrence of *C. kamptneri* below the FAD of *M. murus* in mid-latitude, nearshore sections. An age v. depth plot for the Santee Coastal Reserve core, using both calcareous nannofossil and magnetostratigraphic datum levels, confirms a constant rate of sediment accumulation at this site (Fig. 7; Table 2). This constant-rate interpretation suggests that the lowest occurrence of *M. murus* in the Santee Coastal Reserve core does not represent its true evolutionary first occurrence, but rather represents a delayed arrival as a result of environmental conditions.

Unlike the four cores representing neritic conditions, the FAD of *M. murus* from deep-ocean Hole 1052E is in the stratigraphic order documented by Thierstein (1976), Perch-Nielsen (1985), El Mehaghag (1996) and Burnett (1998). The first appearance of *M. murus* in this hole occurs above the FAD of *L. quadratus* and well below the FAD of *C. kamptneri* (Fig. 6). A hiatus in Hole 1050C has resulted in the delayed first occurrence of both *M. murus* and *C. kamptneri* at this site. As the palaeolatitude of Holes 1050C and 1052E was essentially the same as that of boreholes C-15, Cannon Park, Santee Coastal Reserve and Myrtle Beach during latest Maastrichtian time, it can be concluded that the biostratigraphic range of *M. murus* is dependent on palaeoenvironment as well as palaeolatitude. This situation should be recognized by biostratigraphers: a biostratigraphic sequence that goes from Subzone CC25b to CC26a does not necessarily record missing section; rather, it may simply mean that the defining bioevent is missing but that the correlative chronozone is not (Fig. 6). By applying the use of other stratigraphic methods (e.g. palaeomagnetic data, lithostratigraphy, nannofossils other than the marker species), it should be possible to determine if part of the sedimentary sequence is actually missing or only appears to be gone (Edwards *et al.* 1999). It is herein suggested that *M. murus* may not be a reliable marker for upper Maastrichtian neritic sediments. Further documentation of the relationship between *M. murus* and *C. kamptneri* and the role that palaeoenvironment had on calcareous nannofossil distribution is needed.

Micula praemurus (Fig. 5) is a large, somewhat blocky form having distinct sutures that divide the eight elements into a curved 'pseudo-swastika' pattern. There is considerable range in the size of this species, and specimens are often larger than other members of this genus. *Micula praemurus* is present in rare to frequent abundances in almost every core from South Carolina and is common to frequent in Holes 1050C and 1052E. The FAD of *M. praemurus* is tentatively placed in Subzone CC25a, based on its appearance in 1050C at 462.2 m, above the LAD of *Reinhardtites levis* and below the FAD of *Lithraphidites quadratus*. This placement is further corroborated by data from the Lake City and Myrtle Beach cores, which also indicate a CC25a first occurrence. Most occurrences of this species cited in the literature are from sections lacking basal upper Maastrichtian sediments (e.g. Jafar

Fig. 5. (**a1**) *Lithraphidites grossopectinatus*, cross polars (XP), Santee Coastal Reserve (SCR) core, 119.0 m, Subzone CC26b, South Carolina. (**a2**) Same specimen, phase contrast (PC). (**b**) *Lithraphidites grossopectinatus*, XP, SCR core, 119.0 m, Subzone CC26b, South Carolina. (**c**) *Lithraphidites grossopectinatus*, XP, SCR core, 119.0 m, Subzone CC26b, South Carolina. (**d1**) *Lithraphidites kennethii* (broken), XP, SCR core, 119.0 m, Subzone CC26b, South Carolina. (**d2**) Same specimen, PC. (**e**) *Micula murus*, XP, Hole 1050C, 13R-3, 67–68 cm, Subzone CC26b, Blake Nose. (**f**) *Micula murus*, XP, Hole 1052E, 20R-3, 65–68 cm, Subzone CC26b, Blake Nose. (**g**) *Micula murus*, XP, Hole 1050C, 13R-3, 67–68 cm, Subzone CC26b, Blake Nose. (**h**) *Micula praemurus*, XP, Hole 1050C, 15R-6, 34–35 cm, Subzone CC25c, Blake Nose. (**i**) *Micula praemurus*, XP, Hole 1050C, 15R-6, 34–35 cm, Subzone CC25c, Blake Nose. (**j**) *Micula praemurus*, XP, Hole 1050C, 15R-6, 34–35 cm, Subzone CC25c, Blake Nose. (**k**) *Micula prinsii*, XP, Hole 1050C, 13R-1, 67–68 cm, Subzone CC26b, Blake Nose. (**l1**) *Micula prinsii*, XP, Hole 1050C, 13R-3, 67–68 cm, Subzone CC26b, Blake Nose. (**l2**) Same specimen, PC. (**m**) *Micula swastica*, XP, Hole 1052E, 23R-4, 12 cm, Subzone CC25c, Blake Nose. (**n1**) *Nephrolithus frequens*, XP, SCR core, 119.0 m, Subzone CC26b, South Carolina. (**n2**) Same specimen, PC. (**o1**) *Nephrolithus frequens*, XP, SCR core, 119.0 m, Subzone CC26b, South Carolina. (**o2**) Same specimen, PC. Scale bar represents 5 μm.

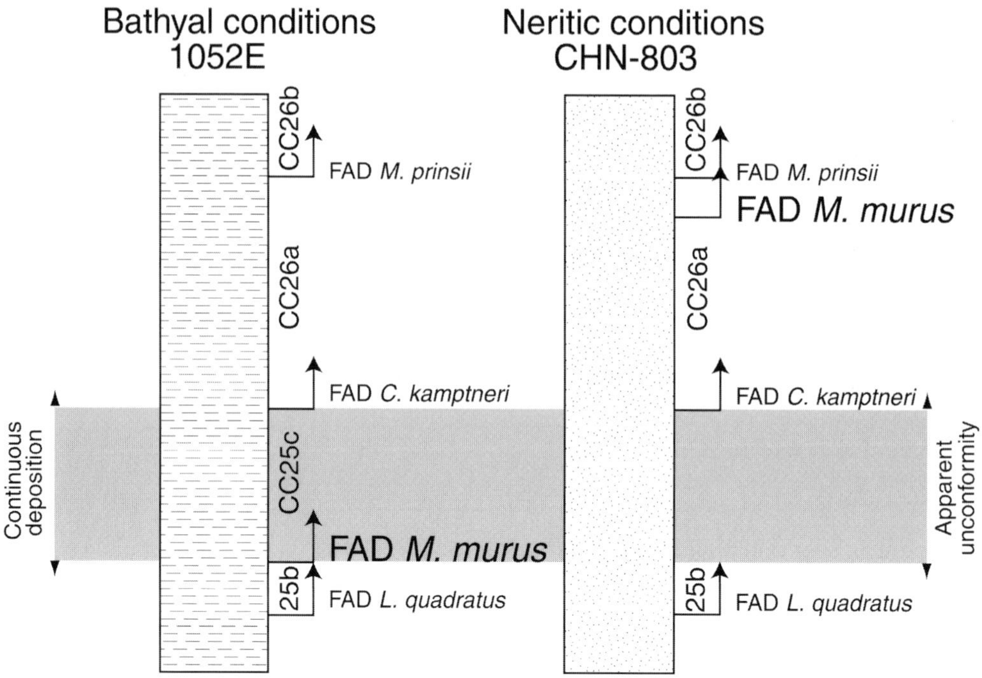

Fig. 6. Schematic diagram showing latitudinal distribution of the first appearance of *Micula murus* across a deep- to shallow-water transect. It should be noted that when there is a complete section with no missing time in bathyal conditions, the same interval will appear to record an apparent unconformity under neritic conditions (grey area).

1994), therefore it is difficult to determine whether the actual FAD of *M. praemurus* more accurately belongs in Zone CC24. However, Proto Decima *et al.* (1978) recorded the first appearance of *M. praemurus* from DSDP Site 364 (southeastern Atlantic) above the last occurrences of *Reinhardtites anthophorus* and *Quadrum trifidum* and below the appearance of *L. quadratus* in samples having a background

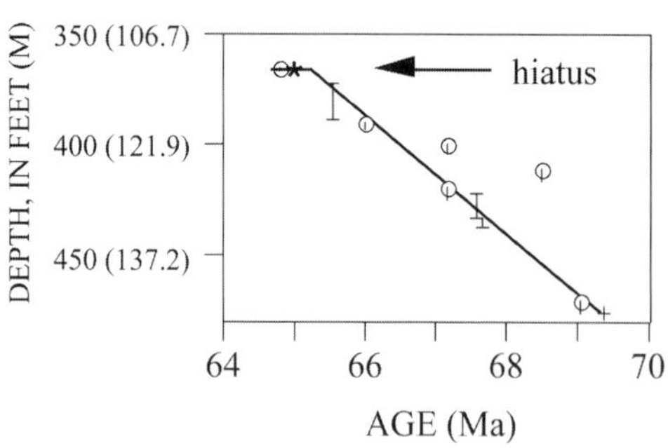

Fig. 7. Age–depth relations for the Santee Coastal Reserve core, SC, using calcareous nannofossil and magnetostratigraphic datum levels. Ages of fossil datum level were assigned by Norris *et al.* (1998). ○, FAD; +, LAD; *, Cretaceous–Tertiary boundary; bar, palaeomagnetic datum. After Edwards *et al.* 1999.

assemblage indicative of the upper lower to lower upper Maastrichtian sediments. The LAD of *M. praemurus* lies within Subzone CC26a, based on data from the Myrtle Beach, Cannon Park and Clubhouse Crossroads cores and from Hole 1052E. It occurs above the FAD of *C. kamptneri*, below the FAD of *M. prinsii*, and around the LAD of *L? charactozorro*. Although *M. praemurus* has a tendency to occur sporadically in neritic sediments, its first appearance and last appearance events are none the less considered to be reliable in both deep-ocean and nearshore settings.

Podorhabdus? elkefensis

One little-used species that is of biostratigraphic importance in this region is *Podorhabdus? elkefensis*. This species occurs sporadically in upper Maastrichtian sediments from all but one core in the study area. The first appearance of *P? elkefensis* is not biostratigraphically useful, but its last appearance shortly above the FAD of *M. prinsii* is a useful datum. The LAD of *P? elkefensis* is in Subzone CC26b above the first occurrence of *M. prinsii* and below the LADs of *C. kamptneri* and *L. kennethii*. The disappearance of this species occurs just above the

Table 2. *Values used in calculations of sediment accumulation rates for the Santee Coastal Reserve core; ages of datum levels are from Erba* et al. *(1995)*

Event	Age (Ma)	Depth (m)	Next sample
FAD *Cruciplacolithus primus*	64.80	−111.50	−111.90
FAD *Micula prinsii*	66.00	−119.00	−119.80
FAD *Ceratolithoides kamptneri*	67.20	−128.00	−129.80
FAD *Nephrolithus frequens*	67.20	−122.10	−123.40
FAD *Micula murus*	68.50	−125.60	−127.00
FAD *Lithraphidites quadratus*	69.00	−143.70	−145.50
FAD *Reinhardtites levis*	69.40	−145.50	−144.50
K–T boundary	65.00	−111.90	
C30n–C29r	65.58	−118.60	−113.50
C30r–C30n	67.61	−132.20	−128.90
C31n–C30r	67.74	−132.20	−133.60

C30n–C29r magnetostratigraphic boundary in the Santee Coastal Reserve core and within C30n at Holes 1050C and 1052E. This species may have been restricted to intermediate- to low-latitude sites, so it is difficult to determine what its true last occurrence is. However, Perch-Nielsen (1981) documented the presence of *P*? *elkefensis* above the FAD of *M. prinsii* in Tunisia. Varol (1991) reported that it commonly occurs in Maastrichtian sediments from Libya, and Burnett (pers. comm.) reported it as frequent in South Africa.

Biostratigraphic sequence of events

On the basis of the information detailed above, a refined biostratigraphic sequence of events, which can be used for the upper Maastrichtian sequences of the northwest Atlantic, is herein documented (Fig. 3). The scheme presented here utilizes both old and new bioevent horizons and is separated into oceanic v. shelf palaeoenvironments. Where possible, the biostratigraphic events are tied to the geochronological time scale of Gradstein *et al.* (1995). Several of the bioevents discussed are shown to be influenced palaeobiogeographically.

Biosubzone CC25a (or equivalent UC19) is an 'absence' zone in neritic and oceanic sediments and is identified by the absence of both *Reinhardtites levis* and *Lithraphidites quadratus*. However, the first common occurrence of *C. amplector* in bathyal sediments can be used to approximate the base of CC25a. Subzone CC25a can be subdivided, in both deep- and shallow-water cores, by the FAD of *M. praemurus*, which occurs near the base of this zone in Chron C31r. Although *M. praemurus* is not present in great abundance in the shallow-water cores, it does occur with enough regularity to be used as a marker species herein.

The base of Subzone CC25b (UC20a) is defined by the FAD of *Lithraphidites quadratus*, a species that is biostratigraphically useful for both shallow- and deep-water sites, virtually globally. However, the base of Subzone CC25c (FAD of *M. murus*) may not be recognizable in sequences from certain palaeoenvironmental settings, but instead must be determined using proxies. The base of this subzone can be approximated using the first appearance of *Lithraphidites*? *charactozorro* and *L. grossopectinatus*, which appear in close proximity near the C31r–C31n boundary and below the FAD of *M. murus* in deep-water settings. Thus, the situation presented herein appears to show that Subzone CC25b is short (0.5 Ma in extent) at bathyal locations but expanded (2.2 Ma) at shallower locations. However, the latter probably chronologically represents Subzones CC25b and CC25c combined (Fig. 6): the delayed first occurrence of *M. murus* in shallow-water sites suggests that CC25c is missing although its corresponding chronozone is present.

The base of subzone CC26a can be approximated at both open-ocean and neritic locations herein, based on the more reliable first occurrence of *Ceratolithoides kamptneri*, rather than on the erratic FAD of *Nephrolithus frequens*. Subdivision of this subzone is possible using the LAD of *L*? *charactozorro*, which occurs below the LAD of *M. praemurus*. This biostratigraphic event is useful at both open-ocean and neritic sites, and corresponds to the first occurrence of *Micula murus* in nearshore settings. Upper Zone CC26a can be even further subdivided based on the last occurrence of *M. praemurus*, which occurs shortly before the first occurrence of *Micula prinsii* in both shallow- and deep-water sites.

The last appearance of *C. kamptneri* shortly after the Chron 30n–29r boundary is an extremely useful biostratigraphic event that has been documented from several sites worldwide. This event further subdivides Zone CC26b into an upper and lower section, each having a 0.5 Ma duration. At deeper-water sites, the LAD of *C. amplector* can be used to further subdivide upper Zone CC26b.

Lower Maastrichtian disconformity

The calcareous nannofossils from all studied South Carolina sites indicate the presence of a lower Maastrichtian disconformity. At each of these sites, upper Maastrichtian sediments of the Peedee Formation rest disconformably on upper Campanian sediments of the Donoho Creek Formation. Sugarman *et al.* (1995) noted the occurrence of this disconformity in sediments from the New Jersey Coastal Plain, and suggested that it is correlative with disconformity UZA4.4 of the Haq sea-level curve (Haq *et al.* 1987). Sediment of this age is also missing from the US Gulf of Mexico Coastal Plain (Bergen, pers. comm). All evidence currently available suggests that this disconformity is ubiquitous throughout the Atlantic and Gulf of Mexico Coastal Plains. However, ODP Holes 1050C and 1052E penetrated upper Cretaceous sediments that contain calcareous nannofossils indicative of an early Maastrichtian age. This sequence at Hole 1050C is condensed (< 5 m thick), occurring just above a package of sediments consisting of a series of hardgrounds separated by thin layers of sediment oozes (Norris *et al.* 1998). The lower Maastrichtian sequence at Hole 1052E is significantly thicker (75 m), although slumping throughout this section has probably distorted the actual thickness somewhat. Norris *et al.* (1998) set forth several possible explanations for the depositional history of this sediment, including mass flow during an interval of changing sedimentation patterns.

Lowered sea level in the Atlantic Ocean during early Maastrichtian time could be the cause of the widespread disconformity along the Western Atlantic seaboard. However, benthic foraminifera at Holes 1050C and 1052E indicate a deepening-upward trend through time for the lower and upper Maastrichtian sequence. This suggests instead that the disconformity could be due to current activity along the shelf–slope for the nearshore locations.

Conclusions

The sedimentary sequences supplied by the Ocean Drilling Program and the US Geological Survey provide a wealth of information concerning upper Maastrichtian calcareous nannofossil biostratigraphy. Detailed study of these sections allows for the subdivision of low- to mid-latitude biozones using a combination of evolution and extinction events that have not been previously utilized. Comparison of bioevent horizons along this shallow-water to open-ocean transect allows for the identification of several species that were influenced by palaeodepth and related factors (possibly nutrients) and whose cosmopolitan utility as marker species is somewhat compromised. Rapid speciation within several calcareous nannofossil genera during late Maastrichtian time coincides with a postulated warming event preceding the K–T boundary (Huber & Watkins 1992; Watkins *et al.* 1996) and provides for a more refined subdivision of upper Maastrichtian sediments in this region. Several species are shown to have been controlled by palaeoenvironmental conditions. Although it is known that certain calcareous nannofossil species show latitudinally induced provincialism, this study documents, for the first time, palaeoenvironmental effects on late Maastrichtian nannofossil assemblages along a shallow- to deep-water transect in this region.

This research was graciously supported by the USAAC-JOI. The author is grateful for the critical reviews supplied by J. A. Burnett (University College, London) and J. J. Pospichal (Bugware). Comments by D. K. Watkins (University of Nebraska) and G. S. Gohn (US Geological Survey) greatly improved the manuscript. E. Seefelt (US Geological Survey) provided valuable assistance in the production of the fossil figures.

Appendix

Cretaceous calcareous nannofossil species cited in text (in alphabetical order by genus).

Ceratolithoides amplector Burnett 1997
Ceratolithoides indiensis Burnett 1997
Ceratolithoides kamptneri Bramlette & Martini 1964
Ceratolithoides pricei Burnett 1997
Ceratolithoides ultimus Burnett 1997
Lithraphidites acutus Verbeek & Manivit, in Manivit *et al.* (1977)
Lithraphidites carniolensis Deflandre 1963
Lithraphidites? charactozorro Self-Trail & Pospichal, in Self-Trail (1999)
Lithraphidites grossopectinatus Bukry 1969
Lithraphidites kennethii Perch-Nielsen 1984
Lithraphidites praequadratus Roth 1978
Lithraphidites quadratus Bramlette & Martini 1964
Micula murus (Martini 1961) Bukry 1973
Micula praemurus (Bukry 1973) Stradner & Steinmetz 1984
Micula prinsii Perch-Nielsen 1979
Micula swastica Stradner & Steinmetz 1984
Nephrolithus frequens Gorka 1957

Podorhabdus? *elkefensis* Perch-Nielsen 1981
Pseudomicula quadrata Perch-Nielsen, in Perch-Nielsen *et al.* (1978)
Quadrum trifidum (Stradner, in Stradner & Papp, 1961) Prins & Perch-Nielsen, in Manivit *et al.* (1977)
Reinhardtites anthophorus (Deflandre 1959) Perch-Nielsen 1968
Reinhardtites levis Prins & Sissingh, in Sissingh (1977)

References

BRALOWER, T. J. & SIESSER, W. G. 1992. Cretaceous calcareous nannofossil biostratigraphy of ODP Leg 122 Sites 761, 762 and 763, Exmouth and Wombat Plateaus, N. W. Australia. *In*: VON RAD, U., HAQ, B. U. *et al.* (eds) *Proceedings of the Ocean Drilling Program, Scientific Results*, **122**. Ocean Drilling Program, College Station, TX, 529–556.

BRALOWER, T. J., LECKIE, R. M., SLITER, W. V. & THIERSTEIN, H. R. 1995. An integrated Cretaceous microfossil biostratigraphy. *In*: BERGGREN, W. A., KENT, D. V., AUBRY, M. P. & HARDENBOL, J. (eds) *Geochronology Time Scales and Global Stratigraphic Correlation. A Unified Temporal Framework for a Historical Geology*. SEPM Special Publications, **54**, 65–79.

BUKRY, D. 1973. Coccolith stratigraphy, Eastern Equatorial Pacific, Leg 16, Deep Sea Drilling Project. *In*: VAN ANDEL, T. H., HEATH, G. R. *et al.* (eds) *Initial Reports of the Deep Sea Drilling Project*, **16**. US Governmnet Printing Office, Washington, DC, 653–711.

BURNETT, J. A. 1997. New species and conjectured evolutionary trends of *Ceratolithoides* Bramlette & Martini, 1964 from the Campanian and Maastrichtian of the Indian Ocean. *Journal of Nannoplankton Research*, **19**(2), 123–132.

BURNETT, J. A. 1998. Upper Cretaceous. *In*: BOWN, P. R. (ed.) *Calcareous Nannofossil Biostratigraphy*. Chapman and Hall, London, 132–199.

BYBELL, L. M., CONLON, K. J., EDWARDS, L. E., FREDERIKSEN, N. O., GOHN, G. S. & SELF-TRAIL, J. M. 1998. *Biostratigraphy and Physical Stratigraphy of the USGS-Cannon Park Core (CHN-800), Charleston County, South Carolina*. US Geological Survey Open-File Report **98-246**.

EDWARDS, L. E., GOHN, G. S., SELF-TRAIL, J. M. *et al.* 1999. *Physical Stratigraphy, Paleontology, and Magnetostratigraphy of the USGS-Santee Coastal Reserve Core (CHN-803), Charleston County, South Carolina*. US Geological Survey Open-File Report **99-308**.

EL MEHAGHAG, A. A. 1996. Cretaceous and Tertiary calcareous nannofossil biostratigraphy of North and Northeast Libya. *In*: SALEM, M. J., MOUZUGHI, A. J. & HAMMUDA, O. S. (eds) *The Geology of Sirt Basin. First Symposium on the Sedimentary Basins of Libya*, **1**, Tripoli 1993, 475–512.

ERBA, E., PREMOLI SILVA, I. & WATKINS, D. K. 1995. Cretaceous calcareous plankton biostratigraphy of Sites 872 through 879. *In*: HAGGERTY, J. A., PREMOLI SILVA, I., RACK, F. & MCNUTT, M. K. (eds) *Proceedings of the Ocean Drilling Program, Scientific Results*, **144**, 157–169.

ESHET, Y., MOSHKOVITZ, S., HABIB, D., BENJAMINI, C. & MAGARITZ, M. 1992. Calcareous nannofossil and dinoflagellate stratigraphy across the Cretaceous/Tertiary boundary at Hor Hahar, Israel. *Marine Micropalaeontology*, **18**, 199–228.

GRADSTEIN, F. M., AGTERBERG, F. P., OGG, J. G., HARDENBOL, J., VAN VEEN, P., THIERRY, J. & HUANG, Z. 1995. A Triassic, Jurassic and Cretaceous time scale. *In*: BERGGREN, W. A., KENT, D. V., AUBRY, M. P. & HARDENBOL, J. (eds) *Geochronology, Time Scales and Global Stratigraphic Correlation*. SEPM Special Publication, **54**, 95–128.

HAQ, B. U., HARDENBOL, J. & VAIL, P. R. 1987. Chronology of fluctuating sea levels since the Triassic (250 million years ago to present). *Science*, **235**, 1156–1167.

HATTNER, J. G. & WISE, S. W., Jr 1980. Upper Cretaceous calcareous nannofossil biostratigraphy of South Carolina. *South Carolina Geology*, **24**, 41–117.

HUBER, B. T. & WATKINS, D. K. 1992. Biogeography of Campanian–Maastrichtian calcareous plankton in the region of the Southern Ocean: palaeogeographic and palaeoclimatic implications. *American Geophysical Union Antarctic Research Series*, **56**, 31–60.

JAFAR, S. A. 1994. Late Maastrichtian calcareous nannofossils from the Lattengebirge (Germany) and the Andaman–Nicobar Islands (India)—Remarks on events around the Cretaceous–Tertiary boundary. *Neues Jahrbuch für Geologie und Mineralogie, Abhandlungen*, **191**(2), 251–269.

JIANG, M. J. & GARTNER, S. 1986. Calcareous nannofossil succession across the Cretaceous–Tertiary boundary in east–central Texas. *Micropalaeontology*, **32**(3), 232–255.

MANIVIT, H., PERCH-NIELSEN, K., PRINS, B. & VERBEEK, J. W. 1977. Mid Cretaceous calcareous nannofossil biostratigraphy. *Koninklijke Nederlandse Akademie van Wetenschappen, Proceedings, Serial B*, **80-3**, 169–181.

MOSHKOVITZ, S. & HABIB, D. 1993. Calcareous nannofossil and dinoflagellate stratigraphy of the Cretaceous–Tertiary boundary, Alabama and Georgia. *Micropalaeontology*, **39**, 167–191.

NORRIS, R. D., KROON, D., KLAUS, A. *et al.* (eds) 1998. *Proceedings of the Ocean Drilling Program, Initial Reports*, **171B**, 93–319.

OWENS, J. P. & GOHN, G. S. 1985. Depositional history of the Cretaceous Series in the United States Atlantic Coastal Plain: stratigraphy, palaeoenvironments, and tectonic controls of sedimentation. *In*: POAG, C. W. (ed.) *Geological evolution of the United States Atlantic Margin*. Van Nostrand Reinhold, New York, 25–86.

PERCH-NIELSEN, K. 1981. New Maastrichtian and Palaeocene calcareous nannofossils from Africa, Denmark, the USA and the Atlantic, and some Palaeocene lineages. *Eclogae Geologicae Helvetiae*, **74-3**, 831–863.

PERCH-NIELSEN, K. 1985. Mesozoic calcareous nannofossils. *In*: BOLLI, H. M., SAUNDERS, J. B. & PERCH-NIELSEN, K. (eds) *Plankton Stratigraphy*. Cambridge University Press, Cambridge, 329–426.

PERCH-NIELSEN, K., SADEK, A., BARAKAT, M. G. & TELEB, F. 1978. Late Cretaceous and Early Tertiary calcareous nannofossil and planktonic foraminifera zones from Egypt. *Annales des Mines et de la Géologie (Tunisia), Actes du VI Colloque Africain de Micropaléontologie, Tunis*, **28-2**, 337–403.

POSPICHAL, J. J. & WISE, S. W. 1990. Maestrichtian calcareous nannofossil biostratigraphy of Maud Rise, ODP Leg 113 Sites 689 and 690, Weddell Sea. *In*: BARKER, P. F., KENNETT, J. P. *et al.* (eds) *Proceedings of the Ocean Drilling Program, Scientific Results*, **113**. Oceab Drilling Program, College Station, TX, 465–487.

PROTO DECIMA, F., MEDIZZA, F. & TODESCO, L. 1978. Southeastern Atlantic Leg 40 calcareous nannofossils. *In*: BOLLI, H. M., RYAN, W. B. F. *et al.* (eds) *Initial Reports of the Deep Sea Drilling Project*, **40**. US Government Printing Office, Washington, DC, 571–632.

RIO, D., VILLA, G. & CANTADORI, M. 1983. Nannofossil dating of helminthoid flysch Units in the Northern Apennines. *Giornale di Geologia*, **45**, 57–86.

RISATTI, J. B. 1973. Nannoplankton biostratigraphy of the Upper Bluffport Marl–Lower Prairie Bluff Chalk interval (Upper Cretaceous), in Mississippi. *In*: SMITH, L. A. & HARDENBOL, J. (eds) *Proceedings of Symposium on Calcareous Nannofossils*. SEPM Gulf Coast Section, Houston, TX, 8–57.

ROTH, P. H. 1978. Cretaceous nannoplankton biostratigraphy and oceanography of the Northwestern Atlantic Ocean. *In*: BENSON, W. E., SHERIDAN, R. E. *et al.* (eds) *Initial Reports of the Deep Sea Drilling Project*, **44**. US Government Printing Office, Washington, DC, 731–760.

ROTH, P. H. & BOWDLER, J. L. 1979. Evolution of the calcareous nannofossil genus *Micula* in the Late Cretaceous. *Micropalaeontology*, **25**, 272–280.

SELF-TRAIL, J. M. 1999. Some new and rarely documented Late Cretaceous calcareous nannofossils from subsurface sediments in South Carolina. *Journal of Palaeontology*, **73-5**, 952–963.

SELF-TRAIL, J. M. & BYBELL, L. M. 1997. *Calcareous Nannofossil Biostratigraphy of the SCDNR Testhole C-15, Jasper County, South Carolina*. US Geological Survey Open-File Report **97-155**.

SELF-TRAIL, J. M. & GOHN, G. S. 1996. *Biostratigraphic Data for the Cretaceous Marine Sediments in the USGS-St. George No. 1 Core (DOR-211), Dorchester County, South Carolina*. US Geological Survey Open-File Report **96-684**.

SISSINGH, W. 1977. Biostratigraphy of Cretaceous calcareous nannoplankton. *Geologie en Mijnbouw*, **56**, 37–65.

SOLLER, D. R. & MILLS, H. H. 1991. Surficial geology and geomorphology. *In*: HORTON, J. W. & ZULLO, V. A. (eds) *The Geology of the Carolinas*. University of Tennessee Press, Knoxville, 290–308.

STRADNER, H. & PAPP, A. 1961. Tertiare Discoasteriden aus Osterreich und deren stratigraphische Bedeutung mit Hinweisen auf Mexico, Rumanien und Italien. *Jahrbuch der Geologischen Bundesanstalt (Wien)*, **7**, 1–159.

SUGARMAN, P. J., MILLER, K. G., BUKRY, D. & FEIGENSON, M. D. 1995. Uppermost Campanian–Maestrichtian strontium isotopic, biostratigraphic, and sequence stratigraphic framework of the New Jersey Coastal Plain. *Geological Society of America Bulletin*, **107**, 19–37.

THEIRSTEIN, H. R. 1976. Mesozoic calcareous nannoplankton biostratigraphy of marine sediments. *Marine Micropalaeontology*, **1**, 325–362.

VAROL, O. 1991. New Cretaceous and Tertiary calcareous nannofossils. *Neues Jahrbuch für Geologie und Paläontologie, Abhandlungen*, **182**, 211–237.

WATKINS, D. K., WISE, S. W., POSPICHAL, J. J. & CRUX, J. 1996. Upper Cretaceous calcareous nannofossil biostratigraphy and palaeoceanography of the Southern Ocean. *In*: MOGUILEVSKY, A. & WHATLEY, R. (eds) *Microfossils and Oceanic Environments*. University of Wales Press, Aberystwyth, 355–381.

WORSLEY, T. R. & MARTINI, E. 1970. Late Maestrichtian nannoplankton provinces. *Nature*, **225**, 1242–1243.

The Maastrichtian record at Blake Nose (western North Atlantic) and implications for global palaeoceanographic and biotic changes

KENNETH G. MACLEOD[1] & BRIAN T. HUBER[2]

[1]*Department of Geological Sciences, University of Missouri, Columbia, MO 65211, USA*

[2]*Department of Paleobiology, Smithsonian Institution, MRC: NHB-121, Washington, DC 20560, USA*

Abstract: Widespread biological, geochemical and sedimentological shifts within the Maastrichtian are well documented, but data are limited for the low-latitude Atlantic. New observations from Ocean Drilling Program (ODP) sites located on Blake Nose in the subtropical western North Atlantic increase information concerning the Maastrichtian history of this critical region. Planktonic $\delta^{18}O$ results suggest up to 6 °C of local surface water warming (or 4‰ decrease in salinity) at the same time as most of the globe was cooling. Benthic $\delta^{13}O$ and $\delta^{13}C$ values of both planktonic and benthic taxa show little if any directional trend or excursions on long time scales; however, planktonic and benthic taxa exhibit strong $\delta^{13}C$ and $\delta^{18}O$ cycles (up to 0.8 and 0.6‰, respectively) across a short interval of high-resolution sampling. Other portions of the cores have not yet been studied at high resolution. The last occurrence of inoceramid shell fragments on Blake Nose matches previously documented global patterns, i.e. a mid-Maastrichtian extinction event that occurred later in low latitudes than in high southern latitudes. Models for Maastrichtian change seem to be converging on variation in intermediate to deep water ocean circulation as a unifying process. Blake Nose data are consistent with this conclusion, but demonstrate new regional patterns and emphasize the importance of precise and accurate chronostratigraphic correlation in understanding Maastrichtian change.

Recognition of a variety of biotic, climatic and oceanographic changes that occurred several million years before the Cretaceous–Tertiary (K–T) boundary has led to considerable interest in the palaeoceanographic and palaeoclimatic history of the Maastrichtian. Widespread cooling has been widely reported for Maastrichtian time (e.g. Douglas & Savin 1975; Boersma & Shackleton 1981; Barrera *et al.* 1987; MacLeod & Huber 1996*a*; Frank & Arthur 1999). Additional Maastrichtian events include the extinction of deep sea inoceramid bivalves (e.g. MacLeod *et al.* 1996, 2000; Chauris *et al.* 1998), extinction of tropical rudist bivalves (Johnson & Kauffman 1990, 1996; Johnson *et al.* 1996), a carbon isotope shift in deep-sea biogenic carbonates (e.g. Barrera 1994; MacLeod & Huber 1996*a*; Barrera & Savin 1999; Li & Keller 1999), and an increasing rate of change of seawater $^{87}Sr/^{86}Sr$ ratios (Nelson *et al.* 1991; Barrera 1994; Barrera & Savin 1999).

Many hypotheses have been advanced to explain Maastrichtian events ranging from long-term Campanian–Maastrichtian climate deterioration (e.g. Dhondt 1983) to rapid, impact-generated perturbation(s) (e.g. Hut *et al.* 1987; Kauffman 1988), but recent explanations have focused on oceanographic changes occurring on intermediate time scales (Barrera 1994; MacLeod 1994*a*; Johnson *et al.* 1996; MacLeod & Huber 1996*a*; Barrera *et al.* 1997; Chauris *et al.* 1998; Barrera & Savin 1999; Frank & Arthur 1999; Li & Keller 1999; Miller *et al.* 1999). These palaeoceanographic models invoke many of the same or similar processes and events to explain similar data (e.g. high-latitude cooling and a reorganization of deep-water circulation). On the other hand, cause and effect, timing and the relative importance ascribed to different processes vary considerably among models, resulting in divergent predictions (both explicit and implicit) about the nature of Maastrichtian change and the behaviour of Maastrichtian climate. Better documentation of the timing and nature of Maastrichtian shifts in different regions is needed to distinguish among

From: KROON, D., NORRIS, R. D. & KLAUS, A. (eds). 2001 *Western North Atlantic Palaeogene and Cretaceous Palaeoceanography*. Geological Society, London, Special Publications, **183**, 111–130.
0305-8719/01/$15.00

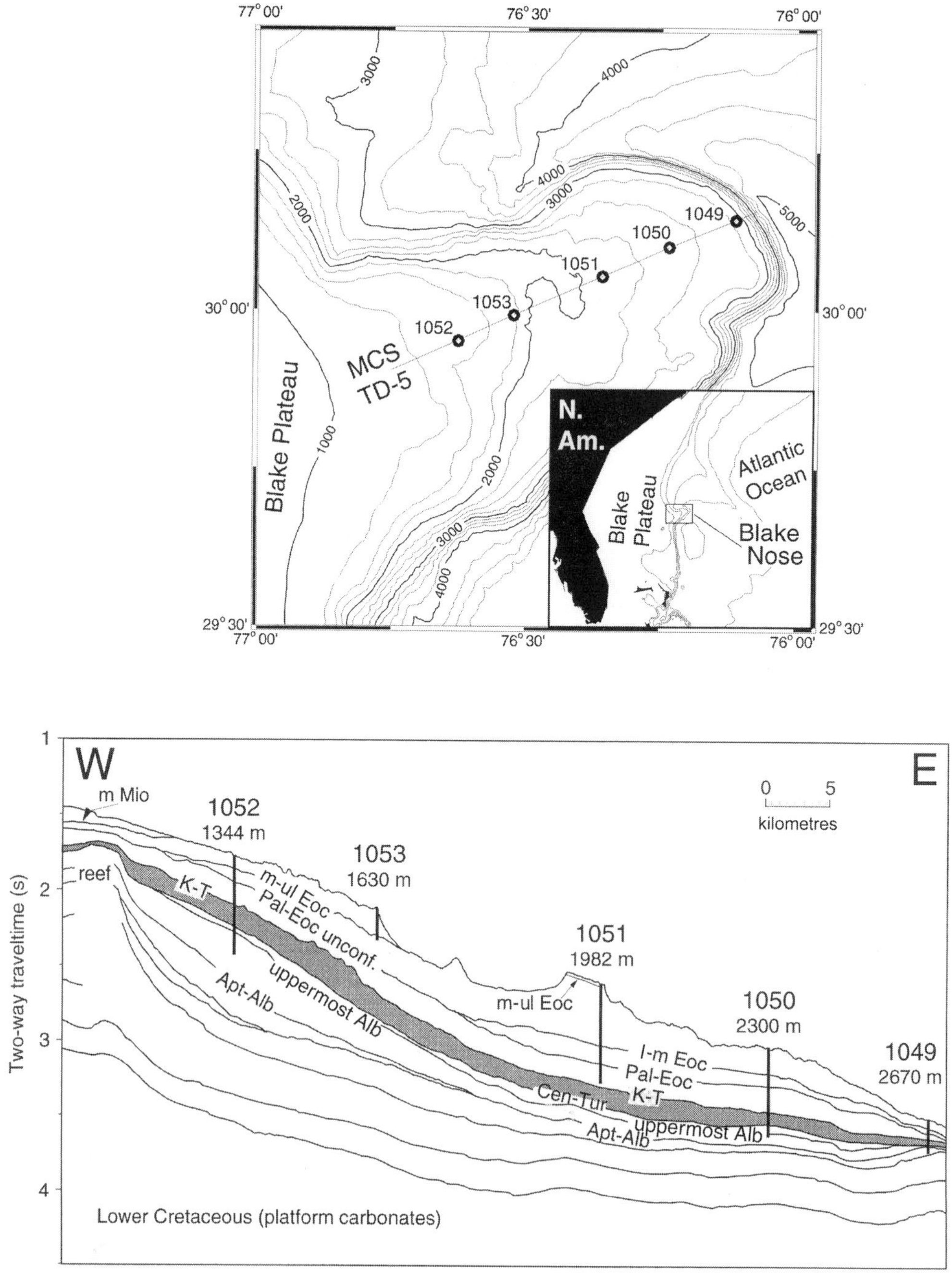

Fig. 1. Topographic map of the Blake Nose (position relative to North America, inset) and an interpretation of a seismic line (MCS-TD-5) shot along the crest of the feature. Location of ODP Leg 171B drill sites is shown in both. The Maastrichtian sequence is shaded on the cross-section.

competing models. This paper presents new isotopic, palaeontological and sedimentological observations from sites drilled on Blake Nose in the subtropical North Atlantic (Fig. 1) and discusses the results relative to previous Maastrichtian studies (Fig. 2).

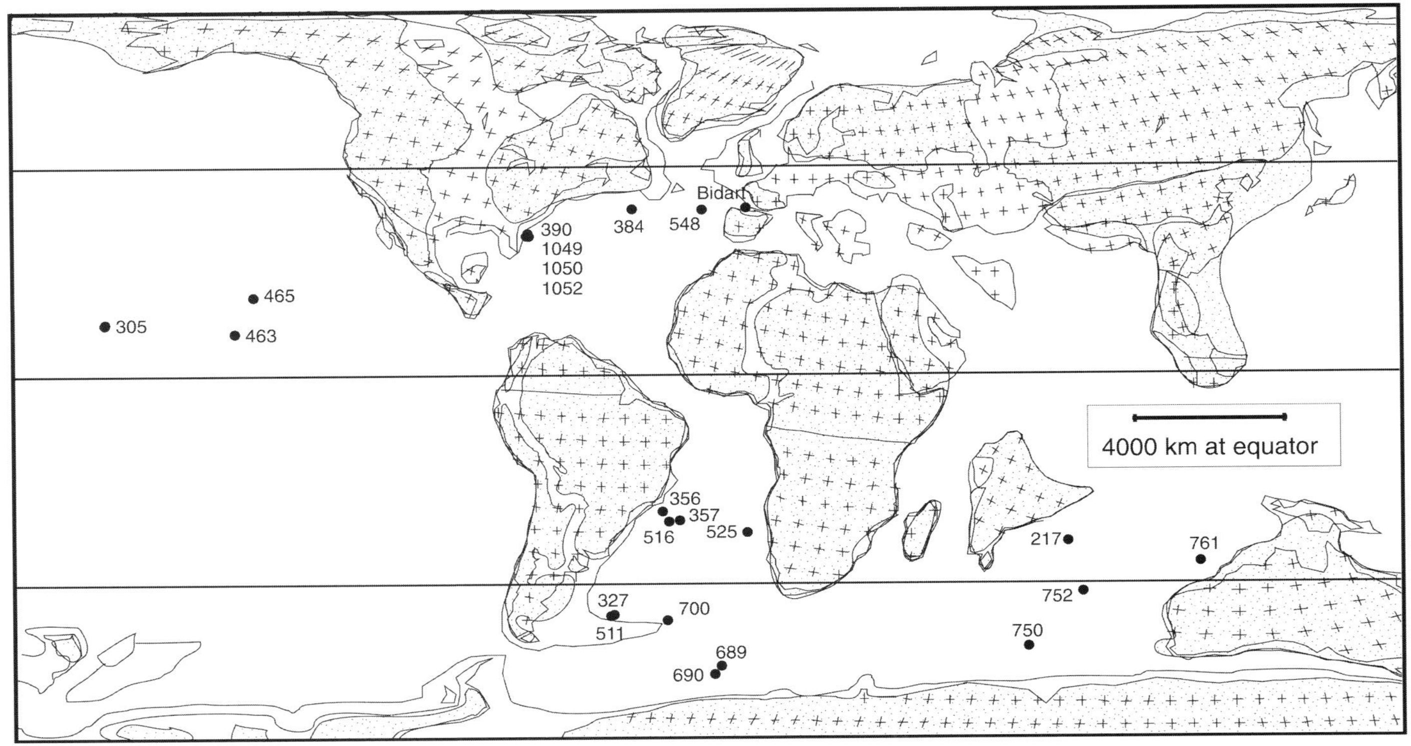

Fig. 2. Palaeogeographical map for 70 Ma showing location of some of the sites that were most important in developing Maastrichtian models discussed in the text. Reconstruction was generated using PGIS/MAC™ software package and land–sea distributions modified from Barron (1987).

Blake Nose

Setting and importance

Blake Nose is a gently sloping topographic high that extends into the western Atlantic from Blake Plateau (Fig. 1). During Maastrichtian time Blake Nose was located at ~30° N. It was drilled during Deep Sea Drilling Project (DSDP) Leg 44 (one site) and Ocean Drilling Program (ODP) Leg 171B (five sites). Maastrichtian strata were recovered at Site 390 during Leg 44 and Sites 1049 (a reoccupation of DSDP Site 390), 1050 and 1052 on Leg 171B. Since Late Cretaceous time, subsidence and deposition have been roughly balanced. Sites 1049/390, 1050 and 1052 are located along the crest of the feature and provide a palaeodepth transect. Benthic foraminiferal assemblages indicate palaeodepths of 1000–2000 m at Site 390 and 600–1000 m at Site 1052, with the difference in depth between the shallowest and deepest site estimated at ~1.5 km based on seismic data (Norris *et al.* 1998).

The circulation of intermediate to deep waters (including the evidence for or against warm, saline water masses), has been discussed in most recent models of Maastrichtian change. Low to mid-latitudes are the most likely locations for formation of saline water masses, but the Maastrichtian record of the tropics and subtropics is not well documented especially in the Atlantic. Thus, data from Blake Nose sites can provide information concerning palaeoceanographic conditions in a critical region.

Materials and methods

Hemipelagic intervals in the Maastrichtian sections on Blake Nose are represented by nannofossil chalk containing abundant well-preserved foraminifera (Norris *et al.* 1998; MacLeod *et al.* 2001). Foraminiferal and nannofossil distributions (Norris *et al.* 1998) and palaeomagnetic studies (Ogg & Bardot 2001) allow both correlation among Blake Nose sections and placement of regional patterns within a global chronostratigraphic framework. Sedimentological observations are based on examination of the archive half of Maastrichtian cores. Bulk samples (~5 cm^3) from each section were disaggregated and washed on a 63 μm screen using standard techniques. Both fine (<63 μm) and coarse (>63 μm) fractions were collected. Coarse fraction residues were examined for foraminiferal and inoceramid content. Isotopic analyses of fine fraction carbonate (Holes 1050C and 1052E) and size-sorted separates of single foraminiferal species (Hole 1050C) were performed at the University of California, Santa Cruz, using a common acid bath system. Preliminary data from Site 1049 confirm patterns observed in Hole 390A (MacLeod *et al.* 2000) and this earlier work at present is more complete than continuing studies of Site 1049. Results are plotted in standard δ-notation relative to the V-PDB standard. Analytical precision is <0.1‰ for both carbon and oxygen. Further details and data tables have been presented by MacLeod *et al.* (2001).

Results

Sedimentology. All three sites (390/1049, 1050 and 1052) show sedimentological changes through the Maastrichtian. At each site lower Maastrichtian hemipelagic sediments are lighter coloured than upper Maastrichtian sediments. At sites 1050 and 1052 the lighter samples contain more $CaCO_3$ than darker samples (Norris *et al.* 1998; MacLeod *et al.* 2001) and we assume the same is true for site 390/1049. Beyond this similarity, sedimentological features at the three sites differ considerably. The outermost site, 390/1049, is composed of massive hemipelagic sediments and a significant hiatus cuts out much of the middle Maastrichtian sequence (Norris *et al.* 1998; MacLeod *et al.* 2000). At Site 1050, ~50 cm thick cyclic colour variations occur throughout the Maastrichtian section, but colour contrast through the cycles increases dramatically in mid-Maastrichtian time near the first appearance (FA) of the planktonic foraminifera *Abathomphalus mayaroensis*, which is here used to mark the base of the upper Maastrichtian (MacLeod *et al.* 2001). At Site 1052 variation in cycle amplitude is difficult to document because cyclicity in Hole 1052E is subtle even at its peak expression. On the other hand, in Hole 1052E there is a marked change in sediment fabric near the FA of *A. mayaroensis*. Most lower Maastrichtian sediments are slightly bioturbated to laminated whereas upper Maastrichtian sediments are thoroughly bioturbated. The abundance of organic carbon is marginally lower in the less bioturbated intervals than in more bioturbated intervals (Norris *et al.* 1998).

Slumping and coring gaps are unfortunate features of the Maastrichtian sections in both Holes 1050C and 1052E. At each site only one hole was drilled deeply enough to recover the Maastrichtian so there is no opportunity to splice across the coring gaps. Fortunately, slumping does not seem to have cut out, repeated or otherwise compromised stratigraphic ordering; expected Maastrichtian Tethyan biozones and magnetozones are present (Klaus *et al.* 2000; Ogg & Bardot 2001). Moreover, a significant deep-sea manifestation of Maastrichtian change, the decline and disappearance of shell fragments of inoceramid bivalves (e.g. MacLeod *et al.* 1996), falls within a recovered interval of hemipelagic sediments at both sites.

Palaeontology. Inoceramid shell fragments are common to abundant in lower Maastrichtian samples and last occur in the lower portions of magnetochron C31N near the FA of *A. mayaroensis* (Fig. 3). Maastrichtian foraminifera and nannofossils are also well represented and well preserved in Blake Nose samples. The sections appear to be biostratigraphically complete and there is no evidence in hemipelagic intervals of reworking or duplication of section. Nine to 13 planktonic foraminifer species first appear within the upper Maastrichtian sequence at different sites (Norris *et al.* 1998; MacLeod *et al.* 2000).

Oxygen isotopes. Planktonic foraminifera (1050C) and fine fraction carbonate (1050C and 1052E) show a decrease in $\delta^{18}O$ of 1‰ to 1.5‰ through the Maastrichtian (Fig. 3). A comparable shift occurs at Site 390, but much of the Maastrichtian may be represented by a hiatus there (MacLeod *et al.* 2000*a*). At Site 1050, all planktonic taxa analysed exhibit similar, correlative $\delta^{18}O$ shifts but benthic $\delta^{18}O$ values are relatively constant. Thus, whereas $\delta^{18}O$ gradients within the planktonic realm do not change greatly, the planktonic to benthic $\delta^{18}O$ gradient increases by ~1‰ through the Maastrichtian.

Carbon isotopes. Carbon isotopes suggest an increased gradient within the planktonic realm but no change in the planktonic to benthic $\delta^{13}C$ gradient during the Maastrichtian (Fig. 3). Presumed surface dwellers (*Pseudogumbelina palpebra* and *Rugoglobigerina* spp.) at Holes 390A and 1050C and the fine fraction from Hole 1052E consistently yield $\delta^{13}C$ values of 2–2.5‰, but $\delta^{13}C$ of deeper-dwelling planktonic species (particularly *Heterohelix globulosa*) decrease by 0.5–0.7‰ through the sections. The youngest benthic sample in Hole 1050C yielded a low $\delta^{13}C$ value and may signal a change in vertical $\delta^{13}C$ gradients near the top of the Maastrichtian, but at present this possibility is supported by a single data point.

Benthic $\delta^{13}C$ values at Hole 1050C show low values in mid-Maastrichtian samples from immediately below the inoceramid extinction, but high-resolution sampling across this interval demonstrates that the low benthic values are the extremes of cyclic variation in $\delta^{13}C$ values rather than a benthic excursion. Further, the benthic $\delta^{13}C$ cycles mirror variation in planktonic $\delta^{13}C$ values and are of the same frequency as the Milankovitch-scale colour cycles (MacLeod *et al.* 2001; Fig. 3). The negative correlation between planktonic and benthic $\delta^{13}C$ values indicates local cyclic variations in surface water productivity and organic carbon export to the bottom (MacLeod *et al.* 2001). It is important to note that only this one short interval has been studied at high resolution.

Discussion

Blake Nose data relative to previous studies and global patterns

Palaeoceanography. During Maastrichtian time the calcite compensation depth (CCD) shifted dramatically in different ocean basins (e.g. Arthur *et al.* 1985; Barrera & Savin 1999; Frank & Arthur 1999). In the Pacific it shallowed by ~3 km, in the South Atlantic it shallowed briefly by *c.* 1 km before deepening by ~1.5 km (Arthur *et al.* 1985), and in the Indian and North Atlantic it deepened by ~1 km (Hemleben & Troester 1984). Colour change and decreased carbonate content across the Maastrichtian at Blake Nose sites is consistent with a deepening of the North Atlantic CCD. However, Blake Nose Maastrichtian sediments all contain >50% $CaCO_3$ (Norris *et al.* 1998) and lack evidence of dissolution, suggesting that changes in productivity or terrigenous dilution rather than the CCD could be proximate causes of changes in the carbonate content. Regardless, colour, productivity, terrigenous dilution and depth of the CCD could all reflect changes brought about by a reorganization of ocean circulation (Frank & Arthur 1999).

Supporting changes in bottom waters, increased bioturbation in Hole 1052E may reflect changes in benthic ecology. Similar to Hole 1052E, in five sections in the Basque region of France and Spain the intensity of bioturbation seems to increase in the mid-Maastrichtian correlative with the decline and disappearance of inoceramids (MacLeod 1994*a*). This change was hypothesized to reflect an increase in bottom-water oxygen concentrations that could have promoted increases among the burrowing population and either directly or indirectly forced the decline of inoceramids. Progressive depletion of Ce relative to other rare earth elements (REE) in Campanian to Eocene samples at several DSDP sites on the Walvis Ridge and Rio Grande Rise (Liu & Schmitt 1984; Wang *et al.* 1986; Hu *et al.* 1988) has also been interpreted as indicating a broadly contemporaneous increase in bottom-water oxygen in the South Atlantic. There are, however, a number of unconstrained complicating factors in REE

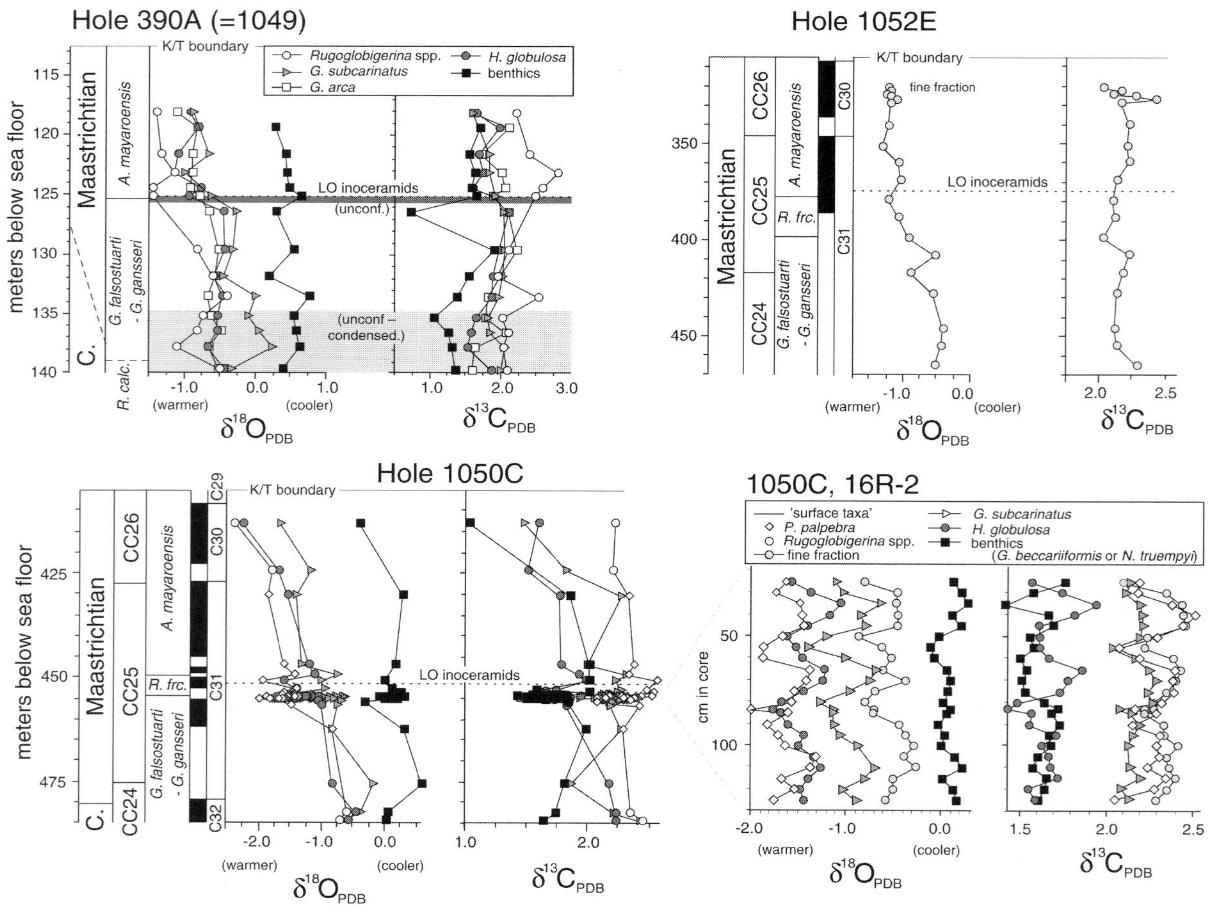

Fig. 3. Stable isotopic results from the Maastrichtian interval of Blake Nose sites. The level of the last occurrence (LO) of inoceramids is shown for each site. Biostratigraphy and magnetostratigraphy from Norris *et al.* (1998), Ogg & Bardot (2001) and J. Self-Trail (pers. comm.). Coring gaps and slumping complicate stratigraphic interpretation, as do multiple reversals in Hole 1050C.

geochemistry that limit utility of Ce anomalies as an independent indicator of changes in Maastrichtian benthic conditions (German & Elderfield 1990; Frank & Arthur 1999).

Inoceramid extinction. The disappearance of inoceramid shell fragments at bathyal depths has been documented in all the world's oceans (MacLeod *et al.* 1996) and their largely Maastrichtian disappearance is a unique and dramatic feature of Late Cretaceous deep-sea cores (MacLeod *et al.* 2000). The globally consistent pattern over thousands of samples (MacLeod *et al.* 1996, 2000; Chauris *et al.* 1998) as well as specific studies examining inoceramid taphonomy (including reworking) support the contention that these shell fragments are usually reliable indicators of local inoceramid occurrences and that reworking can be recognized when present (Saltzman & Barron 1982; MacLeod & Orr 1993; MacLeod 1994*a*; MacLeod & Huber 1996*b*).

The distribution of inoceramid shell fragments in the Blake Nose samples corresponds well to globally established patterns. Inoceramids last occur on the Blake Nose in the lower portion of Chron C31N (mid-Maastrichtian time) and this local disappearance is later than the last occurrence of inoceramids in mid- to high southern latitudes (Chron C31R to Campanian time, respectively) (e.g. MacLeod *et al.* 1996, 2000; Chauris *et al.* 1998). A similar biostratigraphic pattern with higher taxonomic but lower geographical resolution occurs among intact inoceramids (e.g. Dhondt 1992; MacLeod 1994*b*; Voigt 1995; Crame & Luther 1997; Fig. 4).

Diachroneity notwithstanding, the disappearance of bathyal inoceramids appears to represent a coherent global event. After high abundances and stable distributions in the deep sea for ~40 Ma, bathyal inoceramid shell fragments disappeared globally by mid-Maastrichtian time (MacLeod *et al.* 2000). The pattern of decline and disappearance is qualitatively similar in Tethyan, North and South Atlantic, Indian and Southern Ocean sites (MacLeod *et al.* 1996). Inoceramids are absent from a number of sites in the Pacific, but, at sites where they are present, inoceramids disappear well below the K–T boundary (MacLeod *et al.* 1996).

Unfortunately, the ecological tolerances of deep-sea inoceramids are not known well enough to estimate likely palaeoceanographic or palaeoecological differences between inoceramid-bearing and inoceramid-free samples in either time or place (e.g. MacLeod & Hoppe 1992, 1993; Grossman 1993; MacLeod *et al.* 1996). Inoceramid palaeobiogeography and biostratigraphy argue that models for Maastrichtian change should plausibly have effects on the bathyal ocean bottom on a global scale. A single cause with effects that are gradually imposed at any given locality and diachronous across latitudes most simply explains the inoceramid data, but separate proximate causes that resulted in similar patterns of inoceramid decline in different oceans at different times cannot be ruled out.

Oxygen isotopes. Increasing $\delta^{18}O$ values have been documented in planktonic foraminifera through Maastrichtian time at a number of sites, suggesting widespread Maastrichtian cooling (e.g. Douglas & Savin 1975; Boersma & Shackleton 1981; Barrera *et al.* 1987; Barrera & Huber 1990; Barrera 1994; MacLeod & Huber 1996*a*) and perhaps early Maastrichtian glaciation (Barrera & Savin 1999; Miller *et al.* 1999). Thus, the 1–1.5‰ decrease in $\delta^{18}O$ (up to 6 °C warming) in planktonic values on the Blake Nose is surprising (Fig. 3). The decrease in $\delta^{18}O$ is similar in timing, size and absolute value at all three sites. As palaeodepth and burial history vary considerably among these sites, the shift is unlikely to be simply a diagenetic artefact. Additional evidence for regional warming in Maastrichtian time is provided by palaeobotanical evidence from the southern USA (Wolfe & Upchurch 1987) and whole-rock $\delta^{18}O$ results from the European Tethys (e.g. Spicer & Corfield 1992). Further, although tropical Pacific sites show apparent cooling, the increase in $\delta^{18}O$ is <0.5‰ (e.g. Barrera *et al.* 1997) whereas at high latitudes $\delta^{18}O$ values increase up to 1.5‰ (e.g. Barrera & Huber 1990; Barrera & Savin 1999; Frank & Arthur 1999). High-latitude cooling coupled with low-latitude warming or stability suggest that equator-to-pole temperature gradients increased through Maastrichtian time (Barrera & Savin 1999; MacLeod *et al.* 2000).

One intriguing aspect of Maastrichtian $\delta^{18}O$ results on a global scale is relatively low temperatures estimated from low-latitude sites. Blake Nose $\delta^{18}O$ values suggest surface water temperatures 5–10 °C cooler than modern tropical temperatures at the same time as high latitudes were much warmer than at present (D'Hondt & Arthur 1996; Fig. 3). If accurate, the Maastrichtian estimates imply very efficient poleward heat transport (D'Hondt & Arthur 1996). Alternatively, apparent low temperatures might be inaccurate as a result of some combination of diagenetic alteration or salinity effects on $\delta^{18}O_{seawater}$ (e.g. D'Hondt & Arthur 1996; MacLeod *et al.* 2000). If salinity were a major control on $\delta^{18}O$ variation at Blake Nose, the

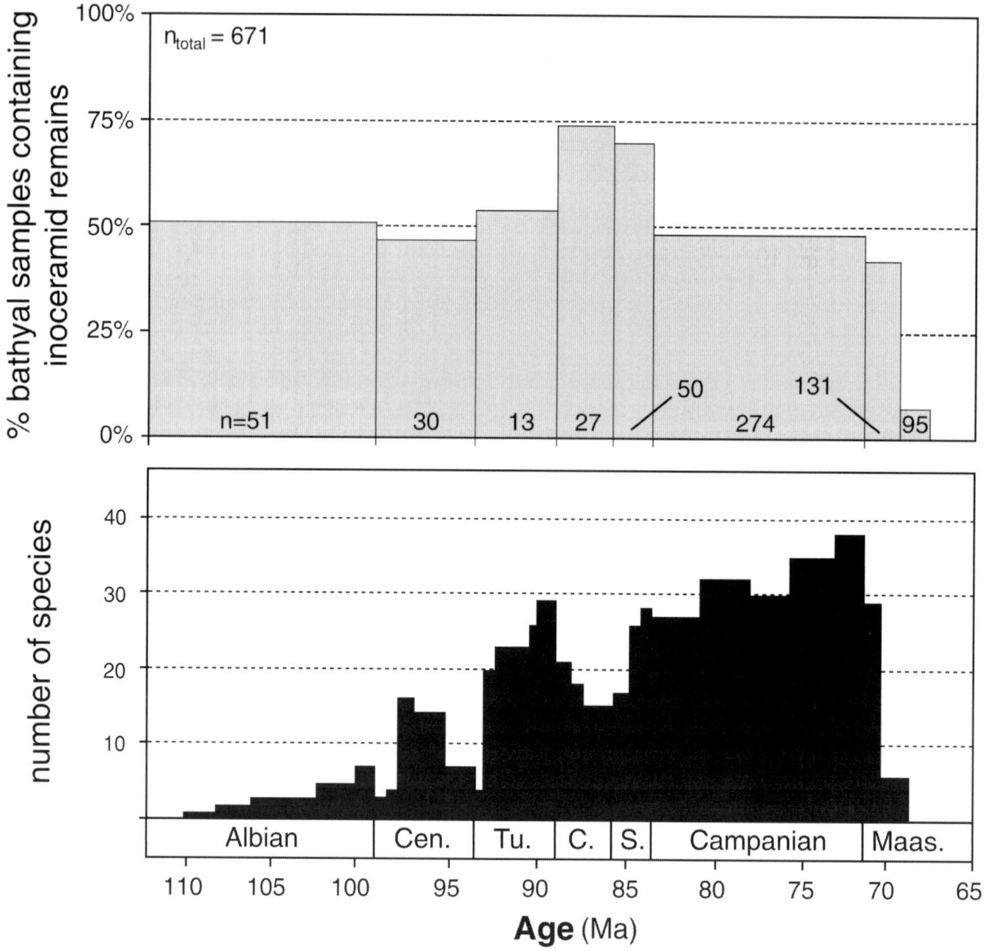

Fig. 4. History of inoceramid diversity and occurrence in bathyal deep-sea sites from Albian to Maastrichtian time (after Voigt 1995; MacLeod *et al.* 2000).

1–1.5‰ $\delta^{18}O$ decrease through Maastrichtian time could reflect up to a 4‰ decrease in surface water salinity (Fig. 5).

Carbon isotopes. A negative ~0.7‰ $\delta^{13}C$ excursion has been reported among benthic foraminifera from sites in the Southern, Indian, South Atlantic and tropical Pacific Oceans (e.g. Barrera & Huber 1990; Barrera 1994; MacLeod & Huber 1996*a*; Barrera *et al.* 1997; Barrera & Savin 1999; Frank & Arthur 1999; Li & Keller 1999). At Sites 689, 690, 525, 752 and 761 the negative $\delta^{13}C$ excursion occurs during Chron C31R; magnetostratigraphy is not available for other sites. At high latitudes ODP Sites 690 and 750 deep-dwelling planktonic taxa show $\delta^{13}C$ trends parallel to the benthic record whereas surface taxa show a smaller excursion (e.g. Barrera 1994; MacLeod & Huber 1996*a*). Subtle shifts among planktonic taxa have also been proposed for the low-latitude Pacific (Barrera 1994; Barrera *et al.* 1997) but are controversial (e.g. MacLeod & Huber 1996*a*; Frank & Arthur 1999). Planktonic $\delta^{13}C$ values at Site 761 from the Indian Ocean show high-frequency variability at the level of the benthic shift without an apparent shift in average values (MacLeod & Huber 1996*a*). At Site 217 the planktonic $\delta^{13}C$ values are fairly stable through the Maastrichtian (Fig. 6). In the North Atlantic neither benthic nor planktonic taxa seem to show an excursion (Frank & Arthur 1999), and results from Blake Nose confirm this conclusion. The single low value reported for Hole 390A, based on a mixed benthic analysis (MacLeod *et al.* 2000*a*), was not reproduced by subsequent analyses (Frank & Arthur 1999) and

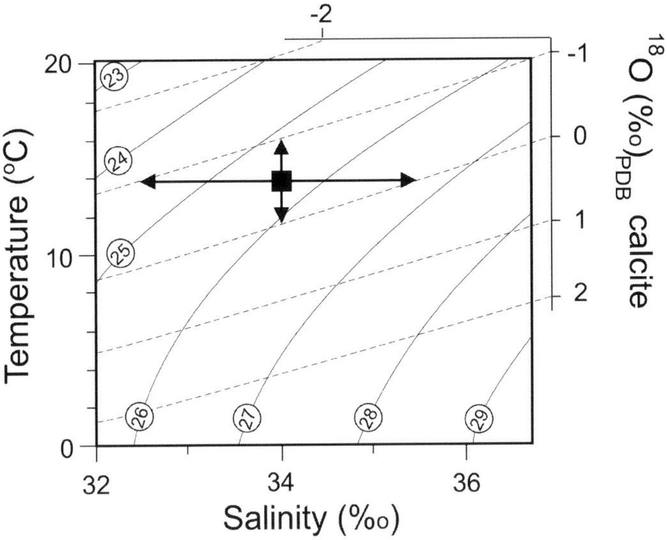

Fig. 5. Temperature–salinity plot contoured in isopycnals (curved continuous lines) with equilibrium $\delta^{18}O$ values shown (dashed line) appropriate for Late Cretaceous oceans (after Woo *et al.* 1992) showing potential salinity (~1.5‰) and temperature (~2 °C) changes (arrows) that could be inferred from a 0.5‰ change in $\delta^{18}O$ values.

low values at Site 1050 are the extremes of cyclic variability best explained by changes in local productivity (MacLeod *et al.* 2001). Thus, although there is a widespread $\delta^{13}C$ excursion that is most commonly documented in benthic taxa, the excursion is not consistently observed across environments and ocean basins.

Maastrichtian history in other environments and groups

As documentation of the Maastrichtian record has improved, models for Maastrichtian change have incorporated observations from an increasing variety of regions, environments and disciplines (e.g. MacLeod 1994*a*; MacLeod *et al.* 1996; Chauris *et al.* 1998; Barrera & Savin 1999; Frank & Arthur 1999; Li & Keller 1999). These changes are an important part of Maastrichtian history and many have not been summarized in other recent papers. In addition, increasing the resolution and reducing uncertainty in some datasets could provide independent tests of different models. Therefore, before discussing models, we briefly summarize additional Maastrichtian studies even though the conclusions are not directly comparable with Blake Nose results.

Rudists. Reef-dwelling rudist bivalves exhibit a well-documented pulse of extinction during mid-Maastrichtian time suggesting significant changes in shallow tropical waters (e.g. Johnson & Kauffman 1990, 1996; Johnson *et al.* 1996; Voigt *et al.* 1999). One cool-water genus (*Gyropleura*) survives to the K–T boundary (e.g. Heinberg 1979; Johnson & Kauffman 1990), but extinctions of reef-dwelling taxa occur much earlier. Caribbean rudist extinctions are considered abrupt and are concentrated in the mid-Maastrichtian near the base of the *A. mayaroensis* planktonic foraminiferal zone (Johnson & Kauffman 1990; 1996; Johnson *et al.* 1996). In the Mediterranean region rudists also range into the lower Maastrichtian (Voigt *et al.* 1999). As discussed by Johnson & Kauffman (1996), a high degree of endemism, differences in the apparent distribution of individuals v. reefal build-ups, a scarcity of biostratigraphically diagnostic taxa, and inconsistencies between macrofossil- and microfossil-based correlations complicate attempts to precisely estimate the timing of rudist extinctions. Depending on the method of correlation and the rudist occurrences discussed, rudist extinctions can be characterized as having occurred between 1 and 3 Ma before the K–T boundary (Johnson & Kauffman 1990, 1996; Johnson *et al.* 1996). Until precise and accurate absolute age estimates are available, incorporating rudist-based insights concerning changes in tropical carbonate platforms into global models must be qualified to the extent that

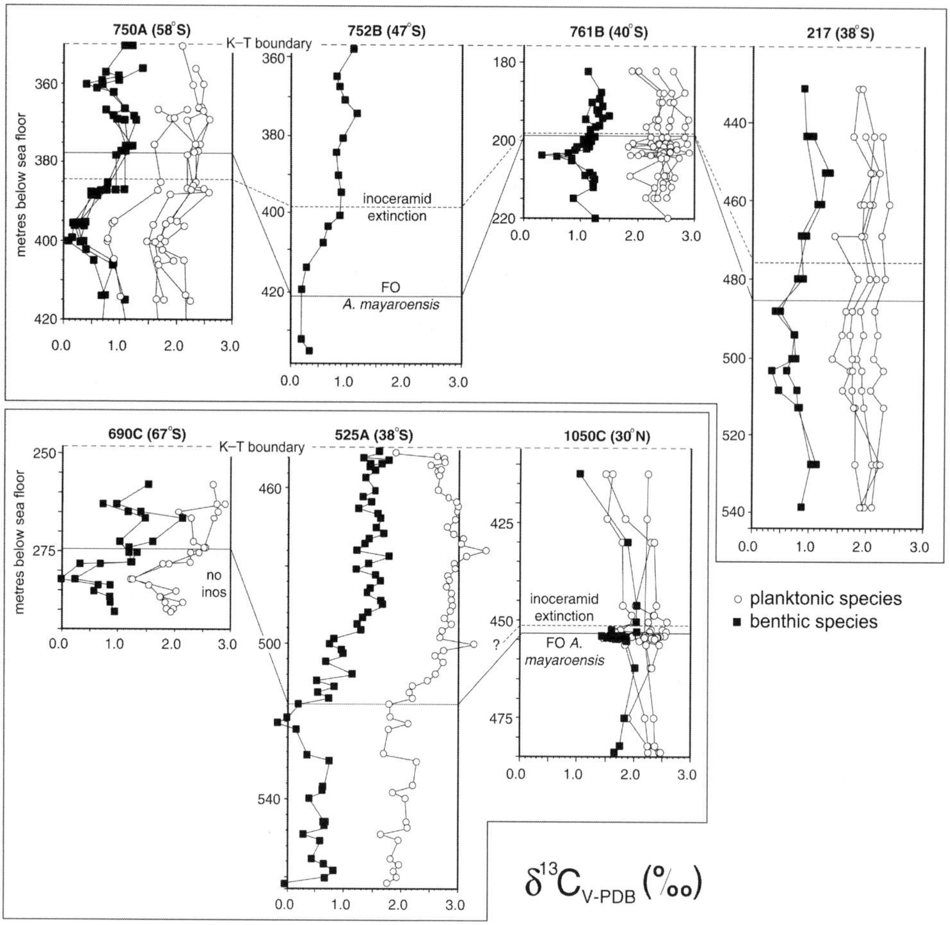

Fig. 6. Plot of $\delta^{13}C$ records for a number of Maastrichtian sequences showing results for a benthic and surface-dwelling planktonic taxa. Horizontal and vertical scales are the same in all plots. Palaeolatitude estimates shown for each site are based on a palaeogeographical reconstruction for 70 Ma by Hay *et al.* (1999). A $\delta^{13}C$ excursion is well expressed at southern high latitudes in both benthic and planktonic taxa but is less well expressed to absent in planktonic taxa in lower-latitude sites. An excursion seems to be absent in both benthic and planktonic taxa in the North Atlantic (see Frank & Arthur (1999) for additional documentation). Data sources: Barrera & Huber (1990) for 690C; Seto *et al.* (1991) for 752B; Barrera (1994) and MacLeod & Huber (1996*a*) for 750A and 761B; MacLeod & Huber (1995) for 217; Li & Keller (1998) for 525; this study for 1050C.

models depend on precise chronostratigraphic correlation.

Benthic foraminifera. Benthic foraminifera could provide important independent tests of the nature of Maastrichtian bathyal changes that have been inferred from benthic $\delta^{13}C$ and $\delta^{18}O$ records, sedimentological changes, and inoceramid distributions. To date, however, changes in deep-sea benthic foraminifera through the Maastrichtian have been documented in only two regions: Maud Rise in the Southern Ocean (Thomas 1990) and Rio Grande Rise in the subtropical South Atlantic (Dailey 1983). The shift in benthic assemblages at Maud Rise occurred at the top of Chron C31R and may be related to changes in food supply (Thomas 1990). Unfortunately, inoceramids do not occur there (MacLeod *et al.* 1996). In the Rio Grande Rise study, sampling density is low. There is a mid-Maastrichtian increase in the relative abundance of *Gavelinella* spp. when only Maastrichtian samples are considered, but the change is no larger than similar shifts occurring in samples from below the Maastrichtian sequence (Dailey 1983).

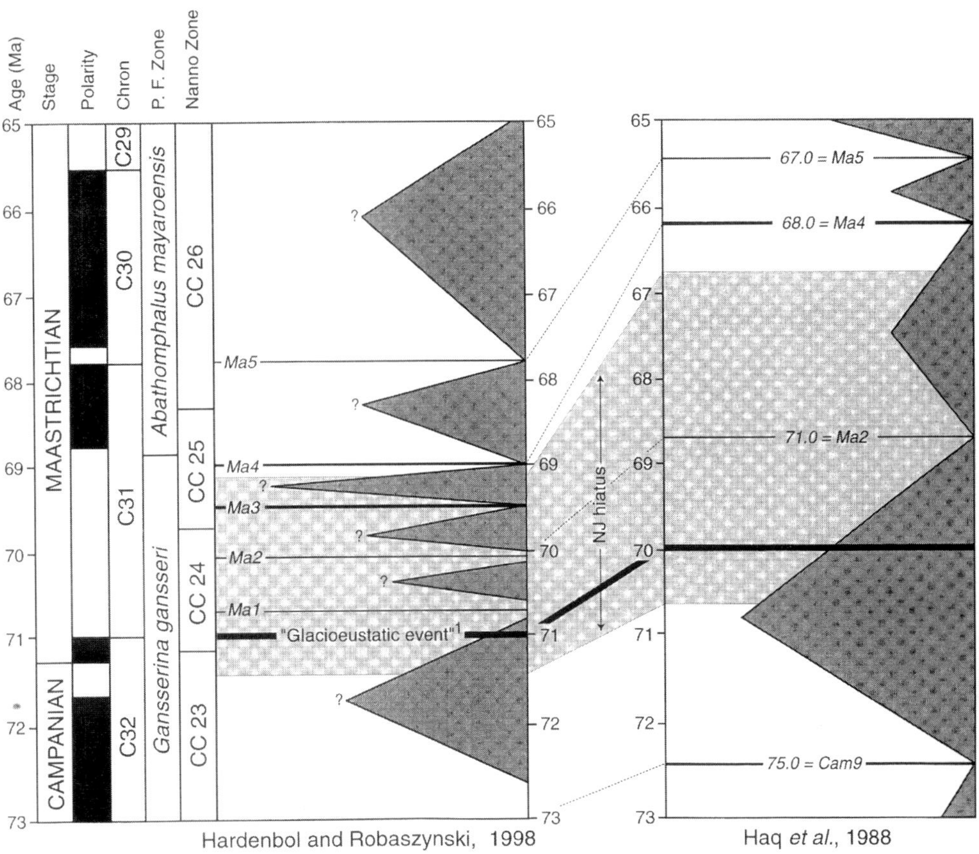

Fig. 7. Sequence chronostratigraphy of Hardenbol & Robaszynski (1998) and sequence stratigraphy of Haq *et al.* (1988) relative to the Gradstein *et al.* (1995) time scale. Vertical axis for Haq *et al.* (1988) curve approximates time on the Gradstein *et al.* (1995) time scale recalibrated by Hardenbol & Robaszynski (1998). Correlation between curves proposed by Hardenbol & Robaszynski (1998) shown as dashed lines between curves (see text). Horizontal lines in each plot represent sequence boundaries, with decreasing thickness of lines denoting major, moderate and minor sequence boundaries, respectively. Left pointing, shaded triangles represent sea-level highstands; their width depicts relative magnitude. '?' in the Hardenbol & Robaszynski (1998) plot indicates that the timing of the maximum highstands is uncertain. The 'Glacioeustatic event' and the estimated time range of hiatus discussed by Miller *et al.* (1999) scale are projected onto the Haq *et al.* (1988) from the Gradstein *et al.* (1995) time scale based on the proposed correlation of sequence boundaries, and do not seem to correlate well with sea-level change inferred from either curve. Recalibrated ages provide worse correspondence between purported glacioeustatic lowstand and the Haq *et al.* (1988) curve.

Planktonic and nektonic organisms. Planktonic foraminiferal and calcareous nannofossil assemblages exhibit different patterns between groups and across latitudes. Beginning in lowermost Chron C31R, a poleward migration of several planktonic foraminiferal species led to increased foraminiferal diversity in the southern high latitudes (Huber 1990; MacLeod *et al.* 2000). From Chron C31R to the end of the Maastrichtian, austral nannofossil diversity declined from 53 species to 20 species (Huber & Watkins 1992). In addition, ~1 Ma after the poleward migrations of foraminifera began, both the planktonic foraminifer *A. mayaroensis* and the calcareous nannofossil *Nephrolithus frequens* extended their range equatorward (Huber & Watkins 1992; Bergen & Sikora 1999). Equatorward range expansion and the decline in nannofossil diversity could be related to mid- to high-latitude cooling but the earlier poleward migrations of planktonic foraminifera are not well understood (Huber & Watkins 1992).

At lower latitudes a pulse of evolutionary diversification occurred among planktonic foraminiferal assemblages during mid- to late Maastrichtian time. At least nine new species first appeared in a narrow interval beginning at ~70 Ma, near the base of the *A. mayaroensis* Zone (e.g. Li & Keller 1998; MacLeod *et al.* 2000). As with the poleward migration of planktonic taxa, it is not clear what could have caused the diversification (MacLeod *et al.* 2000). There are no correlative extinctions that might indicate speciation is part of biotic overturn in the planktonic realm, and evidence for increased stratification of subtropical surface waters leading to the opening of new niche space is contradictory (e.g. compare planktonic $\delta^{13}C$ and $\delta^{18}O$ results for Blake Nose sites, Fig. 3).

Palaeontological data that can be compared with foraminiferal and nannofossil palaeobiogeography are not available for other microfossil and fossil invertebrate groups. In five sections from northern Spain and southwestern France ammonite diversity and abundance appear relatively constant through Maastrichtian time (Ward & Kennedy 1993; MacLeod *et al.* 1996), but global trends in the distribution of ammonites have not been compiled. Similarly, the Maastrichtian distribution of dinoflagellates and siliceous microfossils from deep-sea sites is too poorly known to discern whether or not significant assemblage changes or latitudinal migrations occurred during the mid-Maastrichtian.

Terrestrial ecosystems. The Maastrichtian terrestrial record in North America has been studied across latitudes and groups and suggests considerable regional variations in climatic and biotic trends. Among the floral communities in Wyoming, Montana and North Dakota, there is a shift from an angiosperm–gymnosperm co-dominated assemblage in the lower Maastrichtian Meteesee Formation to an angiosperm dominated assemblage in the upper Maastrichtian Hell Creek Formation, but this transition has not been studied in continuous sections (Johnson & Hickey 1990, 1992; K. Johnson, pers. comm.). In the same general region there are apparent decreases in dinosaur diversity during the Campanian–Maastrichtian (e.g. Clemens 1986; Sloan *et al.* 1986; Archibald 1996), but comparison of diversity within similar facies showed constant within-environment dinosaur diversity through the Maastrichtian followed by a sudden extinction at the K–T boundary (e.g. Sheehan *et al.* 1991; Sheehan & Fastovsky 1992). Floral assemblages from the North Slope of Alaska suggest mean annual temperatures of ~6 °C during the Maastrichtian, 5–8 °C cooler than earlier in the Cretaceous (e.g. Spicer & Parrish 1990; Herman & Spicer 1996), whereas leaf physiognomy indicates 3–5 °C of warming during the early Maastrichtian in the southeastern USA (Wolfe & Upchurch 1987).

$^{87}Sr/^{86}Sr$. Seawater $^{87}Sr/^{86}Sr$ increases during most of the Maastrichtian (e.g. Nelson *et al.* 1991; Barrera 1994; McLaughlin *et al.* 1995; Sugarman *et al.* 1995; Howarth & McArthur 1997). A mid-Maastrichtian increase in the slope of the $^{87}Sr/^{86}Sr$ curve occurs at Bidart (southwestern France, Nelson *et al.* 1991) and Site 690 (Southern Ocean, Barrera 1994). On the other hand, the data for Site 525 (South Atlantic) and a global $^{87}Sr/^{86}Sr$ curve suggest a relatively constant slope through the Maastrichtian (Sugarman *et al* 1995; Howarth & McArthur 1997). Alternative interpretations of these results include a relative increase in continentally derived Sr to oceans during the Maastrichtian (Nelson *et al.* 1991; Barrera 1994) that was not resolved in other studies, or a relatively constant rate of increase in seawater $^{87}Sr/^{86}Sr$ distorted by changes in sedimentation rates at Bidart and Site 690.

Maastrichtian sea level. Sea level figures directly or indirectly in many Maastrichtian models, but estimates of Maastrichtian sea-level history vary widely in the relative timing, magnitude, direction, and cause of proposed transgressive and regressive events. Differences in inferred sea-level histories could result from local tectonic factors (e.g. basinal uplift or subsidence), differing interpretations of geophysical and sedimentological data, and the accuracy and resolution of chronostratigraphic correlation (Lewy 1990; Hallam 1992; Miall 1992; Hancock 1993). Comparison of recent Maastrichtian sea-level curves illustrates the uncertainty in details of the Maastrichtian sea-level record complicated by evolving Maastrichtian time scales (Fig. 7). Haq *et al.* (1988) showed three sequence boundaries in the Maastrichtian, whereas at least five sequence boundaries have been suggested in more recent studies of sequences in the US Gulf coast and northwestern Europe (Hardenbol & Robaszynski 1998) and Tunisia (Li & Keller 1999). Haq *et al.* (1988) proposed that the three recognized sequence boundaries represented eustatic events with maximum regression at 71 Ma, 68 Ma and 67 Ma based on correlation with the Harland *et al.* (1982) time scale. The boundary at 68 Ma was considered major and the other two minor. A recalibration of the Haq *et al.* (1988) curve to the Gradstein *et al.* (1995) time scale suggests ages of 68.8 Ma, 66.2, and 65.4 Ma, respectively, for the three events (Hardenbol & Robaszynski 1998). Using the Gradstein *et al.* (1995) timescale and assuming the proposed correlations between the Hardenbol and Robaszynski (1998) and the Haq *et al.* (1988) curves are accurate and isochronous yields estimated ages of 70.2 Ma, 69.0 Ma and 67.8 Ma for the Haq *et al.* (1988) sequence boundaries with an additional major and an additional minor sequence boundary between the older two.

Despite these uncertainties, the specific transgressive and regressive events identified and dated in the eustatic curve of Haq *et al.* (1988) continue to be invoked to provide a causal link among climatic, oceanic and biotic changes during Maastrichtian time (e.g. Johnson *et al.* 1996; Barrera & Savin 1999; Miller *et al.* 1999). Sea level may be an important factor in the evolution of the Maastrichtian ocean–climate system. However, details of regional to global Maastrichtian sea-level history, as well as chronostratigraphic correlation between sea-level events and other data, are at present poorly known on short time scales.

Maastrichtian models

Given the range of data being integrated and the problems inherent in correlating across large distances and different environments, there is considerable agreement among different Maastrichtian studies. For example, high-latitude cooling (inferred from $\delta^{18}O$ results, foraminiferal and nannofossil distributions, and palaeobotanical data) is well established, and most recently proposed models invoke changes in ocean circulation to explain the Maastrichtian record. Within this palaeoceanographic framework, different studies have advanced a variety of possible causes and effects (e.g. regression, cooling, location(s) of deep-water formation, contraction or elimination of ecospace, productivity, glaciation (Barrera 1994; MacLeod 1994*a*; Johnson *et al.* 1996; MacLeod & Huber 1996*a*; Barrera *et al.* 1997; Chauris *et al.* 1998; Barrera & Savin 1999; Frank & Arthur 1999; Li & Keller 1999; Miller *et al.* 1999)), but models addressing global changes can be lumped into three groups: (1) more or less isochronous changes forced by subsidence of topographic barriers to intermediate water circulation (Frank & Arthur 1999); (2) isochronous changes forced by climate and sea-level changes, the latter perhaps glacioeustatic (Barrera *et al.* 1997; Barrera & Savin 1999; Li & Keller 1999; Miller *et al.* 1999); (3) diachronous changes forced by climatic changes and perhaps sea-level change resulting in a declining importance of mid- to low-latitude saline water masses through the Maastrichtian (MacLeod *et al.* 1996; 2000).

Isochronous model 1: circulation and subsidence. On the basis of a global compilation of new and published benthic $\delta^{13}C$ values, the subsidence model divided the Maastrichtian into three time slices (Frank & Arthur 1999). To minimize diagenetic and biological differences, isotopic data were culled so that only results from selected taxa of well-preserved foraminifera were considered. In time slice 1 (early Maastrichtian) benthic $\delta^{13}C$ values are variable but suggest a decrease from the Pacific and Indian Ocean sites to the southern South Atlantic sites, and show higher values in Atlantic samples from north of the Rio Grande Rise–Walvis Ridge than in the Pacific. This pattern was interpreted to indicate a source of intermediate to deep water in the North Pacific that reached the southern South Atlantic and a second location of deep-water formation in the North Atlantic. During time slice 2, differences in $\delta^{13}C$ among available sites are small, which was interpreted as an indication that the world's oceans were better mixed. Finally, during time 3, relatively low Pacific benthic $\delta^{13}C$ values were proposed to reflect circulation from the Atlantic to the Pacific similar to modern circulation patterns, but high benthic $\delta^{13}C$ in both North and South Atlantic sites precluded proposing a single source region in the Atlantic. Subsidence and breaching of ridges in the South Atlantic was discussed as the primary control on these changing circulation patterns (Frank & Arthur 1999).

The opening or closing of gateways to circulation could be an important palaeoceanographic factor in general (e.g. Maier-Reimer *et al.* 1990; Jackson *et al.* 1993; Mikolajewicz *et al.* 1993) and has long been considered a significant part of the geologic history of the South Atlantic (e.g. Van Andel *et al.* 1977). As discussed by Frank & Arthur (1999), the relative abundance of evaporites, organic-rich shales, and hemipelagic to pelagic carbonates change markedly in the South Atlantic from Early Cretaceous time to early Tertiary time (e.g. Van Andel *et al.* 1977), and economic phosphorite deposits, which are common and widespread in Campanian to lower Eocene deposits, are relatively rare in middle Maastrichtian to middle Palaeocene strata (Cook & McElhinny 1979; Arthur & Jenkyns 1981; Frank & Arthur 1999). These sediment distributions and other factors led previous researchers to conclude that as east–west ridges (e.g. the Walvis Ridge–Rio Grande Rise) subsided, surface-water connections between the North and South Atlantic formed in the Cenomanian at ~95 Ma (Reyment & Tait 1972) and intermediate- then deep-water connections formed from the Campanian to the Eocene between 80 and 55 Ma (e.g. Van Andel *et al.* 1977).

The subsidence model invokes the same process, but uses $\delta^{13}C$ and $\delta^{18}O$ data to propose a precise estimate of the timing and nature of changes in global circulation patterns, as well as a more specific series of consequences related to those changes (Frank & Arthur 1999). The model also invokes regression and tectonism to explain a variety of additional aspects of the Maastrichtian record, including climate change, weathering (reflected by increasing $^{87}Sr/^{86}Sr$ ratios) and palaeontological shifts. The timing of changes (e.g. mid-Maastrichtian rudist reef collapse at $\leqslant$68 Ma v. $^{87}Sr/^{86}Sr$ change at up to 71 Ma) and the considerable diachroneity of other events (e.g. inoceramid extinction and planktonic palaeobiogeographical shifts) are not explored. Variation in geographical coverage of isotopic results among time slices is another possible limitation of this analysis. For example, the most extreme values in time slice 1 (both high and low) come from sites that contained no samples representing time slice 2. If the compilation is limited to data from sites with samples in both time 1 and time 2, the difference between the two intervals becomes less pronounced but does not disappear.

A more significant shortcoming of this model, however, is the expected rate of subsidence. Aseismic features such as the Walvis Ridge and Rio Grande Rise subside at a similar rate to oceanic crust, i.e. $\text{depth}_t \propto 350 \times t^{1/2}$, where t is time (in millions of years) since formation (e.g. Detrick *et al.* 1977; Sclater *et al.* 1977). The Walvis Ridge–Rio Grande Rise system may be as old as 130 Ma and was an important South Atlantic feature by 100 Ma (e.g. Van Andel *et al.* 1977). At 70 Ma the older and deeper parts of the ridge system would conservatively have been 25–50 Ma old and subsiding at a rate of only 25–35 m Ma^{-1}. Thus, although subsidence is likely to be a controlling factor in Cretaceous palaeoceanography on long time scales, on the time scale of Maastrichtian events, the height of topographic barriers seems more likely to be a boundary condition than a variable.

Isochronous model 2: circulation and climate–sea level. Benthic $\delta^{13}C$ gradients also figure prominently in this model. Details differ among studies, but are generally not contradictory (Barrera *et al.* 1997; Barrera & Savin 1999; Li & Keller 1999; Miller *et al.* 1999). Isotopic results discussed are mostly derived from a smaller number of more complete sections than used in the subsidence model. A larger number of taxa are also considered and a correction factor is applied to $\delta^{13}C$ results for some benthic taxa. More specific conclusions regarding inter- and intrasite comparisons (e.g. Li & Keller (1999) proposed five time steps separated by transitional intervals) are reached and different chronostratigraphic assumptions are used (e.g. compare data plotted for Site 750 by Barrera & Savin (1999) and by Frank & Arthur (1999)).

Despite these differences, inferred circulation patterns in the two isochronous models bear notable similarities. In model 2, an early Maastrichtian source region for intermediate waters in the North Pacific was proposed based on the benthic $\delta^{13}C$ gradients between Pacific and Southern Ocean sites and the relatively high benthic $\delta^{18}O$ values in the tropical Pacific. High benthic $\delta^{13}C$ values at North Atlantic DSDP Site 384 (Barrera & Savin 1999) were interpreted as indicative of localized intermediate-water formation isolated from Pacific-sourced water masses (cf. time slice 1 of the subsidence model). Before and after this time, high southern latitudes were interpreted to be closer to the area of intermediate-water formation than Pacific sites (cf. time slice 3 of the subsidence model) with the importance of the North Atlantic source declining in later Maastrichtian time (but see Li & Keller (1999) for mid-Maastrichtian time). Unlike the subsidence model, circulation in the climate model explicitly includes a warm, saline, North Atlantic-sourced water mass and a latest Maastrichtian interval of warming of surface and deep waters.

Climate and sea-level change (rather than subsidence) were forwarded as the primary causes of changes in Maastrichtian circulation patterns in this model. Cooling at high latitudes is well established, and, the uncertainty discussed above notwithstanding, sea-level change in Maastrichtian time has been shown in many studies (e.g. Haq *et al.* 1988; Sugarman *et al.* 1995; Hardenbol & Robaszynski 1998). As in the subsidence model, shifts in the CCD in different ocean basins are consistent with reorganization of ocean circulation patterns; weathering of terrigenous material on coastal plains exposed by sea-level drop could explain the proposed increased flux of ^{87}Sr to the oceans during early Maastrichtian time (e.g. Barrera 1994; MacLeod 1994*a*). The benthic $\delta^{13}C$ excursion seen at a number of sites might similarly be explained by weathering of continental organic matter on continental shelf exposed by regression. Alternatively, changes in productivity and the location of the oxygen minimum zone may have varied locally to regionally as ocean circulation changed as suggested by cyclicity at Site 1050 (Fig. 3).

Although palaeontological changes are not emphasized, the climate model does not simply explain differences in timing of events within and among groups. Shortcomings of the available data include limited coverage of isotopic records relative to detailed regional and bathymetric conclusions and uncertainty regarding Maastrichtian sea-level history (Fig. 7). An interesting feature of some studies grouped in this model is the proposal that a small ice sheet formed during early Maastrichtian time (Barrera *et al.* 1997; Barrera & Savin 1999; Miller *et al.* 1999). Purported temporal correlation among the '71 Ma' eustatic regression (Haq *et al.* 1988; Fig. 7), a sequence boundary in New Jersey dated using $^{87}Sr/^{86}Sr$ (e.g. Sugarman *et al.* 1995; Miller *et al.* 1999), and global cooling inferred from an increase in foraminiferal $\delta^{18}O$ values (e.g. Barrera & Huber 1990; Barrera *et al.* 1997; Barrera & Savin 1999; Miller *et al.* 1999) are the principal evidence cited in favour of glacial ice.

Uncertainty in the sea-level curve and interpretation of $\delta^{18}O$ data are particularly important in evaluating the likelihood of Maastrichtian ice. Miller *et al.* (1999) assumed constant tropical temperatures and used tropical Pacific $\delta^{18}O$ values of planktonic taxa to estimate changes in $\delta^{18}O_{seawater}$ and thereby calculate benthic palaeotemperatures. Blake Nose planktonic $\delta^{18}O$ data do not match Pacific trends or values, but even assuming $\delta^{18}O_{seawater}$ is accurately known, the difference between the estimated palaeotemperatures of 'glacial' Maastrichtian deep water (5–7 °C) and the late Maastrichtian bottom-water temperature of 8 °C (deemed too warm for significant ice accumulation, Miller *et al.* 1999) seems small, especially when potential error in the $\delta^{18}O$ palaeothermometer is considered (e.g. Crowley & Zachos, 2000). Maastrichtian glacial deposits have not been identified at either pole, despite considerable study of many high-latitude cores and near-complete exposure of intensively studied *c.* 120 km^2 of Maastrichtian shallow-marine sediments on Seymour Island, Antarctica (e.g. Askin 1988; Huber 1988; Macellari 1988). Bice (1999) discussed how a small continental ice sheet might not leave much geological evidence except the record of eustatic change and subtle changes in $\delta^{18}O_{seawater}$. Thus, it is difficult to disprove that a small ice sheet existed during Maastrichtian time. However, if sea-level history and $\delta^{18}O$ trends are unclear and if sea-level and $\delta^{18}O$ variations occurred in portions of the Cretaceous agreed to be ice free (Miller *et al.* 1999), there seems to be little support for a Maastrichtian ice sheet.

Diachronous model: circulation and inoceramid extinction. Except for glacioeustacy, the diachronous model invokes similar processes as model 2, with cooling, especially at high latitudes, and reduction of mid- to low-latitude shallow seas as the primary forcing mechanisms for circulation changes during Maastrichtian time. As in the subsidence model, the diachronous model also has advanced the idea that some aspects of climate change (e.g. increased latitudinal temperature gradients) may be partially a result of differences in ocean heat transport brought about by changing circulation patterns. Because the same causes are forwarded, the interpretation of most data in the diachronous model follows that outlined above (e.g. MacLeod 1994*a*; MacLeod *et al.* 1996, 2000). The model differs, however, in emphasizing the progressive and diachronous nature of palaeontological changes and the potential importance of warm, saline intermediate- to deep-water masses formed in low to mid-latitudes.

A low-latitude source region for some intermediate- to deep-water masses during globally warm intervals has long been hypothesized (e.g. Chamberlain 1906; Brass *et al.* 1982; Hay 1988; MacLeod *et al.* 2000). High bottom-water temperatures relative to the modern values are often cited as evidence for saline water masses, but this observation could also reflect decreased latitudinal temperature gradients and elevated temperatures in high-latitude source regions (e.g. Bice *et al.* 2000). The least equivocal evidence for warm, saline deep waters is a reversal with depth in expected $\delta^{18}O$ gradients. In two intervals in the tropical Pacific, late Campanian–early Maastrichtian benthic $\delta^{18}O$ values from a relatively deep site are reported as lower than at two nearby shallower sites (Barrera *et al.* 1997). In addition, at Sites 750 and 761, early Maastrichtian benthic $\delta^{18}O$ values are similar to or slightly lower than $\delta^{18}O$ values of co-occurring deep-dwelling planktonic foraminifera but higher than all planktonic values in late Maastrichtian time (MacLeod and Huber 1996*a*).

Decreasing $\delta^{18}O$ values with increasing depth requires that the warmer deeper waters also had elevated salinity, and such water masses would presumably have originated in mid- to low latitudes. During mid-Maastrichtian time, benthic $\delta^{18}O$ values increased at a number of sites, indicating either cooling in the source region or a shift to a new high-latitude source for intermediate waters. At Sites 750 and 761 the increase in benthic $\delta^{18}O$ is associated with the disappearance of inoceramids and the establishment of a normal planktonic–benthic $\delta^{18}O$ gradient (MacLeod & Huber 1996*a*). At

Site 690 in the Southern Ocean, the increase corresponds to a convergence of planktonic and benthic $\delta^{18}O$ values, perhaps signalling the initiation or intensification of intermediate- to deep-water formation in this region (e.g. Barrera & Huber 1990; Barrera 1994; MacLeod & Huber 1996*a*; Barrera *et al.* 1997; Barrera & Savin 1999; Frank & Arthur 1999; Miller *et al.* 1999). Similarly, divergence of deep planktonic and benthic $\delta^{18}O$ gradients in late Maastrichtian time on Blake Nose could signal the decline or cessation of winter downwelling of saline waters. However, although analyses of the shallowest site (1052) are not complete, benthic $\delta^{13}C$ and $\delta^{18}O$ values on Blake Nose show no apparent change as might be expected if there were a change in the source of local bottom waters.

On a global scale the biggest shortcoming of the diachronous model is the lack of consistent stable isotopic evidence for a change in bottom-water masses associated with the disappearance of inoceramids. Constant benthic $\delta^{13}C$ and $\delta^{18}O$ seen at Blake Nose sites (Fig. 3) and at Sites 327 and 511 in the South Atlantic (MacLeod *et al.* 2000) do not support the contention that bathyal inoceramids were consistently living within saline water masses. Inoceramid abundance at Blake Nose and Sites 327 and 511 is correlated with changes in isotopic gradients among foraminifera, suggesting changes in the water column structure (MacLeod *et al.* 2000; Fig. 3), but the original hypothesis that an equatorward contraction of bathyal regions bathed by warm, saline, water masses is tracked by the diachronous disappearance of inoceramids (e.g. MacLeod 1994*a*; MacLeod & Huber 1996) is clearly too simple. Either inoceramids had regionally specific extinction mechanisms or changes in the distribution of different water masses influenced a number of other factors (e.g. vertical cycling of food or nutrients, oxygen concentrations) that might explain palaeontological patterns and allow for additional complexity.

Conclusions

The Maastrichtian age was a dynamic time. Biological, geochemical and sedimentological shifts occurred in many environments. Although disagreement remains regarding the nature, timing, cause and effects of various Maastrichtian events, consensus seems to be building that shifts in source regions and circulation patterns of intermediate to deep water are directly or indirectly related to most Maastrichtian events. Possible causes for changing circulation patterns include (1) decline or cessation of formation of warm saline water masses; (2) subsidence of topographic barriers to circulation; (3) cooling and formation of glacial ice and associated sea-level change. Possible effects include (1) biological shifts in planktonic, benthic and terrestrial biota; (2) changes in nutrient cycling and productivity; and (3) climate change. At present, different models best explain some aspects of the global dataset, but each also fails to explain some features of the Maastrichtian record. It should be noted that the models are not mutually exclusive, and there is considerable overlap among them. A complete understanding of Maastrichtian evolution, however, will require determining the relative importance of different processes in different regions and at different times. Achieving this goal will require better control on temporal changes in location of intermediate- to deep-water formation, on ocean circulation patterns, and on the relationship of those circulation patterns to Maastrichtian climate and surface and benthic ecology.

We thank the Ocean Drilling Program for providing samples; E. Barrera, C. C. Johnson, A. Nederbragt and R. D. Norris for critical reviews; M. Hogan, P. Koch and J. Zachos for analytical assistance; and the US Science Advisory Council for funding.

References

ARCHIBALD, D. J. 1996. *Dinosaur Extinction and the End of an Era; What the Fossils Say*. Columbia University Press. New York.

ARTHUR, M. A. & JENKYNS, H. C. 1981. Phosphorites and palaeoceanography. *Oceanologica Acta*, **4**, 83–96.

ARTHUR, M. A., DEAN, W. E. & SCHLANGER, S. O. 1985. Variations in the global carbon cycle during the Cretaceous related to climate, volcanism, and changes in atmospheric CO_2. *In*: SUNDQUIST, E. T. & BROECKER, W. S. (eds) *The Carbon Cycle and Atmospheric CO_2; Natural Variations Archean to Present*. Geophysical Monograph, American Geophysical Union, **32**, 504–529.

ASKIN, R. A. 1988. Campanian to Eocene palynological succession of Seymour and adjacent islands, northeastern Antarctic Peninsula. *In*: FELDMANN, R. M. & WOODBURNE, M. O. (eds) *Geology and Paleontology of Seymour Island, Antarctic Peninsula*. Geological Society of America, Memoirs, **169**, 131–153.

BARRERA, E. 1994. Global environmental changes preceding the Cretaceous–Tertiary boundary: early–late Maastrichtian transition. *Geology*, **22**, 877–880.

BARRERA, E. & HUBER, B. T. 1990. Evolution of Antarctic waters during the Maestrichtian: foraminifer oxygen and carbon isotope ratios, ODP Leg 113. *In*: BARKER, P. F. & KENNETT, J. P. (eds) *Proceedings of the Ocean Drilling Program,*

Scientific Results, **113**, Ocean Drilling Program, College Station, TX, 813–823.

BARRERA, E. & SAVIN, S. M. 1999. Evolution of Campanian–Maastrichtian marine climates and oceans. *In*: BARRERA, E. & JOHNSON, C. C. (eds) *Evolution of the Cretaceous Ocean–Climate System*. Geological Society of America, Special Paper, **332**, 245–282.

BARRERA, E., HUBER, B. T., SAVIN, S. M. & WEBB, P. N. 1987. Antarctic marine temperatures: late Campanian through early Palaeocene. *Paleoceanography*, **2**, 21–47.

BARRERA, E., SAVIN, S. M., THOMAS, E. & JONES, C. E. 1997. Evidence for thermohaline-circulation reversals controlled by sea-level change in the latest Cretaceous. *Geology*, **25**, 715–718.

BARRON, E. J. 1987. Global Cretaceous palaeogeography–International Geologic Correlation Program Project 191. *Palaeogeography, Palaeoclimatology, Palaeoecology*, **59**, 207–216.

BERGEN, J. A. & SIKORA, P. J. 1999. Microfossil diachronism in southern Norwegian North Sea chalks: Valhall and Hod fields. *In*: JONES, R. W. & SIMMONS, M. D. (eds) *Biostratigraphy in Production and Development Geology*. Geological Society, London, Special Publications, **60**, 85–111.

BICE, K. 1999. The climate model case for ice in the greenhouse. *Geological Society of America, Abstracts with Programs*, **31**, A311.

BICE, K., SLOAN, L. & BARRON, E. 2000. An early Eocene ocean general circulation model simulation: comparison of the three-dimensional temperature field with isotopic palaeotemperatures. *In*: HUBER, B. T., MACLEOD, K. G. & WING, S. L. (eds) *Warm Climates in Earth History*. Cambridge University Press, Cambridge, 79–131.

BOERSMA, A. & SHACKLETON, N. J. 1981. Oxygen and carbon isotope variations and planktonic foraminiferal depth habitats: Late Cretaceous to Palaeocene, Central Pacific, DSDP Sites 463 and 465, Leg 65. *In*: THIEDE, J. & VALLIER, T. L. (eds) *Initial Reports of the Deep Sea Drilling Project*, **65**. US Government Printing Office, Washington, DC, 513–526.

BRASS, G. W., SOUTHAM, J. R. & PETERSON, W. H. 1982. Warm saline bottom water in the ancient ocean. *Nature*, **296**, 620–623.

CHAMBERLIN, T. C. 1906. On a possible reversal of deep-sea circulation and its influence on geologic climates. *Journal of Geology*, **14**, 363–373.

CHAURIS, H., LEROUSSEAU, J., BEAUDOIN, B., PROPSON, S. & MONTONARI, A. 1998. Inoceramid extinction in the Gubbio basin (northeastern Apennines of Italy) and relations with mid-Maastrichtian environmental changes. *Palaeogeography, Palaeoclimatology, Palaeoecology*, **139**, 177–193.

CLEMENS, W. A. 1986. Evolution of the terrestrial vertebrate fauna during the Cretaceous–Tertiary transition. *In*: ELLIOT, D. K. (ed.) *Dynamics of Extinction*. Wiley, New York, 63–85.

COOK, P. J. & MCELHINNY, M. W. 1979. A reevaluation of the spatial and temporal distribution of sedimentary phosphate deposits in the light of plate tectonics. *Economic Geology and the Bulletin of the Society of Economic Geologists*, **74**, 315–330.

CRAME, J. A. & LUTHER, A. 1997. The last inoceramids in Antarctica. *Cretaceous Research*, **18**, 179–195.

CROWLEY, T. J. & ZACHOS, J. C. 2000. Comparison of zonal temperature profiles for past warm time periods. *In*: HUBER, B. T., MACLEOD, K. G. & WING, S. L. (eds) *Warm Climates in Earth History*, Cambridge University Press, Cambridge, 50–76.

DAILEY, D. H. 1983. Late Cretaceous and Palaeocene benthic foraminifers from Deep Sea Drilling Project Site 516, Rio Grande Rise, western South Atlantic. *In*: HSÜ, K. J., LABRECQUE, J. L. *et al.* (eds). *Initial Reports of the Deep Sea Drilling Project*, **72**. US Government Printing Office, Washington, DC, 757–782.

DETRICK, R. S., SCLATER, J. G. & THIEDE, J. 1977. The subsidence of aseismic ridges. *Earth and Planetary Science Letters*, **34**, 185–196.

DHONDT, A. V. 1983. Campanian and Maastrichtian inoceramids: a review. *Zitteliana*, **10**, 689–701.

DHONDT, A. V. 1992. Cretaceous inoceramid biogeography: a review. *Palaeogeography, Palaeoclimatology, Palaeoecology*, **92**, 217–232.

D'HONDT, S. & ARTHUR, M. A. 1996. Late Cretaceous oceans and the cool tropic paradox. *Science*, **271**, 1838–1841.

DOUGLAS, R. G. & SAVIN, S. M. 1975. Oxygen and carbon isotope analyses of Tertiary and Cretaceous microfossils from the Shatsky Rise and other sites in the North Pacific Ocean. *In*: LARSON, R. L. & MOBERLY, R. (eds) *Initial Reports of the Deep Sea Drilling Project*, **32**, US Government Printing Office, Washington, DC, 509–520.

FRANK, T. & ARTHUR, M. A. 1999. Tectonic forcing of Maastrichtian ocean–climate evolution. *Paleoceanography*, **14**, 103–117.

GERMAN, C. R. & ELDERFIELD, H. 1990. Application of the Ce anomaly as a palaeoredox indicator: the ground rules. *Paleoceanography*, **5**, 823–833.

GRADSTEIN, F. M., AGTERBERG, F. P., OGG, J. G., HARDENBOL, J., VAN VEEN, P., THIERRY, J. & HUANG, Z. 1995. A Triassic, Jurassic and Cretaceous time scale. *In*: BERGGREN, W. A., KENT, D. V., AUBRY, M.-P. & HARDENBOL, J. (eds) *Geochronology, Time Scales and Global Stratigraphic Correlation*. SEPM Special Publications, **54**, 95–126.

GROSSMAN, E. L. 1993. Evidence that inoceramid bivalves were benthic and harbored chemosynthetic symbionts: Comment. *Geology*, **21**, 94–95.

HALLAM, A. 1992. *Phanerozoic Sea-Level Changes: Perspectives in Paleobiology and Earth History Series*. Columbia University Press, New York.

HANCOCK, J. M. 1993. Transatlantic correlations in the Campanian–Maastrichtian stages by eustatic changes of sea-level. *In*: HAILWOOD, E. A. & KIDD, R. B. (eds) *High Resolution Stratigraphy*. Geological Society, London, Special Publications, **70**, 241–256.

HAQ, B. U., HARDENBOL, J. & VAIL, P. R. 1988. Mesozoic and Cenozoic chronostratigraphy and cycles of sea level change. *In*: WILGUS, C. K., HASTINGS, B. S., POSAMENTIER, H., VAN WAGONER, J., ROSS, C. A. & KENDALL, C. G. S. (eds) *Sea Level Changes: an Integrated Approach*. SEPM Special Publications, **42**, 71–108.

HARDENBOL, J. & ROBASZYNSKI, F. 1998. Introduction to the Upper Cretaceous. *In*: DE GRACIANSKY, P.-C., HARDENBOL, J., JACQUIN, T. & VAIL, P. R. (eds) *Mesozoic and Cenozoic Sequence Stratigraphy of European Basins*. Special Publication, Society for Sedimentary Geology, **60**, 329–332.

HARLAND, W. B. *et al.* 1982. *A Geologic Time Scale*. Cambridge University Press, Cambridge.

HAY, W. W. 1988. Paleoceanography; a review for the GSA Centennial. *Geological Society of America Bulletin*, **100**, 1934–1956.

HAY, W. W., DECONTO, R., WOLD, C. N. *et al.* 1999. An alternative global Cretaceous palaeogeography. *In*: BARRERA, E. & JOHNSON, C. (eds) *The Evolution of Cretaceous Ocean/Climate Systems*. Geological Society of America, Special Paper, **332**, 1–47.

HEINBERG, C. 1979. Bivalves from the latest Maastrichtian of Stevns Klint and their stratigraphic affinities. *In*: BIRKELUND, T. & BROMLEY, R. G. (eds) *Symposium on the Cretaceous/Tertiary Boundary Events, I. The Maastrichtian and Danian of Denmark*. University of Copenhagen, Copenhagen, 58–64.

HEMLEBEN, C. & TROESTER, J. 1984. Campanian–Maestrichtian deep-water foraminifers from Hole 543A, Deep Sea Drilling Project. *In*: BIJU-DUVAL, B., MOORE, J. C. *et al.* (eds) *Initial Reports of the Deep Sea Drilling Project*, **78**. US Government Printing Office, Washington, DC, 509–532.

HERMAN, A. B. & SPICER, R. A. 1996. Palaeobotanical evidence for a warm Cretaceous Arctic Ocean. *Nature*, **380**, 330–333.

HOWARTH, R. J. & MCARTHUR, J. M. 1997. Statistics for strontium isotope stratigraphy; a robust LOWESS fit to marine Sr-isotope curve for 0 to 206 Ma, with look-up table for derivation of numeric age. *Journal of Geology*, **105**, 441–456.

HU, X., WANG, Y. L. & SCHMITT, R. A. 1988. Geochemistry of sediments on the Rio Grande Rise and the redox evolution of the South Atlantic Ocean. *Geochimica et Cosmochimica Acta*, **52**, 201–207.

HUBER, B. T. 1988. Upper Campanian–Palaeocene foraminifera from the James Ross Island region (Antarctic Peninsula). *In*: FELDMANN, R. M. & WOODBURNE, M. O. (eds) *Geology and Paleontology of Seymour Island, Antarctica*. Geological Society of America, Memoirs, **169**, 163–252.

HUBER, B. T. 1990. Maestrichtian planktonic foraminifer biostratigraphy of the Maud Rise (Weddell Sea, Antarctica): ODP Leg 113 Holes 689B and 690C. *In*: BARKER, P. F. & KENNETT, J. P. (eds) *Proceedings of the Ocean Drilling Program, Scientific Results*, **113**. Ocean Drilling Program, College Station, TX, 489–513.

HUBER, B. T. & WATKINS, D. K. 1992. Biogeography of Campanian–Maastrichtian calcareous plankton in the region of the Southern Ocean: palaeogeographic and palaeoclimatic implications. *In*: KENNETT, J. P. & WARNKE, D. A. (eds) *The Antarctic Paleoenvironment: a Perspective on Global Change*. Antarctic Research Series, American Geophysical Union, **56**, 31–60.

HUT, P., ALVAREZ, W., ELDER, W. P. *et al.* 1987. Comet showers as a cause of mass extinctions. *Nature*, **329**, 118–125.

JACKSON, J. B. C., PETER, J., COATES, A. G. & COLLINS, L. S. 1993. Diversity and extinction of tropical American mollusks and emergence of the Isthmus of Panama. *Science*, **260**, 1624–1626.

JOHNSON, C. C. & KAUFFMAN, E. G. 1990. Originations, radiations and extinctions of Cretaceous rudistid bivalve species in the Caribbean. *In*: KAUFFMAN, E. G. & WALLISER, O. H. (eds) *Extinction Events in Earth History*. Springer, Berlin, 305–324.

JOHNSON, C. C. & KAUFFMAN, E. G. 1996. Maastrichtian extinction patterns of Caribbean Province rudistids. *In*: MACLEOD, N. & KELLER, G. (eds) *Cretaceous–Tertiary Mass Extinctions: Biotic and Environmental Changes*. Norton, New York, 231–273.

JOHNSON, C. C., BARRON, E. J., KAUFFMAN, E. G., ARTHUR, M. A., FAWCETT, P. J. & YASUDA, M. K. 1996. Middle Cretaceous reef collapse linked to ocean heat transport. *Geology*, **24**, 376–380.

JOHNSON, K. R. & HICKEY, L. J. 1990. Megafloral change across the Cretaceous/Tertiary boundary in the northern Great Plains and Rocky Mountains, USA. *In*: SHARPTON, V. L. & WARD, P. D. (eds) *Global Catastrophes in Earth History: an Interdisciplinary Conference on Impacts, Volcanism, and Mass Mortality*. Geological Society of America, Special Papers, **247**, 433–444.

JOHNSON, K. R. & HICKEY, L. J. 1992. Foliar physiognomy of Maastrichtian leaf floras from the northern Great Plains: implications for palaeoclimate. *SEPM 1992 Theme Meeting, Mesozoic of the Western Interior, Ft Collins, Colorado*. Abstracts, 36.

KAUFFMAN, E. G. 1988. The dynamics of marine stepwise extinction. *Revista Española de Paleontologia, Extraordinario*, 57–71.

KLAUS, A., NORRIS, R. D., KROON, D. & SMIT, J. 2000. Impact-induced mass wasting at the K–T boundary: Blake Nose, western North Atlantic. *Geology*, **28**, 319–322.

LEWY, Z. 1990. Transgressions, regressions and relative sea level changes on the Cretaceous shelf of Israel and adjacent countries. A critical evaluation of Cretaceous global sea level correlations. *Paleoceanography*, **5**, 619–637.

LI, L. & KELLER, G. 1998. Maastrichtian climate, productivity and faunal turnovers in planktonic foraminifera in South Atlantic DSDP sites 525A and 21. *Marine Micropalaeontology*, **33**, 55–86.

LI, L. & KELLER, G. 1999. Variability in Late Craetceous climate and deep waters: evidence from stable isotopes. *Marine Geology*, **161**, 171–190.

LIU, Y.-G. & SCHMITT, R. A. 1984. Chemical profiles in sediment and basalt samples from Deep Sea Drilling Project Leg 74, hole 525A, Walvis Ridge. *In*: MOORE, T. C., Jr & RABINOWITZ, P. D. (eds) *Initial Reports of the Deep Sea Drilling Project*, **74**. US Government Printing Office, Washington, DC, 713–730.

MACELLARI, C. E. 1988. Stratigraphy, sedimentology and palaeoecology of Late Cretaceous/Palaeocene shelf–deltaic sediments of Seymour Island (Antarctic Peninsula). *In*: FELDMANN, R. M. & WOODBURNE, M. O. (eds) *Geology and Paleontology of Seymour Island, Antarctica*. Geological Society of America, Memoirs, **169**, 25–53.

MACLEOD, K. G. 1994*a*. Bioturbation, inoceramid extinction, and mid-Maastrichtian ecological change. *Geology*, **22**, 139–142.

MACLEOD, K. G. 1994*b*. The extinction of inoceramid bivalves in Maastrichtian strata of the Bay of Biscay region of France and Spain. *Journal of Paleontology*, **68**, 1048-1066.

MACLEOD, K. G. & HOPPE, K. A. 1992. Evidence that inoceramid bivalves were benthic and harbored chemosynthetic symbionts. *Geology*, **20**, 117–120.

MACLEOD, K. G. & HOPPE, K. A. 1993. Evidence that inoceramid bivalves were benthic and harbored chemosynthetic symbionts: Reply. *Geology*, **21**, 95–96.

MACLEOD, K. G. & HUBER, B. T. 1995. Testing for subtle alteration of calcitic microfossils: caveats and inferences from analyses of suspect samples. *Geological Society of America, Abstracts with Programs*, **27**, A266.

MACLEOD, K. G. & HUBER, B. T. 1996*a*. Reorganization of ocean structure and deep circulation correlated with Maastrichtian global change. *Nature*, **380**, 422–425.

MACLEOD, K. G. & HUBER, B. T. 1996*b*. Strontium isotopic evidence for extensive reworking in sediments spanning the Cretaceous–Tertiary boundary at ODP Site 738. *Geology*, **24**, 463–466.

MACLEOD, K. G. & ORR, W. N. 1993. The taphonomy of Maastrichtian inoceramids in the Basque region of France and Spain and the pattern of their decline and disappearance. *Paleobiology*, **19**, 235–250.

MACLEOD, K. G., HUBER, B. T. & DUCHARME, M. L. 2000. Paleontological and geochemical constraints on changes in the deep ocean during the Cretaceous greenhouse interval. *In*: HUBER, B. T., MACLEOD, K. G. & WING, S. L. (eds) *Warm Climates in Earth History*. Cambridge University Press, Cambridge, 241–274.

MACLEOD, K. G., HUBER, B. T., PLETSCH, T., RÖHL, U. & KUCERA, M. 2001. Maastrichtian foraminiferal and palaeoceanographic changes on Milankovitch time scales. *Paleoceanography*, in press.

MACLEOD, K. G., HUBER, B. T. & WARD, P. D. 1996. The biostratigraphy and palaeobiogeography of Maastrichtian inoceramids. *In*: RYDER, G., GARTNER, S. & FASTOVSKY, D. (eds) *New Developments Regarding the KT Event and Other Catastrophes in Earth History*. Geological Society of America, Special Papersm **307**, 361–374.

MAIER-REIMER, E., MIKOLAJEWICZ, U. & CROWLEY, T. J. 1990. Ocean General Circulation Model sensitivity experiment with an open Central American isthmus. *Paleoceanography*, **5**, 349–366.

MIKOLAJEWICZ, U., MAIER-REIMER, E., CROWLEY, T. J. & KIM, K.-Y. 1993. Effect of Drake and Panamanian gateways on the circulation of an ocean model. *Paleoceanography*, **8**, 409–426.

MCLAUGHLIN, O., M., MCARTHUR, J. M., THIRLWALL, M. F., HOWARTH, R., BURNETT, J., GALE, A. S. & KENNEDY, W. J. 1995. Sr evolution of Maastrichtian seawater, determined from the chalk of Hemmoor, NW Germany. *Terra Nova*, **7**, 491–499.

MIALL, A. D. 1992. Exxon global cycle chart: an event for every occasion? *Geology*, **20**, 787–790.

MILLER, K. G., BARRERA, E., OLSSON, R. K., SUGARMAN, P. J. & SAVIN, S. M. 1999. Does ice drive Maastrichtian eustasy? *Geology*, **27**, 783–786.

NELSON, B. K., MACLEOD, K. G., & WARD, P. D. 1991. Rapid change in strontium isotopic composition of sea water before the Cretaceous/Tertiary boundary. *Nature*, **351**, 644–647.

NORRIS, R. D., KROON, D., KLAUS, A. *et al.* (eds) 1998. *Proceedings of the Ocean Drilling Program, Initial Reports*, **171B**. Ocean Drilling Program, College Station, TX.

OGG, J. G. & BARDOT, L. 2001. Aptian through Eocene magnetostratigraphic correlation of Blake Nose transect (ODP Leg 171), Florida continental margin. *In*: KROON, D., NORRIS, R. D. et al. (eds) *Proceedings of the Ocean Drilling Program, Scientific Results*, **171**, in press.

REYMENT, R. A. & TAIT, E. A. 1972. Biostratigraphic dating of the early history of the South Atlantic Ocean. *Philosophical Transactions of the Royal Society of London, Series B*, **264**, 55–95.

SALTZMAN, E. & BARRON, E. J. 1982. Deep circulation in the Late Cretaceous: oxygen isotope palaeotemperatures from *Inoceramus* remains in DSDP cores. *Palaeogeography, Palaeoclimatology, Palaeoecology*, **40**, 167–181.

SCLATER, J. G., HELLINGER, S. & TAPSCOTT, C. 1977. The palaeobathymetry of the Atlantic Ocean from the Jurassic to the present. *Journal of Geology*, **85**, 509–552.

SETO, K., NOMURA, R. & NIITSUMA, N. 1991. Data report: oxygen and carbon isotope records of the upper Maastrichtian to Lower Eocene benthic foraminifers at site 752 in the eastern Indian Ocean. *In*: WEISSEL, J., PEIRCE, J., TAYLOR, E. & ALT, J. (eds) *Proceedings of the Ocean Drilling Program, Scientific Results*, **121**. Ocean Drilling Program, College Station, TX, 885–889.

SHEEHAN, P. M. & FASTOVSKY, D. E. 1992. Major extinctions of land-dwelling vertebrates at the Cretaceous–Tertiary boundary, eastern Montana. *Geology*, **20**, 556–560.

SHEEHAN, P. M., FASTOVSKY, D. E., HOFFMAN, R. G., BERGHAUS, C. B. & GABRIEL, D. L. 1991. Sudden extinction of the dinosaurs: Latest Cretaceous, upper Great Plains, USA. *Science*, **254**, 835–839.

SLOAN, L. C., WALKER, J. C. G. & MOORE, T. C., Jr 1995. Possible role of oceanic heat transport in the early Eocene climate. *Paleoceanography*, **10**, 347–356.

SLOAN, R. E., RIGBY, J. K., Jr, VAN VALEN, L. M. & GABRIEL, D. 1986. Gradual dinosaur extinction and simultaneous ungulate radiation in the Hell Creek Formation. *Science*, **232**, 629–633.

SPICER, R. A. & CORFIELD, R. M. 1992. A review of terrestrial and marine climates in the Cretaceous with implications for modelling the 'Greenhouse Earth'. *Geological Magazine*, **129**, 169–180.

SPICER, R. A. & PARRISH, J. T. 1990. Late Cretaceous–early Tertiary palaeoclimates of northern high latitudes; a quantitative view. *Paleoclimates*, **147**, 329–341.

SUGARMAN, P. J., MILLER, K. G., BURKY, D. & FEIGENSON, M. D. 1995. Uppermost Campanian–Maastrichtian strontium isotopic, biostratigraphic, and sequence stratigraphic framework of the New Jersey Coastal Plain. *Geological Society of America Bulletin*, **107** 19–37.

THOMAS, E. 1990. Late Cretaceous through Neogene deep-sea benthic foraminifers (Maud Rise, Weddell Sea, Antarctica): ODP leg 113 holes 689B and 690C. *In*: BARKER, P. F. & KENNETT, J. P. (eds) *Proceedings of the Ocean Drilling Program, Scientific Results*, **113**. Ocean Drilling Program, College Station, TX, 571–594.

VAN ANDEL, T. H., THIEDE, J. R., SCLATER, J. G. & HAY, W. W. 1977. Depositional history of the South Atlantic Ocean during the last 125 million years. *Journal of Geology*, **85**, 651–698.

VOIGT, S. 1995. Paleobiogeography of early Late Cretaceous inoceramids in the context of a new global palaeogeography. *Cretaceous Research*, **16**, 343–356.

VOIGT, S., HAY, W. W., HÖFLING, R. & DECONTO, R. M. 1999. Biogeographic distribution of late Early to Late Cretaceous rudist-reefs in the Mediterranean as climate indicators. *In*: BARRERA, E. & JOHNSON, C. C. (eds) *Evolution of the Cretaceous Ocean–Climate System*. Geological Society of America, Special Papers, **332**, 91–103.

WANG, Y. L., LIU, Y.-G. & SCHMITT, R. A. 1986. Rare earth element geochemistry of South Atlantic deep sea sediments: Ce anomaly change at ~54 My. *Geochimica et Cosmochimica Acta*, **50**, 1337–1355.

WARD, P. D. & KENNEDY, W. J. 1993. Maastrichtian ammonites from the Biscay region (France, Spain). *Journal of Paleontology, Memoir*, **34**, 1–58.

WOLFE, J. A. & UPCHURCH, G. R., Jr 1987. North American nonmarine climates and vegetation during the Late Cretaceous. *Palaeogeography, Palaeoclimatology, Palaeoecology*, **61**, 33–77.

WOO, K.-S., ANDERSON, T. F., RAILSBACK, L. B. & SANDBERG, P. A. 1992. Oxygen isotopic evidence for high-salinity surface seawater in the mid-Cretaceous Gulf of Mexico: implications for warm, saline deepwater formation. *Paleoceanography*, **7**, 673–685.

Geochemistry of the Cretaceous–Tertiary boundary at Blake Nose (ODP Leg 171B)

F. MARTÍNEZ-RUIZ[1], M. ORTEGA-HUERTAS[2], D. KROON[3], J. SMIT[4], I. PALOMO-DELGADO[2] & R. ROCCHIA[5]

[1]*Instituto Andaluz de Ciencias de la Tierra, CSIC-Universidad de Granada, Facultad de Ciencias, Avda. Fuentenueva s/n, 18002 Granada, Spain (e-mail: fmruiz@goliat.ugr.es)*

[2]*Departamento de Mineralogía y Petrología, Facultad de Ciencias, Universidad de Granada, Avda. Fuentenueva s/n, 18002 Granada, Spain*

[3]*Department of Geology and Geophysics, University of Edinburgh, Grant Institute, West Mains Road, Edinburgh EH9 3JW, UK*

[4]*Department of Sedimentary Geology, Vrije Universiteit, 1081 HV Amsterdam, Netherlands*

[5]*Centre des Faibles Radioactivités, Laboratoire CEA–CNRS, 91198 Gif-sur-Yvette Cedex, France*

Abstract: The Cretaceous–Tertiary (K–T) boundary at Blake Nose, in the NW Atlantic, is recorded by a coarse, poorly graded and poorly cemented layer mostly consisting of green spherules that are mainly composed of smectite. Geochemical patterns across the boundary are governed by the source material of the spherule bed and postdepositional processes. The chemical composition and the nature of this bed indicate that it derived from melted target rocks from the Chicxulub impact structure. Ir and other typical extraterrestrial elements do not present significant enrichments, which suggests that the spherule bed material derived from crustal rocks. Ir instead reaches its highest concentration in the burrow-mottled calcareous ooze above the spherule bed, suggesting that it is associated to the finest fraction deposited after the target-rock-derived material. Only the Ni and Co content show slight enrichments within the upper part of the spherule layer, although most of the trace element profiles resulted from diagenetic alteration. During the alteration of glass to smectite, the concentrations of certain trace elements, such as the rare earth elements, were severely changed. In addition, oxygen-poor conditions also led to the remobilization of redox-sensitive elements, which show enhanced concentration at the top or above the spherule bed. Diagenetic remobilization may have also affected the Ir concentration.

Since Alvarez *et al.* (1980) first proposed a meteorite impact as the probable cause of the mass extinction at the end of the Cretaceous period, several lines of evidence have been used to support this model, one of the main arguments being the enhanced concentration of typical extraterrestrial elements in the boundary material, such as Ir and other platinum-group elements (PGE). These elements have been sought in sediments from this age to constrain the boundary in relation to the faunal turnover. In the last two decades, Ir anomalies at the Cretaceous–Tertiary (K–T) boundary have been found at many K–T sites throughout the world. In addition to the PGEs, anomalous concentrations of other chemical elements have also been reported in the boundary sediments (e.g. Kyte *et al.* 1980, 1985; Schmitz 1985; Strong *et al.* 1987; Zachos *et al.* 1989; Smit 1990; Martínez-Ruiz *et al.* 1992; Bhandari *et al.* 1993). The vast literature on this topic prevents our mentioning all the sites and geochemical anomalies reported thus far, although all the data published until now provide consistent evidence supporting first the extraterrestrial hypothesis (in the 1980s) and later also the impact site at the Chicxulub crater (e.g. Hildebrand *et al.* 1991; Koeberl & Sigurdsson 1992, Sharpton *et al.* 1993; Smit 1999). Despite all the advances made in the knowledge of this devastating impact and its environmental consequences, more work is undeniably needed for complete understanding. Ocean Drilling Program (ODP) Leg 171B contributed a substantial

From: KROON, D., NORRIS, R. D. & KLAUS, A. (eds). 2001 *Western North Atlantic Palaeogene and Cretaceous Palaeoceanography*. Geological Society, London, Special Publications, **183**, 131–148.
0305-8719/01/$15.00

advance in this field by adding a new site (Blake Nose, NW Atlantic) offering the possibility of studying the composition of the Cretaceous, Tertiary and K–T boundary sediments in a succession recording the mass extinction and the proximal ejecta fallout from the Chicxulub crater, and providing also further evidence for a bolide impact at Chicxulub (Norris *et al.* 1998), sometimes questioned in the literature (e.g. Keller *et al.* 1993). Cretaceous and Tertiary boundary sediments recovered by ODP Leg 171 at Blake Nose are similar to those from other marine K–T sections in the North Atlantic margin (Klaver *et al.* 1987; Olsson *et al.* 1997), with the exception of the remarkable thickness of the spherule bed at Site 1049. The thickest bed (17 cm) was recovered at Hole 1049A, whereas at Holes 1049B and 1049C it is around 9 cm thick. The variable thickness of the spherule bed at the three holes drilled at this site, as well as the presence of Cretaceous foraminifera and Cretaceous clasts in the spherule bed, point to reworking of the spherule bed material (Norris *et al.* 1999; Klaus *et al.* 2000). Despite the absence of a precise stratigraphy, its geochemical composition is evidence of the nature of the impact-generated material and further supports the Chicxulub impact model. The main focus of this paper is therefore to study the geochemical anomalies from Blake Nose sediments to constrain the terrestrial–extraterrestrial contribution at this location, the patterns for extraterrestrial material distribution, as well as the chemical element distribution associated with the impact event.

Samples and analytical methods

Samples from the K–T boundary interval at Holes 1049A, 1049B, 1050C and 1052E, Blake Nose (NW Atlantic) (Fig. 1), were analysed for major and trace element concentrations. In the studied interval, a spherule bed occurs at the biostratigraphic boundary between the Cretaceous and the Palaeocene sediments at Site 1049 (Fig. 2). It sharply overlies slumped uppermost Cretaceous foraminiferal–nannofossil ooze and is overlain by Tertiary clay-rich ooze with planktonic foraminiferal assemblages indicative of Early Danian Foraminiferal Zone P-alpha (Norris *et al.* 1998, 1999). This layer mainly consists of green spherical and oval-shaped spherules mostly composed of smectite, which derives from alteration of the original material (Martínez-Ruiz *et al.* this volume) and is capped by a 3 mm-thick orange Fe-oxide layer that initially appeared to be the fireball layer (Norris, Kroon, Klaus *et al.* 1998) equivalent to the uppermost layer in non-marine sections (Pollastro & Bohor 1993). The spherule bed also contains some lithic fragments, Cretaceous foraminifera and clasts of Cretaceous material, suggesting reworking of the spherule bed material, which is further supported by the variable thickness. Notwithstanding this reworking, which limits the interpretation of the possible original stratigraphy of the ejecta deposit at Blake Nose, the ejecta fallout from the Chicxulub impact is recorded. The spherule bed was not preserved at Holes 1050C and 1052E, although at Hole 1052E some burrows are filled with ejecta material (Norris *et al.* 1998; Klaus *et al.* 2000). Samples were collected from cores 16X, 17X and 18X at Hole 1049A, taking samples continuously every 2 cm in an interval from 125.69 to 126.23 m below sea floor (mbsf) in section 17X-2 in order to analyse continuously and with higher resolution the K–T boundary layer and nearby sediments above and below. At Hole 1049B, a

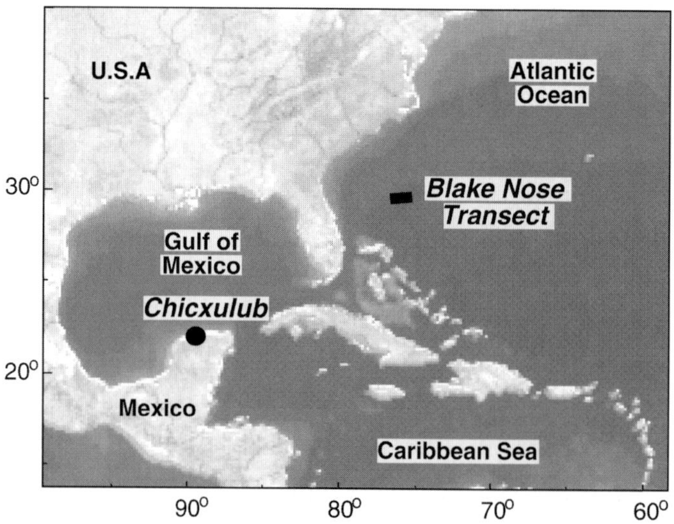

Fig. 1. Location of the ODP Leg 171B Blake Nose drilling transect.

continuous sampling of the K–T boundary interval was also carried out in section 8H-2 every 2 cm in an interval from 111.00 to 111.32 mbsf. Other scattered samples were also taken at these holes (see Tables 1 and 2 for sample position). At Holes 1050C and 1052E, where the spherule bed is not well preserved, lower-resolution sampling was carried out. Section 10R-2 was sampled at Hole 1050C, and sections 18R-1, 18R-2, 18R-3 and 19R-1 were sampled at Hole 1052E (see Tables 3 and 4 for sample positions).

Samples from the spherule bed were cleaned of Cretaceous clasts under a stereomicroscope and dried, homogenized and ground in an agate mortar for chemical analyses using inductively coupled plasma-mass spectrometry (ICP-MS) and atomic absorption spectrometry (AAS). Rb, Sr, Ba, V, Cr, Co, Ni, Cu, Zr, Hf, Mo, Pb, U, Th and rare earth elements (REE) were analysed by ICP-MS, and Al, K, Fe, Mn, Ca and Mg were analysed by AAS. Analyses were performed on bulk samples after sample digestion with $HNO_3 + HF$ of 0.100 g of sample powder in a Teflon-lined vessel at high temperature and pressure, evaporation to dryness and subsequent dissolution in 100 ml of 4 vol. % HNO_3. ICP-MS instrument measurements were performed in

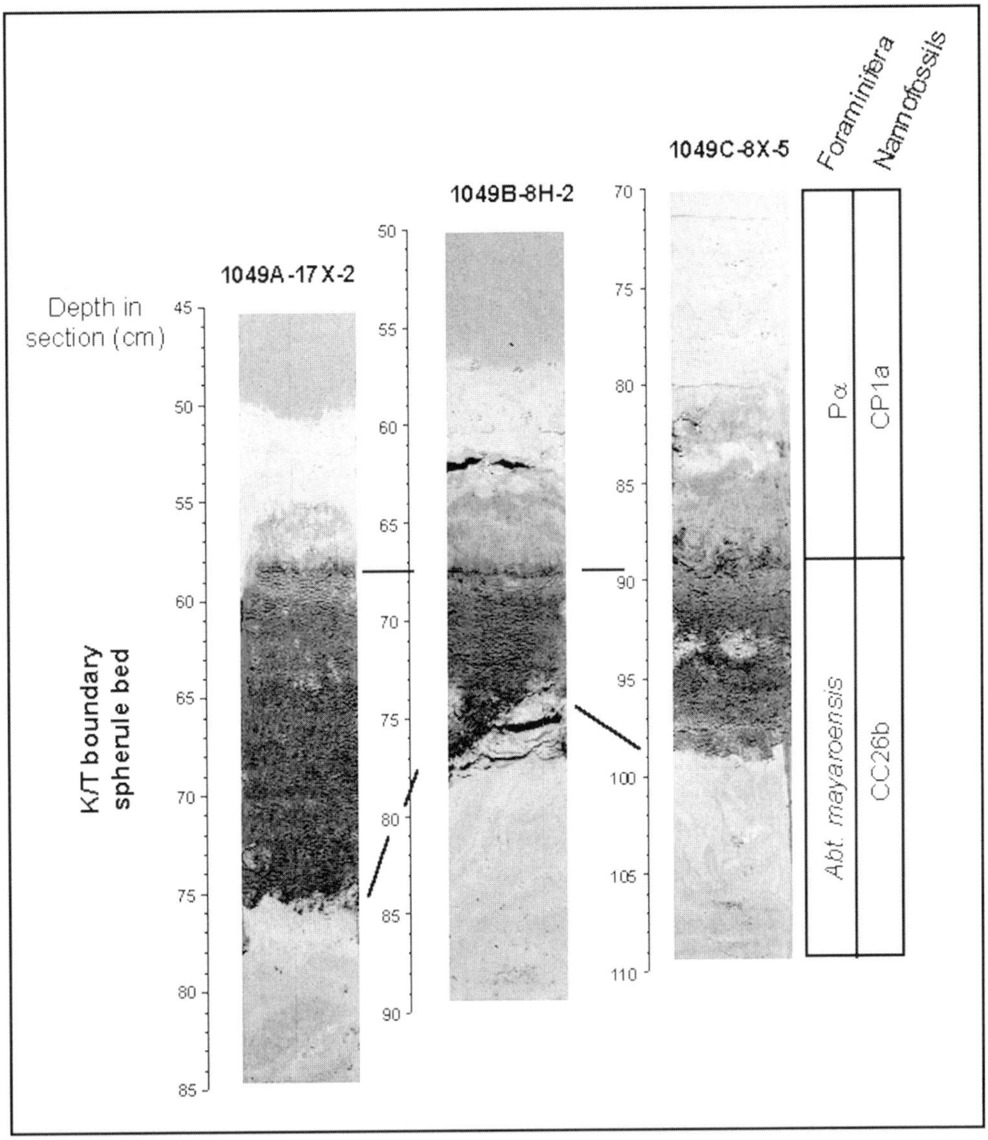

Fig. 2. Core photographs of the K–T boundary interval at Site 1049 showing the spherule bed marking the boundary at the three holes drilled at this site and its variable thickness (biostratigraphy from Norris *et al.* (1999)).

Table 1. *Geochemical data from the K–T boundary interval at Hole 1049A*

Sample	Depth (mbsf)	Al	K	Fe	Mn	Ca	Mg	Rb	Sr	Ba	V	Cr	Co	Ni	Cu	Zr	Hf	Mo	Pb	U	Th	REE	Ua
16X 01 002–004	115.02	3.14	1.09	1.40	270	21.0	0.80	46.6	607	101	64.8	134	7.97	29.0	24.2	25.6	0.60	0.29	7.37	0.47	2.75	73.33	0.52
16X 01 042–044	115.42	3.30	1.07	1.46	200	17.7	0.80	49.0	583	97	80.5	153	9.23	27.8	25.9	24.8	0.51	0.17	3.89	0.41	2.32	67.09	0.53
16X 01 093–095	115.93	3.68	1.10	1.60	590	16.5	0.92	53.2	669	101	82.3	141	8.39	29.6	20.6	25.9	0.64	0.20	4.90	0.50	3.12	74.15	0.48
16X 01 135–137	116.35	3.43	1.01	1.54	190	17.9	0.73	50.0	663	85	96.6	184	10.2	31.4	22.8	24.5	0.50	0.16	3.98	0.43	2.36	66.98	0.55
16X 02 025–025	116.75	2.80	0.89	1.29	260	20.5	0.80	39.4	637	83	48.2	93.8	6.71	21.1	12.7	20.3	0.58	0.11	4.86	0.54	2.74	76.07	0.59
16X 02 077–079	117.27	2.48	0.74	1.20	180	21.9	0.74	32.4	654	70	38.5	76.9	5.09	13.9	10.5	16.2	0.45	0.56	4.73	0.47	2.66	65.81	0.53
16X 02 122–124	117.72	2.60	0.77	1.27	230	21.8	0.74	36.8	687	73	42.7	81.1	5.52	16.3	10.5	16.3	0.50	0.28	4.96	0.43	2.78	64.61	0.46
16X 03 002–004	118.02	1.77	0.60	0.97	230	24.3	0.62	25.8	734	47	35.6	64.5	4.63	13.5	9.78	11.9	0.32	1.11	3.09	0.34	1.59	51.32	0.64
16X 03 044–046	118.44	1.55	0.50	0.80	220	14.6	0.57	24.8	788	36	41.0	66.7	4.95	13.2	9.11	10.9	0.24	0.02	2.15	0.24	1.18	44.28	0.62
16X 03 093–095	118.93	2.00	0.62	1.17	520	23.7	0.64	28.7	734	58	42.4	68.3	10.2	21.1	11.56	15.8	0.41	0.83	4.85	0.40	2.09	63.19	0.58
16X 03 132–134	119.32	1.92	0.60	1.08	320	22.7	0.62	28.8	684	59	38.8	64.3	9.55	13.7	9.59	14.7	0.40	0.79	7.18	0.39	2.05	56.53	0.57
16X 04 041–043	119.91	2.90	0.78	1.43	230	18.6	0.87	44.4	627	74	85.2	150	8.43	24.4	17.0	19.9	0.43	0.07	3.40	0.57	2.00	56.35	0.86
16X 04 099–101	120.49	3.38	0.94	1.76	280	14.1	0.99	51.3	615	104	90.4	159	7.76	23.5	20.4	23.1	0.57	0.32	4.62	0.76	2.97	73.23	0.77
16X 05 012–014	121.12	3.90	0.94	1.73	180	15.5	1.02	56.6	660	96	124	165	8.19	22.4	33.6	23.0	0.47	0.35	4.61	0.75	2.57	67.16	0.87
16X 05 069–071	121.69	2.35	0.61	1.05	230	20.4	0.70	31.1	744	73	41.5	77.2	4.31	10.6	21.5	14.7	0.39	0.36	4.36	0.76	2.29	56.88	0.99
16X 05 129–131	122.29	1.62	0.45	0.70	270	24.6	0.87	24.7	788	41	48.9	86.5	5.11	9.62	13.3	10.0	0.23	0.82	2.06	0.51	1.09	35.66	1.40
16X 06 033–035	122.83	3.13	0.85	1.42	230	17.9	0.94	48.1	564	96	94.8	113.4	11.0	22.2	10.1	22.3	0.63	0.38	5.54	1.20	3.47	82.89	1.04
16X 06 083–085	123.33	3.53	1.03	1.71	180	15.4	0.99	59.8	540	127	93.7	149	28.8	42.2	29.4	28.1	0.73	0.44	7.54	1.52	3.92	90.60	1.17
16X CC 015–017	123.82	2.82	0.89	1.38	250	19.5	0.85	46.8	557	94	69.3	117	13.1	24.8	9.44	21.0	0.55	0.20	4.92	1.38	3.37	81.44	1.22
17X 01 069–071	125.29	1.97	0.58	0.92	230	23.9	0.69	32.2	722	63	46.2	77.9	14.1	19.7	12.0	15.1	0.38	0.11	3.61	0.92	2.26	54.70	1.22
17X 02 001–003	125.32	2.05	0.61	0.56	288	31.6	0.75	39.0	764	73	39.6	75.4	12.2	24.2	<0.5	16.2	0.43	0.25	2.99	0.91	2.19	57.10	1.24
17X 02 004–006	125.35	1.89	0.54	0.52	256	31.8	0.76	35.8	763	60	36.5	66.0	10.0	20.0	<0.5	13.8	0.40	0.27	2.68	0.81	1.91	52.07	1.27
17X 02 008–010	125.39	2.46	0.76	1.03	295	30.3	0.77	42.3	722	74	44.9	75.8	13.6	26.1	2.86	16.5	0.47	0.23	3.81	0.91	2.37	61.96	1.15
17X 02 012–014	125.43	1.89	0.57	0.73	262	33.8	0.72	28.0	737	54	33.4	62.3	11.1	21.4	<0.5	12.2	0.34	0.24	3.61	0.89	1.80	49.88	1.49
17X 02 016–018	125.47	2.26	0.68	0.89	264	30.7	0.80	30.9	729	68	39.5	73.0	13.4	25.2	<0.5	14.9	0.43	0.22	3.79	1.05	2.20	57.42	1.43
17X 02 020–022	125.51	3.35	0.95	1.32	286	25.8	0.97	43.2	671	112	58.7	122	33.7	71.3	5.29	25.0	0.67	0.64	7.15	1.27	3.37	83.46	1.13
17X 02 024–026	125.55	3.48	0.91	1.34	213	23.6	0.92	55.6	663	115	67.8	117	58.1	76.1	<0.5	27.7	0.82	0.34	6.36	1.49	3.48	82.74	1.29
17X 02 028–030	125.59	2.48	0.69	0.99	249	28.8	0.75	38.2	716	79	46.9	83.0	17.2	26.5	<0.5	19.3	0.51	0.42	4.49	1.04	2.48	64.99	1.26
17X 02 032–034	125.63	2.19	0.62	0.90	247	20.7	0.72	29.2	727	69	43.5	71.0	13.6	22.2	<0.5	17.4	0.47	0.23	3.84	0.95	2.39	62.41	1.19
17X 02 035–037	125.66	2.31	0.64	0.97	271	29.6	0.75	37.4	731	71	44.5	73.6	12.8	23.8	<0.5	18.3	0.47	0.25	3.74	0.99	2.26	64.33	1.32
17X 02 038–040	125.69	2.24	0.69	0.95	304	29.9	0.75	35.8	722	69	44.2	73.4	11.0	23.1	<0.5	18.0	0.43	0.20	3.71	0.94	2.32	64.64	1.21
17X 02 040–042	125.71	2.18	0.64	0.91	300	28.6	0.71	37.5	723	70	44.1	73.2	11.0	25.5	<0.5	19.0	0.46	0.20	3.55	0.89	2.24	64.41	1.19
17X 02 042–044	125.73	2.19	0.67	0.99	305	27.5	0.74	36.1	712	70	43.7	73.0	11.0	20.6	<0.5	19.1	0.45	0.27	3.60	0.92	2.33	67.04	1.18
17X 02 044–046	125.75	2.25	0.64	0.89	285	28.4	0.75	39.9	705	73	44.3	73.4	10.0	19.2	<0.5	19.4	0.49	0.26	3.57	0.90	2.37	68.93	1.15
17X 02 046–048	125.77	2.16	0.64	1.02	298	28.0	0.72	35.7	695	70	44.0	72.2	10.3	20.7	<0.5	18.6	0.43	0.21	3.32	0.87	2.25	68.48	1.16
17X 02 048–050	125.79	2.17	0.65	1.03	291	25.4	0.68	37.5	713	71	46.4	75.8	18.5	34.5	<0.5	19.5	0.42	0.23	4.08	0.92	2.39	73.63	1.16
17X 02 050–052	125.81	1.47	0.53	0.77	350	31.2	0.61	27.0	506	48	46.3	48.6	6.88	20.1	3.50	14.7	0.36	0.17	2.44	1.04	1.68	62.72	1.86
17X 02 052–054	125.83	1.55	0.47	0.70	487	31.7	0.54	24.5	388	42	34.4	36.8	2.61	9.58	<0.5	13.9	0.38	0.13	1.81	1.11	1.65	65.39	2.01
17X 02 054–056	125.85	2.03	0.62	0.86	757	29.9	0.61	19.1	421	50	36.6	44.8	2.97	10.8	1.55	16.9	0.46	0.24	2.15	0.90	1.90	79.66	1.43
17X 02 056–058	125.87	1.80	0.61	0.91	341	29.7	0.65	30.6	424	54	36.6	44.0	3.58	14.2	2.13	19.3	0.48	0.16	4.55	0.96	2.00	86.07	1.45
17X 02 058–060	125.89	5.13	0.71	4.66	100	11.3	2.00	36.7	298	17	53.8	49.0	38.9	91.3	<0.5	103	2.46	2.34	6.33	1.03	3.38	24.85	0.92
17X 02 060–062	125.91	6.15	0.82	2.14	100	6.8	1.92	29.9	252	338	61.3	65.0	155	247	24.0	123	3.82	0.40	9.75	2.00	3.64	12.52	1.65

17X 02 062–064	125.93	6.45	0.80	2.13	100	6.1	1.88	34.5	264	198	56.1	75.7	207	293	15.2	143	4.11	0.51	11.66	2.88	4.61	14.40	1.88
17X 02 064–066	125.95	7.28	0.98	2.37	100	7.5	2.44	31.7	234	56	69.0	59.9	36.4	54.3	3.86	139	3.62	0.30	4.93	2.14	4.26	22.81	1.51
17X 02 066–068	125.97	7.64	0.79	2.12	100	4.2	2.10	30.4	142	9	74.8	61.5	26.5	39.0	6.67	144	3.83	0.32	4.24	2.41	4.80	28.60	1.51
17X 02 068–070	125.99	8.25	0.97	2.25	100	3.1	2.12	30.3	132	9	73.1	76.8	34.3	47.7	20.7	163	4.74	0.38	5.47	4.18	5.10	20.20	2.45
17X 02 070–072	126.01	8.63	1.11	1.77	100	3.4	2.00	26.4	133	43	76.0	84.3	27.3	44.9	53.0	168	4.83	0.39	5.53	4.85	4.78	24.49	3.04
17X 02 072–074	126.03	8.00	0.88	2.20	100	3.8	2.21	23.7	142	192	80.4	57.6	13.3	26.4	45.3	146	4.36	0.47	6.37	3.82	5.14	15.58	2.23
17X 02 074–076	126.05	9.25	1.04	1.86	100	5.1	1.55	64.3	167	46	68.6	92.8	91.4	86.6	89.1	165	5.10	0.60	13.16	8.18	6.03	16.01	4.07
17X 02 076–078	126.07	2.57	0.57	0.86	220	30.3	0.55	35.9	864	66	41.6	43.6	13.9	30.2	6.57	22.8	0.50	0.69	3.49	0.96	2.12	58.23	1.36
17X 02 078–080	126.09	2.54	0.62	0.87	248	29.9	0.50	37.2	880	69	36.9	44.7	38.1	71.3	13.4	19.3	0.41	0.18	3.39	0.55	2.20	63.45	0.75
17X 02 080–082	126.11	2.55	0.64	1.09	254	29.5	0.51	42.4	895	74	40.7	48.5	38.8	77.3	7.21	19.8	0.41	0.28	3.46	0.53	2.25	63.65	0.71
17X 02 082–084	126.13	2.98	0.69	1.01	288	27.2	0.53	43.8	920	92	49.9	55.8	49.7	97.9	8.27	20.5	0.46	0.26	4.33	0.54	2.64	69.76	0.62
17X 02 084–086	126.15	2.62	0.65	0.95	329	30.2	0.51	42.2	915	76	42.1	51.5	31.1	50.2	<0.5	18.6	0.39	0.22	3.05	0.52	2.45	61.54	0.64
17X 02 086–088	126.17	2.31	0.60	0.89	358	31.5	0.46	39.2	962	70	37.7	46.7	36.6	52.4	<0.5	15.0	0.33	0.20	2.70	0.46	2.16	56.75	0.65
17X 02 088–090	126.19	2.24	0.72	1.16	340	29.7	0.47	35.1	890	73	36.5	51.6	17.9	30.8	5.40	14.9	0.36	0.22	3.10	0.47	2.40	59.24	0.59
17X 02 090–092	126.21	2.08	0.73	1.05	337	29.2	0.47	25.7	881	74	35.5	42.9	34.4	44.2	5.78	14.9	0.35	0.16	4.71	0.60	2.46	59.36	0.73
17X 02 092–094	126.23	2.51	0.77	1.15	322	29.8	0.49	42.5	909	79	39.1	48.1	23.4	33.8	0.92	15.3	0.36	0.16	3.00	0.47	2.38	59.00	0.60
17X 02 096–098	126.27	2.67	0.77	1.26	326	29.9	0.52	42.6	911	85	40.5	51.1	5.31	16.9	0.83	15.8	0.38	0.15	3.36	0.46	2.54	58.81	0.54
17X 02 100–102	126.31	2.73	0.67	0.81	296	27.5	0.54	42.9	938	87	44.8	57.4	5.80	16.9	<0.5	15.7	0.39	0.21	3.39	0.46	2.80	59.48	0.50
17X 02 104–106	126.35	2.78	0.64	1.10	315	28.8	0.53	40.2	947	89	41.2	51.1	5.50	17.7	<0.5	14.0	0.37	0.24	4.32	0.45	2.59	56.52	0.52
17X 02 108–110	126.39	2.71	0.64	1.23	315	28.1	0.52	36.4	914	88	41.2	50.8	5.45	16.5	<0.5	14.8	0.40	0.26	3.15	0.44	2.70	57.02	0.48
17X 02 112–114	126.43	2.87	0.69	1.31	266	26.0	0.50	45.0	928	91	44.4	58.0	5.97	18.9	<0.5	15.2	0.39	0.27	4.28	0.44	2.74	57.53	0.48
17X 02 116–118	126.47	3.00	0.79	1.42	296	29.3	0.53	30.3	798	92	40.2	50.8	5.73	16.5	<0.5	14.4	0.43	0.25	3.96	0.49	3.04	59.29	0.49
17X 02 120–122	126.51	2.92	0.87	1.39	267	28.5	0.55	40.3	864	86	41.6	56.5	5.55	17.5	2.36	14.4	0.44	0.28	5.62	0.46	2.77	57.63	0.50
17X 02 124–126	126.55	3.05	0.89	1.45	283	27.1	0.55	48.9	870	96	43.9	57.9	5.93	21.5	1.66	15.8	0.41	0.40	6.48	0.50	3.00	58.78	0.50
17X 02 128–130	126.59	2.74	0.73	1.30	289	27.9	0.51	41.4	870	84	38.6	51.3	5.63	17.0	0.31	13.1	0.35	0.18	3.66	0.45	2.65	51.75	0.51
17X 02 132–134	126.63	2.30	0.76	1.19	314	29.8	0.46	42.3	912	77	36.1	50.0	52.5	16.8	<0.5	11.9	0.34	0.20	3.88	0.41	2.39	52.70	0.52
17X 02 136–138	126.67	2.36	0.73	1.22	293	29.0	0.46	34.1	880	77	35.6	50.3	5.36	16.0	<0.5	11.9	0.34	0.22	3.50	0.42	2.46	47.86	0.51
17X 02 148–150	126.79	2.06	0.60	1.03	260	24.3	0.31	35.1	940	68	33.8	49.8	5.12	12.1	5.95	11.0	0.31	0.05	5.05	0.47	2.06	44.49	0.68
17X 03 024–026	127.05	2.55	0.65	1.30	220	23.2	0.46	43.2	949	85	42.8	63.6	7.18	14.9	8.12	14.4	0.40	0.12	4.26	0.47	2.79	53.55	0.51
17X 03 054–056	127.35	2.22	0.66	1.17	210	32.4	0.42	42.1	951	77	38.5	61.6	7.12	13.7	7.12	12.8	0.35	0.09	370	0.45	2.51	46.98	0.53
17X 03 075–077	127.56	2.32	0.67	1.31	210	21.7	0.41	42.9	921	77	43.2	59.3	6.88	17.7	7.24	12.6	0.31	0.05	4.14	0.65	2.42	45.76	0.81
17X 03 082–084	127.63	2.27	0.66	1.24	220	23.2	0.42	41.5	937	78	44.6	66.6	7.22	15.3	5.89	12.7	0.40	0.09	3.73	0.46	2.64	47.63	0.52
17X 03 110–113	127.91	1.99	0.63	1.10	230	26.0	0.37	39.7	961	73	40.6	58.9	6.71	12.8	7.22	11.7	0.32	0.08	3.98	0.50	2.28	45.19	0.65
17X CC 026–028	128.36	2.47	0.76	1.18	260	23.5	0.46	49.5	912	84	39.6	51.3	7.87	14.3	8.89	13.8	0.39	0.06	6.20	0.60	2.59	52.77	0.69
18X 01 035–037	134.65	2.33	0.72	1.02	230	24.1	0.39	48.5	919	84	44.1	57.6	7.90	15.0	9.77	13.6	0.36	0.10	4.01	0.67	2.73	53.14	0.74
18X 01 115–117	135.45	2.39	0.73	1.11	250	26.6	0.38	49.1	914	84	42.1	57.7	7.50	15.0	10.3	13.4	0.35	0.72	4.31	0.72	25.1	50.97	0.86
18X 02 006–008	135.86	2.69	0.81	1.29	280	23.4	0.40	51.8	940	95	43.2	57.7	7.81	15.6	9.34	15.5	0.44	0.19	4.36	0.69	2.87	56.79	0.73
18X 02 057–059	136.37	2.38	0.73	1.13	240	23.4	0.39	48.0	947	82	45.2	57.3	8.04	15.2	13.2	14.1	0.39	0.07	4.38	0.83	2.71	54.21	0.92
18X 02 130–132	137.10	2.31	0.71	1.09	270	26.3	0.38	47.7	953	84	42.6	54.7	7.76	15.5	11.2	13.9	0.38	0.05	3.76	0.68	2.47	53.26	0.83
18X 03 048–050	137.78	2.43	0.76	1.18	250	24.1	0.40	52.8	956	88	47.1	58.9	8.06	15.0	11.0	15.0	0.39	0.11	3.73	0.81	2.71	54.79	0.90
18X 03 124–126	138.54	2.39	0.76	1.18	250	23.7	0.41	46.4	955	86	47.7	61.0	8.96	16.9	15.7	15.9	0.40	0.57	6.22	0.95	2.66	54.56	1.08
18X 04 011–013	138.91	2.61	0.78	1.20	220	23.9	0.39	53.1	985	89	47.9	59.4	10.6	19.1	141	16.0	0.40	0.11	8.66	22.19	2.78	53.71	23.92
18X 04 095–097	139.75	1.06	0.33	0.46	260	31.5	0.40	17.1	870	33	15.5	17.1	4.91	8.41	6.25	8.26	0.22	0.00	2.60	0.34	1.12	31.76	0.91

Ua, authigenic uranium (total U – Th/3). Al, K, Fe, Ca and Mg in wt %; others in ppm. (Shaded area corresponds to the spherule bed.)

Table 2. *Geochemical data from the K–T boundary interval at Hole 1049*

Sample	Depth (mbsf)	Al	K	Fe	Mn	Ca	Mg	Rb	Sr	Ba	V	Cr	Co	Ni	Cu	Zr	Hf	Mo	Pb	U	Th	REE	Ua
8H 01 015–017	109.15	2.44	0.67	1.03	290	22.1	0.77	35.1	717	64	82.2	100	13.4	27.2	14.8	15.9	0.38	0.18	3.47	0.74	1.90	53.07	0.11
8H 01 035–037	109.25	2.11	0.58	0.58	243	29.7	1.36	34.9	740	66	44.6	73.3	10.1	23.2	<0.5	14.9	0.41	0.40	3.20	0.81	2.03	55.62	0.13
8H 01 055–057	109.55	2.21	0.61	0.94	280	23.1	0.70	32.9	746	57	60.8	97.3	14.9	26.0	12.9	14.8	0.30	0.29	2.98	0.71	1.73	49.06	0.13
8H 01 075–077	109.75	2.30	0.63	0.97	250	23.0	0.71	31.9	713	67	41.5	66.6	12.4	17.8	11.9	14.8	0.41	0.35	4.07	0.90	2.30	58.58	0.13
8H 01 095–097	109.95	2.88	0.74	0.73	238	26.1	0.89	35.4	752	65	44.9	68.0	10.1	21.7	<0.5	15.2	0.46	0.26	3.15	0.83	2.09	54.73	0.13
8H 01 115–117	110.15	2.38	0.79	0.62	249	28.1	0.82	39.6	749	76	40.8	78.7	8.78	21.2	<0.5	17.2	0.50	0.24	3.64	0.88	2.28	57.76	0.12
8H 01 125–127	110.25	2.08	0.58	0.60	229	29.1	0.71	38.8	708	65	41.1	67.0	10.9	21.1	<0.5	14.3	0.43	0.24	2.88	0.83	2.12	56.94	0.12
8H 01 135–137	110.35	2.23	0.70	0.62	268	30.0	0.77	35.0	698	69	41.6	69.3	16.8	31.1	3.65	15.6	0.46	0.22	3.40	0.89	2.27	58.67	0.12
8H 01 145–147	110.45	2.16	0.69	0.62	263	30.2	0.75	25.9	697	69	41.9	68.1	12.7	27.2	1.67	14.8	0.42	0.27	3.88	1.00	2.35	61.26	0.21
8H 02 003–005	110.53	2.07	0.69	0.65	266	30.2	0.73	35.9	718	68	42.8	71.9	12.3	24.5	<0.5	15.9	0.41	0.26	3.56	0.90	2.16	59.36	0.18
8H 02 010–012	110.60	2.18	0.77	0.64	272	29.7	0.72	34.3	700	70	42.1	70.8	12.7	25.0	<0.5	15.6	0.45	0.24	3.80	0.94	2.29	61.47	0.17
8H 02 016–018	110.66	2.33	0.73	0.70	297	30.9	0.76	34.8	707	69	42.7	70.6	13.5	25.8	2.56	15.8	0.43	0.33	3.56	0.97	2.34	62.04	0.19
8H 02 023–025	110.73	2.21	0.75	0.67	264	32.4	0.75	36.5	721	69	43.8	72.4	13.1	24.1	<0.5	16.7	0.50	0.33	3.48	0.89	2.30	63.03	0.12
8H 02 031–033	110.81	2.17	0.75	0.63	262	21.0	0.77	33.4	677	68	41.4	68.1	23.8	35.9	<0.5	16.4	0.47	0.25	3.94	0.90	2.33	64.66	0.12
8H 02 038–040	110.88	2.71	0.70	0.65	244	28.7	0.78	41.6	711	82	51.8	83.2	12.7	24.4	<0.5	20.7	0.56	0.29	4.32	1.03	2.68	73.20	0.14
8H 02 045–047	110.95	2.41	0.65	0.60	231	29.4	0.78	37.2	706	72	44.7	73.7	11.3	28.5	<0.5	19.1	0.47	0.19	3.61	0.92	2.35	69.81	0.13
8H 02 050–052	111.00	2.42	0.64	1.07	254	29.6	0.72	41.7	710	74	46.4	76.6	14.1	27.1	<0.5	20.3	0.44	0.24	3.48	0.90	2.34	74.41	0.12
8H 02 052–054	111.02	2.40	0.63	1.05	273	30.4	0.70	38.3	722	75	45.4	76.8	16.5	32.6	<0.5	22.3	0.53	0.39	3.83	0.88	2.33	75.90	0.10
8H 02 054–056	111.04	2.29	0.65	0.95	255	29.7	0.69	30.6	693	71	52.1	72.4	22.0	44.6	<0.5	20.8	0.50	0.22	4.27	0.90	2.30	75.30	0.13
8H 02 056–058	111.06	1.97	0.54	0.87	315	31.2	0.61	30.7	526	57	52.7	55.4	11.8	20.6	7.69	17.7	0.44	0.24	2.76	0.88	1.98	77.39	0.22
8H 02 058–060	111.08	1.50	0.43	0.71	295	32.6	0.57	19.1	403	42	31.7	38.0	3.5	10.9	<0.5	15.6	0.40	0.23	1.53	1.05	1.63	73.18	0.50
8H 02 060–062	111.10	2.03	0.51	0.91	239	30.9	0.72	23.4	454	51	36.2	44.4	3.8	13.0	<0.5	19.8	0.48	0.15	2.05	0.95	1.92	79.76	0.31
8H 02 062–064	111.12	3.52	0.75	1.59	206	22.7	1.53	43.8	442	60	52.8	52.8	12.1	46.6	14.8	43.2	1.06	0.32	3.35	1.08	3.11	77.67	0.05
8H 02 064–066	111.14	3.83	0.75	0.99	197	18.8	1.65	39.4	418	54	55.6	46.5	22.3	93.7	18.8	47.3	1.27	0.34	4.07	0.99	4.17	71.49	−0.40
8H 02 066–068	111.16	3.61	0.57	4.91	50	7.8	1.94	20.0	258	13	64.1	47.1	75.4	115	16.7	67.2	1.42	3.51	8.62	0.75	2.94	11.76	−0.23
8H 02 068–070	111.18	7.33	0.69	1.13	<50	7.6	1.85	37.8	277	293	58.1	95.7	287	433	19.9	130	3.87	0.63	9.43	2.18	4.36	9.36	0.72
8H 02 070–072	111.20	7.48	0.66	1.96	<50	7.3	1.82	37.8	288	135	65.4	76.2	160	225	24.3	118	2.30	0.40	7.70	2.29	6.07	12.77	0.26
8H 02 072–074	111.22	6.89	0.77	2.07	<50	8.1	1.72	36.4	311	322	63.1	122	277	295	58.9	134	3.76	0.59	12.15	3.40	4.13	10.26	2.02
8H 02 074–076	111.24	6.34	0.91	2.63	<50	8.2	1.78	44.6	312	15	89.6	88.7	44.6	60.9	53.6	125	3.73	0.47	7.33	2.26	3.29	13.86	1.16
8H 02 076–078	111.26	2.25	0.54	0.96	199	31.3	0.47	34.5	895	62	31.7	42.1	12.5	23.4	<0.5	18.2	0.36	0.17	2.97	0.50	2.21	56.57	−0.24
8H 02 078–080	111.28	1.97	0.61	0.90	229	33.8	0.42	30.4	907	66	33.8	41.1	8.9	21.5	<0.5	18.2	0.36	0.17	3.56	0.38	1.98	55.18	−0.28
8H 02 080–082	111.30	2.17	0.65	1.10	218	31.0	0.42	43.1	917	71	35.1	44.8	11.0	20.8	<0.5	18.9	0.41	0.17	2.65	0.40	2.15	59.28	−0.32
8H 02 082–084	111.32	2.11	0.74	1.07	226	32.5	0.39	46.4	924	69	34.1	43.6	15.6	24.6	<0.5	17.8	0.39	0.16	2.69	0.42	2.16	58.32	−0.30

Ua, authigenic uranium (total U – Th/3). Al, K, Fe, Ca and Mg in wt %; others in ppm. (Shaded area corresponds to the spherule bed.)

Table 3. *Geochemical data from the K–T boundary interval at Hole 1050C*

Sample	Depth (mbsf)	Al	K	Fe	Mn	Ca	Mg	Rb	Sr	Ba	V	Cr	Co	Ni	Cu	Zr	Hf	Mo	Pb	U	Th	REE	Ua
10R 02 010–012	405.70	1.96	0.49	1.02	340	25.4	0.62	25.1	548	56	42.0	57.4	6.11	12.6	18.0	13.7	0.42	0.22	8.95	1.12	2.29	56.70	0.35
10R 02 025–027	405.85	2.18	0.55	1.34	440	23.2	0.62	31.0	510	65	40.2	57.6	6.15	15.3	8.27	18.1	0.56	0.24	4.56	0.77	2.73	69.44	−0.14
10R 02 033–035	405.93	1.90	0.47	1.15	160	24.8	0.58	29.0	518	53	31.6	43.6	5.01	13.8	6.59	14.9	0.49	0.24	3.75	0.59	2.43	63.34	−0.22
10R 02 040–042	406.00	1.76	0.45	1.99	350	27.9	0.37	25.9	956	66	34.5	56.2	4.76	10.6	5.06	13.6	0.40	0.09	2.75	0.49	2.03	50.52	−0.19
10R 02 044–046	406.04	2.71	0.58	1.47	160	21.1	0.47	38.4	920	86	45.5	73.4	8.02	19.4	7.63	17.4	0.47	0.14	6.45	0.56	3.19	57.42	−0.51
10R 02 048–050	406.08	3.25	0.61	1.55	100	20.2	0.56	40.0	957	97	39.4	64.7	8.23	24.7	8.26	17.3	049	0.23	5.42	0.52	3.43	58.22	−0.62
10R 02 055–057	406.15	2.28	0.56	1.20	440	23.2	0.44	35.2	988	90	36.3	59.8	5.58	15.1	12.4	15.9	0.46	0.21	7.74	0.56	2.82	60.88	−0.37
10R 02 060–062	406.20	2.32	0.59	1.38	340	22.4	0.45	36.9	1004	96	39.5	63.6	6.41	16.3	26.3	17.1	0.49	0.32	6.12	0.60	2.98	61.45	−0.39
10R 02 068–070	406.28	2.81	0.61	1.48	120	21.2	0.54	38.0	989	96	41.7	70.8	7.42	18.8	27.8	16.4	0.47	0.28	5.14	0.62	3.06	60.55	−0.40
10R 02 078–080	406.38	2.31	0.56	1.31	120	22.7	0.45	33.7	1005	96	40.2	65.0	7.48	18.6	24.2	15.5	0.46	0.93	5.92	0.65	2.95	57.79	−0.33

Ua, authigenic uranium (total U – Th/3). Al, K, Fe, Ca and Mg in wt %; others in ppm.

Table 4. *Geochemical data from the K–T boundary interval at Hole 1052E*

Sample	Depth (mbsf)	Al	K	Fe	Mn	Ca	Mg	Rb	Sr	Ba	V	Cr	Co	Ni	Cu	Zr	Hf	Mo	Pb	U	Th	REE	Ua
18R 01 065–068	300.75	2.30	0.59	1.10	120	22.5	0.84	31.0	1003	89	38.9	78.5	5.68	13.1	11.3	16.3	0.48	0.17	4.55	1.00	2.59	57.91	0.13
18R 01 105–108	301.15	1.84	0.51	0.93	160	23.4	0.87	27.4	940	70	36.4	69.9	4.00	11.0	6.17	13.7	0.41	0.11	3.78	1.01	2.45	50.16	0.19
18R 01 145–148	301.55	1.88	0.51	0.98	130	24.9	0.85	28.1	949	69	36.1	66.1	5.54	15.1	10.8	14.2	0.40	0.17	4.43	1.12	2.32	48.51	0.34
18R 02 013–017	301.73	1.81	0.48	0.88	240	25.4	0.78	25.5	938	61	35.3	64.2	5.09	13.0	12.4	14.4	0.40	0.16	4.06	1.08	2.12	50.62	0.37
18R 02 029–032	301.89	1.94	0.52	0.93	340	24.5	0.82	27.5	935	66	35.2	66.5	5.12	14.5	7.76	15.4	0.42	0.15	3.79	1.25	2.19	55.36	0.51
18R 02 043–046	302.03	2.29	0.58	1.05	480	22.0	0.89	32.7	894	81	42.3	78.6	6.70	19.0	12.2	19.3	0.54	0.16	4.72	1.36	2.78	65.67	0.43
18R 02 048–051	302.08	2.61	0.65	1.20	460	20.5	0.88	36.7	907	91	43.9	81.0	10.1	24.9	8.06	22.9	0.61	0.17	5.04	1.50	2.98	72.07	0.50
18R 02 051–054	302.11	1.66	0.43	0.94	70	25.3	0.86	21.2	584	46	32.1	41.5	7.10	18.5	8.28	17.0	0.47	0.21	4.09	1.06	2.05	56.40	0.38
18R 02 057–059	302.16	1.16	0.32	0.71	100	25.6	0.88	14.3	518	28	19.9	25.6	3.93	11.1	4.97	12.1	0.34	0.11	3.63	0.82	1.36	39.29	0.37
18R 03 000–004	302.21	2.27	0.53	1.16	80	22.3	0.59	32.3	1079	79	35.5	53.3	3.57	8.19	16.8	16.3	0.55	0.11	4.20	0.60	2.78	55.65	−0.32
18R 03 027–030	302.48	2.12	0.49	1.10	150	22.2	0.62	30.3	1074	78	34.8	51.7	4.27	10.5	9.81	14.2	0.39	0.10	5.40	0.52	2.70	53.04	−0.38
18R 03 052–055	302.73	2.31	0.54	1.24	120	24.8	0.65	31.6	1060	99	36.7	54.9	4.46	10.0	6.34	14.9	0.41	0.08	4.71	0.52	2.96	56.37	−0.46
19R 01 024–027	309.94	2.65	0.64	1.28	160	22.8	0.56	46.1	1146	117	42.5	66.0	7.29	15.8	11.1	14.6	0.44	0.23	6.19	0.99	3.39	59.39	−0.14
19R 01 053–056	310.23	2.69	0.65	1.23	130	21.6	0.52	43.0	1143	121	43.0	63.0	5.96	12.7	12.0	14.5	0.48	0.24	5.99	1.02	3.28	56.33	−0.07
19R 01 099–102	310.69	2.34	0.61	1.21	1.60	21.4	0.51	41.6	1099	112	40.0	60.7	6.98	15.4	10.3	14.6	0.46	0.19	5.41	0.93	3.23	53.81	−0.14

Ua, authigenic uranium (total U – Th/3). Al, K, Fe, Ca and Mg in wt %; others in ppm.

triplicate using a Perkin Elmer Sciex Elan-5000 spectrometer with Rh as internal standard, and AAS analyses were carried out with a Perkin Elmer 5100 ZL spectrometer. The quality of the analyses was monitored with laboratory and international standards from the United States Geological Survey (USGS). ICP-MS precision and accuracy were better than $\pm 2\%$ and $\pm 5\%$ relative for analyte concentrations of 50 and 5 ppm in the rock, respectively. AAS analytical error was $<2\%$.

Ir analyses of bulk samples were carried out by neutron activation at Centre des Faibles Radioactivés, Gif-sur-Yvette. Iridium was counted with a γ–γ spectrometer detecting the 316–468 keV γ-ray coincidence resulting from the decay of ^{192}Ir (Rocchia *et al.* 1990).

Element stratigraphy across the K–T boundary

The chemical data from analyses performed on all selected samples from the K–T boundary interval at Blake Nose are presented in Tables 1–4, corresponding to Holes 1049A, 1049B, 1050C and 1052E, respectively, and are plotted in Figs 3–7. Figure 3 represents the entire analysed interval at Hole 1049A, although the closely spaced samples at the K–T boundary bed and nearby sediments above and below are not clearly visible in this graph. Therefore, a selected interval showing the K–T boundary chemical composition in greater detail has been plotted in Fig. 4. When analysing sediments with considerable variations in carbonate content, as is the case for K–T boundary interval sediments, some fluctuations in elemental contents may result from variations in the carbonate/aluminosilicate ratio. Whole-rock concentrations of elements have, therefore, been normalized to Al as an index of the relative abundance of detrital phases and plotted in Figs 3–7. Figures 8 and 9 show the Ir profiles at the K–T boundary interval in Holes 1049A and 1049B.

Al and Ca are plotted in Figs 3–7 to give an idea of the abundances of detrital and biogenic material. As occurs in other K–T sections, the boundary is marked by a significant decrease in carbonate content as a consequence of the mass extinction and deposition of the impact fallout material. Figure 4 shows these Al and Ca variations and also the element abundance variations across the K–T boundary at Hole 1049A. As the boundary layer is better preserved at this hole, our discussion will focus on this interval and the geochemical record will then be compared with the interval analysed in the other holes.

Sr and Mg

Because Sr is closely related to Ca and carbonates, the Sr and Ca contents across the K–T boundary show a similar decrease in concentration (Tables 1 and 2), and, therefore, the Sr/Ca ratio has been considered and plotted in Figs 3–7. Even considering whole-rock analysis data, the observed decrease in Sr content in relation to Ca content across the boundary suggests a response to a global Sr decrease in early Tertiary time, reported also at other K–T boundary sections (Renard 1986). When whole-rock Mg data are considered, the profile would be expected to show a significant increase in the Mg content of the spherule bed (Tables 1 and 2) as a consequence of the decrease in carbonates and an increase in smectites. However, the Mg content normalized to Al shows an increase in Mg across the boundary which is also consistent with a global trend (Renard 1986). Variations in the Sr and Mg contents are also in good agreement with data from Hole 1049C (Speed & Kroon 2000). Although further analyses on the carbonate fraction are needed, these variations in Mg and Sr concentration seem to correlate with sea-level changes and global fluctuations.

Redox-sensitive elements and diagenetic remobilization

In the analysed interval, the Fe and Mn profiles are mainly governed by diagenetic remobilization. At Hole 1049A, where a thicker interval has been analysed, the Fe concentration is similar in Cretaceous and in Tertiary sediments (Fig. 3), suggesting similar detrital fluxes and palaeoceanographic conditions. However, the K–T boundary is marked by a significant Fe increase in the top part of the boundary bed (Fig. 4) and lower Fe and Mn concentrations within the boundary bed; the enhanced concentrations of these elements indicate diagenetic remobilization. After the mass extinction, oxygen consumption would have been enhanced by oxidation of the accumulated organic matter, and suboxic or reducing conditions would be expected after deposition. In such conditions, elements sensitive to changes in redox conditions, such as Fe and Mn, would have undergone severe redistribution. These oxygen-depleted conditions led to diffusion of Fe and Mn, precipitated upon encountering oxygenated pore waters, being oxidized and immobilized as oxyhydroxides. The two elements became decoupled during diagenesis, the Mn peak being located above the Fe peak (Fig. 4), which constrains the Eh conditions at which Fe was immobilized and Mn continued to diffuse upward upon encountering more oxygenated pore waters. Such Fe diagenetic remobilization

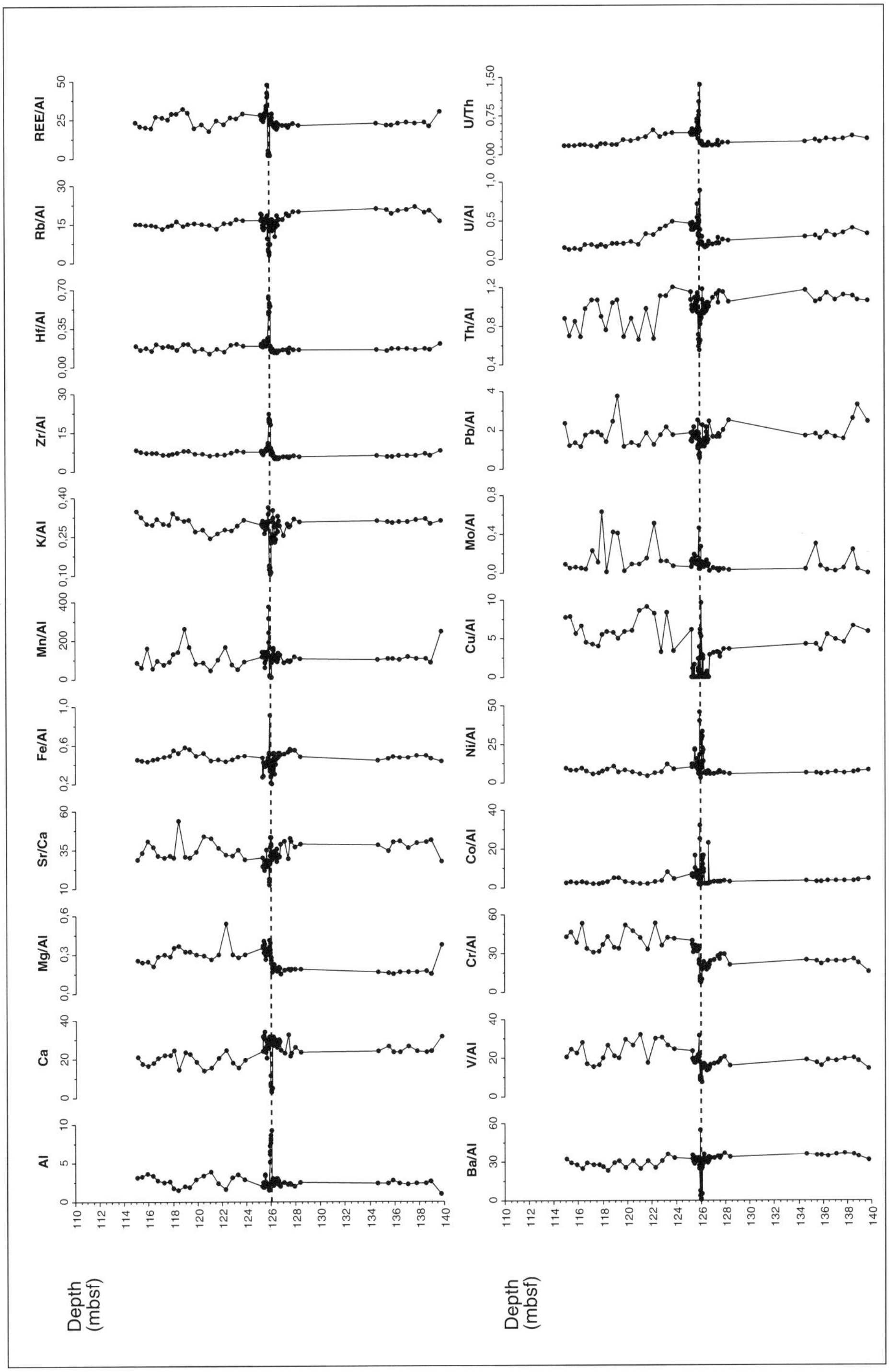

Fig. 3. Geochemical data from the K–T boundary interval at Hole 1049A. Plots show the Ca and Al concentrations (wt %), Sr/Ca and Th/U ratios, Fe, K and Mg concentrations normalized to Al and trace-element/Al weight ratio ($\times 10^4$) v. depth profiles.

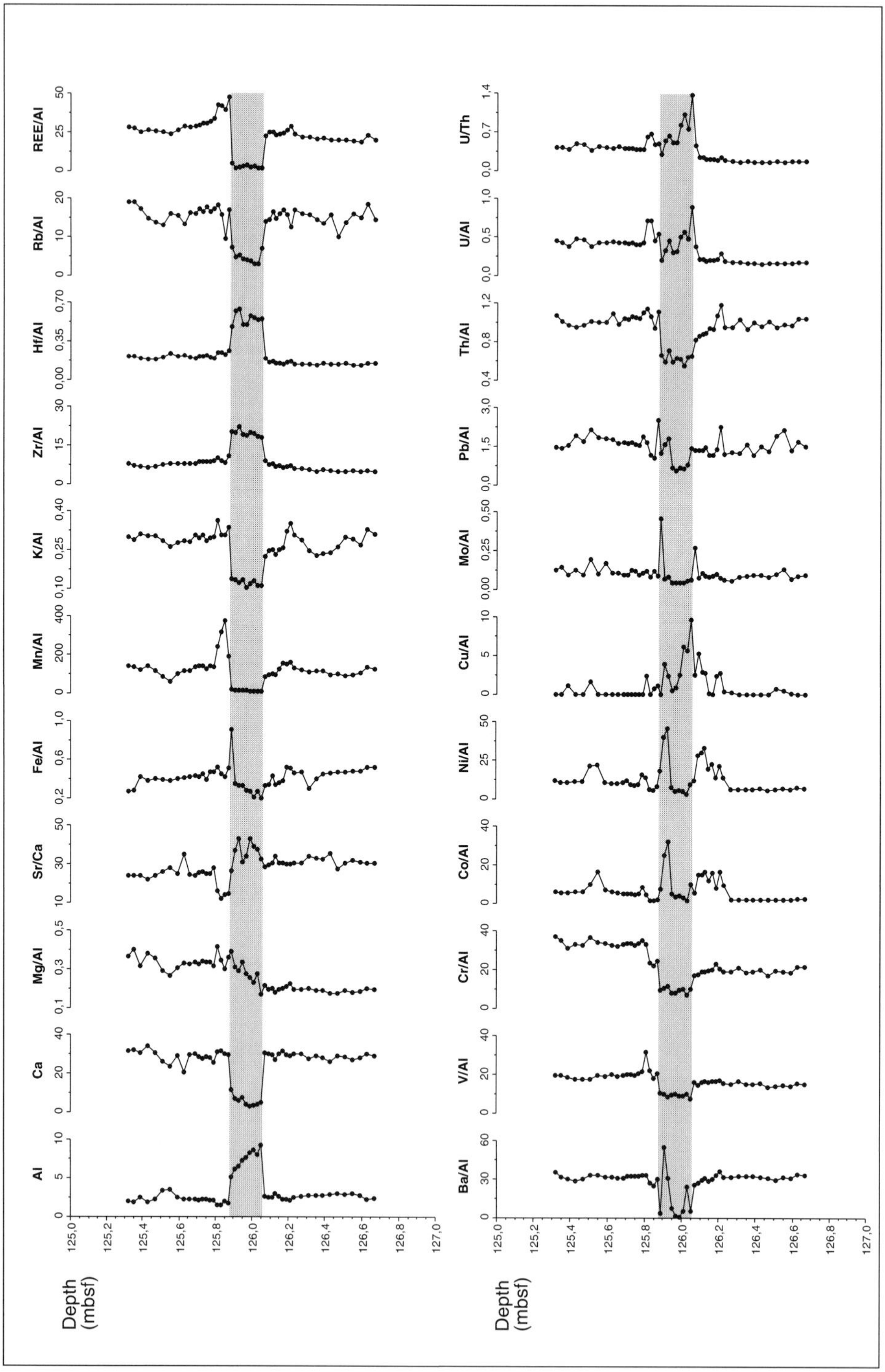

Fig. 4. Enlarged detail of K–T boundary interval showed in Fig. 3 presenting the K–T chemical profiles of the spherule bed and adjacent sediments at Hole 1049A. (Shaded area represents the spherule bed.)

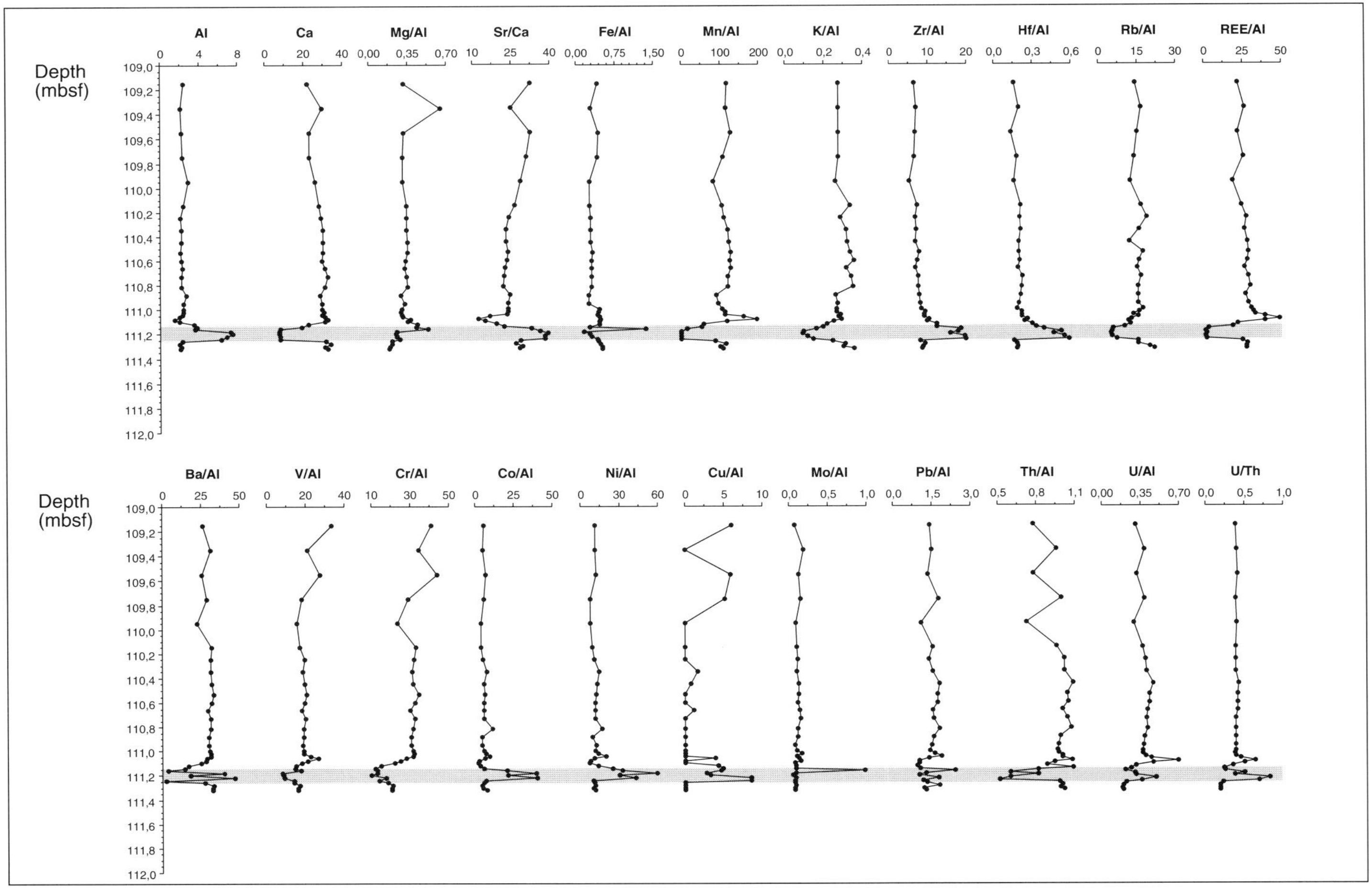

Fig. 5. Geochemical data from the K–T boundary interval at Hole 1049B. Plots show the Ca and Al concentrations (wt %), Sr/Ca and Th/U ratios, Fe, K and Mg concentrations normalized to Al and trace-element/Al weight ratio ($\times 10^4$) v. depth profiles. (Shaded area represents the spherule bed.)

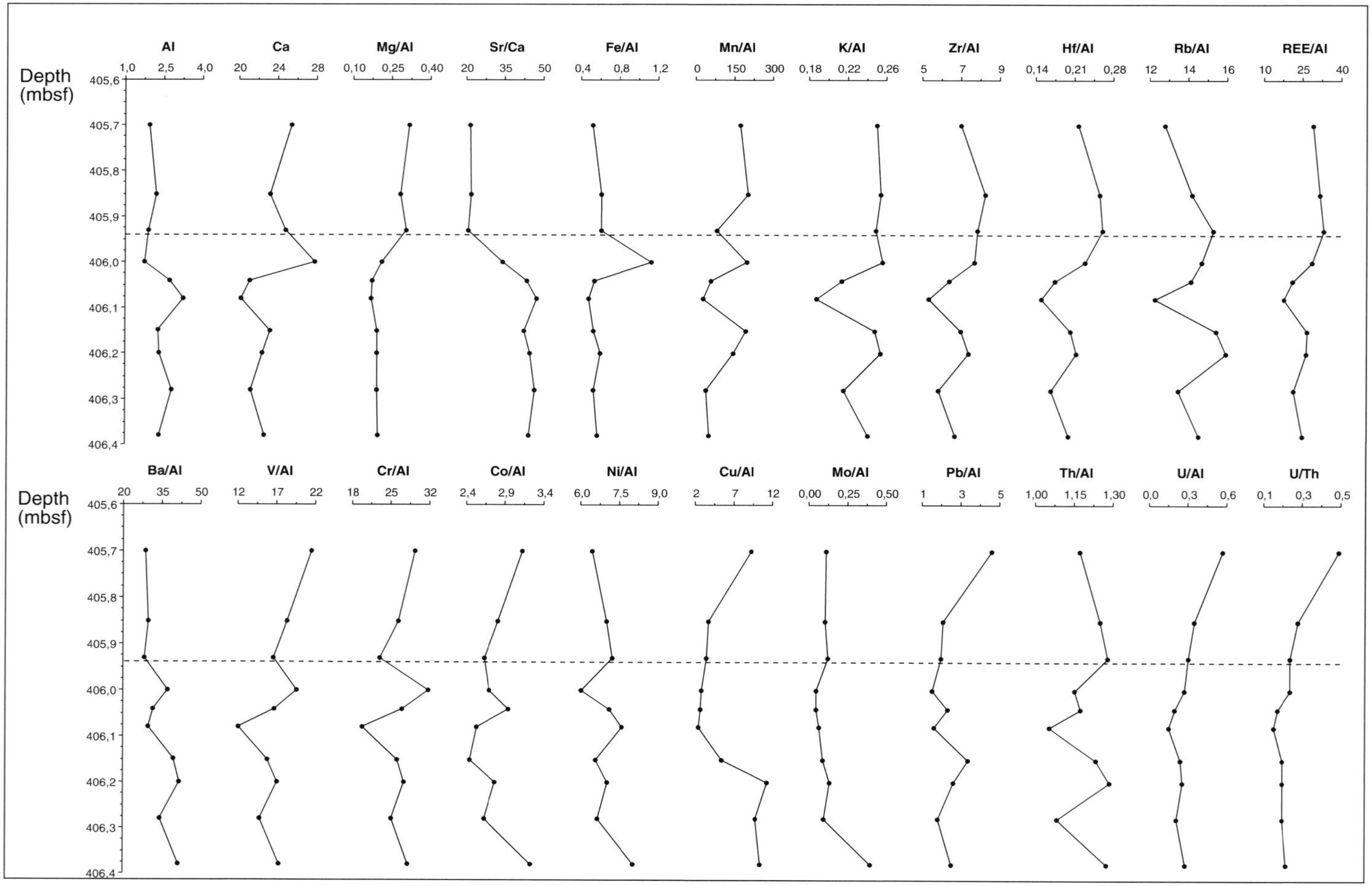

Fig. 6. Geochemical data from the K–T boundary interval at Hole 1050C. Plots show the Ca and Al concentrations (wt %), Sr/Ca and Th/U ratios, Fe, K and Mg concentrations normalized to Al and trace-element/Al weight ratio ($\times 10^4$) v. depth profiles. (The dashed line marks the K–T boundary.)

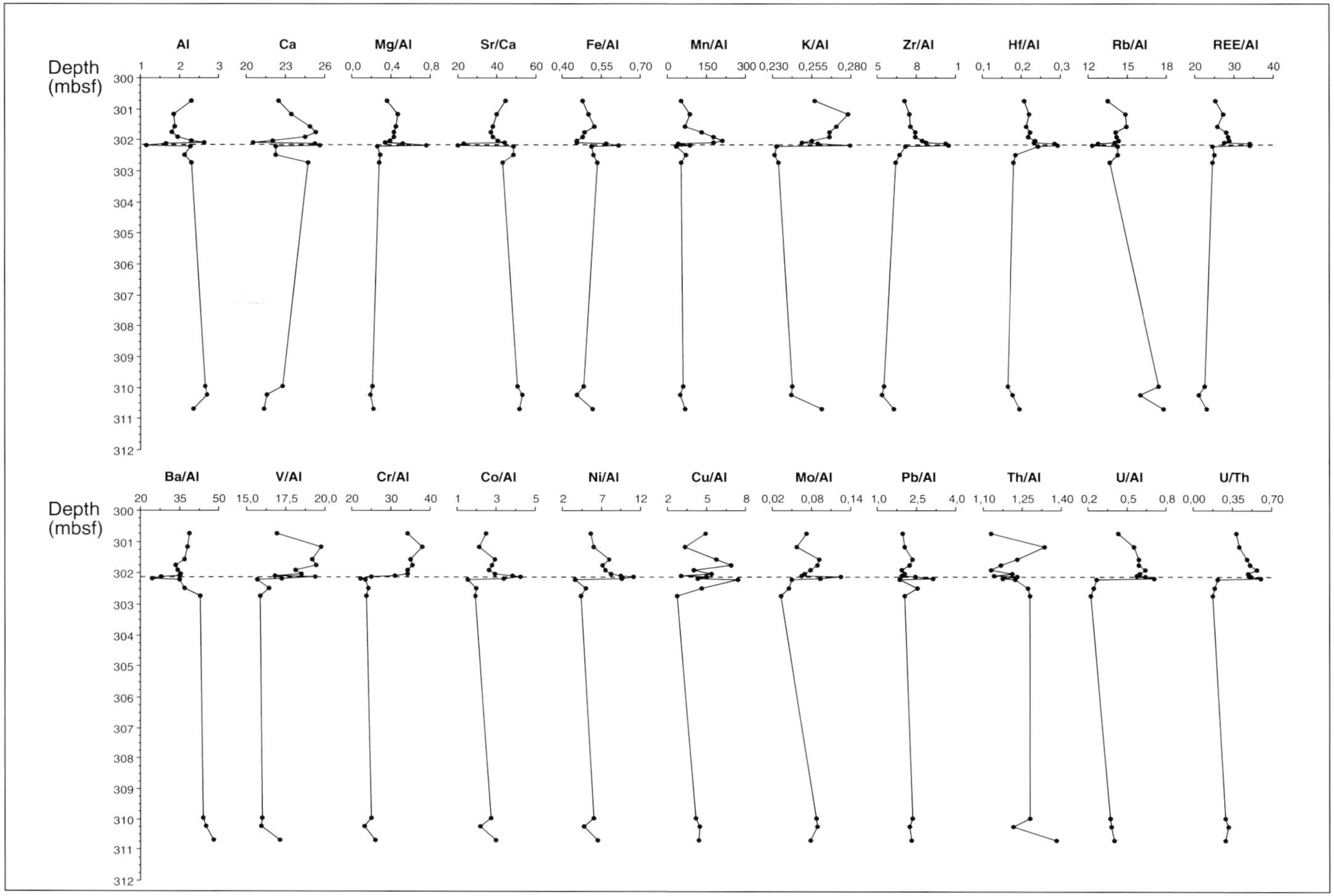

Fig. 7. Geochemical data from the K–T boundary interval at Hole 1052E. Plots show the Ca and Al concentrations (wt %), Sr/Ca and Th/U ratios, Fe, K and Mg concentrations normalized to Al and trace-element/Al weight ratio ($\times 10^4$) v. depth profiles. (The dashed line marks the K–T boundary.)

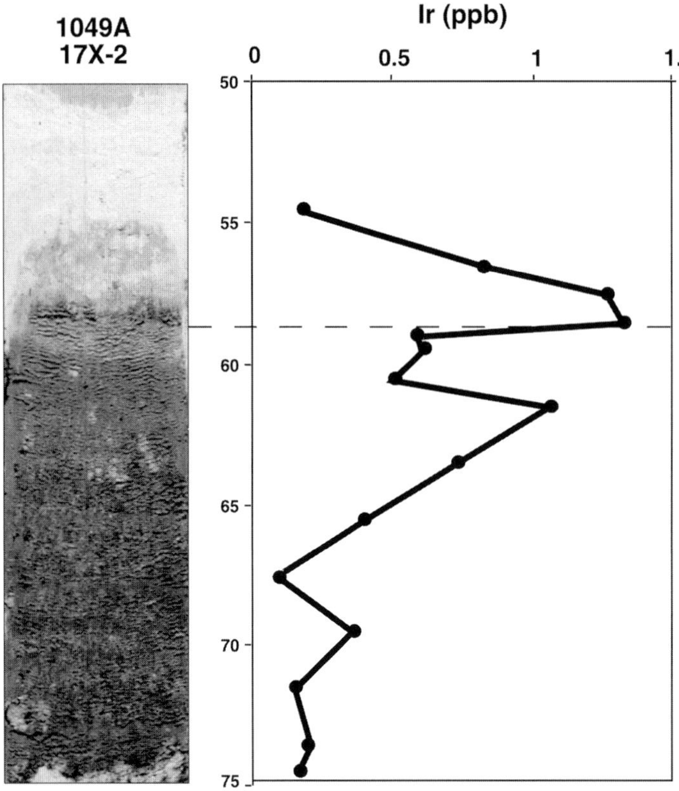

Depth (cm in core)	Ir (ppb)
54.0	0.20 ± 0.05
56.0	0.82 ± 0.10
57.0	1.26 ± 0.10
58.0	1.32 ± 0.12
58.5	0.60 ± 0.08
59.0	0.62 ± 0.10
60.0	0.50 ± 0.08
61.0	1.05 ± 0.12
63.0	0.71 ± 0.10
65.0	0.38 ± 0.10
69.0	0.28 ± 0.10
71.0	0.12 ± 0.10
73.0	0.20 ± 0.10
74.0	0.18 ± 0.10

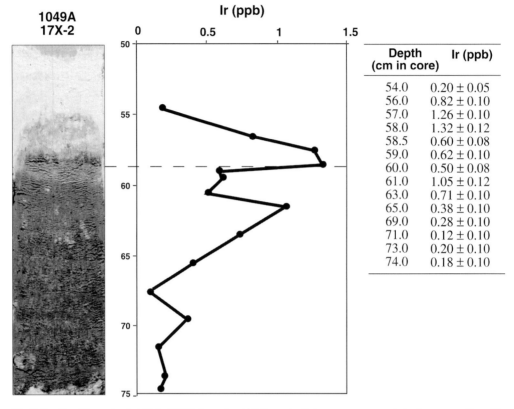

Fig. 8. Profile of Ir content at Hole 1049A (sample at 58.5 cm depth in core corresponds to the orange Fe-oxide layer).

is responsible for the 3 mm thick rusty layer capping the spherule bed (Fig. 2) that initially appeared to be the fireball layer (Norris *et al.* 1998), but only represents in fact a diagenetic concentration of Fe, with no enhanced extra-terrestrial-element flux observed, as discussed below. At Hole 1049B where the spherule bed is preserved, and Hole 1052E, with burrows containing spherules, the same Fe and Mn patterns are observed.

Reducing conditions are also constrained by the U/Th ratio, which is a potential palaeoredox indicator. Th is relatively immobile in the sedimentary environment and is concentrated in the detrital fraction associated with heavy minerals or clays. In contrast, the U concentration is dependent on redox conditions. U^{6+} is soluble, but U^{4+} is precipitated, therefore, concentrating in a reducing environment and raising the U/Th ratio. Wignall & Myers (1988) proposed the authigenic U content (Ua) as an index of bottom-water oxygenation calculated as (authigenic U) (total U) – Th/3, which can be a reliable redox proxy (Jones & Manning 1994). The U/Th ratio and Ua content (Figs 4 and 5; Tables 1–3) reach higher values in the spherule bed, indicating more reducing conditions.

Ba also shows some fluctuations in its content in the spherule bed, governed by diagenetic remobilization. In deep-sea sediments, bulk Ba is associated with the detrital fraction, with carbonates, or it occurs as authigenic or biogenic barite. It is known that Ba has enhanced concentrations in pelagic sediments that are deposited in high-productivity areas, as it has a strong biogenic association, and its content increases with higher biogenic production (Goldberg & Arrhenius 1958; Schmitz, 1987; Dymond *et al.* 1992; Paytan *et al.* 1993; Paytan 1995). The Ba/Al ratio gives a general idea of the Ba associated with carbonates or biogenic barite, and although any palaeoproductivity interpretation would require an evaluation of biogenic barite concentration (e.g. Paytan *et al.* 1993; Paytan 1995), the slight decreases in Ba above the spherule bed could be related to lower

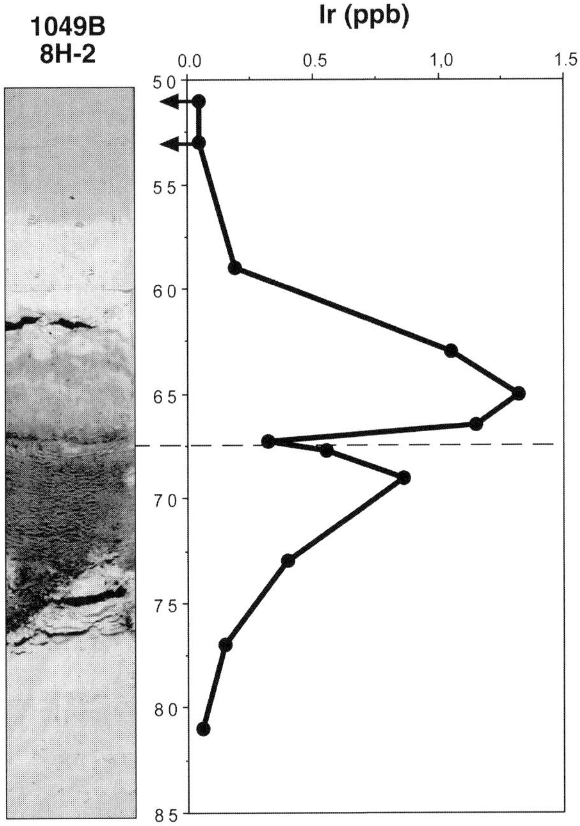

Depth (cm in core)	Ir (ppb)
51.00	≤ 0.05
53.00	≤ 0.05
59.00	0.19 ± 0.03
63.00	1.05 ± 0.07
65.00	1.32 ± 0.06
66.50	1.15 ± 0.08
67.25	0.32 ± 0.03
67.75	0.55 ± 0.09
69.00	0.86 ± 0.08
73.00	0.40 ± 0.07
77.00	0.15 ± 0.03
81.00	0.06 ± 0.03

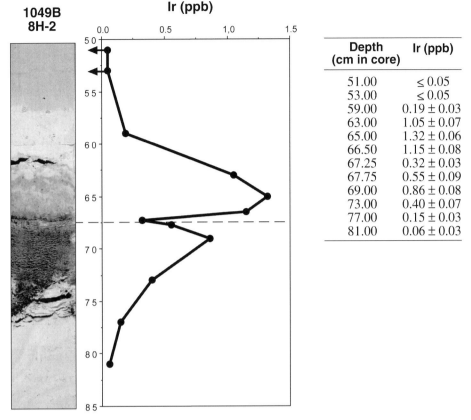

Fig. 9. Profile of Ir content at Hole 1049B (sample at 67.25 cm depth in core corresponds to the orange Fe-oxide layer).

productivity. The low carbonate content and the absence of biogenic barite in the spherule bed led to low Ba contents (Tables 1 and 2). Ba in this bed is probably associated with the aluminosilicate fraction and secondary barite, as the enhanced Ba content within the upper part of the spherule bed also suggests diagenetic remobilization of Ba during suboxic diagenesis.

The V, Mo and Cu profiles are significantly affected by changes in redox conditions and diagenetic remobilization as well. The V content increases above the spherule bed, similar to Mn, and it may have been mobilized upward and precipitated with oxidized phases. A similar profile for Mo with an enrichment at the top of the spherule bed also indicates redistribution of this element.

Detrital elements

Detrital element profiles are governed by source material and mineral composition. The decrease in K content is related to changes in the composition of the clay mineral assemblages and the absence of illite in the spherule layer. In the Cretaceous and Tertiary sediments, clay mineral associations are composed of smectite, illite, kaolinite and minor proportions of chlorite (Martínez-Ruiz *et al.* 2000), whereas the spherule bed is mainly composed of smectite (Martínez-Ruiz *et al.* this volume) of authigenic origin. Typical detrital elements associated with detrital clays have lower concentrations in this bed. In addition, diagenetic alteration of the ejecta material led to REE depletion (Tables 1 and 2). These circumstances would explain the decrease in K, Rb, REE, Pb and Th. In fact, severe changes in major and trace element composition may have occurred during the alteration of the ejecta material to smectite. Izett (1991) examined Haitian spherules, and showed that some trace elements are severely depleted in smectite relative to the impactite glass (Izett 1991, Table 5). Thus, Cr or Rb could have been severely depleted. Izett's analyses also showed depletion in Zr and Hf, although both these elements are abundant in the spherule bed at Blake Nose, probably derived from rutile and zircon, which have been reported by scanning electron microscopy analyses.

Extraterrestrial elements

Because diagenetic alteration limits the interpretation of extraterrestrial Fe, other typical extraterrestrial elements are considered to evaluate the possible extraterrestrial contribution in Blake Nose sediments. The Ir concentration is very low at Blake Nose (Smit *et al.* 1997), reaching the highest concentration above the spherule bed in the burrow-mottled calcareous ooze (Figs 8 & 9). The contents of Cr, Co and Ni present different profiles at this location. The Cr content decreases considerably in the spherule bed, which points to the absence of a significant extraterrestrial Cr contribution. Although Co and Ni concentrations are not as high as in distal sections, both elements are enriched in the upper part of the spherule bed, suggesting a possible enrichment derived from extraterrestrial material. Although the main contribution to the Blake Nose spherule bed is target rock material from Chicxulub it seems to have been associated with the finest fraction and therefore deposited after the spherule bed material, although diagenetic remobilization during suboxic diagenesis could also have been an important factor in Ir distribution at this location. Ir reaches its maximum content just above the orange Fe-oxide layer (Figs 8 & 9).

Comparison with other K–T ejecta deposits

The Chicxulub impact resulted in ejecta deposits identified in outcrops and drill cores world-wide (e.g. Alvarez *et al.* 1995; Pierazzo & Melosh 1999; Smit 1999), with melted target rocks deposited closer to the crater site. A major terrestrial contribution is, therefore, expected at Blake Nose. The vertically-expanding hot vapour plume of vaporized bolide with some entrained melted target rocks was dispersed and deposited globally (the fireball layer). In some K–T boundary sections, such as North American nonmarine sections, a dual nature of the K–T boundary interval is clearly evident (e.g. Pollastro & Bohor 1993), and in distal marine sections, such as those in SE Spain, only a single clay layer, equivalent to the uppermost layer in nonmarine sections, records the boundary event. In the basinal Gulf of Mexico and the Caribbean, the K–T boundary is marked by a mixture of reworked microfossils, impact-derived material and lithic fragments deposited by giant gravity flows, the K–T boundary 'cocktail' (Bralower *et al.* 1998). Impact signatures and geochemical anomalies are, therefore, different in proximal and distal locations, also as a function of different diagenetic environment. At Blake Nose, the major contribution of ejecta material is derived from target rocks, and the geochemical composition of the spherule bed reflects the composition of the precursor material modified by diagenetic alteration and presents minor extraterrestrial contribution. At distal locations, the nature of the spherules, which are microkrystites (Smit *et al.* 1992), and other impact signatures such as Ni-rich spinels (Bohor *et al.* 1986), indicate a different nature of the impact-derived material. Broadly speaking, after an impact the projectile material is ejected first and attains the highest speed, whereas the target material is ejected later and is slower (e.g. Melosh & Vickery 1991). The deposition of the impact-melted target rocks from Chicxulub, mainly consisting of impact glasses (e.g. Koeberl & Sigurdsson 1992), would have occurred quickly after the impact and more rapidly than deposition of the finest fraction derived from a cloud of vaporized bolide and entrained target-rock material. This material, usually termed the fireball layer, was deposited by gravitational processes on top of the recently deposited target-rock-derived material (e.g. Pollastro & Bohor 1993). Reworking of the K–T boundary material at Blake Nose (Klaus *et al.* 2000) prevented the preservation of the ejecta layer stratigraphy, although enhanced concentrations of Ir, Co and Ni in the upper part of the spherule bed suggest a more significant extraterrestrial contamination within this part. Nevertheless, no clear evidence for significant contribution of extraterrestrial material is observed at Blake Nose. The nature of the spherules and geochemical composition of the spherule bed suggest it is composed mostly of target-rock material, with very low amounts of bolide-derived material.

Conclusions

The K–T boundary at Blake Nose is marked by a spherule bed up to 17 cm thick at ODP Site 1049 containing green spherules, composed mostly of smectite. Reworking of the spherule material limits the interpretation of the possible original stratigraphy, but this layer none the less provides further evidence for the deposition of the Chicxulub impact-generated material, and it represents the impact-melted target rocks. Ir and other extraterrestrial elements show lower concentrations than at distal locations. Only Ni and Co have higher concentrations within the upper part of the spherule bed. Original concentrations have been severely modified after deposition. Low Eh conditions led to trace-element remobilization. Fe, Mn, V and Mo mobilized, diffusing upward and precipitating upon encountering the oxygenated pore waters required for their precipitation. As different oxygen conditions are required for precipitation of these elements, their concentrations became decoupled, showing peaks at different depths. Major chemical changes also accompanied the diagenetic alteration of glass to smectite. REE, and possibly other associated elements, were significantly depleted during this alteration. Eh and alteration of glass are therefore the main factors controlling the geochemical profiles across the K–T boundary at Blake Nose.

We thank the ODP Leg 171B Shipboard Scientific Party and the crew of the *JOIDES Resolution* for assistance with the samples and data, and the Bremen ODP Core Repository for assistance during the sampling party. One of the authors (F. Martínez-Ruiz) thanks the University of Granada and 'Junta de Andalucía' for financial support for participation on ODP Leg 171B. This work was partially supported by Project PB96-1429 (DGES, MEC, Spain) and Research Group RNM0179 (Junta de Andalucía, Spain). We thank the C. I. C. (University of Granada, Spain) for the use of the analytical facilities. We also thank P. Sánchez-Gómez, I. Nieto and E. Abarca for their help in the laboratory. This paper benefited from the careful revision of C. Koeberl.

References

ALVAREZ, L. W., ALVAREZ, W., ASARD, F. & MICHEL, N. V. 1980. Extraterrestrial cause for Cretaceous/Tertiary extinction. *Science*, **208**, 1095–1108.

ALVAREZ, W., CLAEYS, P. & KIEFFER, S. 1995. Emplacement of Cretaceous–Tertiary boundary shocked quartz from Chicxulub crater. *Science*, **269**, 930–935.

BHANDARI, N., SHUKLA, P. N. & CASTAGNOLI, G. C. 1993. Geochemistry of some K–T sections in India. *Palaeogeography, Palaeoclimatology, Palaeoecology*, **104**, 199–211.

BOHOR, B. F., FOORD, E. E. & GANAPATHY, R. 1986. Magnesioferrite from the Cretaceous–Tertiary boundary, Caravaca, Spain. *Earth and Planetary Science Letters*, **81**, 57–66.

BRALOWER, T. J., PAULL, C. K. & LECKIE, R. M. 1998. The Cretaceous–Tertiary boundary cocktail: Chicxulub impact triggers margin collapse and extensive sediment gravity flows. *Geology*, **26**, 331–334.

DYMOND, J., SUESS, E. & LYLE, M. 1992. Barium in deep-sea sediment: a geochemical proxy for palaeoproductivity. *Palaeoceanography*, **7**, 163–181.

GOLDBERG, E. D. & ARRHENIUS, G. 1958. Chemistry of Pacific pelagic sediments. *Geochimica et Cosmochimica Acta*, **13**, 153–212.

HILDEBRAND, A. R., PENFIELD, G. T., KRING, D. A. *et al.* 1991. Chicxulub crater: a possible Cretaceous/Tertiary boundary impact crater on the Yucatan Peninsula, Mexico. *Geology*, **19**, 867–871.

IZETT, G. A. 1991. Tektites in Cretaceous–Tertiary boundary rocks on Haiti and their bearing on the Alvarez impact extinction hypothesis. *Journal of Geophysical Research*, **96**, 20879–20905.

JONES, B. & MANNING, D. A. C. 1994. Comparison of geochemical indices used for the interpretation of palaeoredox conditions in ancient mudstones. *Chemical Geology*, **111**, 111–129.

KELLER, G., MACLEOD, N., LYONS, J. B. & OFFICER, C. B. 1993. Is there evidence for Cretaceous–Tertiary boundary age deep water deposits in the Caribbean and Gulf of Mexico? *Geology*, **21**, 776–780.

KLAUS, A., NORRIS, R. D., KROON, D. & SMIT, J. 2000. Impact-induced mass wasting at the K–T boundary: Blake Nose, western North Atlantic. *Geology*, **28**, 319–322.

KLAVER, G. T., VAN KEMPEN, T. M. G., BIANCHI, F. R. & VAN DER GAAST, S. J. 1987. Green spherules as indicators of the Cretaceous/Tertiary boundary in Deep Sea Drilling Project Hole 603B. *In*: VAN HINTE, J. E., WISE, S. W., Jr *et al.* (eds). *Initial Reports of the Deep Sea Drilling Project*, **93**. US Government Printing Office, Washington, DC, 1039–1056.

KOEBERL, C. & SIGURDSSON, H. 1992. Geochemistry of impact glasses from the K–T boundary in Haiti: relation to smectites and a new type of glass. *Geochimica et Cosmochimica Acta*, **56**, 2113–2129.

KYTE, F. T., SMIT, J. & WASSON, J. 1985. Siderophile interelement variations in the Cretaceous–Tertiary boundary sediments from Caravaca, Spain. *Earth and Planetary Science Letters*, **73**, 183–195.

KYTE, F. T., ZHOU, Z. & WASSON, J. 1980. Siderophile-enriched sediments from the Cretaceous–Tertiary boundary. *Nature*, **288**, 651–656.

MARTÍNEZ-RUIZ, F., ORTEGA-HUERTAS, M. & PALOMO, I. 2000. Climate, tectonics and meteoritic impact expressed by clay mineral sedimentation across the Cretaceous–Tertiary boundary at Blake Nose, Northwestern Atlantic. *Clay Minerals*, in press.

MARTÍNEZ-RUIZ, F., ORTEGA-HUERTAS, M., PALOMO, I. & BARBIERI, M. 1992. The geochemistry and mineralogy of the Cretaceous–Tertiary boundary at Agost (southeast Spain). *Chemical Geology*, **95**, 265–281.

MARTÍNEZ-RUIZ, F., ORTEGA-HUERTAS, M., PALOMO-DELGADO, I. & SMIT, J. 2001. K–T boundary spherules from Blake Nose (ODP Leg 171B) as a record of the Chicxulub ejecta deposits. *This volume*.

MELOSH, H. J. & VICKERY, A. M. 1991. Melt droplet formation in energetic impact events. *Nature*, **350**, 494–496.

NORRIS, R. D., HUBER, B. T. & SELF-TRAIL, J. 1999. Synchroneity of the K–T oceanic mass extinction and meteorite impact: Blake Nose, Western North Atlantic. *Geology*, **27**, 419–422.

NORRIS, R. D., KROON, D., KLAUS, A. *et al.* (eds) 1998. *Proceedings of the Ocean Drilling Program, Initial Reports*, **171B**. Ocean Drilling Program, College Station, TX.

OLSSON, R. K., MILLER, K. G., BROWNING, J. V., HABIB, D. & SUGARMAN, P. J. 1997. Ejecta layer at the Cretaceous–Tertiary boundary, Bass River, New Jersey (Ocean Drilling Program Leg 174AX). *Geology*, **25**, 759–762.

PAYTAN, A. 1995. *Marine barite, a recorder of oceanic chemistry, productivity, and circulation*. Ph.D thesis, Scripps Institute of Oceanography, University of California, San Diego.

PAYTAN, A., KASTNER, M., MARTIN, E. D., MACDOUGALL, J. D. & HERBERT, T. 1993. Marine barite as monitor of seawater strontium isotope composition. *Nature*, **366**, 445–449.

PIERAZZO, E. & MELOSH, J. H. 1999. Hydrocode modeling of Chicxulub as an oblique impact event. *Earth and Planetary Science Letters*, **165**, 163–176.

POLLASTRO, R. M. & BOHOR, B. F. 1993. Origin and clay-mineral genesis of the Cretaceous/Tertiary boundary unit, Western Interior of North America. *Clays and Clay Minerals*, **41**, 7–25.

RENARD, M. 1986. Pelagic carbonate chemostratigraphy (Sr, Mg, ^{18}O, ^{13}C). *Marine Micropalaeontology*, **10**, 117–164.

ROCCHIA, R., BOCLET, D., BONTÉ, Ph. *et al.* 1990. The Cretaceous–Tertiary boundary at Gubbio revisited: vertical extent of the Ir anomaly. *Earth and Planetary Science Letters*, **99**, 206–219.

SCHMITZ, B. 1985. Metals precipitation in the Cretaceous–Tertiary boundary clay at Stevns Klint, Denmark. *Geochimica et Cosmochimica Acta*, **49**, 2361–2370.

SCHMITZ, B. 1987. Barium, equatorial high productivity, and the northward wandering of the Indian continent. *Palaeoceanography*, **2**, 63–78.

SHARPTON, V. L., BURKE, K., CAMARGO, A. *et al.* 1993. Chicxulub multiring impact basin: size and other characteristics derived from gravity analysis. *Science*, **261**, 1564–1567.

SMIT, J. 1990. Meteorite impact, extinctions and the Cretaceous–Tertiary boundary. *Geologie en Mijnbouw*, **69**, 187–204.

SMIT, J. 1999. The global stratigraphy of the Cretaceous–Tertiary boundary impact ejecta. *Annual Reviews of Earth and Planetary Sciences*, **27**, 75–113.

SMIT, J., MONTANARI, A., SWINBURNE, N. H. M. *et al.* 1992. Tektite-bearing, deep-water clastic unit at the Cretaceous–Tertiary boundary in northeastern Mexico. *Geology*, **20**, 99–103.

SMIT, J., ROCCHIA, R., ROBIN, E. & ODP Leg 171B Shipboard Party. 1997. Preliminary iridium analyses from a graded spherule layer at the K–T boundary and late Eocene ejecta from ODP Sites 1049, 1052, 1053, Blake Nose, Florida. *Geological Society of America, Abstracts with Programs*, **29**, A141.

SPEED, C. D. & KROON, D. 2000. Inorganic geochemistry and mineralogy of the Cretaceous–Tertiary boundary section in Hole 1049C. *In*: KROON, D., NORRIS, R. D. & KLAUS, A. (eds). *Ocean Drilling Program, Scientific Results*, **171B**. Ocean Drilling Program, College Station, TX, 1–26.

STRONG, C. P., BROOKS, R. R., WILSON, S. M. *et al.* 1987. A new Cretaceous–Tertiary boundary site at Flasboumern River, New Zealand: biostratigraphy and geochemistry. *Geochimica et Cosmochimica Acta*, **51**, 2769–2777.

WIGNALL, P. B. & MYERS, K. J. 1988. Interpreting the benthic oxygen levels in mudrocks: a new approach. *Geology*, **16**, 452–455.

ZACHOS, J. C., ARTHUR, M. A. & DEAN, W. E. 1989. Geochemical and palaeoenvironmental variations across the Cretaceous/Tertiary boundary. *Palaeogeography, Palaeoclimatology, Palaeoecology*, **69**, 245–266.

K–T boundary spherules from Blake Nose (ODP Leg 171B) as a record of the Chicxulub ejecta deposits

F. MARTÍNEZ-RUIZ[1], M. ORTEGA-HUERTAS[2], I. PALOMO-DELGADO[2] & J. SMIT[3]

[1]*Instituto Andaluz de Ciencias de la Tierra, CSIC-Universidad de Granada, Facultad de Ciencias, Avda. Fuentenueva, s/n. 18002 Granada, Spain (e-mail: fmruiz@goliat.ugr.es)*

[2]*Departamento de Mineralogía y Petrología, Facultad de Ciencias, Universidad de Granada, Avda. Fuentenueva, s/n. 18002 Granada, Spain*

[3]*Department of Sedimentary Geology, Vrije Universiteit, 1081 HV Amsterdam, Netherlands*

Abstract: The Cretaceous–Tertiary (K–T) boundary interval recovered by the ODP Leg 171 at Site 1049 (Blake Nose, NW Atlantic) contains a thick (9–17 cm) spherule bed marking the boundary. The spherules are mainly perfect spheres with a lesser proportion of oval spherules. They usually range from 100 to 1000 µm. This bed represent the diagenetically altered impact ejecta from Chicxulub and further supports this structure as the site of the K–T impact. Mineralogical and geochemical investigations indicate that impact-generated glass was altered to smectite. Transmission electron microscopy observations revealed in some spherules that smectite is forming from a Si-rich or Ca-rich material, which could suggest a precursor similar to Haitian glasses. The variable thickness and the presence of some Cretaceous planktonic foraminifera and clasts of Cretaceous chalk suggest reworking of the ejecta material. However, the spherule bed confirms that a large volume of the Chicxulub ejecta material reached the Blake Nose Plateau.

Introduction

The Ocean Drilling Program (ODP) Leg 171B recovered an excellent Cretaceous–Tertiary (K–T) boundary interval at three locations along Blake Nose (NW Atlantic): Sites 1049, 1050 and 1052. At Site 1049 a spectacular boundary layer was recovered within this interval, providing further evidence supporting the Chicxulub structure as the site for the K–T impact. OPD Site 1049 is located on the eastern margin of Blake Nose and represents the deepest site of the Blake Nose transect, at a present depth of 2671 m below sea level. The boundary layer, mostly consisting of green spherules interpreted to be of impact origin (Norris *et al.* 1998), shows a variable thickness in the three holes drilled at Site 1049. These circumstances suggest reworking of the ejecta material down-slope after deposition. The green spherules are none the less a record of impact-generated material.

The Chicxulub impact originated distinct dispersal phases (e.g. Alvarez *et al.* 1995; Pierazzo & Melosh 1999) mostly derived from: (1) the turbulent front of melted target rocks deposited in the proximity of the crater site, and (2) the vertically expanding hot vapour plume of vaporized bolide with entrained melted target rocks that was dispersed and deposited globally (the fireball layer). Thus, locations proximal to Chicxulub show greater contributions of the ejecta blanket derived from the crater rocks. Drilled holes at Site 1049 would have been located *c.* 2000 km NE of Chicxulub and the boundary spherules would represent proximal ejecta material. The purpose of this paper is therefore to present mineralogical and geochemical data from the K–T boundary spherules from Blake Nose to constrain the source material and postdepositional processes that altered the original composition and the original record of the impact event.

Samples and methods

Mineralogical and geochemical analyses of the K–T boundary spherules were carried on samples from Holes 1049A and 1049B drilled by ODP Leg 171B along the Blake Nose transect in the NW Atlantic (Fig. 1). In the studied interval, the boundary layer occurs at the biostratigraphic boundary between the Cretaceous and the Palaeocene sediments. It sharply

From: KROON, D., NORRIS, R. D. & KLAUS, A. (eds). 2001 *Western North Atlantic Palaeogene and Cretaceous Palaeoceanography*. Geological Society, London, Special Publications, **183**, 149–161.
0305-8719/01/$15.00

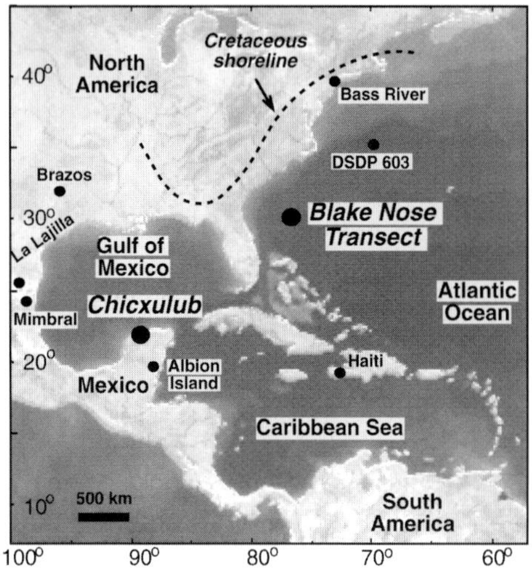

Fig. 1. Location of the Blake Nose transect and some other K–T boundary sections where spherules have been reported.

overlies slumped uppermost Cretaceous foraminiferal–nannofossil ooze (*Abathomphalus mayaroensis* Zone and *Micula prinsii* Zone) and is overlain by Tertiary clay-rich ooze with planktonic foraminiferal assemblages indicative of Early Danian Foraminiferal Zone P-alpha (Norris *et al.* 1998, 1999). This layer, 9–17 cm thick, consists of green spherical and oval-shaped spherules composed of clay as well as some Cretaceous planktonic foraminifera and clasts of Cretaceous chalk (Fig. 2). This spherule bed is capped by a 3 mm thick orange limonite layer that was initially assumed to be the fireball layer (Norris *et al.* 1998). However, the geochemical composition of this limonite layer indicates that it is similar to the rest of the spherule bed except for an enrichment in iron (Martínez-Ruiz *et al.* 2000). This suggests that it is related to only a diagenetic remobilization of iron and that extraterrestrial contribution is not significantly higher.

The spherule bed and sediments above and below were sampled in sections 1049A-17X-2 and 1049B-8H-2, with continuous sampling every 2 cm of the complete interval. Spherules were hand-picked under a stereomicroscope with a dry brush, as spherules in contact with water disintegrate completely. Bulk samples and representative hand-picked spherules were subjected to mineralogical and geochemical analyses using the following methods.

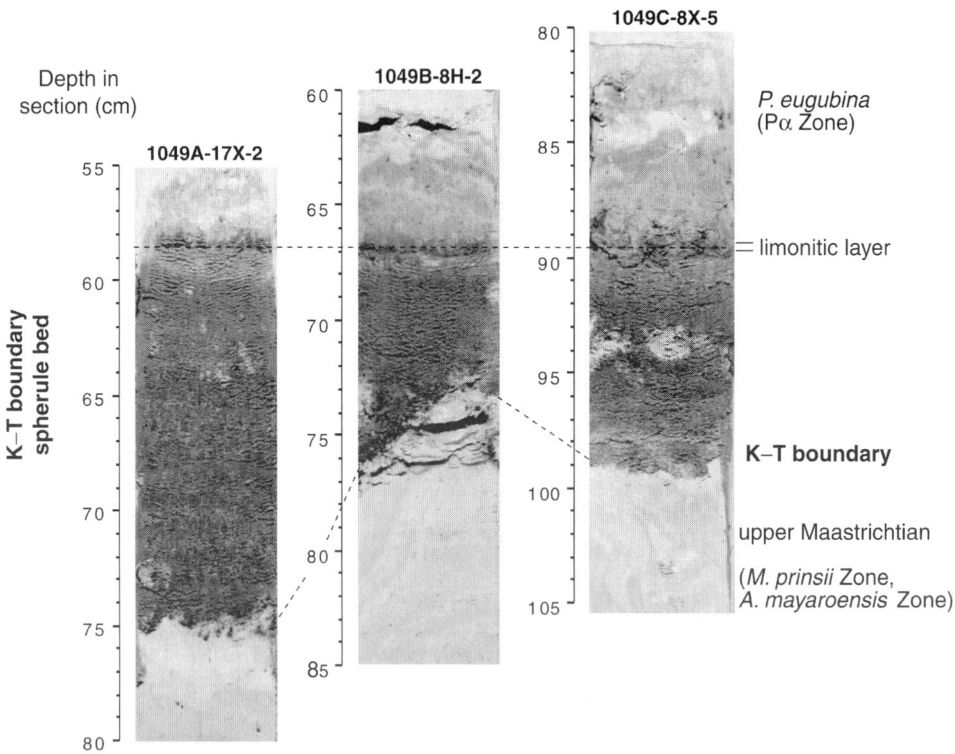

Fig. 2. Core photographs from the K–T boundary layer at Site 1049 for the three holes drilled at this site, showing the variable thickness of the spherule bed and the presence of Cretaceous material within this bed (modified after Norris *et al.* (1998, p. 56)).

X-ray diffraction (XRD)

For bulk mineralogy analyses, samples were packed in Al holders for XRD. For clay mineral analyses, the carbonate fraction was removed using acetic acid, starting the reaction at very low concentration (0.1 mol.$^{-1}$) and increasing to 1 mol.$^{-1}$. Clays were deflocculated by successive washing and the <2 μm fraction was separated by centrifugation. The clay fraction was smeared onto glass slides for XRD. Diffractograms were obtained using a Philips PW 1710 diffractometer with Cu-K_α radiation. Scans were run from 2 to 64° 2θ for bulk samples and untreated clay preparations, and from 2 to 30° 2θ for glycolated, heated and dimethyl-sulphoxide treated samples. Semiquantitative analyses were performed considering the integrated peak area using a specific computer program for the diffractometer used (Nieto *et al.* 1989). The estimated semiquantitative analysis error for bulk-mineralogy absolute values is 5%. For clay-mineral proportions in absolute values, the error ranges from 5 to 10%, but the main aim of the semiquantitative analyses here is to reveal trends in mineral abundance.

Electron microscopy

Morphological studies on bulk samples and hand-picking of spherules were performed using binocular microscopy and scanning electron microscopy (SEM; Zeiss DSM 950). Internal textures were investigated after breaking selected spherules. Quantitative micro-analyses of clay minerals were obtained by transmission electron microscopy (TEM, Philips CM-20 equipped with an EDAX microanalysis system). Quantitative analyses were obtained in scanning TEM mode only from edge particles using a 7 nm diameter beam and 20 × 100 nm scanning area and a short counting time to avoid alkali loss (Nieto *et al.* 1996). Smectite formules were normalized to 11 oxygens.

Blake Nose spherule bed

The spherule bed at Blake Nose (Fig. 2) consists of a coarse, poorly graded and poorly cemented unit. Mineralogical analyses reveal that it is mostly composed of clays and minor proportions of carbonates (Table 1). Other minerals also present in lower proportions are quartz, zeolites and minor amounts of rutile, biotite and some lithic fragments. Clays are mostly smectites, with occasional traces of illite and kaolinite in some of the samples taken from this bed. The contact of the spherule bed with sediments above and below is very sharp. The contact surface of the Cretaceous sediments with the spherule bed contains some spherical impressions (Fig. 3j and k) that seem to be bubbles or deformations by deposition of glassy spherules in soft sediments. The nature of this contact suggests that deposition occurred very rapidly. Cretaceous sediments are slump-folded; however, the overlying stratigraphy is undisturbed. The presence of Cretaceous materials within the spherule bed also suggests downslope transport of the spherule bed material. Deformation of Cretaceous sediments is a general feature at proximal ejecta sites. Deformation and large-scale slope failures were related to the seismic energy input from Chicxulub impact, some of it induced before the emplacement of the ejecta from the same impact (Smit 1999).

Table 1. *XRD semiquantitative data on the main mineral components from the K–T boundary interval at Hole 1049A (see Fig. 2 for location of samples)*

Samples	Depth (mbsf)	Clays	Quartz	Calcite	Dolomite
17X 02 046–048	125.77	9	<5	87	<5
17X 02 048–050	125.79	11	<5	85	<5
17X 02 050–052	125.81	12	<5	84	<5
17X 02 052–054	125.83	11	<5	85	<5
17X 02 054–056	125.85	14	<5	83	<5
17X 02 056–058	125.87	12	<5	85	<5
17X 02 060–062	125.91	75	<5	12	10
17X 02 062–064	125.93	92	<5	6	<5
17X 02 064–066	125.95	90	<5	8	<5
17X 02 066–068	125.97	91	<5	7	<5
17X 02 068–070	125.99	90	<5	8	<5
17X 02 072–074	126.03	91	<5	7	<5
17X 02 074–076	126.05	91	<5	7	<5
17X 02 076–078	126.07	13	<5	84	<5
17X 02 078–080	126.09	23	<5	75	<5
17X 02 080–082	126.11	22	<5	75	<5
17X 02 082–084	126.13	22	<5	74	<5
17X 02 086–088	126.17	21	<5	76	<5
17X 02 088–090	126.17	21	<5	76	<5

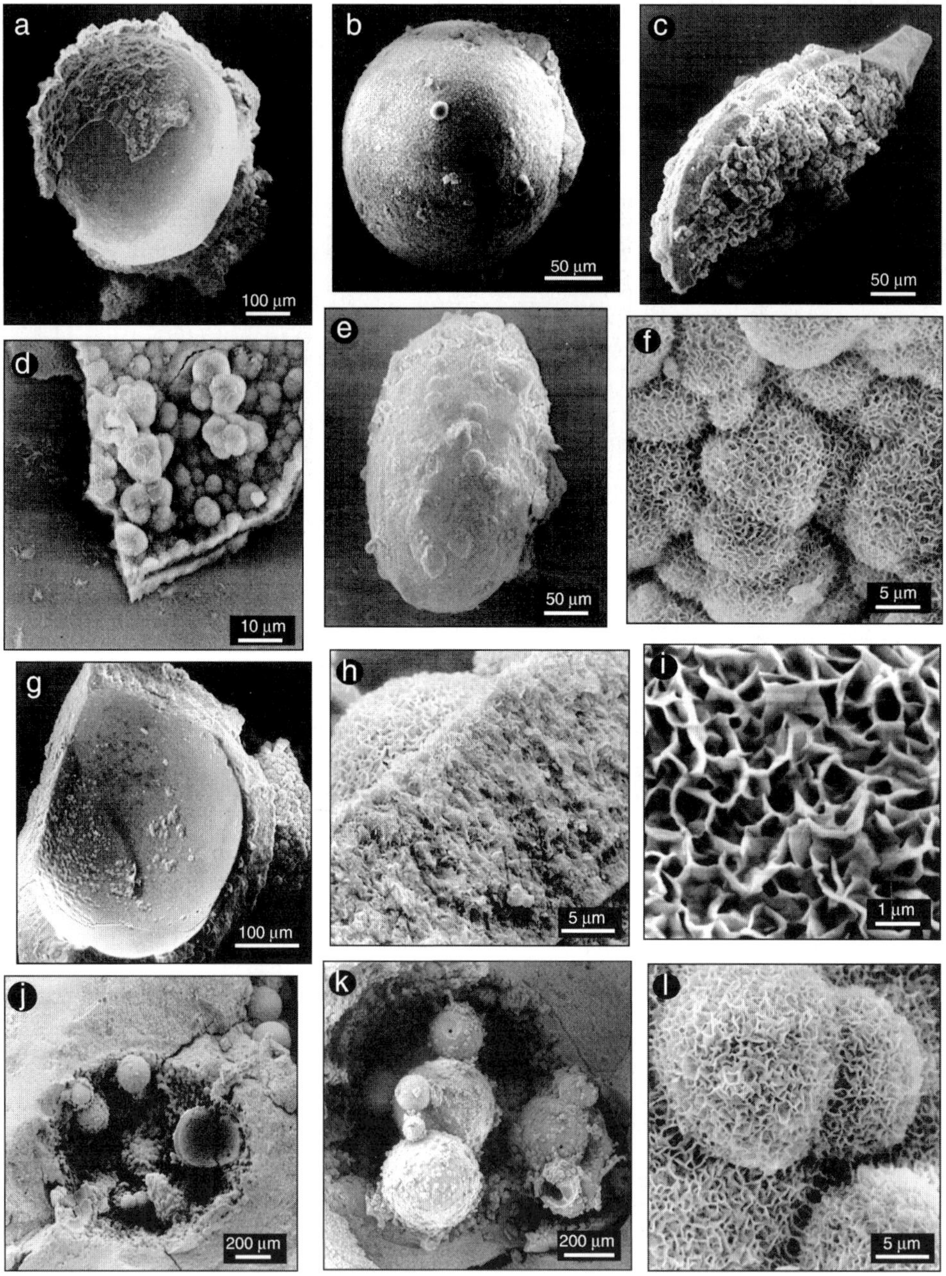

Fig. 3. Scanning electron micrographs of smectite dark green spherules from Blake Nose at Site 1049. (**a**, **b**) View of spherical spherules showing a nodular surface shell ranging from a lumpy surface shell (**a**) to scarce globules on surface (**b**). (**c**) View of a drop-shaped spherule fill with smectite aggregates. (**d**) Enlarged detail of the surface of the spherule shown in (**a**). (**e**) View of an oval spherule with nodular surface. (**f**) Higher-resolution micrographs of the globules from surface spherules showing smectite morphologies. (**g**) Fragment of a hollow spherule showing the inner nodular side and outer lumpy surface. (**h**) Detail of the spherule wall shown in (**g**). (**i**) Higher-resolution micrograph of the smectite wall. (**j**, **k**) Spherical impressions on top of the Cretaceous material filled with spherules with lumpy surfaces. (**l**) Enlarged detail of the lumpy surface of spherules shown in (**j**) and (**k**).

Table 2. *Color, morphology and surface texture of Blake Nose spherules*

	Spherical	Oval	Smooth surface	Rough surface	Nodular surface	Hollow
Light green	X	X		X		
Dark green	X	X		X	X	X
Pale yellow	X		X			X

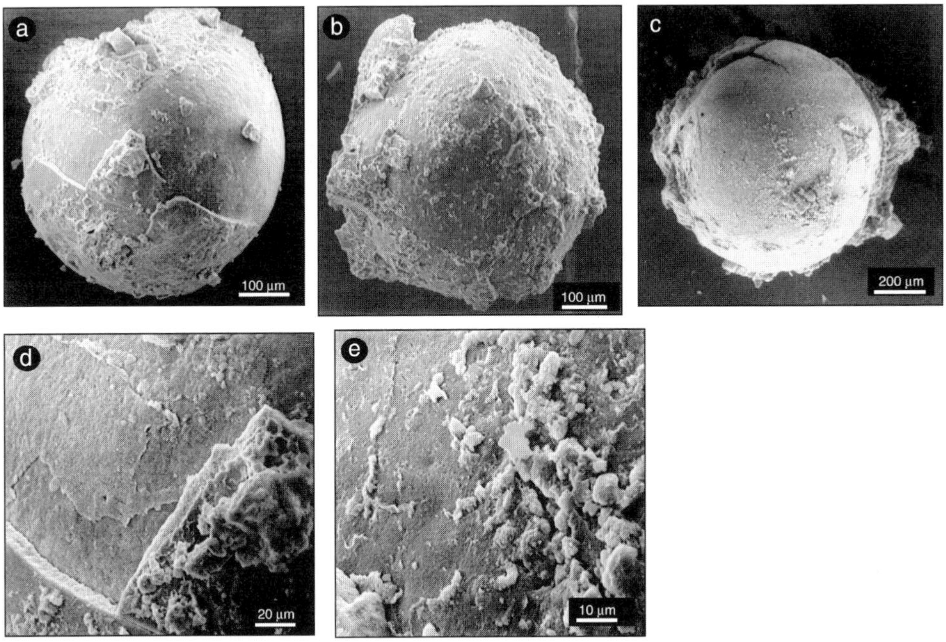

Fig. 4. Scanning electron micrographs of smectite pale yellow spherules from Blake Nose at Site 1049 showing their spherical morphologies (**a**–**c**) and enlarged details of the surface (**d**, **e**).

Spherules and diagenetic alteration

Stereomicroscope and SEM observations reveal that the morphologies of the Blake Nose spherules are tektite like and they mainly correspond to perfect spheres and lesser proportions of oval spherules. Size usually ranges from 100 to 1000 μm. Some hollow spherules and spherical voids are filled with smaller spherules (Figs. 3 j and k) that may represent diagenetic infills of original bubbles. Figures 3–5 show some examples of the spherules from the K–T boundary layer at Blake Nose. The surface of the spherules is nodular (Fig 3a, b and e), smooth (Fig 4a–c) or rough (Fig. 5). Different types of spherules have been distinguished based on colour, morphology and surface texture, which are summarized in Table 2.

The smallest dark green spherules (Fig. 3), and occasionally pale yellow ones (Fig. 4), also occur as aggregates (Fig. 3k).The surface of dark green spherules is usually nodular (Fig. 3b), from a few nodules on the surface (Fig. 3b) to a lumpy surface with a nodular aspect (Fig 3d and k). When hollow they are partially or completely filled with aggregates (Fig. 3c). The pale yellow spherules have a smoother outer surface and either are filled with a fine matrix or are hollow with aggregates covering the internal walls, as also occurs in the dark green spherules (Fig. 3g). The light green spherules are massive and present rough surfaces; they are the most abundant (Fig. 5).

Although the morphologies or surfaces of the spherules are different, the XRD scans on oriented samples reveal that spherules are mainly composed of smectite, and no clear evidence for glass relicts has been observed. TEM microanalyses reveal that the smectite corresponds to a dioctahedral type. Compositions are presented in Tables 3 and 4, and Fig. 6. Some compositional variations are observed between dark green spherules and pale yellow spherules (Table 3). Dark green spherules are richer in Fe and pale

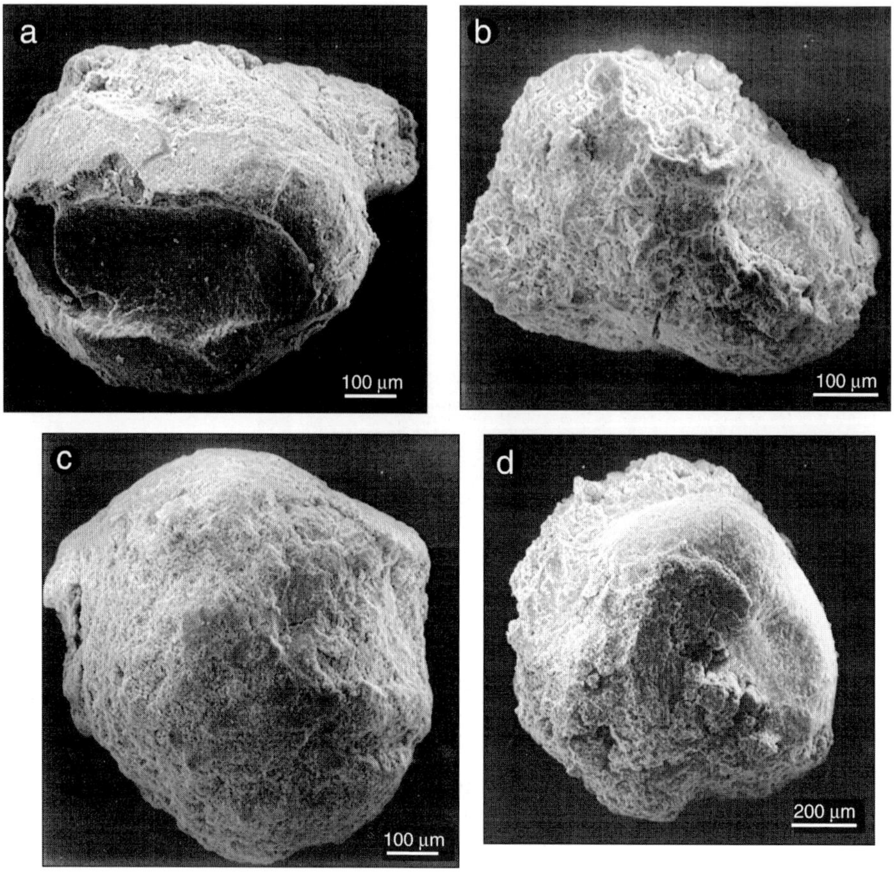

Fig. 5. Scanning electron micrographs of smectite light green spherules from Blake Nose at Site 1049 showing their morphology, size and rough surface.

yellow ones are richer in Ca. These differences may derive from different precursor glass types. In fact, two types of glass occur in the proximal K–T ejecta, black andesitic glass and honey-coloured CaO-rich glass (Izzet 1991; Sigurdsson *et al.* 1991; Smit *et al.* 1992). Precursors of the Blake Nose spherules could then have been similar to those impact glasses. TEM analyses on pale yellow spherules also reveal they contain a very rich-Ca matrix altering to smectite. Some calcite crystals are also observed in this matrix. This calcite could be an original and unaltered phase, but it may also be derived from diagenetic reactions leading to clay authigenesis, such as, for instance, reactions including the formation of palygorskite from smectite and dolomite (Jones & Galán 1988):

$$Ca_{0.25}(Mg_{0.5}Al_{1.5})Si_4O_{10}(OH)_2 + 1.25CaMg(CO_3)_2 + 2SiO_2 = 2Mg_{0.87}Al_{0.75}Si_3O_{7.5}(OH) + 1.5CaCO_3 + CO_2$$

Palygorskite is observed by TEM, suggesting that such reactions could have taken place. Moreover, dolomite and calcite could have been abundant in the ejecta material. The high Ca content could probably derive from carbonate target rocks. In fact, carbonate material is abundant in other K–T ejecta deposits, such as those from Albion Island, Belize, where the ejecta come from the outer portion of the ejecta blanket of the Chicxulub crater. In this formation, abundant clay spheroids are altered impact glass and dolomite spheroids are accretionary lapilli (Pope *et al.* 1999). The differences in composition between spherules suggest therefore that they derive from alteration of a variety of impact glasses and tektite-like material, similar to those reported in the Gulf of Mexico region (Sigurdsson *et al.* 1991; Smit *et al.* 1992; Koeberl & Sigurdsson 1992).

Smectite compositions are also variable within the same spherule; smectites vary from having

Table 3. *Representative AEM data from smectites of the K–T boundary bed at Hole 1049A normalized to* $O_{10}(OH)_2$

Sample	Si	Al^{IV}	Al^{VI}	Mg	Fe	Ti	Σ^{VI}	K	Ca	Na	$\Sigma^{Int.}$
17X-2 58–60	3.78	0.22	1.24	0.49	0.29	0.03	2.05	0.13	0.13	–	0.26
	3.77	0.23	1.73	0.25	0.12	0.00	2.10	0.19	0.01	–	0.20
	3.82	0.18	1.18	0.62	0.40	0.02	2.22	0.08	0.01	–	0.09
	3.96	0.04	1.15	0.43	0.48	0.01	2.07	0.13	0.02	–	0.15
	3.89	0.11	1.31	0.52	0.33	0.00	2.16	0.03	0.04	–	0.07
	3.88	0.12	1.19	0.49	0.43	0.03	2.14	0.08	0.04	–	0.12
	3.90	0.10	1.01	0.45	0.67	0.00	2.13	0.05	0.04	–	0.09
	3.84	0.16	1.22	0.51	0.45	0.00	2.18	0.10	0.02	–	0.12
	3.85	0.15	1.29	0.44	0.42	0.00	2.15	0.13	0.05	–	0.18
17X-2 62–64	3.81	0.19	1.55	0.46	0.16	0.00	2.17	0.07	0.03	–	0.10
	3.78	0.22	1.54	0.52	0.16	0.00	2.22	0.07	0.02	–	0.09
	3.85	0.15	1.64	0.43	0.05	0.00	2.12	0.07	0.03	–	0.10
	3.84	0.16	1.46	0.44	0.23	0.03	2.16	0.06	0.01	–	0.07
	3.87	0.13	1.50	0.43	0.14	0.05	2.12	0.07	0.01	–	0.08
17X-2 68–70	3.63	0.37	1.31	0.84	0.17	0.00	2.32	0.08	0.06	–	0.14
	3.99	0.01	1.63	0.40	0.09	0.00	2.12	0.08	0.00	–	0.08
	3.94	0.06	1.64	0.38	0.11	0.00	2.13	0.02	0.01	–	0.03
	3.95	0.05	1.52	0.35	0.16	0.06	2.09	0.07	0.01	–	0.08
	3.75	0.25	1.44	0.53	0.20	0.03	2.20	0.09	0.06	–	0.15
17X-2 70–72	3.80	0.20	1.47	0.42	0.23	0.03	2.15	0.04	0.02	0.05	0.11
	3.80	0.20	1.55	0.43	0.10	0.04	2.12	0.05	0.05	0.10	0.20
	3.71	0.29	1.53	0.55	0.09	0.06	2.23	0.06	0.02	0.21	0.29
	3.89	0.11	1.44	0.44	0.15	0.05	2.08	0.05	0.03	0.12	0.20
	3.65	0.35	1.36	0.61	0.16	0.12	2.25	0.12	0.02	0.37	0.51
	3.76	0.24	1.28	0.51	0.35	0.00	2.14	0.16	0.01	0.15	0.32
	3.70	0.30	1.50	0.47	0.19	0.00	2.16	0.07	0.04	0.15	0.26
	3.71	0.29	1.21	0.57	0.33	0.00	2.11	0.09	0.04	0.25	0.38
	3.75	0.25	1.38	0.47	0.27	0.02	2.14	0.06	0.19	0.19	0.44
	3.53	0.47	1.28	0.58	0.17	0.09	2.12	0.11	0.15	0.19	0.45
17X-2 74–76	3.58	0.42	1.41	0.33	0.39	0.00	2.13	0.23	0.10	–	0.33
	3.60	0.40	1.42	0.46	0.36	0.00	2.24	0.07	0.03	–	0.10
	3.90	0.10	1.52	0.32	0.27	0.00	2.11	0.05	0.01	–	0.06
	3.80	0.20	1.60	0.31	0.17	0.03	2.11	0.05	0.02	–	0.07
	3.72	0.28	1.65	0.29	0.19	0.00	2.13	0.13	0.03	–	0.16
Dark green spherules	3.57	0.43	1.71	0.49	0.81	0.04	2.05	0.49	0.02	0.18	0.69
	3.67	0.35	1.77	0.41	0.81	0.04	2.03	0.41	0.00	0.12	0.53
	3.66	0.34	1.84	0.48	0.71	0.03	2.06	0.22	0.02	0.25	0.49
	3.56	0.42	1.83	0.53	0.68	0.04	2.08	0.29	0.03	0.27	0.59
	3.77	0.23	1.91	0.44	0.64	0.03	2.02	0.29	0.02	0.08	0.39
Pale yellow spherules	3.64	0.36	1.35	0.50	0.37	0.00	2.22	0.13	0.00	0.07	0.20
	3.64	0.36	1.18	0.39	0.48	0.02	2.07	0.20	0.02	0.16	0.38
	3.69	0.31	1.20	0.39	0.48	0.03	2.10	0.21	0.00	0.16	0.37
	3.56	0.44	1.04	0.48	0.48	0.00	2.00	0.06	0.11	0.48	0.65
	3.76	0.24	1.24	0.40	0.40	0.01	2.05	0.05	0.05	0.13	0.23

–, not determined.
Analyses on single dark green and pale yellow spherules are also presented (see Fig. 2 for location of samples).

normal percentages of Si to being very Si enriched. TEM observations also reveals very high silica areas that do not correspond to a real smectite composition, suggesting that Si-rich glass could have been the precursor. Octahedral cations also present wide ranges of abundance (Tables 3 and 4). The original material is expected to be compositionally variable, as impact-generated glass is only briefly molten and there is not enough time for mixing and homogenizing of the composition (e.g. Alvarez *et al.* 1992).

TEM observations reveal smectite morphologies similar to those of smectites originated from the alteration of volcanic glass (Fig. 7), such as hair-like smectites (Fig. 7a and b) and broad

Table 4. *Representative AEM data from smectites of the K–T boundary bed at Hole 1049B normalized to $O_{10}(OH)_2$ (see Fig. 2 for location of samples)*

Sample	Si	Al^{IV}	Al^{VI}	Mg	Fe	Ti	Σ^{VI}	K	Ca	$\Sigma^{Int.}$
8H-2 68–70	3.76	0.24	1.52	0.53	0.09	0.00	2.14	0.11	0.09	0.20
	3.88	0.12	1.54	0.50	0.14	0.00	2.18	0.02	0.03	0.05
	3.73	0.27	1.37	0.36	0.23	0.00	1.96	0.08	0.32	0.40
	3.86	0.14	1.47	0.35	0.26	0.02	2.10	0.05	0.04	0.09
	3.80	0.20	1.64	0.32	0.13	0.03	2.12	0.03	0.03	0.06
	3.77	0.23	1.44	0.42	0.28	0.02	2.18	0.01	0.03	0.04
	3.91	0.09	1.14	0.18	0.62	0.04	1.98	0.07	0.10	0.17
	3.88	0.12	1.43	0.38	0.28	0.02	2.11	0.04	0.05	0.09
	3.96	0.04	1.63	0.34	0.08	0.03	2.08	0.02	0.02	0.04
	3.95	0.05	1.46	0.37	0.22	0.03	2.08	0.05	0.01	0.06
	3.89	0.11	1.55	0.39	0.17	0.01	2.12	0.04	0.03	0.07
	3.68	0.32	1.09	0.26	0.65	0.07	2.07	0.12	0.08	0.20
	3.79	0.21	1.11	0.13	0.53	0.09	1.86	0.21	0.23	0.44
	3.39	0.61	0.57	0.17	0.09	0.34	1.17	0.25	0.17	0.42
	3.22	0.78	0.46	0.24	1.19	0.10	1.99	0.37	0.23	0.60
	3.35	0.65	0.44	0.16	1.28	0.08	1.96	0.39	0.21	0.60
	3.82	0.18	1.46	0.39	0.25	0.03	2.13	0.06	0.05	0.11
8H-2 70–72	3.84	0.16	1.87	0.14	0.06	0.00	2.07	0.01	0.03	0.04
	3.77	0.23	1.72	0.32	0.03	0.00	2.07	0.02	0.11	0.13
	3.73	0.27	1.43	0.52	0.22	0.05	2.22	0.03	0.02	0.05
	3.94	0.06	1.65	0.29	0.16	0.00	2.10			
	3.75	0.25	1.50	0.45	0.25	0.00	2.20	0.43	0.02	0.45
	3.88	0.12	1.51	0.38	0.21	0.01	2.11	0.07	0.02	0.09
	3.84	0.16	1.53	0.45	0.14	0.02	2.14	0.09	0.02	0.11
	3.83	0.17	1.63	0.34	0.07	0.07	2.11	0.05	0.02	0.07
	3.88	0.12	1.59	0.39	0.12	0.02	2.12	0.05	0.02	0.07
	3.84	0.16	1.40	0.37	0.26	0.02	2.05	0.13	0.09	0.22
8H-2 72–74	3.84	0.16	1.51	0.44	0.20	0.00	2.15	0.07	0.05	0.12
	3.88	0.12	1.45	0.47	0.12	0.06	2.10	0.10	0.05	0.15
	3.79	0.35	1.30	0.41	0.25	0.04	2.00	0.10	0.04	0.14
	3.88	0.12	1.59	0.49	0.07	0.00	2.15	0.05	0.04	0.09
	3.89	0.11	1.64	0.44	0.02	0.00	2.10	0.05	0.09	0.14
	3.75	0.25	1.32	0.53	0.36	0.00	2.21	0.10	0.02	0.12
	3.89	0.11	1.55	0.46	0.16	0.00	2.17	0.04	0.01	0.05

board-shaped smectites (Fig. 7c) (e.g. Chamley 1989). TEM morphological observations have also shown the smectite growing from Si-rich material, probably the altering glass (Fig. 7d) and palygorskite forming from a smectite precursor (Fig. 7e and f). This suggests that smectite directly replaced the original glass phase. At Hole 603B, also located in the NW Atlantic, Klaver *et al.* (1987) proposed that smectite directly replaced the original phase; however, they did not exclude the hypothesis that the smectite represents a second stage of alteration of K–T spherules, after initial alteration to illite. Those workers also considered that the high K_2O content of the smectite might derive from illite. Nevertheless, the TEM observations are in favour of the hypothesis of direct replacement of the original glass. On the other hand, the existence of authigenic palygorskite indicates a high-silica source and alkaline conditions. Couture (1977) reported the association of palygorskite with high-silica minerals such as opal and clinoptilolite and environments with mobilization of silica. In general, high-silica environments favour the precipitation of chain-structure silicates (Beck & Weaver 1978) and smectite can be a precursor of fibrous clays (Singer 1979). Low-temperature alteration of basalts also involves the expulsion of the necessary elements for both fibrous clays and zeolites (Velde 1985). In this case, when the original glass material was altered to smectite, not all the available silica from the former glass was incorporated within the smectite, which favoured the high-silica environments required for the formation of such high-silica minerals. The existence of Ca-rich and Si-rich precursors in some spherules agrees with the geochemistry of Haitian glasses (Izett 1991; Koeberl & Sigurdsson 1992), linked to the Chicxulub crater

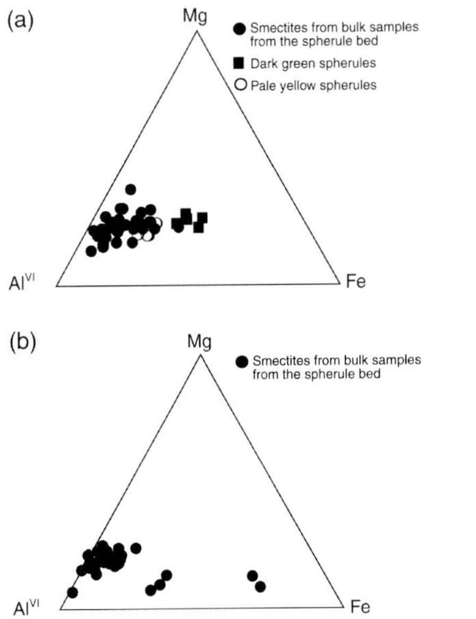

Fig. 6. Al–Fe–Mg diagrams (based on Güven (1988)) showing the smectite composition from spherule bed samples from Holes 1049A (**a**) and 1049B (**b**).

(Swisher *et al.* 1992; Blum *et al.* 1993; Hough *et al.* 1998); however, the pre-impact target stratigraphy would require the presence of Si-rich rocks near the surface (Koeberl 1993) to account for an Si-rich precursor.

Other components

In addition to smectite, the spherule bed also contains minor amounts of calcite, zeolites (Fig. 8), quartz and scarce rutile grains, biotite and some lithic fragments. Calcite mostly derives from the presence of carbonate material in the spherule bed, represented by Cretaceous clast and Cretaceous foraminifera, although some of it could also be represented by the original unaltered ejecta material or authigenic phases. In fact, Ca-rich material has also been observed by TEM in the spherules as discussed above, and stereomicroscope observations also reveal that spherules are macroscopically included in a matrix containing some white patches. SEM analysis confirms that this white matrix is composed of smectite, but also contains calcite and dolomite. Quartz is detrital in origin and shocked quartz is rare in this layer. Rutile and other minor mineral components probably derived from target rock material in a mixture of microfossils, lithic fragments and impact-generated material, labelled by Bralower *et al.* (1998) 'the K–T boundary cocktail'. Zeolites resulted from authigenesis during alteration of the spherule material. Zeolitization requires specific conditions, with temperature, pressure and pH being the most important factors (Hall 1998). The diagenetic conditions of the spherule bed and the geological context preclude high temperature or pressure; pH is thus the most important factor controlling the formation of zeolites during the alteration of the ejecta material. Although not very abundant ($<5\%$), zeolites are not common in other K–T boundary layers and therefore provide in this case some constraints on the diagenetic evolution of the boundary material. The Chicxulub impact resulted in acid conditions, as reported by many workers (Hildebrand *et al.* 1991; Sigurdsson *et al.* 1992; Sharpton *et al.* 1993). Increased weathering induced by acid rain has been proposed to explain the crustal Sr enrichment in the oceans at the K–T boundary (MacDougall 1988; Martin & MacDougall 1991). However, large volumes of acid were consumed by reactions with the impact ejecta (e.g. Retallack 1996) and in the boundary deposits, reactions within the ejecta buffered the pH to alkaline conditions. The large contribution of carbonate material from the impact site also favoured such conditions and during alteration of the ejecta material raised pH, allowed zeolitization to take place.

Interpretations and comparison with other K–T ejecta deposits

The K–T material from Blake Nose derives from the fallout of the material generated by the Chicxulub impact. Reworking of the impact-generated material downslope prevents a precise determination of the original thickness and limits evidence for inferring precise ejecta material distribution NE of Chicxulub. It does, however, provide evidence of the large volume of ejecta material reaching Blake Nose Plateau.

When compared with proximal ejecta, the spherules from Blake Nose are similar to spherules from different locations on the North America Atlantic margin, such as Bass River and Deep Sea Drilling Project (DSDP), 603B (Fig. 1). At Hole 603B, Klaver *et al.* (1987) reported green spherules in the lower part of a turbidite section that were also composed of smectite. The green spherules from Blake Nose (Site 1049) represent the same diagenetically altered impact ejecta from the Chicxulub crater. Spherules recovered at DSDP Hole 603B are spherical, with smooth or nodular morphologies, and they are massive or hollow (Klaver *et al.* 1987). At Bass River,

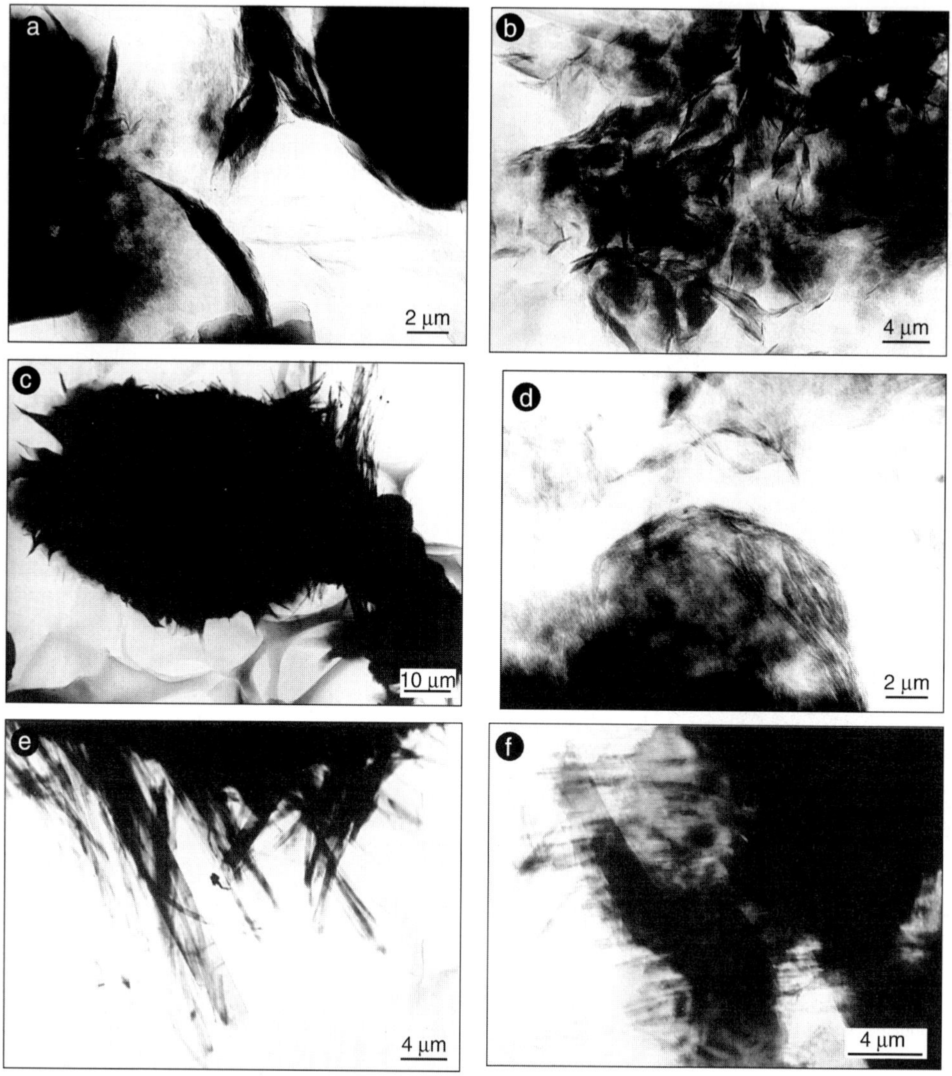

Fig. 7. Selected transmission electron micrographs of smectites from the spherule bed at Site 1049. (**a**, **b**) Hair-like smectite, which has replaced the original spherule glass. (**c**) Authigenic smectite replacing glass with some filmy veils and palygorskite fibres forming from smectite. (**d**) Smectites forming from a high-silica material, probably the altering glass. (**e**, **f**) Higher-resolution micrographs showing the authigenesis of palygorskite fibres.

New Jersey, the ODP Leg 174AX recovered a K–T interval with a 6 cm thick spherule bed (Miller *et al.* 1998). The spherule bed represents the altered impact ejecta and consists of a coarse and poorly graded unit containing spherical and oval-shaped spherules. These spherules have a smooth outer surface, a thin solid rim and a fine matrix containing in turn rounded spherules (Olsson *et al.* 1997). Spherules from the El Mimbral and La Lajilla sections (Smit *et al.* 1992; Keller *et al.* 1994) in Mexico are also similar but often contain a preserved impact glass core. Clay spheroids from Albion Island (Fig. 1) are also advocated to be impact glass (Ocampo *et al.* 1996; Pope *et al.* 1999). In Haiti, exposures of the ejecta deposits reveal spherules that are not completely altered, consisting of impact glasses and glass spherules (e.g. Izett 1991; Koeberl & Sigurdsson 1992).

The spherule bed recovered in the NW Atlantic at Bass River, DSDP Hole 603B and Blake Nose represent similar ejecta blanket deposits from the Chicxulub impact. The thickness of the spherule bed at Bass River is consistent with the southeast

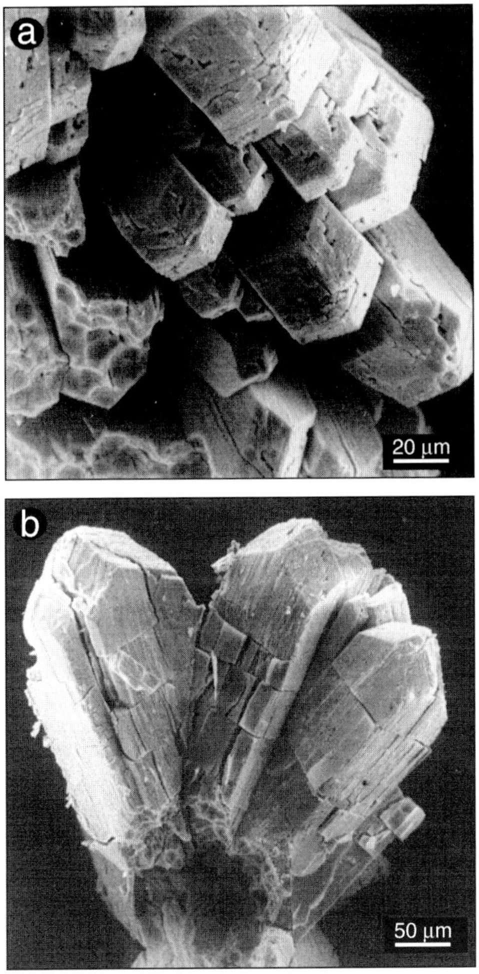

Fig. 8. Scanning electron micrographs of zeolites from the spherule bed at Site 1049.

low-angle impact hypothesis (Schultz & D'Hondt 1996) according to Olsson *et al.* (1997).

When comparing Blake Nose spherules and others from proximal sites (tektites and impact glasses) with those found at distal sites, it is found that their nature is very different. At distal sites spherules show evidence of a crystalline phase and quench-crystal textures, and, altered or not, are considered as microkrystites (Smit *et al.* 1992). At distal sections such as those at Agost and Caravaca (SE Spain), which are some of the most complete K–T boundary outcrops, the ejecta layer is mostly composed of smectite with abundant spherules, which are microkrystites according to the term proposed by Glass & Burns (1988). The ejecta deposits at these sections record a significant extraterrestrial contamination (Smit 1990; Martínez-Ruiz *et al.* 1997). Thus, the spherule bed at Blake Nose records mainly the terrestrial material involved in the impact. In contrast, distal ejecta deposits contain microkrystites and higher extraterrestrial contributions. The existence of tektites and microkrystites in the K–T boundary layer at proximal and distal sequences (Smit *et al.* 1992; Smit 1999) further supports the Chicxulub impact and patterns of global distribution of the impact-generated materials.

Conclusions

Spherules from the K–T ejecta deposits at Blake Nose derive from the impact material generated at the Chicxulub crater site. The original material has been altered to smectite. Mineralogical and geochemical evidence indicates that the smectites derive from the alteration of Si-rich and Ca-rich glass. During diagenetic processes, other mineral phases such as zeolites and palygorskite originated. Although the spherule bed resulted from reworking of the impact-generated material downslope, the ejecta deposit at ODP Site 1049 confirms that a large volume of this material that reached the Blake Nose Plateau.

We thank the ODP Leg 171B Shipboard Scientific Party and the crew of the *JOIDES Resolution* for assistance with the samples and data, and the Bremen ODP Core Repository for assistance during the sampling party. One of the authors (F. Martínez-Ruiz) thanks the University of Granada and 'Junta de Andalucía' for financial support for participation on ODP Leg 171B. This work was partially supported by Project PB96-1429 (DGES, MEC, Spain) and Research Group RNM0179 (Junta de Andalucía, Spain). We thank the C.I.C. (University of Granada, Spain) for the use of the analytical facilities. We also thank P. Sánchez-Gómez, I. Nieto and E. Abarca for their help in the laboratory. We also thank M. Dubois and P. Claeys for constructive reviews and the improvement of this paper.

References

Alvarez, W., Claeys, P. & Kieffer, S. 1995. Emplacement of Cretaceous–Tertiary boundary shocked quartz from Chicxulub crater. *Science*, **269**, 930–935.

Alvarez, W., Smit, J., Lowrie, W. *et al.* 1992. Proximal impact deposits at the Cretaceous–Tertiary boundary in the Gulf of Mexico: a restudy of DSDP Leg 77 Sites 536 and 540. *Geology*, **20**, 697–700.

Beck, K. C. & Weaver, C. E. 1978. Miocene of the S.E. United States: a model for chemical sedimentation in a peri-marine environment. Reply. *Sedimentary Geology*, **21**, 154–157.

BLUM, J. D., CHAMBERLAIN, C. P., HINGSTON, M. P. & KOEBERL, C. 1993. Isotopic composition of K–T boundary impact glass with melt rock from Chicxulub and Manson impact structures. *Nature*, **364**, 325–327.

BRALOWER, T. J., PAULL, C. K. & LECKIE, R. M. 1998. The Cretaceous–Tertiary boundary cocktail: Chicxulub impact triggers margin collapse and extensive sediment gravity flows. *Geology*, **26**, 331–334.

CHAMLEY, H. 1989. *Clay Sedimentology*. Springer, Berlin.

COUTURE, R. A. 1977. Composition and origin of palygorskite-rich and montmorillonite-rich zeolite-containing sediments from the Pacific Ocean. *Chemical Geology*, **21**, 149–153.

GLASS, B. P. & BURNS, C. A. 1988. Microkrystites: a new term for impact-produced glassy spherules containing primary crystallites. *In*: RYDER, S. (ed.) *Lunar and Planetary Science Conference, 18*, Pergamon, New York, 455–458.

GÜVEN, N. 1988. Smectites. *In*: BAILEY, S. W. (ed.) *Hydrous Phyllosilicates (Exclusive of Micas)*. Mineralogical Society of America, Reviews in Mineralogy, **19**, 497–559

HALL, A. 1998. Zeolitization of volcaniclastic sediments: the role of temperature and pH. *Journal of Sedimentary Research*, **68**, 739–745.

HILDEBRAND, A. R., PEWNFIELD, G. T., KRING, D. A., PILKINGTON, M., CAMARGO, Z. A., JACOBSEN, S. B. & BOYNTON, W. V. 1991. Chicxulub Crater: a possible Cretaceous–Tertiary boundary impact crater on the Yucatan Peninsula, Mexico. *Geology*, **19**, 867–871.

HOUGH, R. M., WRIGHT, I. P., SIGURDSSON, H., PILLINGER, C. T. & GILMOUR, I. 1998. Carbon content and isotopic composition of K–T impact glasses from Haiti. *Geochimica et Cosmochimica Acta*, **62**, 1285–1291.

IZETT, G. A. 1991. Tektites in Cretaceous/Tertiary boundary rocks on Haiti and their bearing on the Alvarez impact extinction hypothesis. *Journal of Geophysical Research*, **96**, 20879–20905.

JONES, B. F. & GALÁN, E. 1988. Sepiolite and palygorskite. *In*: Bailey, S. W. (ed.) *Hydrous Phyllosilicates (Exclusive of Micas)*. Mineralogical Society of America, Reviews in Mineralogy, **19**, 631–674.

KELLER, G., STINNESBECK, W. & LÓPEZ-OLIVA, J. G. 1994. Age, deposition, and biotic effects of the Cretaceous/Tertiary boundary event at Mimbral, NE Mexico. *Palaios*, **9**, 144–157.

KLAVER, G. T., VAN KEMPEN, T. M. G., BIANCHI, F. R. & VAN DER GAAST, S. J. 1987. Green spherules as indicators of the Cretaceous/Tertiary boundary in Deep Sea Drilling Project Hole 603B. *In*: VAN HINTE, J. E., WISE, S. W. Jr, *et al.* (eds) *Initial Reports of the Deep Sea Drilling Project*, **93**, US Government Printing Office, Washington, DC, 1039–1056.

KOEBERL, C. 1993. Chicxulub Crater, Yucatan: tektites, impact glasses, and the geochemistry of target rocks and breccias. *Geology*, **21**, 211–214.

KOEBERL, C. & SIGURDSSON, H. 1992. Geochemistry of impact glasses from the K–T boundary in Haiti: relation to smectites and a new type of glass. *Geochimica et Cosmochimica Acta*, **56**, 2113–2129.

MACDOUGALL, J. D. 1988. Seawater strontium isotopes, acid rain, and the Cretaceous–Tertiary boundary. *Science*, **239**, 485–487.

MARTIN, E. E. & MACDOUGALL, J. D. 1991. Seawater Sr isotopes at the Cretaceous–Tertiary boundary. *Earth and Planetary Science Letters*, **104**, 166–180.

MARTÍNEZ-RUIZ, F., ORTEGA-HUERTAS, M., KROON, D., SMIT, J., PALOMO-DELGADO, I. & ROCCHIA, R. 2001. Geochemistry of the Cretaceous–Tertiary boundary at Blake Nose (ODP Leg 171B). *This volume*.

MARTÍNEZ-RUIZ, F., ORTEGA-HUERTAS, M., KROON, D. & ACQUAFREDDA, P. 1997. Quench textures in altered spherules from the Cretaceous–Tertiary boundary layer at Agost and Caravaca, SE Spain. *Sedimentary Geology*, **113**, 137–147.

MILLER, K. G., SUGARMAN, P. J., BROWNING, J. V. *et al.* (eds) 1998. *Proceedings of the Ocean Drilling Program, Initial Reports,* **174AX**. Ocean Drilling Program, College Station, TX.

NIETO, F., LÓPEZ-GALINDO, A. & PEINADO-FENOLL, E. 1989. *Programa de recogida de datos del difractómetro de rayos X*. Departamento de Mineralogía y Petrología, Universidad de Granada.

NIETO, F., ORTEGA-HUERTAS, M., PEACOR, D. R. & AROSTEGUI, J. 1996. Evolution of illite/smectite from early diagenesis through incipient metamorphism in sediments of the Basque–Cantabrian basin. *Clays and Clay Minerals*, **44**, 304–323.

NORRIS, R. D., HUBER, B. T. & SELF-TRAIL, J. 1999. Synchroneity of the K–T oceanic mass extinction and meteorite impact: Blake Nose, Western North Atlantic. *Geology*, **27**, 419–422.

NORRIS, R. D., KROON, D., KLAUS, A. *et al.* 1998. *Proceedings of the Ocean Drilling Program, Initial Reportsn*, **171B**. Ocean Drilling Program, College Station, TX, 47–91.

OCAMPO, A. C., POPE, K. O. & FISCHER, A. G. 1996. Ejecta blanket deposits of the Chicxulub crater from Albion Island, Belize. *In*: RYDER, G., FASTOVSKY, D. & GARTNER, S. (eds) *The Cretaceous–Tertiary Event and Other Catastrophes in Earth History*. Geological Society of America, Special Papers, **397**, 75–88.

OLSSON, R. K., MILLER, K. G., BROWNING, J. V., HABIB, D. & SUGARMAN, P. J. 1997. Ejecta layer at the Cretaceous–Tertiary boundary, Bass River, New Jersey (Ocean Drilling Program Leg 174AX). *Geology*, **25**, 759–762.

PIERAZZO, E. & MELOSH, J. H. 1999. Hydrocode modeling of Chicxulub as an oblique impact event. *Earth and Planetary Science Letters*, **165**, 163–176.

POPE, K. O., OCAMPO, A. C., FISCHER, A. G. *et al.* 1999. Chicxulub impact ejecta from Albion Island, Belize. *Earth and Planetary Science Letters*, **170**, 351–364.

RETALLACK, G. J. 1996. Acid trauma at the Cretaceous–Tertiary boundary in Eastern Montana. *GSA Today*, **6**, 1–7.

SCHULTZ, P. H. & D'HONDT, S. 1996. Cretaceous–Tertiary (Chicxulub) impact angle and its consequences. *Geology*, **24**, 963–967.

SHARPTON, V. L., BURKE, K., CAMARGO-ZANOGUERA, A. *et al.* 1993. Chicxulub multiring impact basin: size and other characteristics derived from gravity analyses. *Science*, **261**, 1564–1567.

SIGURDSSON, H., D'HONT, S., ARTHUR, M. A., BRALOWER, T. J., ZACHOS, J. C., VAN FOSSEN, M. & CHANNELL, E. T. 1991. Glass from the Cretaceous/Tertiary boundary in Haiti. *Nature*, **349**, 482–487.

SIGURDSSON, H., D'HONT, S. & CAREY, S. 1992. The impact of the Cretaceous/Tertiary bolide on evaporite terrane and generation of major sulfuric acid aerosol. *Earth and Planetary Science Letters*, **109**, 543–559.

SINGER, A. 1979. Palygorskite in sediments: detrital, diagenetic or neoformed. A critical review. *Geologsche Rundschau*, **68**, 996–1008.

SMIT, J. 1990. Meteorite impact, extinctions and the Cretaceous–Tertiary boundary. *Geologie en Mijnbouw*, **69**, 187–204.

SMIT, J. 1999. The global stratigraphy of the Cretaceous–Tertiary boundary impact ejecta. *Annual Review of Earth and Planetary Sciences*, **27**, 75–113.

SMIT, J., MONTANARI, A., SWINBURNE, N. H. S. *et al.* 1992. Tektite-bearing, deep-water clastic unit at the Cretaceous–Tertiary boundary in northeastern Mexico. *Geology*, **20**, 99–103.

SWISHER, C. C., GRAJALES, N. J. M., MONTANARI, A. *et al.* 1992. Coeval $^{40}Ar/^{39}Ar$ ages of 65.0 million years from Chicxulub melt rock and Cretaceous–Tertiary boundary tektites. *Science*, **257**, 954–958.

VELDE, B. 1985. *Clay Minerals. A Physical-Chemical Explanation of their Occurrence.* Elsevier, Amsterdam.

Astronomical calibration of the Danian time scale

URSULA RÖHL[1], JAMES G. OGG[2], TRICIA L. GEIB[2] & GEROLD WEFER[1]

[1]*Geosciences Department, Bremen University, D-28334 Bremen, Germany (e-mail: uroehl@allgeo.uni-bremen.de)*

[2]*Department of Earth and Atmospheric Sciences, Purdue University, West Lafayette, IN 47907-1397, USA*

Abstract: Ocean Drilling Program Sites 1001A (Caribbean Sea) and 1050C (western North Atlantic) display obliquity and precession cycles throughout polarity zone C27 of the late Danian stage (earliest Cenozoic time). Sliding-window spectra analysis and direct cycle counting on downhole logs and high-resolution Fe variations at both sites yield the equivalent of 35–36 obliquity cycles. This cycle-tuned duration for polarity chron C27 of 1.45 Ma (applying a modern mean obliquity period of 40.4 ka) is consistent with trends from astronomical tuning of early Danian polarity chron C29 and $^{40}Ar/^{39}Ar$ age calibration of the Campanian–Maastrichtian magnetic polarity time scale. The cycle-tuned Danian stage (*sensu* Berggren *et al.* 1995, in SEPM Special Publications, **54**, 129–212) spans 3.65 Ma (65.5–61.85 Ma). Spreading rates on a reference South Atlantic synthetic profile display progressive slowing during the Maastrichtian to Danian stages, then remained relatively constant through late Palaeocene and early Eocene time.

Milankovitch climate cycles are created by oscillations in the annual and seasonal distribution of insolation as a result of quasi-periodic changes in the obliquity or tilt of the Earth's spin axis, in the precession or relative tilt of the Earth upon closest approach to the Sun, and in the eccentricity of the Earth's orbit. Geochemical and other proxies in deep-sea and other sediments record the resulting climatic and associated oscillations in ocean circulation, productivity and other surface conditions.

Obliquity oscillations (40.4 ka dominant period) shift the latitude of the polar circle and cause significant climatic variation in high latitudes. Such effects can be transmitted throughout the ocean via thermal–haline circulation. However, obliquity also produces significant changes in evaporation–precipitation balance, and other climatic factors in mid- to low latitudes. Even under ice-free conditions, Park & Oglesby (1990) concluded that 'sensitivity to obliquity can be important at low, as well as high, latitudes'. Precession cycles (modern mean periods of 19 and 23 ka) alter the seasonal contrast at a given latitude, and are a major control on monsoon intensity and associated continental precipitation and coastal upwelling (e.g. deMenocal *et al.* 1993; Olsen & Kent 1996). The main role of eccentricity (95, 123 and 403 ka) is to modulate the amplitude of the precession effect.

The relative importance of obliquity v. precession within the deep-sea sedimentary record is governed by many factors, including palaeogeographical location, geological time and the selected climate proxy. For example, uppermost Cretaceous to lowest Danian sediments of the mid-latitude South Atlantic faithfully record precession cycles modulated by eccentricity (Herbert *et al.* 1995; Herbert 1999). In contrast, Miocene–Oligocene pelagic sediments of the equatorial Atlantic and Pacific record a dominant obliquity cycle with secondary precession effects (e.g. Keigwin 1987; Shackleton *et al.* 1999, 2000).

The astronomical factors influencing obliquity, precession and eccentricity can be projected into mid-Cenozoic time (e.g. Laskar *et al.* 1993; Laskar 1999), and these orbital solutions provide a target curve for high-resolution correlation of sedimentary cycles. This tuning of deep-sea sediment cycles has allowed the development of a high-resolution time scale of biostratigraphy and geomagnetic polarity reversals for Oligocene to Pleistocene time (e.g. Shackleton *et al.* 1990, 1995, 1999; Hilgen, 1991; Hilgen *et al.* 1995; Shackleton & Crowhurst, 1997).

An accurate orbital solution and target curve is not yet available for pre-Oligocene sediments (Laskar 1999). However, the regular dominant periods of obliquity, precession and eccentricity

From: KROON, D., NORRIS, R. D. & KLAUS, A. (eds). 2001 *Western North Atlantic Palaeogene and Cretaceous Palaeoceanography*. Geological Society, London, Special Publications, **183**, 163–183.

allow computation of elapsed duration if these Milankovitch components are demonstrated in sedimentary successions. This form of astronomical calibration has been applied to selected intervals in Mesozoic time (e.g. Huang *et al.* 1993; Gale 1995; Herbert *et al.* 1995; Gale *et al.* 1999; Hinnov & Park 1999; Kent & Olsen 1999; Weedon *et al.* 1999) and to earliest Danian time (Herbert *et al.* 1995).

Palaeogene time scale and the Danian stage

A current time scale 'CK95' for latest Cretaceous and early Cenozoic time was derived from an age model for magnetic polarity chrons derived from a cubic-spline fit of the marine magnetic anomaly pattern in the South Atlantic to selected radiometric ages (Cande & Kent 1992, 1995). Magnetobiostratigraphic correlations allow projection of ages from the 'CK95' model to biostratigraphic datum points and associated geological stages (Berggren *et al.* 1995) (Fig. 1).

In this paper, polarity chron (time) and polarity zone (stratigraphy) nomenclature is the system of Cande & Kent (1992), with suffix n denoting normal polarity or r denoting the previous (underlying) reversed polarity interval. The relative timing (position) of an event (level) within a polarity chron (zone) is 'defined as the relative position in time or distance between the younger and older chronal boundaries' (system of Hallam *et al.* 1985, p. 126). In this proportional stratigraphic convention, the location of the Cretaceous–Palaeogene boundary at Gubbio (Alvarez *et al.* 1977) occurs at C29r.75, indicating that 75 % of reversed-polarity zone C29r is below the event. (Cande & Kent (1992) used an inverted stratigraphic placement relative to present; therefore, C29r.3 in their notation indicated that 30 % of reversed-polarity chron C29r followed the event. This system mirrors the convention of measuring geological time and the numbering magnetic anomalies backwards from the present.)

The Palaeocene–Eocene portion of this 'CK95' age model for magnetic polarity chrons is calibrated by four age control points within a 30 Ma span: 65 Ma (C29r.7; base of Cenozoic sequence), 55 Ma (C24r.34; base of Eocene sequence), 46.8 Ma (C21n.67; lower middle Eocene sequence), and 33.7 Ma (C13r.86; base of Oligocene sequence). Three of these age controls require minor to moderate revision (Berggren *et al.* 1995; Herbert *et al.* 1995; Wei, 1995; Hicks *et al.* 1999). The associated age–distance model for the South Atlantic implied a dramatic slowing by two-thirds of spreading rates from Campanian to Danian time, followed by a doubling of spreading rates to mid-Eocene time, then relative stability until late Miocene time.

Cycle stratigraphy and associated astronomical calibration of the durations of selected Palaeocene–Eocene polarity chrons would help constrain the spreading rates of this 'CK95' model. When coupled with revised age calibrations of polarity chrons, an improved Palaeocene–Eocene time scale can be assembled. Our objectives are to obtain a cycle stratigraphy and associated duration for polarity chron C27 and to derive a more accurate spreading rate history of the South Atlantic magnetic anomaly pattern.

The termination of the Danian stage has not yet been fixed by international agreement on a boundary stratotype or primary marker. Indeed, there is a debate on whether the succeeding stage should be the Selandian stage ('mid'-Palaeocene time), or whether the Thanetian stage of the late Palaeocene period should be extended earlier to create a two-fold division of the Palaeocene period (e.g. Schmitz *et al.* 1998). Berggren *et al.* (1995) utilized the base of the planktonic foraminifer Zone P3 (first appearance datum ('FAD') of *Morozovella angulata*) as a provisional Danian–Selandian boundary, and they estimated a placement near the top of polarity zone C27n (approximately C27n.8), although a precise calibration was not possible (Fig. 1). A more significant global event occurs in Denmark and Spain near the FAD of calcareous nannofossil *Neochiastozygus perfectus* (near the base of nannofossil Zone NP 5) and is projected to be near the base of planktonic foraminifer Zone P3b (FAD of *Igorina albeari*). This placement, which correlates with the major lithological shift in the original earliest Selandian stage in Denmark, is being considered by the Palaeocene Working Group (Schmitz *et al.* 1998; B. Schmitz, pers. comm.). Such a boundary definition, which is in the lower middle of polarity chron C26r, would add *c.* 1 Ma to the duration of the 'Danian' stage in the Cenozoic time scale of Berggren *et al.* (1995). Another possibility for the Danian–Selandian boundary is the FAD of planktonic foraminifer *Praemurica uncinata* (base of foraminifer Zone P2) (H.-P. Luterbacher, pers. comm.), which is near the base of polarity zone C27n (Berggren *et al.* 1995).

In ODP Holes 1050C and 1001A, the magnetic polarity zones are well defined, but the ranges of planktonic microfossils are relatively less constrained. In Hole 1050C, the base of planktonic foraminifer Zone P3a and entire Zone P2 are constrained to be within Cores 1050C-4R to lower 7R (345–380 m below sea floor (mbsf)), but cannot be assigned because of the absence of

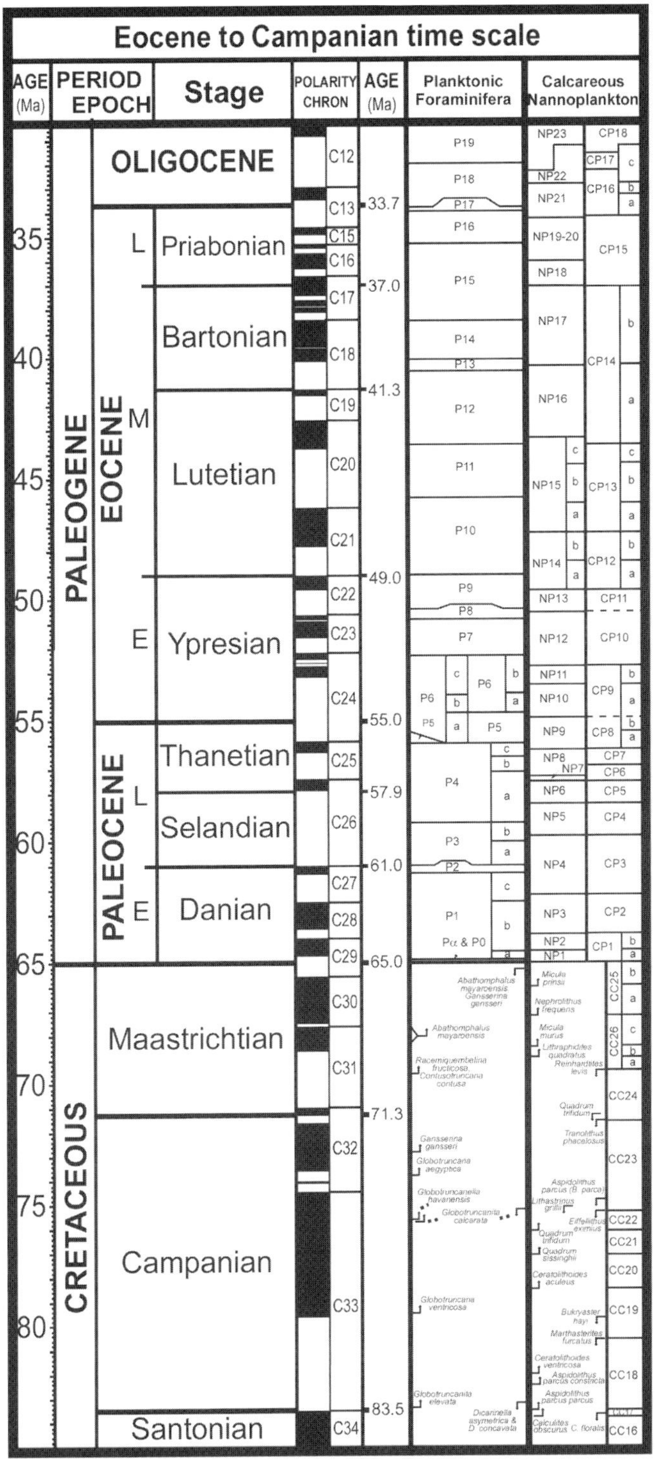

Fig. 1. Late Cretaceous and Palaeogene biochronology and magnetic polarity time scale. Palaeogene portion is modified after Berggren *et al.* (1995), and Late Cretaceous section is modified from Erba *et al.* (1995) and Shipboard Scientific Party (1998*a*).

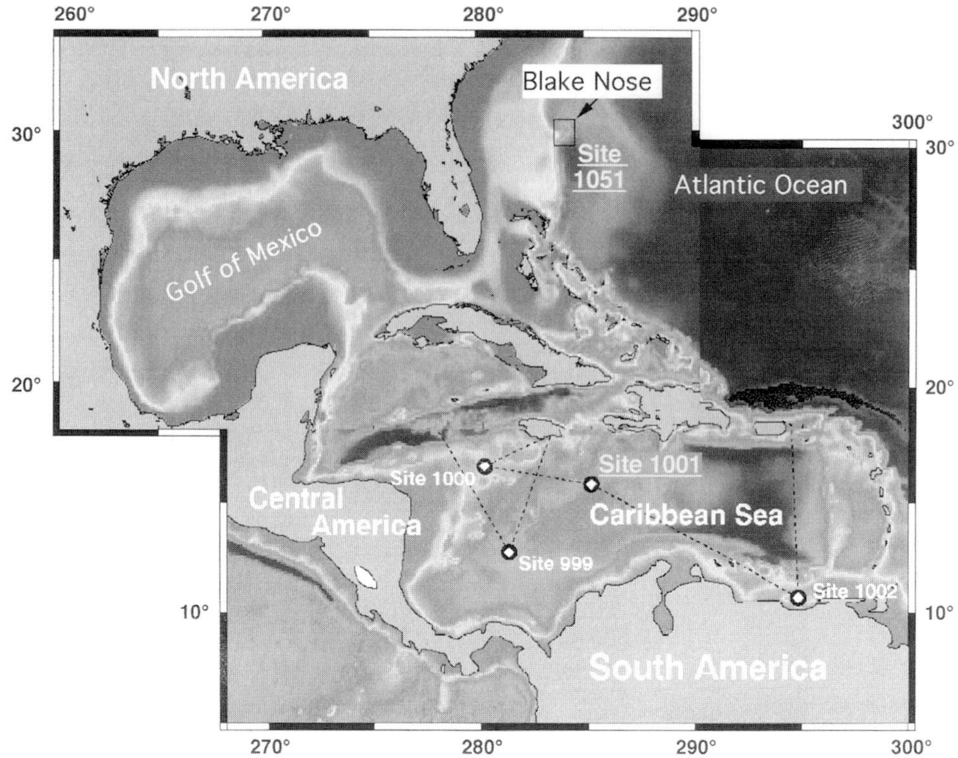

Fig. 2. Location of ODP Site 1050 at the Blake Nose in the western North Atlantic and Site 1001 at the Lower Nicaraguan Rise in the Colombian Basin of the Caribbean Sea (bathymetry from IOC–IHO–BDOC 1994).

the biomarkers (Shipboard Scientific Party 1998*b*). Similarly, Zones P2–P4 cannot be differentiated in Hole 1001A (Shipboard Scientific Party 1997). Therefore, in this paper, we use the unofficial temporary assignment of the Danian–Selandian boundary to near the top of polarity zone C27n. This allows a direct comparison with the duration of the 'Danian' concept in the widely distributed Cenozoic time scale of Berggren *et al.* (1995). However, to clarify the lack of an international definition of the 'Danian' stage, we denote this usage as 'Danian stage (*sensu* Berggren *et al.* 1995)'.

Upper Danian stratigraphy at ODP Sites 1050 and 1001

Magnetic polarity zones

We analysed the cyclicity of iron (Fe) variations in sediments spanning upper Danian polarity zone C27 at two Ocean Drilling Program (ODP) sites (Fig. 2). Upper Danian magnetic polarity zones C28n–C26r were identified by palaeomagnetic measurements on discrete minicores enhanced by pass-through cryogenic magnetometer data. Polarity chron assignments were based upon the biostratigraphic constraints coupled with the distinctive pattern of normal- and reversed-polarity zones.

The base of polarity zone C27r and the top of zone C27n occur in cores with complete recovery at both sites. In both holes, the ratio of the thickness of polarity zone C27n to C27r is slightly greater than the ratio of the widths of marine magnetic anomaly C27n to C27r in the South Atlantic model of Cande & Kent (1995). We suggest that this systematic thickening of polarity zone C27n may indicate an underestimate of the width of anomaly C27n. Anomaly C27n has an average width of 5 km on South Atlantic synthetic flow line, but may have an approximate uncertainty of about 1 km (as estimated from 95 % confidence intervals in table 4 in Cande & Kent (1992), although C27n is not tabulated individually). To reduce the influence of these uncertainties in the duration of individual brief polarity chrons, we analysed only the total interval spanned by combined polarity zones C27r and C27n.

Hole 1050C

ODP Site 1050 was drilled on the Blake Nose (Atlantic margin of northern Florida). The upper Danian sediments range from greenish grey clayey siliceous chalk to slightly darker nannofossil-rich claystone. Oscillations on a 1 m of calcium carbonate, silica and clay content are pronounced features in the shipboard multisensor core logs and downhole geophysical logs, and correspond to muted light–dark variations in the sediments (Shipboard Scientific Party 1998*b*). In contrast, the more calcareous middle Danian (upper polarity zone C28n and lowermost C27r) chalk displays striking oscillations between greenish and pinkish coloration, indicating changing iron oxidation states within the sediment, with an average wavelength of about 50 cm.

An initial estimate of the periodicity of the light–dark variations in the upper Danian sequence can be made from the palaeomagnetic-derived accumulation rates. Polarity chron C27 (C27n plus C27r) spans *c.* 1500 ka (Cande & Kent 1995). In Hole 1050C, polarity zone C27 spans 37.0 ± 0.7 m: the top of C27n is at 332.7 ± 0.7 mbsf from palaeomagnetic analysis of discrete minicores, and the base of C27r is tightly constrained at 370.69 ± 0.03 mbsf from high-resolution pass-through cryogenic magnetometer records (Shipboard Scientific Party 1998*b*; Ogg & Bardot 2000). Therefore, the sedimentation rate for late Danian time (C27) averages 24 m Ma^{-1}, implying that the 1 m banding corresponds to *c.* 40 ka periodicity. Their monotonous regularity and approximate periodicity suggest that these light–dark cycles in Hole 1050C are caused by the obliquity component of Milankovitch variations.

Compared with the upper Danian sequence (polarity zone C27), the underlying middle Danian (polarity zone C28n) chalk has a slower average accumulation of about 15 m Ma^{-1}. This implies that the 50 cm variations between greenish and pinkish hues have an approximate periodicity of 30–35 ka. However, the wavelength of these colour variations is irregular, which suggests the possibility of superimposed precession and obliquity controls.

The overlying Selandian to lower Thanetian (polarity zone C26r to C25r) chalks have an average accumulation of about 15–16 m Ma^{-1}. The coincidence of clay enrichment and relatively higher accumulation rate for the upper Danian sequence (C27) indicates an increased influx of clay.

Accumulation rates slow to about 8–9 m Ma^{-1} for the uppermost Palaeocene units (C25n). At nearby Site 1051, the uppermost Palaeocene sequence to the carbon-isotope shift that heralds the Palaeocene–Eocene transition (lower part of C24r) contains superimposed 29 and 23 cm cycles in magnetic susceptibility and iron abundance. These have been interpreted as the pair of precession components (23 and 19 ka) of Milankovitch climate cycles (Norris & Röhl 1999).

Hole 1001A

ODP Site 1001 on the lower Nicaraguan Rise (Caribbean Sea) recovered an Upper Cretaceous to Palaeocene succession (Fig. 2). The upper Palaeocene to middle Eocene sediments are primarily composed of foraminifer-bearing chalk with clay, minor chert and volcanic ash layers (Shipboard Scientific Party 1997).

In ODP Hole 1001A, polarity zone C27 spans about 12 m (328.0–339.9 mbsf) (Shipboard Scientific Party 1997; Louvel & Galbrun 2000; see Fig. 7), therefore the mean accumulation rate for the upper Danian sequence is 8 m Ma^{-1}, which is about one-third the sediment accumulation rate of Hole 1050C.

X-ray fluorescence (XRF) scanning

The Danian sediments were measured with the X-ray fluorescence (XRF) core scanner at Bremen University (Röhl & Abrams 2000). The XRF scanner allows closely spaced (or even continuous), non-destructive analyses of major and minor elements at the surfaces of the split cores (Jansen *et al.* 1998). XRF measurements of the entire suite of elements between potassium (K, atomic number 19) and strontium (Sr, atomic number 38) were recorded at 2 cm (Hole 1001A) or 4 cm (Hole 1050C) intervals, in accordance with the different accumulation rates. The intensities in counts per second (c.p.s.) were calibrated between each core by measuring a suite of lithological standards.

Fe was selected as a geochemical proxy for climate–palaeoceanographic cycles. This major element mirrors changes in carbonate/clay ratios, and is in significant abundance throughout the sediments, therefore it has a high signal-to-noise ratio. Fe is located in the centre of the measurable element range (atomic number 26), and therefore provides a stronger signal than the heavy or light ends of the XRF range. Calcium (Ca) was simultaneously measured by the XRF scanner, and generally varied inversely with Fe. Quantitative documentation of cyclicity at other ODP sites has utilized magnetic susceptibility (e.g. Shackleton *et al.* 1995, 1999), which is an indirect indicator of the combined influences of Fe content, Fe oxidation state and Ca abundance. In general, the concentration of Fe is less affected by post-burial diagenetic alteration than is the oxidation state of Fe and associated magnetic susceptibility (Norris & Röhl, 1999; Röhl & Abrams, 2000).

Spectral analysis and direct cycle counts

Methods

Two related methods were used to ascertain the presence of Milankovitch cycles in the XRF Fe intensity measurements. The first was to apply spectral analysis to a series of sliding windows through the Fe record to identify the continuity and ratios of significant peaks in cyclicity with depth. Continuity of peaks among windows requires an implicit assumption that sedimentation rates are not changing rapidly.

Spectral analysis can generate artefacts, but the dominant peaks should also be visually apparent in the raw data. Therefore, the second method was to directly count the regular oscillations in the Fe record and downhole logs. In intervals where both obliquity and precession appeared to contribute to the cyclicity, then the distinction of obliquity v. precession counts were partially guided by their relative spacing.

Spectral analyses of selected windows used the AnalySeries 1.1 software package (Paillard *et al.* 1996). The XRF Fe intensity measurements within each window were detrended with a 51-point moving average, normalized to unit variance, and pre-whitened before spectral analysis by the Blackman–Tukey method. Anomalous concentrations in Fe associated with rare volcanic ash horizons in Hole 1001A were filtered to reduce spectral artefacts.

Continuous downhole geophysical logging of Hole 1050C included a measurement of natural gamma radiation (NGT log) at 15 cm intervals. The uranium component was visually cyclic and provided an independent record for spectral analysis that was not affected by core recovery. Spectral analysis of the uranium gamma radiation log in Hole 1050C yielded the same wavelengths for the dominant cycles, and most of the individual uranium excursions could be directly matched to intensity peaks in the XRF Fe record. In contrast, downhole logs in Hole 1001A lacked the resolution to resolve precession–obliquity cycles, as a result of the much slower accumulation rates.

There are a few minor gaps in core recovery. Correlation of characteristics of the XRF Fe record with downhole logs of magnetic susceptibility seemed to indicate that all gaps in the core recovery were caused by failure to obtain the basal portion of the cored interval, rather than a partial concatenation within the core. Therefore, these recovery gaps do not affect determination of average wavelengths of component cycles and associated estimates of sediment accumulation rates from spectral analysis. However, these recovery gaps preclude using progressive overlapping sliding windows through the entire C27 interval.

Therefore, for one series of spectral analyses, we constructed a synthetic XRF-depth scale that adjusted for significant spacing between core recovery. Overlapping windows of 5 m Hole 1001A) or 10 m (Hole 1050C) were applied to monitor the continuity and relative importance of spectral peaks and the variations in accumulation rate. These rates were then applied to the true thickness of polarity zone C27 in each hole. A similar sliding-window spectral analysis was also performed with the Joint Time-Frequency Analysis Tool (JTFA) of LabVIEW 4.1 from National Instruments. The JTFA produced a 3D, short-time Fourier transform spectrogram by performing a full-bandwidth spectral analysis of a 256-point sliding window through the data (window size of 10 m for Hole 1050C and 5 m for Hole 1001A). However, even though these colourful JTFA plots provided a useful glimpse of the main trends, the stacked series of selected windows from the AnalySeries output allowed a more direct interpretation of shifting peak positions as visually displayed in the original records. The spectral results from overlapping sliding windows through this synthetic no-gap record yielded identical trends in peak positions to the spectra of the non-overlapping intervals from the individual cores.

However, these gaps in recovery do require making an estimate of the number of 'missing cycles' when using the direct counting method. Estimates of 'missing cycles' were obtained by assuming that the mean wavelength of major oscillations within the overlying metre was maintained in the underlying recovery gap. Only in Hole 1001A, where a 1.2 m interval was not recovered between Cores 36R and 37R, was a significant interpolation required in counting 'missing cycles'. These gaps in core recovery do not apply to the downhole geophysical logs, and cycle counts of the uranium component of natural gamma radiation in Hole

1050C provided an independent verification of the cyclostratigraphy estimates from the high-resolution Fe record.

Hole 1050C

Hole 1050C sediments display the full suite of obliquity, precession and eccentricity cycles.

A pronounced 1 m oscillation is present in both the scanning XRF Fe records from cores and the downhole log of natural gamma radiation (Figs. 3–6). Its mean wavelength from spectral analysis of both records is 0.92 m (Fig. 4). The average 24 m Ma^{-1} sedimentation rate for the upper Danian sequence implies that this 1 m monotonous cycle is obliquity, which has a modern mean period of 40.4 ka (Herbert *et al.* 1995). Many of the broader and major oscillations in high-resolution Fe record have corresponding features in the downhole logs of magnetic susceptibility, total natural gamma radiation, and the uranium component of natural gamma radiation (Fig. 5). Relatively high Fe content generally correlates with a higher magnetic susceptibility, a darker coloration in the recovered sediments, a lower carbonate content (reflected in a lower Ca intensity in the XRF scans), and a higher total and uranium component of natural gamma radiation. The position of this 'obliquity' spectral peak shifts through the C28n–C27 interval (Fig. 3). As previously estimated from the relative widths of polarity chrons, there is an increase in accumulation rate near the top of polarity zone C28n and an interpreted trend toward slower accumulation in the lowermost portion of C26r (Fig. 3). The ubiquitous repetition of this 1 m cycle was used to count 'obliquity cycles' in the XRF Fe record (Fig. 3).

Spectral analysis of the full interval of polarity zone C27 yields a pair of higher-frequency peaks with mean wavelengths of *c.* 0.53 and 0.45 cm (Fig. 4). Relative to the assigned obliquity peak (40.4 ka), their respective projected periods are 23 and 20 ka, implying that these are the pair of precession cycles (modern periods of 23 and 19 ka). The sliding window spectra and associated short-wavelength oscillations in the Fe record indicate that this precession component has variable importance. Dual precession peaks are resolved in the progressive 5 m spectral windows in the lower portion of the analysed interval (upper part of polarity zone C28n). The two precession components are interpreted to merge into a single '21 ka' peak within polarity zone C27n, where the higher accumulation rates imply fewer precession cycles within each 5 m window. The precession component is relatively enhanced within the basal portion of polarity zone C27r (Cores 6R and 7R), which has 50 cm alternations from reddish to greenish coloration. As the accumulation rate increases upward within this lower interval, the wavelengths of the obliquity and dual precession peaks maintain constant ratios in the progressive spectral windows (Fig. 3). The precession component (and dual components in some intervals) is superimposed upon the 1 m obliquity oscillations, thereby creating a degree of ambiguity in attempting direct counts of precession-induced Fe peaks (Fig. 6).

Spectra analysis of both the Fe and uranium gamma radiation records indicate a pronounced long-wavelength oscillation centred near 2.2 m wavelength (Fig. 4). The detailed Fe record has an additional subdued spectral peak near 2.75 m. Relative to the assigned obliquity peak, these wavelengths correspond to *c.* 100 ka, indicating the partial resolution of short-period eccentricity (mean average period is *c.* 109 ka). This eccentricity component is only partially resolved within the 5 m spectral windows, except within the basal interval of slower accumulation rates (Fig. 3).

Spectra of the uranium component of the gamma radiation log through the entire C27 interval yield an additional long-wavelength peak at 9.35 m (not shown in figures). This long-wavelength oscillation is displayed as sets of relatively intense uranium gamma radiation peaks relative to the background mean levels at *c.* 351, 361 and 369 mbsf (Fig. 5). This long-wavelength oscillation has a projected period of 410 ka relative to the average obliquity (40.4 ka) wavelength of 0.92 m, therefore it may reflect the long-period (403 ka) eccentricity cycle. However, because of the inadequate length of record, this spectral peak may be fortuitous.

Therefore, spectral analysis of both the high-resolution Fe records and the downhole logs of uranium gamma radiation through the C27 interval indicates the presence of the entire suite of Milankovitch cycles of precession, obliquity and short-period eccentricity, and perhaps long-period eccentricity. A precise 'tuning' conversion of the signal–depth records of Hole 1050C to a signal–time record can be achieved by adjusting the visually apparent obliquity peaks to a fixed even spacing (e.g. method used by Shackleton *et al.* (1999)). Spectral analysis of such an 'obliquity tuned' signal–time record would probably enhance the resolution of the suite of precession and eccentricity cycles, and eventually allow calibration to a target curve derived from orbital solutions. However, the importance of obliquity in the untuned Fe record is adequate to

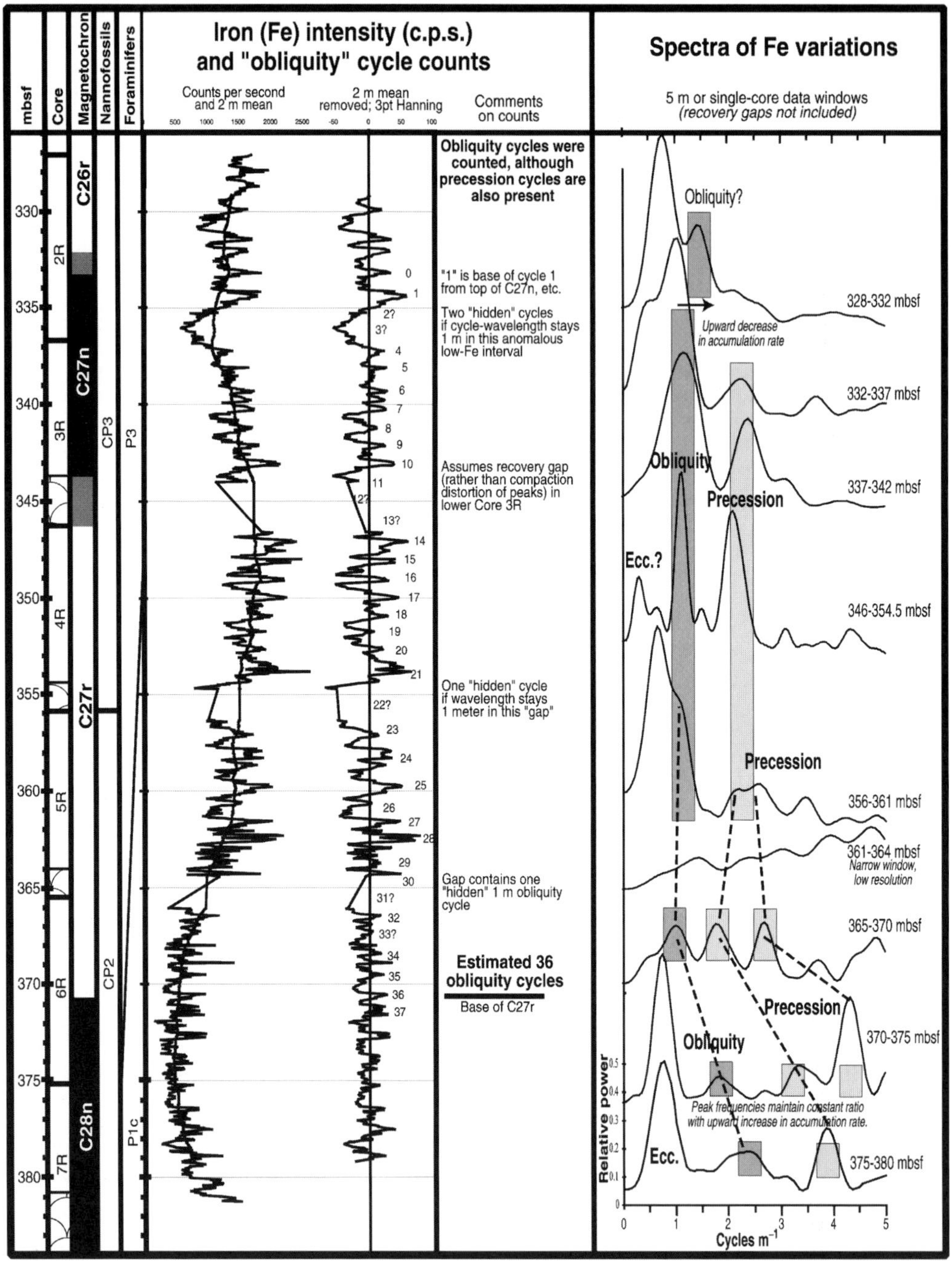

Fig. 3. Composite of bio- and magnetostratigraphy, Fe intensity curves (bold line indicates 2 m moving average), obliquity cycle counts, and selected interval-spectra for ODP Hole 1050C. Depth is in metres below seafloor (mbsf), gaps in coring recovery are shaded with curved lines, planktonic foraminifer and calcareous nannofossil zones (P and CP zonations) are from Shipboard Scientific Party (1998*b*). Spectral windows generally span 5 m intervals. The assignment of obliquity and precession to selected peaks and interpretations of changing accumulation rates are explained in the text.

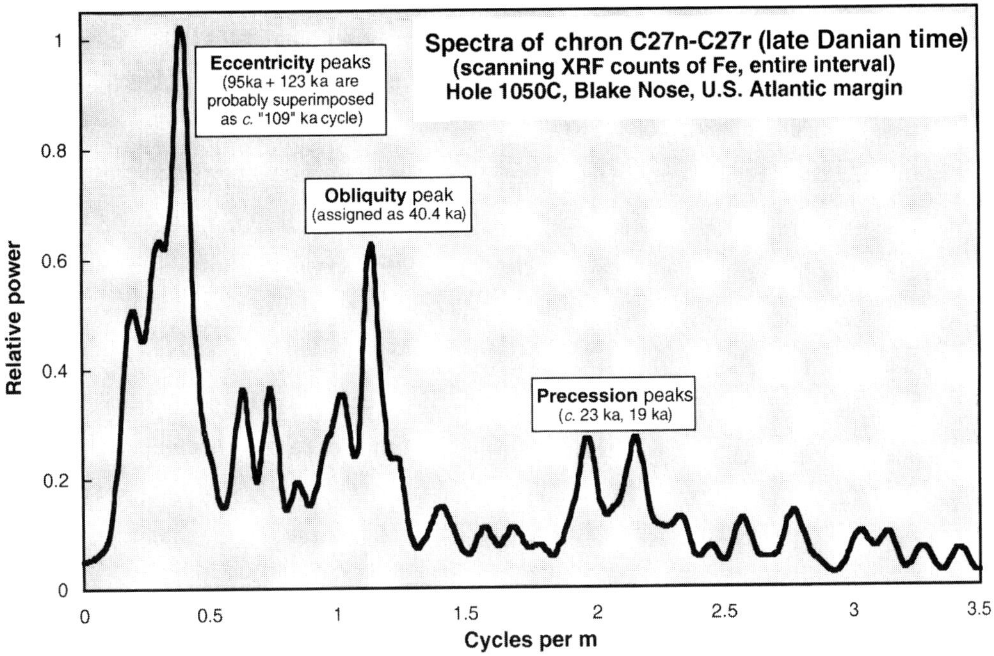

Fig. 4. Spectra from the XRF Fe record of entire polarity zone C27 interval within ODP Hole 1050C. Recovery gaps were removed to create a continuous synthetic record. The wider window allows resolution of eccentricity and the double peak of precession (23 and 19 ka modern periods).

estimate the duration of polarity zone C27 (Fig. 3).

In Hole 1050C, polarity zone C27 spans 38.0 ± 0.7 m. Spectral results from the entire interval (Fig. 4) cannot be applied, because of the slower accumulation rate in the lower portion of polarity zone C27r. Therefore, applying the obliquity wavelengths from the succession of overlapping 10 m windows yields a total span of polarity chron C27 of 36.4 ± 0.7 obliquity cycles.

Direct counting of the major 'obliquity' oscillations in the Fe record requires interpolations across gaps, adjusting for the slower sedimentation rate in the lowermost portion of polarity zone C27r, and some subjective assignments in some intervals. This count, despite these problems, also yielded an estimate of 36 cycles (Fig. 3). Independent visual counting of 'obliquity' cycles in a filtered uranium gamma radiation record that used a sine curve of wavelength 0.92 m as an additional guide yielded between 34 and 35 cycles.

Therefore, all four methods (spectral analysis and direct counting of monotonous peaks in both the Fe and uranium gamma radiation records) converge to yield an estimated duration of polarity zone C27n equivalent to 36 obliquity cycles.

Hole 1001A

Hole 1001A sediments also display superimposed precession and obliquity cycles, but precession cycles dominate over most of polarity zone C27.

Cycle stratigraphy in ODP Hole 1001A has complications introduced by slower sediment accumulation rates and by random high-Fe spikes associated with volcanic horizons. In ODP Hole 1001A, polarity zone C27 spans about 12 m (328.0–339.9 mbsf) (Shipboard Scientific Party 1997; Louvel & Galbrun 2000; Fig. 7), therefore 1 m of sediment encompasses about 125 ka, which is about one-third of the sediment accumulation rate in Hole 1050C.

The suite of spectra through progressive 3 m windows displays three main peaks that remain semi-stable: one in the 2.5–3 cycles m^{-1} range, a second in the 5–6 cycles m^{-1} range, and the third at about 7–8 cycles m^{-1} (Fig. 7). Applying the mean sedimentation rate yields estimated average durations of *c.* 45, 23 and 17 ka. On the basis of these estimates, these three peaks are considered to represent Milankovitch cycles of obliquity and both precession components (23 and 19 ka modern periods). Because of the narrow 3 m windows, no unambiguous spectra evidence for the equivalent of eccentricity wavelengths (expected at about 1 cycle m^{-1}).

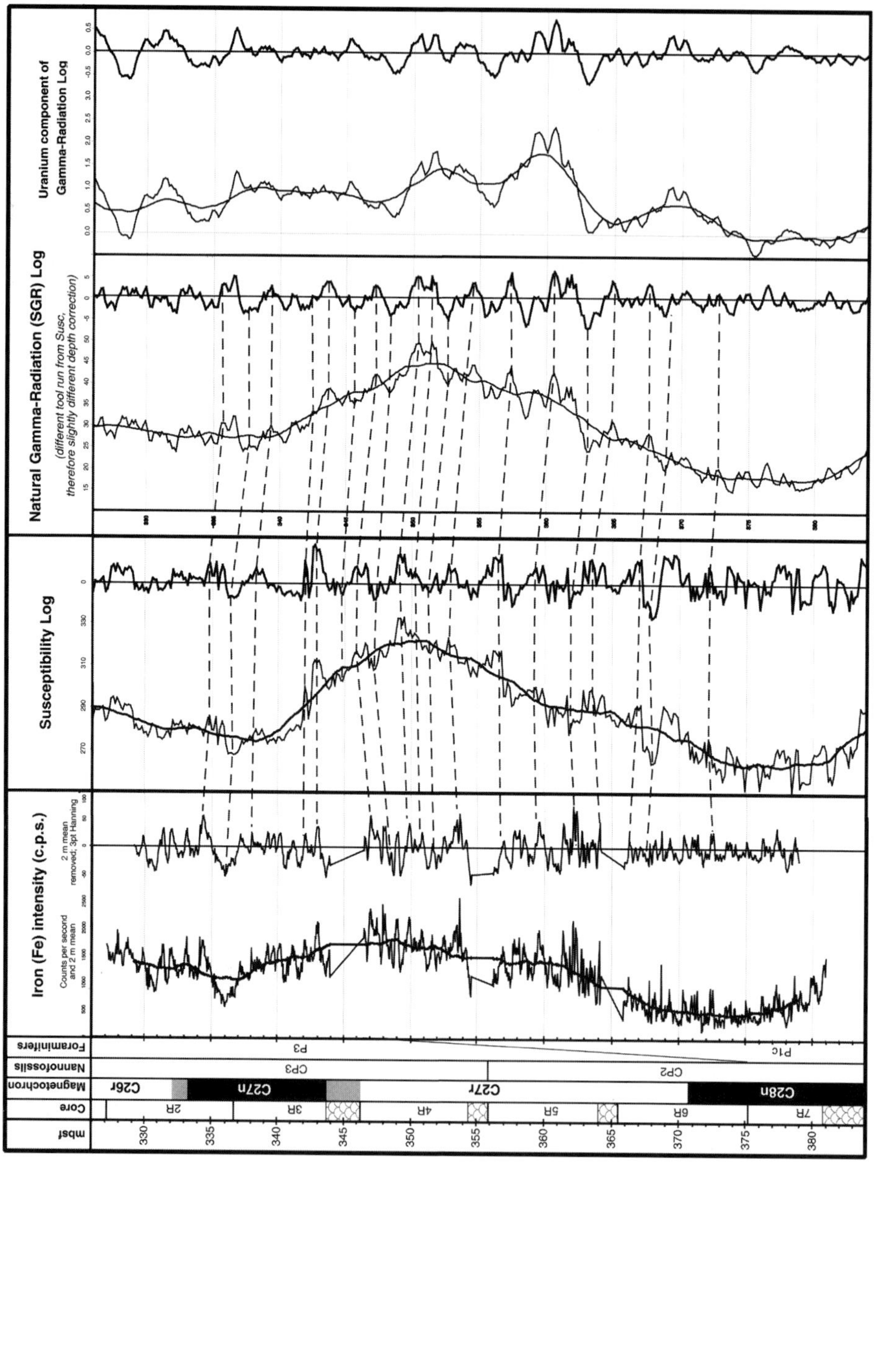

Fig. 5. Composite of bio- and magnetostratigraphy, Fe intensity curves (bold line indicates 2 m moving average) for ODP Hole 1050C. Downhole logs of susceptibility (arbitrary dimensionless unit), natural gamma ray (API units) and the uranium component of this gamma-ray log are from Shipboard Scientific Party (1998*b*). (See Fig. 3 for explanation of other abbreviations and patterns.)

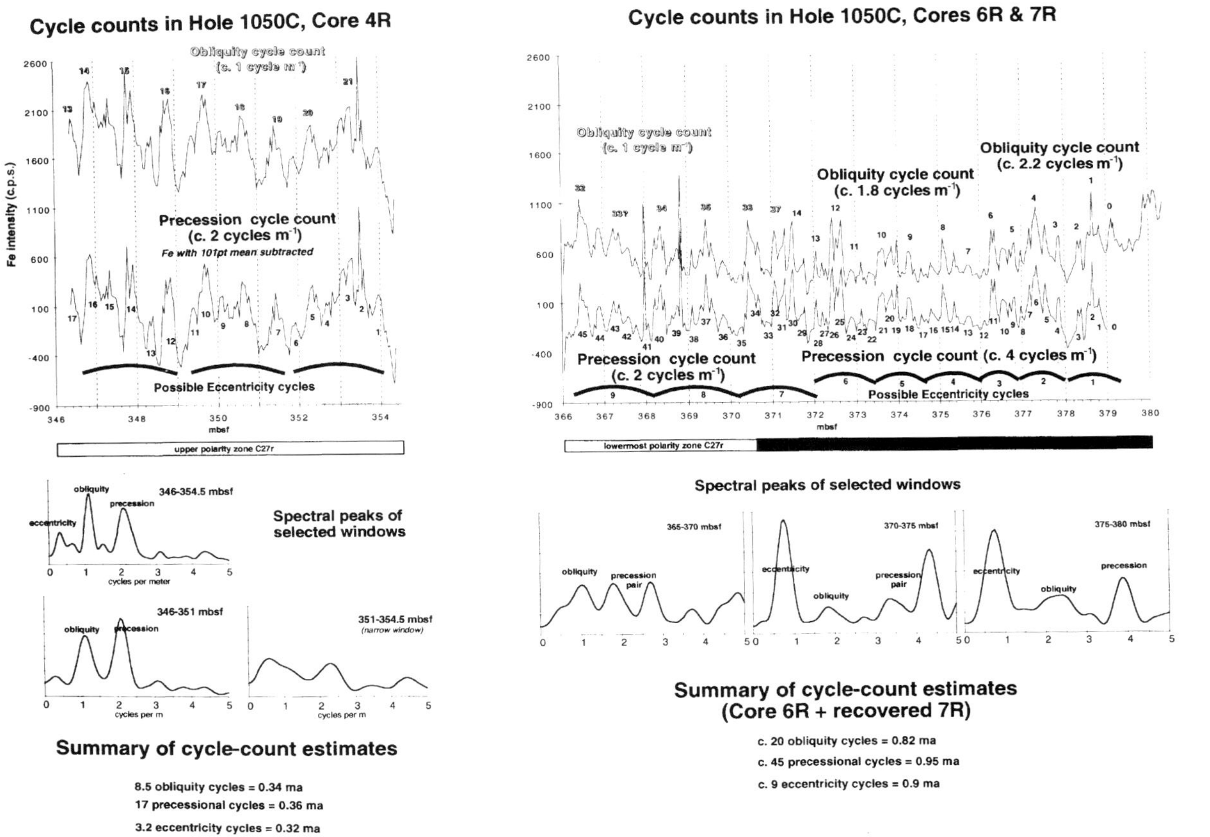

Fig. 6. Expanded record of Fe intensity variations within two intervals of ODP Hole 1050C that illustrate superimposed obliquity and precession cycles. Numbered 'obliquity cycles' within polarity zone C27 are from Fig. 3. The record of Core 4R is interpreted as a main set of obliquity cycles of 1 m wavelength with precession adding secondary peaks and eccentricity shifting the baseline at about 2.5 m intervals. Estimated durations of the record from Core 4R are similar from trial cycle counts of each Milankovitch component. In contrast, Cores 6R and 7R have a lesser obliquity component relative to a pronounced precession (lower set of peak counts); therefore, assignment of peaks to individual components becomes ambiguous. The lower set of obliquity counts within polarity zone C28n is arbitrarily numbered upward from 379 mbsf.

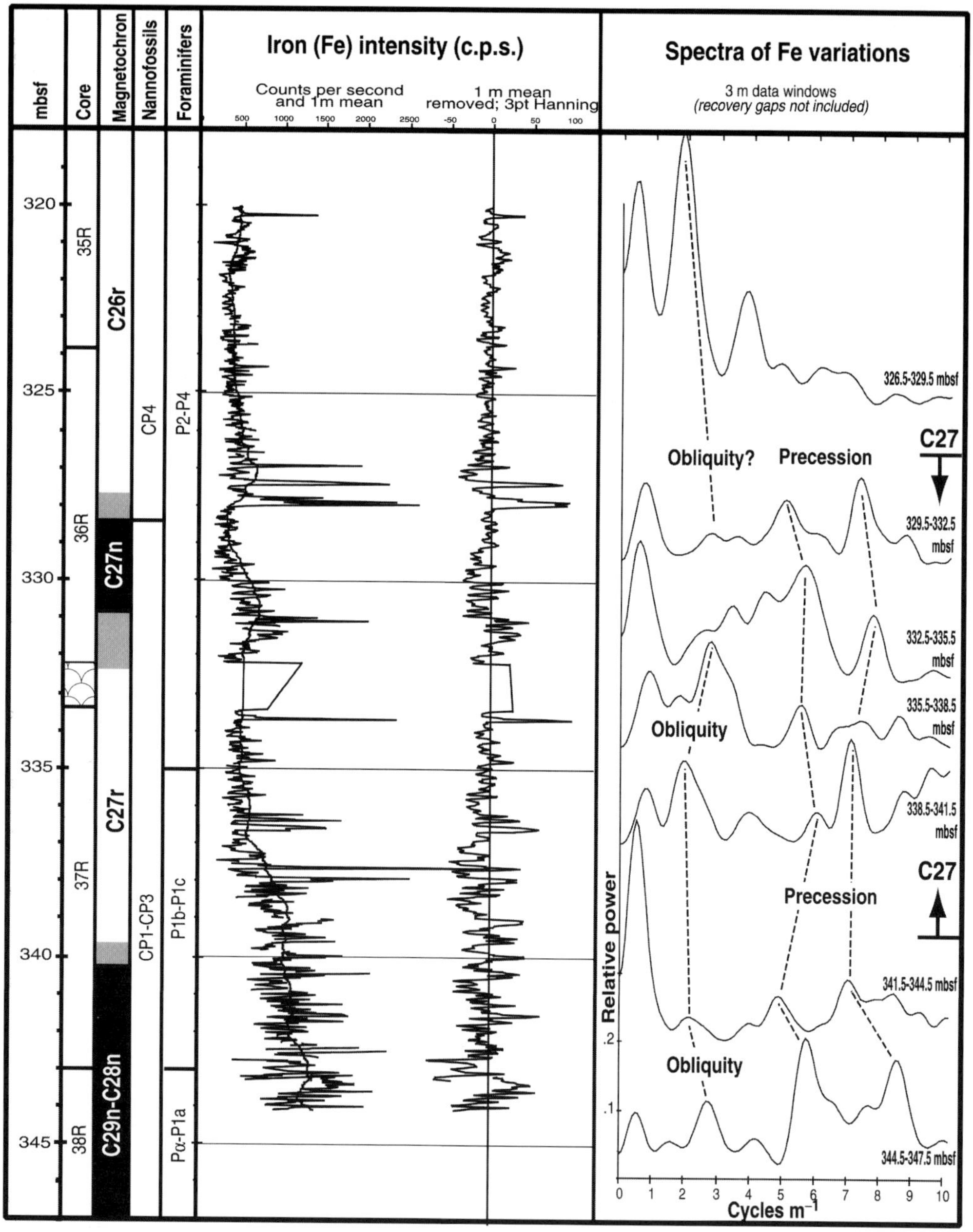

Fig. 7. Composite of bio- and magnetostratigraphy, Fe intensity curves (bold line indicates 1 m moving average), and selected interval-spectra for ODP Hole 1001A. Interval spectra for an XRF scan of Fe intensity in upper Core 38R (344.5–347.5 mbsf) is included, although the Fe curve from this poorly dated core is not included. (See Fig. 3 for explanation of other abbreviations and patterns.).

The relative importance of the assigned obliquity and precession components changes through the analysed interval. Precession is dominant below polarity zone C27r, especially within the poorly dated Core 38R. The obliquity component assumes an important role in the lower half of zone C27r, then appears to vanish in upper C27r and within C27n, where precession again becomes dominant. Visual examination of the Fe record suggests a sudden transition at 336.5 mbsf, with monotonous obliquity dominating perhaps 11 major oscillations in the lower portion of C27r. Irregular amplitudes of these short-wavelength oscillations in the upper portion of the Fe record add considerable ambiguity to direct counting of regular variations (Fig. 8).

Even through the assigned precession cycles appear to shift their wavelengths slightly within the suite of individual spectra, there are no significant trends in accumulation rate from *c.* 345 mbsf to 329 mbsf. Therefore, applying the mean 0.34 m obliquity wavelength within the lower portion of polarity zone C27r to the entire polarity zone implies that polarity zone C27 (12 m) spans 35 obliquity cycles.

Attempts to assign and count precession and obliquity cycles were made independently by two of the authors (Fig. 8). One attempt counted 41'precession' cycles within the upper portion of C27 overlying 11 'obliquity' cycles within lower portion of C27. The other attempt concentrated on a single cycle band, and obtained either 70 precession cycles or 35 obliquity cycles. In both cases, the counts of 'precession' had implicitly assumed a single '21 ka' cycle, rather than trying to distinguish the superimposed '23 ka' and '19 ka' cycles. Even though such direct counting procedures are highly subjective, these independent trials yielded nearly identical estimates of the duration of polarity zone C27 as equivalent to 1.45 Ma using modern Milankovitch cycle durations.

Therefore, with either an assumption of quasi-stable accumulation rate or direct counting of either precession or obliquity cycles, the entire polarity zone C27 in Hole 1001A spans the equivalent of 35 obliquity cycles.

Relative importance of obliquity and precession

In summary, obliquity cycles are identified and counted throughout polarity zone C27 in Hole 1050C off Florida even though the full suite of obliquity, precession and eccentricity Milankovitch cycles are present. Assignment of the high-amplitude oscillations, of 1 m wavelength, to obliquity (40.4 ka modern period) is consistent with the estimated sediment accumulation rate of polarity zone C27. The obliquity assignment also agrees with the relative wavelengths of superimposed cycles of lesser amplitude, which are consistent with the ratios to precession and eccentricity periods. In the lowest portion of polarity zone C27 within Hole 1050C, precession is associated with redox oscillations (reddish v. greenish coloration) and a relatively low abundance of Fe. In contrast, precession dominates through most of Hole 1001A in the Caribbean, where obliquity cycles are significant only in the lower portion of polarity zone C27.

The relative importance of obliquity cycles in the upper Danian sequence of Hole 1050C is interesting. In general, climate cycles associated with precession are considered to be most important in tropical to subtropical regions, whereas obliquity is mainly important at high to polar latitudes. Site 1050C was at *c.* 20° N latitude (Ogg & Bardot 2000) during Danian time. There has been no identification of continental ice during Danian time; therefore it was expected that the influence of obliquity would not be significant in governing relative influxes of carbonate and clay. However, the importance of obliquity cycles in governing facies variations is not restricted to high latitudes or to glacial periods. General circulation models of the regional impact of different Milankovitch cycles during the ice-free late Cretaceous period suggest that low latitudes are sensitive to obliquity variations (Park & Oglesby 1990). It is also possible that the precession signal may devolve to near-zero during intervals of geological time, leaving Earth's obliquity as the dominant forcing parameter (L. Hinnov, pers. comm.). An important role of obliquity cycles has been recorded at other tropical to subtropical sites, such as in the Oligocene sequence at Ceara Rise in near-equatorial South Atlantic (Shackleton *et al.* 1999), several intervals in the Valanginian–Hauterivian–Barremian sequence of Southeastern France at 30° N palaeolatitude (Huang *et al.* 1993; Giraud 1995), and the Kimmeridge Clay and Rhaetian–Sinemurian sequence of southern England at respective 30° N and near-equatorial palaeolatitudes (Weedon *et al.* 1999). An upward shift from dominant precession–eccentricity cyclicity to dominant obliquity cyclicity occurs across the Early to Mid-Jurassic boundary within tropical pelagic carbonates in northern Italy (Hinnov & Park 1999).

In the case of coeval Sites 1001 and 1050, it appears that obliquity was generally much less significant in the semi-enclosed Caribbean than in the eastern margin of the central Atlantic during late Danian time. Therefore, regional palaeoceanographic factors contribute to the local relative expression of precession v. obliquity cycles, in addition to probable global palaeoclimatic factors.

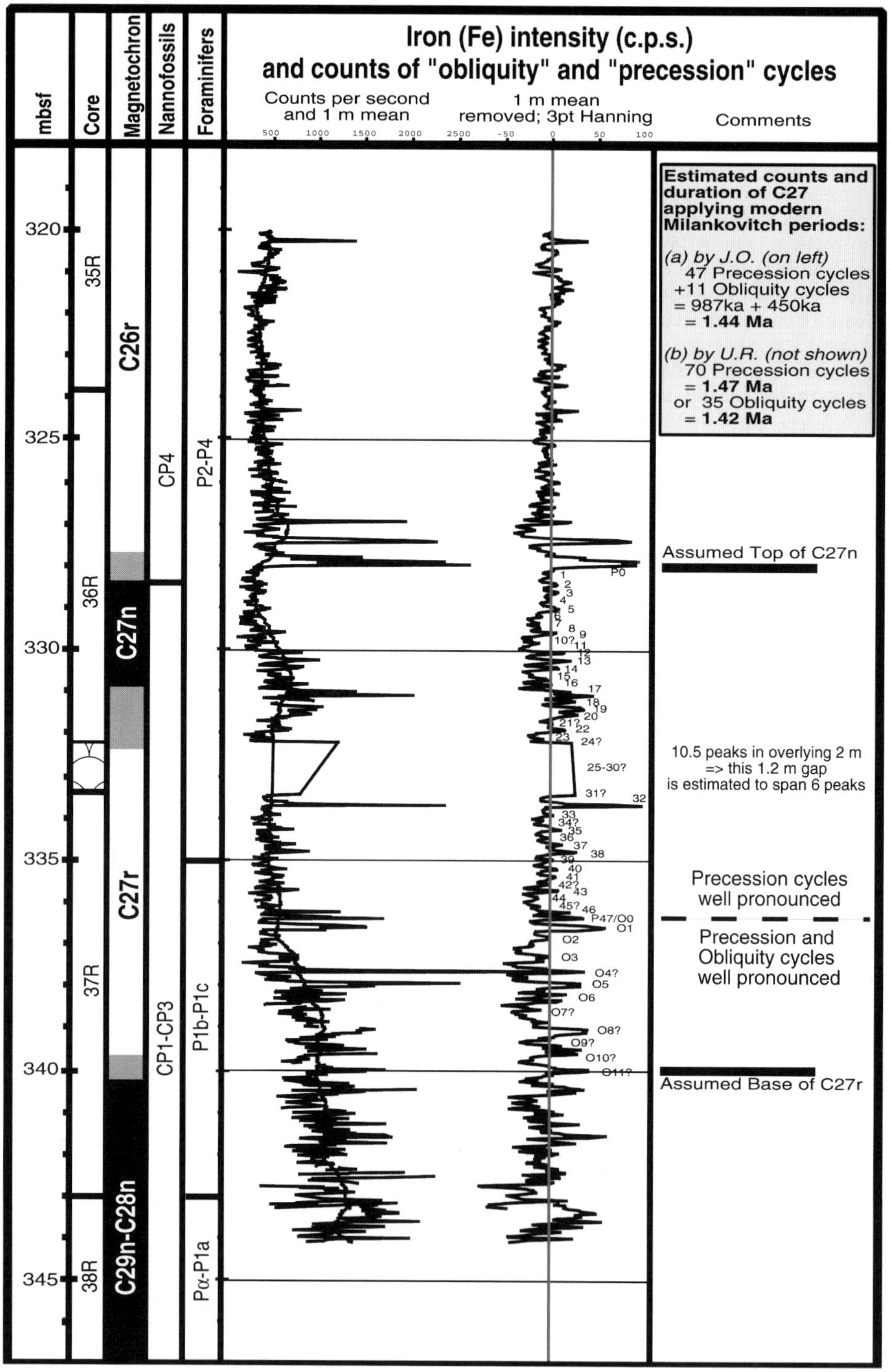

Fig. 8. Composite of bio- and magnetostratigraphy, Fe intensity curves (bold line indicates 1 m moving average), and cycle counts for ODP Hole 1001A. Attempts to assign and count precession and obliquity cycles were carried out independently by two of the authors (J.O. and U.R.), and the assignments by J.O. are shown adjacent to the Fe intensity curve. These independent trials yielded identical estimates of the duration of polarity zone C27.

Astronomical calibration of the Danian time scale

Periods of obliquity and precession cycles during Danian time

Modern obliquity cycles have a 40.4 ka period and precession cycles average 20.8 ka (Herbert *et al.* 1995). A model incorporating progressive lunar recession suggests that these two periods may have been about 1.3 ka and 0.4 ka shorter, respectively, at 65 Ma (Berger *et al.* 1992). The main periods of eccentricity are relatively invariant through Cenozoic–Mesozoic time; therefore measuring the relative ratio of precession cycles to eccentricity in ancient sediments can test this model. Spectral analysis by Olsen & Kent (1996) of upper Triassic lake sediments, in which the same astronomical model predicts shortening of the 19 ka and 23 ka precession periods to *c.* 18 and 21.5 ka, respectively did not show this theoretical effect. Olsen & Kent concluded that, considering the uncertainties associated with projections of the recession rate of the Moon and of the length of the day, it 'is, in fact, plausible that the climatic precession cycles in the Triassic were not . . . different from present day values'. A similar test is not yet available to test the associated prediction of shortening of the obliquity cycle.

We shall take the conservative approach and apply modern values of Milankovitch precession and obliquity cycles to estimate the duration of polarity chrons and the Danian stage. As will be demonstrated, these modern values still imply significantly shorter durations than are estimated from the CK95 spreading rate model. Applying the theoretical shorter periods for precession and obliquity will only magnify this difference.

Estimated durations of polarity chrons C27 and C29, and intervening chron C28

Polarity zone C27 at both sites spans the equivalent of 35–36 obliquity cycles. The conservative estimate of 36 obliquity cycles at 40.4 ka implies that polarity chron C27 spans 1.45 Ma.

Anomaly C27 spans 21.90 km in the synthetic composite flow line of the South Atlantic magnetic anomalies that serves as a standard for scaling the Late Cretaceous and Cenozoic geomagnetic polarity pattern (Cande & Kent 1992). Therefore, the average spreading rate was 15.05 km Ma^{-1} on this flow line during chron 27.

Polarity chron C29 of latest Maastrichtian and earliest Danian time spans 65 ± 2 precession cycles (Herbert *et al.* 1995). Applying the modern average precession period of 20.8 ka implies that the duration of chron C29 is 1.352 ± 0.042 Ma.

Anomaly C29 spans 24.81 km in the South Atlantic flow line (Cande & Kent 1992). Therefore, the average spreading rate was 18.35 km Ma^{-1} on this flow line during chron 29.

Intervening anomaly C28 spans 21.32 km on this flow line (Cande & Kent 1992). As C28 has a similar width to both C29 and C27, the spreading rate during chron C28 is assigned to the intermediate value of 16.7 km Ma^{-1}. This yields an estimated duration for Chron C28 of 1.28 Ma.

Therefore, from chron C29 to chron C27, spreading rates on the synthetic composite flow line of the South Atlantic are modelled as a smooth deceleration through Danian time.

Absolute ages of the Danian stage and magnetic polarity time scale

Analysis of cycle-magnetostratigraphy from the South Atlantic (Deep Sea Drilling Project (DSDP) Holes 516F and 528) and from Spanish sections led Herbert *et al.* (1995) and Herbert (1999) to conclude that reversed-polarity chron C29r spans 33 ± 2 precession cycles, of which 18.5 ± 1 are in the Maastrichtian stage. Therefore, in the stratigraphic convention, the Cretaceous–Palaeogene boundary (inappropriately called the 'K–T' boundary in some publications) is placed at C29r.56 in the magnetic polarity time scale.

The age of the Cretaceous–Palaeogene boundary has been examined by several isotopic dating studies using $^{39}Ar/^{40}Ar$ laser fusion methods; however, the computed age depends upon the calibration of the neutron fluence monitor standard. Normalizing the suite of dates to an age of 28.32 Ma for the Taylor Creek Rhyolite (TCR) standard yields an average Cretaceous–Palaeogene boundary age of *c.* 65.5 Ma (compiled by Obradovich, in Hicks *et al.* (1999)). In contrast, Swisher *et al.* (1993) published a K–T boundary age of 65.16 ± 0.04 Ma. However, Renne *et al.* (1998), who obtained a recalibrated age of 28.34 ± 0.16 Ma for the TCR standard (and associated 523.1 ± 2.6 Ma for the MMhb-1 standard), recomputed this K–T date as 65.46 ± 0.63 Ma. A slight adjustment of published $^{39}Ar/^{40}Ar$ ages to older values is also implied by astronomical ages of radiometrically dated ash beds (Hilgen *et al.* 1997). Therefore, we utilize a K–T age of 65.5 Ma.

The array of two cycle-calibrated durations of polarity chrons C29 and C27, an interpolated

Table 1. Magnetic polarity time scale for the Danian stage derived from cycle stratigraphy

Polarity chron	Age (Ma)	Geological stage
C27n (end)	61.80	Earliest Selandian
c. C27n.8	61.85	Danian–Selandian stage boundary
C28n (end)	63.25	Mid-Danian
C29n (end)	64.53	Early Danian
C29n (base)	65.20	Earliest Danian
C29r.56	65.50	Start of Danian ('K–T' boundary)
C30n (end)	65.88	Latest Maastrichtian

duration of polarity chron C28 from the synthetic South Atlantic magnetic anomaly profile, and the cycle-calibrated position of the Cretaceous–Palaeogene boundary at C29r.56 with an age of 65.5 Ma establish the Danian magnetic polarity time scale (Table 1).

The projected age of the base (older limit) of polarity chron C28r from the precession cycle data of Herbert *et al.* (1995) is 64.53 Ma. Recomputing the radiometric date published by Swisher *et al.* (1993) from the W-Coal near the polarity zone C28r–C29n boundary to the revised monitor standard yields a date of 64.52 ± 0.02 Ma (J. D. Obradovich, pers. comm.), nearly identical to the estimate derived from the cycle calibration.

The Danian–Selandian stage boundary has not yet been formally defined by a Global Stratotype and Stratigraphic Point (GSSP). As explained in the introductory section, a common oceanic definition is the lowest occurrence of planktonic foraminifer *Morozovella angulata* (base of planktonic foraminifer Zone P3) (Berggren *et al.* 1995). This datum is estimated to occur within the uppermost portion of polarity zone C27n (Berggren *et al.* 1995). Because of the imprecise planktonic foraminifer biostratigraphy at these ODP sites, we have used this polarity-chron assignment to place the top of the Danian sequence (*sensu* Berggren *et al.* 1995).

Applying this potential correlation of the Danian–Selandian stage boundary to magnetostratigraphy projects the corresponding age of the top of the Danian sequence as *c*. 61.85 Ma. The Danian stage (*sensu* Berggren *et al.* 1995) spans 3.65 Ma, from 65.5 to 61.85 Ma.

Revised spreading-rate model for the South Atlantic

This cycle-calibrated duration of the Danian stage (*sensu* Berggren *et al.* 1995) is nearly 10% shorter than the 4.0 Ma duration derived by Berggren *et al.* (1995). This previous estimate was derived from computing ages and duration of polarity chrons from fitting a natural cubic spline to selected calibration ages applied to widths of magnetic anomalies in the synthetic composite flow line of the South Atlantic (Cande & Kent 1992, 1995). The Late Cretaceous–Palaeogene portion of this spline-fit model utilized six calibration points of age-chron–distance. Of these six ages, only the youngest estimate (Eocene–Oligocene boundary has an extrapolated age of 33.7 ± 0.4 Ma and is positioned at C13r.86 (Odin *et al.* 1991)) has not been revised by later data, revision of monitor standards, and re-evaluation of biostratigraphic calibrations.

A revised spreading rate model for this same synthetic South Atlantic profile is constrained by approximately 15 calibrations on age–chron–distance relationships (Table 2, Fig. 9). The intervals between each pair of age–distance calibration points yield the average spreading rate for the synthetic South Atlantic magnetic anomaly profile compiled by Cande & Kent (1992).

In the revised age–distance model, spreading rates remained constant at about 27 km Ma^{-1} during Campanian time (85–70 Ma), then decelerated smoothly through Maastrichtian and Danian time to reach a minimum at about 60 Ma (Fig. 9). The minimum rate of about 13 km Ma^{-1} was maintained through late Palaeocene and early Eocene time (60 to *c*. 50 Ma). Spreading underwent rapid acceleration through mid-Eocene time (*c*. 50 Ma to 42 Ma) to reach a plateau of about 25–27 km Ma^{-1} during late Eocene time and into Oligocene time.

These general trends were identified by Cande & Kent (1995), but the revised curve yields younger ages for the main acceleration and deceleration surges and suggests quasi-stable spreading rates during three time intervals (Campanian, late Palaeocene–early Eocene and late Eocene–early Oligocene time). Compared with the previous curve (Cande & Kent 1995), the shorter duration of the Danian stage (*sensu* Berggren *et al.* 1995) in this new model implies faster average spreading rates through Danian time, the same rate for the late Palaeocene epoch (and same durations for the associated stages), and much slower spreading rates for early Eocene time (implying a longer duration). Essentially, the shortening of the Danian stage required by cycle calibration and the shift of the Cretaceous–Palaeogene boundary to the older

Table 2. *Calibration array for Late Cretaceous and Palaeogene magnetic polarity time scale*

Age (Ma)	Placement in polarity chron	Geological age	Distance (km) in South Atlantic profile	Calibration type*	Source or review references
28.1 ± 0.3	C9n (base)	Mid-Oligocene (earliest Chattian)	607.96	K–Ar and ^{40}Ar–^{39}Ar ages and magnetostratigraphy in Italy	Odin *et al.* (1991); Wei (1995)
33.7 ± 0.4	C13r.86	Eocene–Oligocene stage boundary	759.49	K–Ar and ^{40}Ar–^{39}Ar dating in Italy	Odin *et al.* (1991); Cande & Kent (1992)
35.2 ± 0.27	C15n (base)	Latest Eocene	791.78	K–Ar and ^{40}Ar–^{39}Ar ages and magnetostratigraphy in Italy	Odin *et al.* (1991); Wei (1995)
45.60 ± 0.38	C21n.67	Mid-Eocene (early Lutetian)	1071.62	^{40}Ar–^{39}Ar dating of magnetostratigraphy in DSDP Hole 516F	Berggren *et al.* (1995) (postscript)
52.8 ± 0.3	C24n.1n (base)	Early Eocene (mid-Ypresian)	1184.03	^{40}Ar–^{39}Ar ages and magnetostratigraphy in Wyoming	Tauxe *et al.* (1994); Wei (1995)
55.07 ± 0.5	C24r.5	Near Palaeocene–Eocene boundary	1214.93	^{40}Ar–^{39}Ar ages and magnetostratigraphy in DSDP Hole 550	Berggren *et al.* (1995) (age by Obradovich in postscript)
61.80 ± 0.02	C27n (top)	Early to late Palaeocene boundary	1303.81	Cycles (central Atlantic and Caribbean)	This study
63.25 ± 0.1	C28n (top)	Mid-Danian	1325.71	Cyclo-magnetostratigraphy (central Atlantic and Caribbean)	This study
64.53 ± 0.04	C29n (top)	Early Danian	1347.03	Cyclo-magnetostratigraphy (South Atlantic and Spain)	Herbert *et al.* (1995)
65.20 ± 0.04	C29n (base)	Earliest Danian	1358.66	Cyclo-magnetostratigraphy (South Atlantic and Spain)	Herbert *et al.* (1995)
65.5 ± 0.1	C29r.56	Start of Cenozoic ('K–T' boundary)	1364.45	^{40}Ar–^{39}Ar ages and cyclostratigraphy of polarity zone C29r	Herbert *et al.* (1995); Hicks *et al.* (1999); Renne *et al.* (1998)
65.88 ± 0.02	C30n (top)	Latest Maastrichtian	1371.84	Cyclo-magnetostratigraphy (South Atlantic and Spain)	Herbert *et al.* 1995
70.44 ± 0.65	C32n (top)	Early Maastrichtian	1481.12	^{40}Ar–^{39}Ar ages and magnetostratigraphy of US western interior	Hicks & Obradovich (1995); Hicks *et al.* 1999
79.34 ± 0.5	C33n (base)	Mid-Campanian	1723.76	^{40}Ar–^{39}Ar ages and magnetostratigraphy of US western interior	Hicks *et al.* (1995)
84.4 ± 0.5	C33r (base)	Mid–late Santonian substage boundary	1862.32	^{40}Ar–^{39}Ar age constraints on biostratigraphy associated with magnetostratigraphy	Montgomery *et al.* (1998); Obradovich (1993)

*Uncertainties are 2σ for ^{40}Ar–^{39}Ar ages, unless estimates in assignment within polarity chron suggested a wider range (e.g. relation of 55 Ma ash bed within polarity chron C24r after adjusting for hiatus in Hole 550 magnetostratigraphy). Uncertainty estimates for cycle-tuned ages (top of C30n to top of C27n) are relative to fixed Cretaceous–Palaeogene ('K–T') boundary. A suggested age of 69.01 ± 0.5 Ma for the base of late Maastrichtian polarity chron C31n by Hicks *et al.* (1999) was not used for determining spreading rates because it is derived by projecting a constant sedimentation rate derived from ^{40}Ar–^{39}Ar ages in marine shale upward by 100 m across a sharp contact into the nearshore clastic-rich facies containing the polarity reversal.

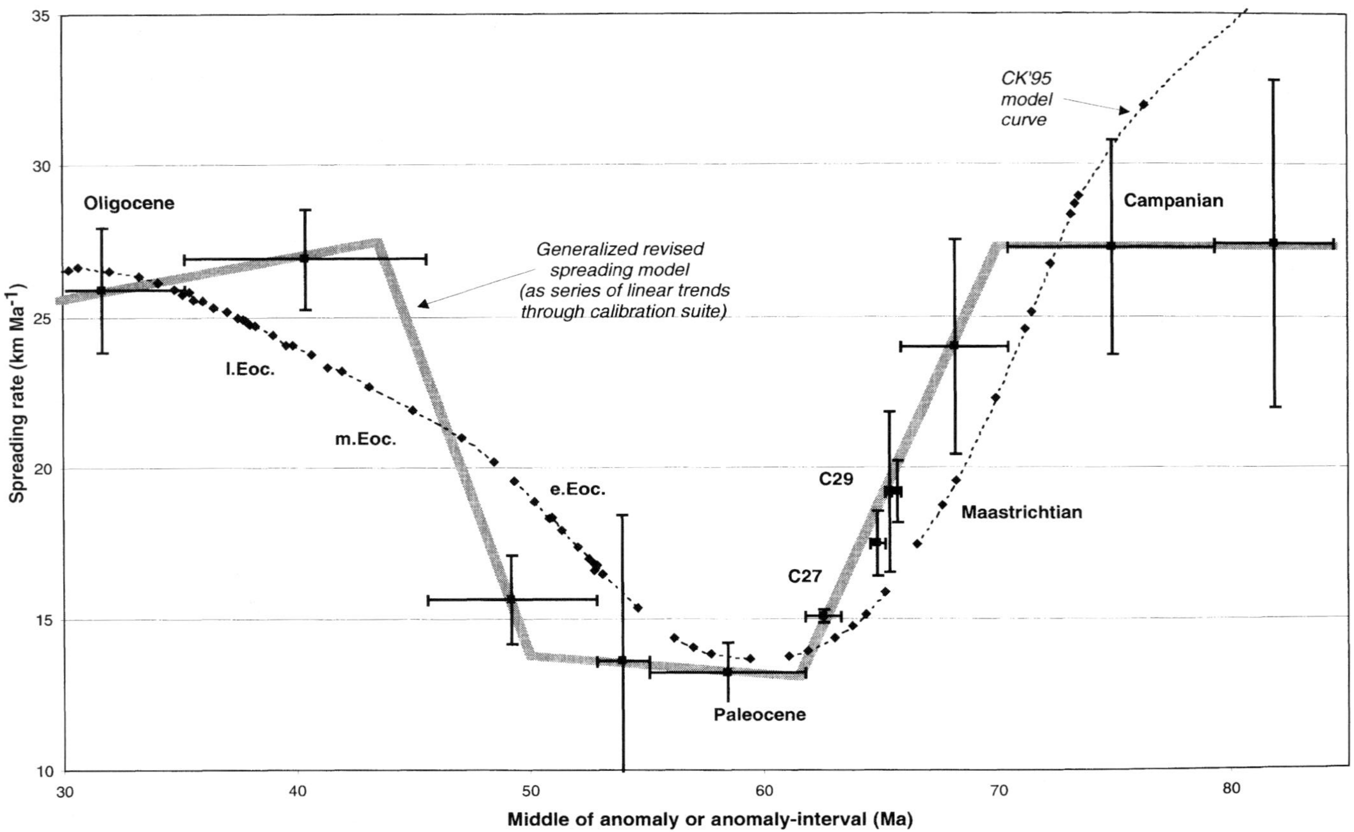

Fig. 9. Revised spreading rates for the South Atlantic. This synthetic profile was compiled by Cande & Kent (1992) to estimate ages of magnetic polarity chrons and associated biostratigraphic datums. Interval spreading rates (■) from an array of new age–chron calibrations (Table 2) indicate five general trends (grey line): a stable and rapid spreading rate during Campanian time, a slowing in Maastrichtian and Danian time (early Palaeogene period), a stable slow rate in late Palaeogene and early Eocene time, a rapid acceleration during mid-Eocene time, and a stable rate during late Eocene and Oligocene time. Previous calibration curve (dashed line with ◆) used by Cande & Kent (1995) to scale magnetic anomalies had similar trends but was displaced to older ages. Bars on average spreading rates indicate the distance (time) interval between calibration ages (horizontal bar) and the combined rate uncertainty associated with confidence levels on those ages (vertical bar).

65.5 Ma age is partially compensated by an expansion of the early Eocene interval.

Summary

High-resolution geochemical profiles through Danian pelagic sediments from the Caribbean and the Blake Nose margin of Florida display regular variations of Fe abundance associated with Milankovitch orbital–climate cycles of obliquity and precession. Spectral analysis of overlapping windows through the interval spanned by magnetic polarity zones C27r and C27n indicate that polarity chron C27 is equivalent to 35–36 obliquity cycles, or a duration of 1.45 Ma. Combining the obliquity-tuned duration of C27 with the precession-tuned 1.35 Ma duration of polarity chron C29 (Herbert *et al.* 1995), the interpolated duration for chron C28, and the cycle-scaled placement of the Cretaceous–Palaeogene boundary (65.5 Ma) within chron C29r implies that the Danian stage (*sensu* Berggren *et al.* 1995) spans 3.65 Ma (65.5–61.85 Ma). For the synthetic South Atlantic marine magnetic profile that serves as a standard for scaling magnetic polarity chrons, spreading rates were decreasing smoothly through Maastrichtian and Danian time, and maintained a constant rate during the late Palaeogene and early Eocene epochs before accelerating during mid-Eocene time.

We are grateful to the Ocean Drilling Program for inviting us to participate in Leg 165 and 171B, and to our shipmates, especially the palaeomagnetic downhole (logging) and physical properties teams. Financial support was provided by the Deutsche Forschungsgemeinschaft (DFG) to U. Röhl and G. Wefer, and by JOI-USSAC and the New Zealand Institute of Geological and Nuclear Sciences to J. Ogg. We thank B. Koster, A. Vaars (both at Netherlands Institute for Sea Research (NIOZ), Texel), and W. Bevern (Bremen) for technical assistance. B. Schmitz and H.-P. Luterbacher enlightened us on the status of defining the Danian and Selandian stages, and B. Oglesby provided background on the importance of obliquity for low-latitude climate. Valuable suggestions for alternative data interpretations, enhancing the graphics and clarifying the text were received from T. Herbert, B. Berggren, D. Norris and an anonymous reviewer.

References

ALVAREZ, W., ARTHUR, M. A., FISCHER, A. G., LOWRIE, W., NAPOLEONE, G., PREMOLI SILVA, I. & Roggenthen, M. W. 1977. Upper Cretaceous–Palaeocene magnetic stratigraphy at Gubbio, Italy—V. Type section for the Late Cretaceous–Palaeocene geomagnetic reversal time scale. *Geological Society of America Bulletin*, **88**, 383–389.

BERGER, A., LOUTRE, M. F. & LASKAR, J. 1992. Stability of the astronomical frequencies over the Earth's history for palaeoclimate studies. *Science*, **255**, 560–566.

BERGGREN, W. A., KENT, D. V., SWISHER, C. C., III. & AUBRY, M.-P. 1995. A revised Cenozoic geochronology and chronostratigraphy. *In*: BERGGREN, W. A., KENT, D. V. & HARDENBOL, J. (eds) *Geochronology, Time Scales and Global Stratigraphic Correlations: a Unified Temporal Framework for a Historical Geology*. SEPM Special Publications, **54**, 129–212.

CANDE, S. C. & KENT, D. V. 1992. A new geomagnetic polarity time scale for the Late Cretaceous and Cenozoic. *Journal of Geophysical Research*, **97**, 13917–13951.

CANDE, S. C. & KENT, D. V. 1995. Revised calibration of the geomagnetic polarity timescale for the Late Cretaceous and Cenozoic. *Journal of Geophysical Research*, **100**, 6093–6095.

DEMENOCAL, P. B., RUDDIMAN, W. F. & POKRAS, E. M. 1993. Influences of high- and low-latitude processes on African terrestrial climate: Pleistocene eolian records from equatorial Atlantic Ocean Drilling Program Site 663. *Paleoceanography*, **89**, 209–242.

ERBA, E., PREMOLI-SILVA, I. & WATKINS, D. K. 1995. Cretaceous calcareous plankton biostratigraphy of Sites 872 through 879. *In*: HAGGERTY, J. A., PREMOLI-SILVA, I., RACK, F. & MCNUTT, M. K. (eds) *Proceedings of the Ocean Drilling Program*, Scientific. Results, **144**. Ocean Drilling Program, College Station, TX, 157–169.

GALE, A. S. 1995. Cyclestratigraphy and correlation of the Cenomanian of Europe, *In*: HOUSE, M. R. & GALE, A. S. (eds) *Orbital Forcing Timescales and Cyclostratigraphy*. Geological Society, London, Special Publications, **85**, 177–197.

GALE, A. S., YOUNG, J. R., SHACKLETON, N. J., CROWHURST, S. J. & WRAY, D. S. 1999. Orbital tuning of Cenomanian marly chalk successions: towards a Milankovitch time-scale for the Late Cretaceous. *Philosophical Transactions of the Royal Society of London*, Series A, **357**, 1815–1829.

GIRAUD, F. 1995. *Recherche des périodicités astronomiques et des fluctuations du niveau marin à partir de l'étude du signal carbonaté des séries pélagiques alternantes*. Documents des Laboratoires de Géologie Lyon, **134**.

HALLAM, A., HANCOCK, J. M., LABRECQUE, J. L., LOWRIE, W. & CHANNELL, J. E. T. 1985. Jurassic and Cretaceous geochronology and Jurassic to Palaeogene magnetostratigraphy, *In*: SNELLING, N. J. (ed.) *The Chronology of the Geological Record*. Geological Society, London, Memoirs, **10**, 118–140.

HERBERT, T. D. 1999. Toward a composite orbital chronology for the Late Cretaceous and Early Palaeocene GPTS. *Philosophical Transactions of the Royal Society of London, Series A*, **357**, 1891–1905.

HERBERT, T. D., D'HONDT, S. L., PREMOLI-SILVA, I., ERBA, E. & FISCHER, A. G. 1995. Orbital chronology of Cretaceous–Early Palaeocene marine sediments, *In*: BERGGREN, W. A., KENT, D. V. & HARDENBOL, J. (eds) *Geochronology, Time Scales and Global Stratigraphic Correlations: a Unified Temporal Framework for a Historical Geology*. SEPM Special Publicatiions, **54**, 81–94.

HICKS, J. F. & OBRADOVICH, J. D.,1995. Isotopic age calibration of the GRTS from C33N to C31N: Late Cretaceous Pierre Shale, Red Bird section, Wyoming, USA. *Geological Society of America, Abstracts with Programs*, **27**, A174.

HICKS, J. F., OBRADOVICH, J. D. & TAUXE, L. 1995. A new calibration point for the Late Cretaceous time scale: the $^{40}Ar/^{39}Ar$ isotopic age of the C33r/C33n geomagnetic reversal from the Judith River Formation (Upper Cretaceous), Elk Basin, Wyoming, USA. *Journal of Geology*, **103**, 243–256.

HICKS, J. F., OBRADOVICH, J. D. & TAUXE, L. 1999. Magnetostratigraphy, isotopic age calibration and intercontinental correlation of the Red Bird section of the Pierre Shale, Niobrara County, Wyoming, USA. *Cretaceous Research*, **20**, 1–27.

HILGEN, F. J. 1991. Astronomical calibration of Gauss to Matuyama sapropels in the Mediterranean and implication for the geomagnetic polarity time scale. *Earth and Planetary Science Letters*, **104**, 226–244.

HILGEN, F. J., KRIJGSMAN, W., LANGEREIS, C. G., LOURENS, L. J., SANTARELLI, A. & ZACHARIASSE, W. J. 1995. Extending the astronomical (polarity) time scale into the Miocene. *Earth and Planetary Science Letters*, **136**, 495–510.

HILGEN, F. J., KRIJGSMAN, W. & WIJBRANS, J. R. 1997. Direct comparison of astronomical and $^{40}Ar/^{39}Ar$ ages of ash beds: potential implications for the age of mineral dating standards. *Geophysical Research Letters*, **24**, 2043–2046.

HINNOV, L. A. & PARK, J. J. 1999. Strategies for assessing Early–Middle (Pliensbachian–Aalenian) Jurassic cyclochronologies. *Philosophical Transactions of the Royal Society of London*, Series A, **357**, 1831–1859.

HUANG, Z., OGG, J. G. & GRADSTEIN, F. M. 1993. A quantitative study of Lower Cretaceous cyclic sequences from the Atlantic Ocean and the Vocontian Basin (SE France). *Palaeoceanography*, **8**, 275–291.

IOC–IHO–BDOC (Intergovernmental Oceanographic Comission–International Hydrographic Organization–British Oceanographic Data Centre) 1994. *General Bathymetric Chart of the Ocean (GEBCO)*. British Oceanographic Data Centre, Birkenhead.

JANSEN, J. H. F., VAN DER GAAST, S. J., KOSTER, B. & VAARS, A. J. 1998. CORTEX, a shipboard XRF-scanner for element analyses in split sediment cores. *Marine Geology*, **151**, 143–153.

KEIGWIN, L. D. 1987. Toward a high-resolution chronology for latest Miocene palaeoceanographic events. *Palaeoceanography*, **2**, 639–660.

KENT, D. V. & OLSEN, P. E. 1999. Astronomically tuned geomagnetic polarity timescale for the Late Triassic. *Journal of Geophysical Research*, **104**, 12831–12841.

LASKAR, J. 1999. The limits of Earth's orbital calculations for geological time-scale use. *Philosophical Transactions of the Royal Society of London, Series A*, **357**, 1735–1759.

LASKAR, J., JOUTEL, F. & BOUDIN, F. 1993. Orbital, precessional, and insolation quantities for the Earth from −20 Myr to +10 Myr. *Astronomy and Astrophysics*, **270**, 522–533.

LOUVEL, V. & GALBRUN, B. 2000. Magnetic polarity sequences from downhole measurements in ODP Holes 998B and 1001A, Leg 165, Caribbean Sea. *Marine Geophysical Research*, in press.

MONTGOMERY, P., HAILWOOD, E. A., GALE, A. S. & BURNETT, J. A. 1998. The magnetostratigraphy of Coniacian-Late Campanian chalk sequences in southern England. *Earth and Planetary Science Letters*, **156**, 209–224.

NORRIS, R. D. and RÖHL, U. 1999. Astronomical chronology for the Palaeocene/Eocene transient global warming and carbon isotope anomaly. *Nature*, **401**, 775–778.

OBRADOVICH, J. D. 1993. A Cretaceous time scale. *In*: CALDWELL, W. G. E. & KAUFFMANN, E. G. (eds) *Evolution of the Western Interior Basin*. Geological Association of Canada, Special Papers, **39**, 379–396.

ODIN, G. S., MONTANARI, A., Deino, A., DRAKE, R., GUISE, P. G., KEUZER, H. & REX, D. C. 1991. Reliability of volcano-sedimentary biotite ages across the Eocene–Oligocene boundary (Apennines, Italy). *Chemical Geology*, **86**, 203–224.

OGG, J. G. & BARDOT, L. (eds) 2000. Aptian through Eocene magnetostratigraphic correlation of the Blake Nose transect (ODP Leg 171), Florida continental margin. *Proceedings of the Ocean Drilling Program, Scientific Results*, **171B**, Ocean Drilling Program, College Station, TX, in press.

OLSEN, P. E. & KENT, D. V. 1996. Milankovitch climate forcing in the tropics of Pangaea during the Late Triassic. *Palaeogeography, Palaeoclimatology, Palaeoecology*, **122**, 1–26.

PAILLARD, D., LABEYRIE, L. & YIOU, P. 1996. Macintosh program performs time-series analysis. *EOS Transactions, American Geological Union*, **77**, 379.

PARK, J. & OGLESBY, R. J. 1990. A comparison on precession and obliquity effects in a Cretaceous palaeoclimate simulation. *Geophysical Research Letters*, **17**, 1929–1932.

RENNE, P. R., SWISHER, C. C., DEINO, A. L., KARNER, D. B., OWNS, T. L. & DEPAOLO, D. J. 1998. Intercalibration of standards, absolute ages and uncertainties in $^{40}Ar/^{39}Ar$ dating. *Chemical Geology*, **145**, 117–152.

RÖHL, U. & ABRAMS, L. J. 2000. High-resolution, downhole and non-destructive core measurements from Sites 999 and 1001 in the Caribbean Sea: application to the Late Palaeocene Thermal Maximum. *In*: LECKIE, R. M., SIGURDSSON, H., ACTON, G. D. & DRAPER, G. (eds). *Proceedings of the Ocean Drilling Program, Scientific Results*,

165. Ocean Drilling Program, College Station, TX, 191–203.

SCHMITZ, B., MOLINA, E. & VON SALIS, K. 1998. The Zumaya section in Spain: a possible global stratotype section for the Selandian and Thanetian Stages. *Newsletters on Stratigraphy*, **36**, 35–42.

SHACKLETON, N. J. & CROWHURST, S. 1997. Sediment fluxes based on an orbitally tuned time scale 5 to 14 Ma, Site 926. *In*: SHACKLETON, N. J., CURRY, W. B., RICHTER, C. & BRALOWER, T. J. (eds). *Proceedings of the Ocean Drilling Program, Scientific Results*, **154**. Ocean Drilling Program, College Station, TX, 69–82.

SHACKLETON, N. J., BERGER, A. & PELTIER, W. A. 1990. An alternative astronomical calibration of the lower Pleistocene timescale based on ODP Site 677. *Transactions of the Royal Society of Edinburgh: Earth Science*, **81**, 251–261.

SHACKLETON, N. J., CROWHURST, S., HAGELBERG, T., PISIAS, N. G. & SCHNEIDER, D. A. 1995. A new late Neogene time scale: application to Leg 138 sites, *In*: PISIAS, N. G., MAYER, L. A., JANECEK, T. R., PALMER-JULSON, K. A. & VAN ANDEL, T. H. (eds). *Proceedings of the Ocean Drilling Program, Scientific Results*, **138**, 73–101.

SHACKLETON, N. J., CROWHURST, S., WEEDON, G. P. & LASKAR, J. 1999. Astronomical calibration of Oligocene–Miocene time, *Philosophical Transactions of the Royal Society of London*, Series A, **357**, 1907–1929.

SHACKLETON, N. J., HALL, M. A., RAFFI, I., TAUXE, L. & ZACHOS, J. 2000. Astronomical calibration age for the Oligocene–Miocene boundary. *Geology*, **28**, 447–450.

SHIPBOARD SCIENTIFIC PARTY, 1997. Site 1001. *In*: SIGURDSSON, H., LECKIE, R. M., ACTON, G. D. *et al.* (eds) *Proceedings of the Ocean Drilling Program, Initial Reports*, **165**. Ocean Drilling Program, College Station, TX, 291–357, 677–763.

SHIPBOARD SCIENTIFIC PARTY, 1998*a*. Explanatory notes. *In*: NORRIS, R. D., KROON, D., KLAUS, A. *et al.* (eds) *Proceedings of the Ocean Drilling Program, Initial Reports*, **171B**. Ocean Drilling Program, College Station, TX, 11–44.

SHIPBOARD SCIENTIFIC PARTY, 1998*b*, Site 1050. *In*: NORRIS, R. D., KROON, D., KLAUS, A. *et al.* (eds) *Proceedings of the Ocean Drilling Program, Initial Reports*, **171B**. Ocean Drilling Program, College Station, TX, 93–169, 401–487.

SWISHER, C. C., DINGUS, L. & BUTLER, R. F. 1993. $^{40}Ar/^{39}Ar$ dating and magnetostratigraphic correlation of the terrestrial Cretaceous–Palaeogene boundary and Puercan Mammal Age, Hell Creek–Tullock formations, eastern Montana. *Canadian Journal of Earth Sciences*, **30**, 1981–1996.

TAUXE, L., GEE, J., GALLET, Y., PICK, T. & BOWN, T. 1994. Magnetostratigraphy of the Willwood Formation, Bighorn Basin, Wyoming: new constraints on the location of the Palaeocene/Eocene boundary. *Earth and Planetary Science Letters*, **125**, 159–172.

WEEDON, G. P., JENKYNS, H. C., COE, A. L. & HESSELBO, S. P. 1999. Astronomical calibration of the Jurassic time-scale from cyclostratigraphy in British mudrock formations, *Philosophical Transactions of the Royal Society of London, Series A*, **357**, 1787–1813.

WEI, W. 1995. Revised age calibration points for the geomagnetic polarity time scale. *Geophysical Research Letters*, **22**, 957–960.

Biostratigraphic implications of mid-latitude Palaeocene–Eocene radiolarian faunas from Hole 1051A, ODP Leg 171B, Blake Nose, western North Atlantic

ANNIKA SANFILIPPO[1] & CHARLES D. BLOME[2]

[1]*Scripps Institution of Oceanography, University of California, San Diego, La Jolla, CA 92093-0244, USA (e-mail: asanfilippo@ucsd.edu)*

[2]*US Geological Survey, MS 913, Box 25046, Federal Center, Denver, CO 80225, USA*

Abstract: Abundant well-preserved radiolarians were recovered from Ocean Drilling Program Leg 171B Hole 1051A, western North Atlantic, and range from upper middle Eocene radiolarian Zone RP16 through lower Palaeocene Zone RP6. This mid-latitude fauna contrasts with tropical faunas in lacking many tropical zonal markers and in its high proportion of diachronous first and last occurrences. The sequence from Hole 1051A contains the lower Eocene–middle Eocene and Palaeocene–Eocene (P–E) boundaries, and the only known record of a well-preserved Late Palaeocene Thermal Maximum (LPTM) radiolarian assemblage. There is no gross change observed in the composition of the fauna, only a minor increase in the number of first and last occurrences across the LPTM interval and P–E boundary. Calcareous evidence indicates two hiatuses, each 1–2 Ma long, one in the lowermost middle Eocene sequence and a second in the upper Palaeocene sequence. Presence of the middle Eocene hiatus is corroborated by an abnormally large radiolarian turnover. Twenty-six events are documented and show that most of radiolarian Zone RP10 and a substantial part of Zone RP9 are missing. Seven new species are described: *Spongatractus klausi*, *Calocyclas aphradia*, *Lychnocanoma (?) parma*, *Sethocyrtis austellus*, *Sethocyrtis chrysallis*, *Thyrsocyrtis (Pentalacorys) krooni* and *Thyrsocyrtis (Thyrsocyrtis) norrisi*.

During Ocean Drilling Program (ODP) Leg 171B five sites were drilled on the Blake Plateau and Blake Nose in the western North Atlantic Ocean (Fig. 1). The sediments offer an ideal record for reconstructing variability in the Cretaceous and early Cenozoic sedimentation and deep-water circulation, which is closely linked to climate change. The sites were positioned so as to give a depth transect of cores along the palaeoslope to provide information on depth-dependent sedimentation, deep-ocean chemistry and biota that can be used to reconstruct the past circulation of the North Atlantic.

The Palaeogene Blake Nose sediments span numerous short-term events of palaeoceanographic and biological significance, including the Cretaceous–Palaeocene extinction, the late Palaeocene extinction and the late Eocene impact horizons. A thicker than anticipated Palaeogene section (644.6 m) was drilled in the middle part of the Blake Nose at Site 1051 (Fig. 2), located at 30°03.1741′ N, 76°21.4580′ W, water depth 1980.6 m. Drilling through the lower Palaeocene units to the Cretaceous–Tertiary (K–T) boundary was not attempted because of slow drilling rates. Detailed biostratigraphic analyses indicate that an almost complete sequence was recovered from the uppermost part of the middle Eocene to the upper part of the lower Palaeocene units. The Palaeocene–Eocene boundary was completely recovered, and consists predominantly of laminated sediments in the lower Eocene sequence and in parts of the upper Palaeocene sequence, indicating decreased bioturbation. Two hiatuses of approximately 1–2 Ma at the lower–middle Eocene boundary and in the upper Palaeocene sequence are recorded.

The Eocene and Palaeocene sequences contain mainly oozes and chalks composed predominantly of nannofossils, foraminifera, siliceous microfossils and clay. The siliceous microfossils in Hole 1051A largely consist of radiolarians and sponge spicules, with diatoms occurring sporadically. Clay content increases downward in the lower Eocene and Palaeocene parts of 1051A, and more than 25 ash layers are found throughout the Eocene part of the sequence (Norris *et al.* 1998).

The sedimentary sequences drilled at Site 1051 are unique in containing a near-continuous

From: KROON, D., NORRIS, R. D. & KLAUS, A. (eds). 2001 *Western North Atlantic Palaeogene and Cretaceous Palaeoceanography*. Geological Society, London, Special Publications, **183**, 185–224.
0305-8719/01/$15.00

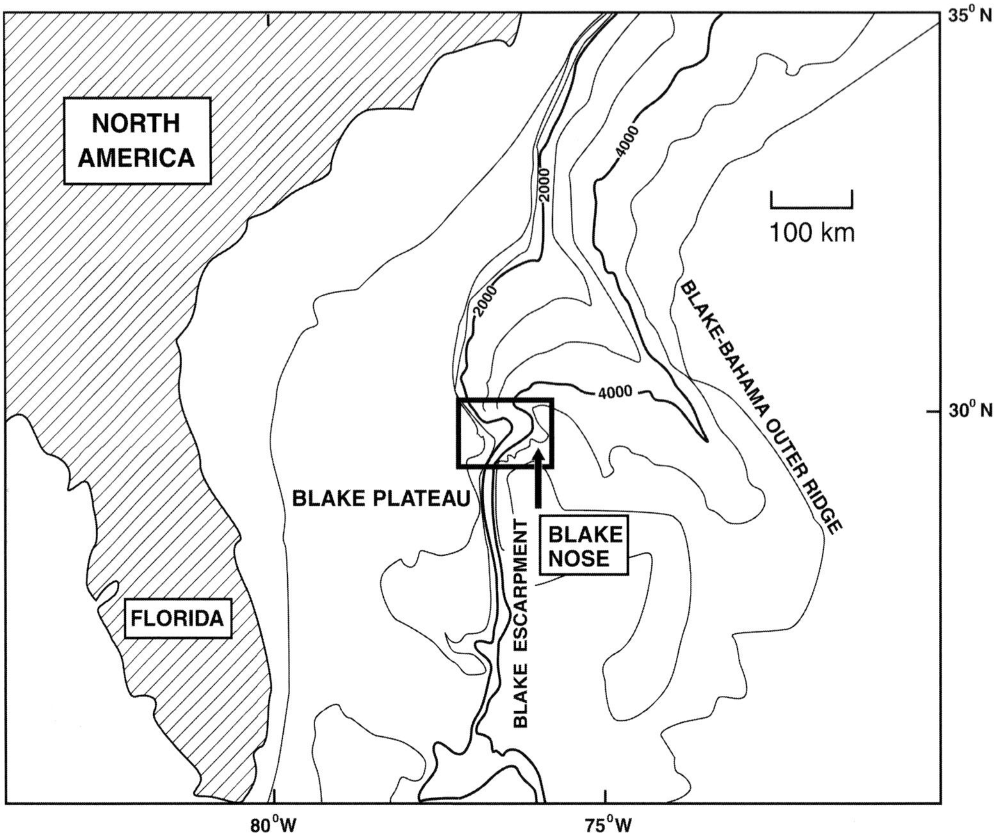

Fig. 1. Location of the ODP Leg 171B Blake Nose palaeoceanographic transect, western North Atlantic Ocean (from Norris *et al.* 1998). Contours in metres.

record of well-preserved radiolarian faunas of late mid-Eocene to late early Palaeocene age. This study aims to (1) document Palaeogene radiolarian faunas from mid-latitude intermediate waters, (2) demonstrate that the radiolarian zonal schemes developed for tropical assemblages are difficult to apply to mid-latitude Atlantic faunas, (3) document 200 radiolarian events (first and last occurrences and evolutionary transitions) occurring within a 23 Ma period in the western North Atlantic based on calcareous nannofossil datum levels, (4) discuss the 2 Ma hiatus and other coeval hiatuses that occur near the lower–middle Eocene boundary in relation to radiolarian events, and (5) document a continuous sedimentary record across the Palaeocene–Eocene boundary and Late Palaeocene Thermal Maximum (LPTM) interval to show that radiolarian assemblages cross these intervals without major faunal change.

Methods

Samples were sieved at 44, 63 and 150 μm and prepared following procedures documented by Sanfilippo *et al.* (1985). The preservation of the assemblage has been noted as very good (VG): no dissolution observed; good (G): individual specimens exhibit little dissolution and delicate parts of the skeleton are preserved; moderate (M): dissolution and breakage of individual specimens apparent, but identification of species not impaired; poor (P): individual specimens exhibit considerable dissolution and breakage, and identification of some species is not possible. Abundances of individual taxa have been tabulated as abundant A(bundant) >10 %, C(ommon) >1–10 %, F(ew) 0.5–1 %, VF (very few) 0.1–0.5 %, MR (moderately rare) 0.05–0.1 %, R(are) 0.01–0.05 %, VR (very rare) two specimens, ' + ' one specimen, '–' looked for but not found, and (!) indicates suspected reworking. Total abundance of the assemblage on a slide has been estimated as abundant (A) >10 001, common (C) 5001–10 000, few (F) 1001–5000, rare (R) 11–1000; very rare (VR) 1–9, or barren (B).

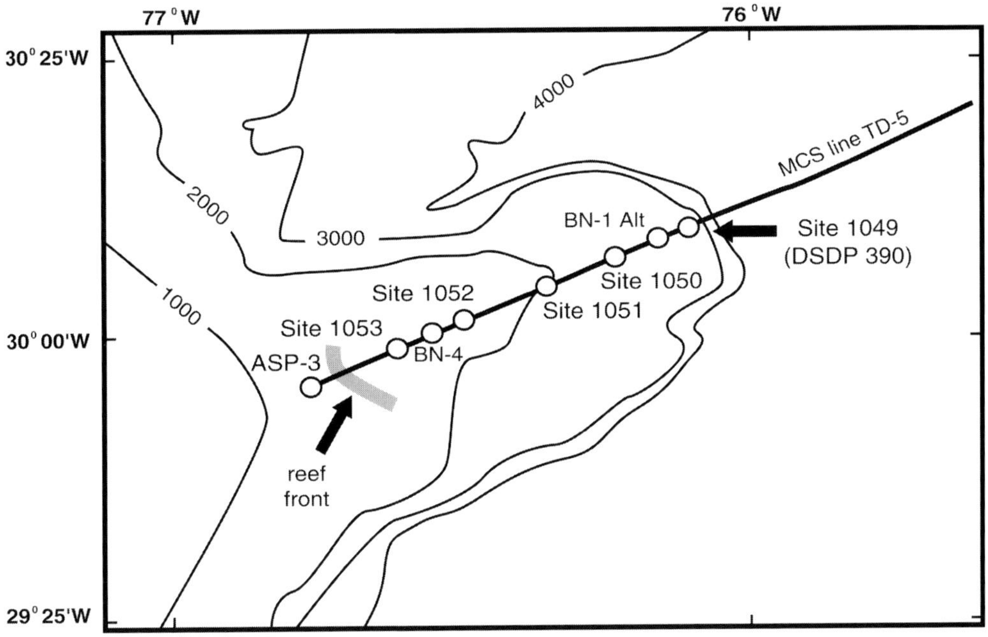

Fig. 2. Location of individual sites drilled during ODP Leg 171B. The investigated Site 1051 occurs near the centre of the palaeoceanographic transect (from Norris *et al.* 1998). Contours in metres.

The occurrence of stratigraphically useful radiolarians is recorded in Table 1 (provided in the envelope at the back of this volume), arranged in the order of first appearance. A summary of the most significant first and last occurrences and evolutionary transitions recorded in Table 1 is presented in Table 2. The table includes the top and bottom intervals for each event, depth in metres below sea floor (mbsf) and mean depth (in mbsf) for each event. The events are listed according to their mean depth in the core and numbered sequentially from one to 200. (For an explanation of evolutionary transitions, see Sanfilippo & Nigrini (1998*b*, p. 111).)

Lithostratigraphy

Two holes were drilled at Site 1051; Hole 1051A was drilled to a depth of 644.6 mbsf and Hole 1051B was drilled to 526.6 mbsf. Lithologies at Hole 1051A have been subdivided into four lithological units (Units I–IV) on the basis of colour, microfossil content and lithology.

Unit I consists of an uppermost layer (Subunit IA) of 2.6 m of manganese nodules, Subunit IB, 62 m thick, characterized by yellow, middle Eocene siliceous nannofossil ooze with foraminifers and clay, a 66 m thick interval (Subunit IC) of greenish grey siliceous nannofossil ooze, and a basal 241.3 m thick interval (Subunit ID) of siliceous nannofossil chalk with siliceous microfossils.

Unit II is somewhat condensed and was poorly recovered, and contains strongly altered, dark green porcellanitic smectitic clay overlain by silicified foraminifer porcellanite. This unit, which represents the transition from middle to lower Eocene sediments, contains bentonitic clay and a 2 Ma hiatus based on calcareous nannofossil datum levels. This disconformity and the large radiolarian turnover above and below the hiatus is discussed in the section entitled 'Hiatuses, upper lower and lower middle Eocene radiolarian record'.

The transition from Unit II to Unit III is marked by a change from laminated chert at the base of Unit II to siliceous nannofossil chalk in Unit III. Lithologies in Unit III are described as 178 m of lower Eocene to upper Palaeocene dark, siliceous nannofossil chalk with some clay. A complete Eocene–Palaeocene boundary is recorded within Unit III and is characterized by laminated sediments in the lowermost Eocene interval.

Unit IV is the oldest unit and, was recovered only in Hole 1051A. This unit contains *c.* 76 m of dark green siliceous claystone and siliceous nannofossil chalk. All four units are characterized by high silica content. At least 25 discrete vitric ash layers have been used to correlate lithostratigraphy between Holes 1051A and

Table 2. *Summary of first and last occurrences and evolutionary transitions recorded in Table 1*

Zone	Event No	Datum	Species	Core, Section, Interval (cm) top	Core, Section, Interval (cm) bottom	Depth top (mbsf)	Depth bot (mbsf)	Mean (mbsf)
RP16	1	LO	*Sethochytris triconiscus*	1051A-6H-2, 40-42	1051A-6H-5, 40-42	45.70	50.20	47.95
	2	LO	*Spongatractus pachystylus*	1051A-6H-2, 40-42	1051A-6H-5, 40-42	45.70	50.20	47.95
	3	LO	*Lithochytris vespertilio*	1051A-6H-2, 40-42	1051A-6H-5, 40-42	45.70	50.20	47.95
	4	FO	*Lithocyclia aristotelis* gr.	1051A-6H-CC, 21-24	1051A-7H-2, 40-42	53.62	55.20	54.41
RP15	5	LO	*Dictyophimus craticula*	1051A-7H-CC, 18-21	1051A-8H-2, 38-39	62.90	63.50	63.20
	6	LO	*Podocyrtis (Podocyrtopsis) apeza* var. A	1051A-9H-2, 40-42	1051A-9H-5, 40-42	74.20	78.70	76.45
	7	FO	*Podocyrtis (Podocyrtopsis) apeza*	1051A-9H-2, 40-42	1051A-9H-5, 40-42	74.20	78.70	76.45
	8	FO	*Artophormis barbadensis*	1051A-9H-2, 40-42	1051A-10H-2, 40-42	74.20	83.70	78.95
	9	FO	*Spongatractus klausi*	1051A-9H-5, 40-42	1051A-10H-2, 40-42	78.70	83.70	81.20
	10	LO	*Lophocyrtis biaurita*	1051A-9H-CC, 0-1	1051A-10H-2, 40-42	81.09	83.70	82.40
	11	LO	*Dictyoprora amphora*	1051A-12H-2, 41-43	1051A-13H-2, 40-42	102.71	112.20	107.46
	12	EVOL	*Podocyrtis (Lampterium) mitra - P. (L.) chalara*	1051A-12H-5, 39-41	1051A-13H-2, 40-42	107.19	112.20	109.70
RP14	13	LO	*Dictyoprora pirum* var. A	1051A-14H-CC, 29-31	1051A-15H-2, 42-44	129.81	131.22	130.52
	14	LO	*Periphaena tripyramis triangula*	1051A-15H-2, 42-44	1051A-15H-5, 41-43	131.22	135.71	133.47
	15	LO	*Podocyrtis (Lampterium) trachodes*	1051A-15H-5, 41-43	1051A-16H-2, 41-43	135.71	140.71	138.21
	16	FO	*Calocyclas aphradia*	1051A-15H-5, 41-43	1051A-16H-2, 41-43	135.71	140.71	138.21
	17	FO	*Podocyrtis (Podocyrtopsis) apeza* var. A	1051A-15H-5, 41-43	1051A-16H-2, 41-43	135.71	140.71	138.21
	18	FO	*Thyrsocyrtis (Thyrsocyrtis) norrisi*	1051A-15H-5, 41-43	1051A-16H-2, 41-43	135.71	140.71	138.21
	19	FO	*Podocyrtis (Lampterium) chalara*	1051A-15H-5, 41-43	1051A-16H-2, 41-43	135.71	140.71	138.21
	20	EVOL	*Dictyoprora pirum* var. A - *D. pirum*	1051A-16H-2, 41-43	1051A-17X-2, 42-45	140.71	150.22	145.47
	21	FO	*Lychnocanoma lucerna*	1051A-16H-5, 40-42	1051A-16H-CC, 25-27	145.20	148.88	147.04
	22	FO	*Calocyclas turris*	1051A-16H-5, 40-42	1051A-16H-CC, 25-27	145.20	148.88	147.04
	23	FO	*Lychnocanoma amphitrite*	1051A-16H-5, 40-42	1051A-16H-CC, 25-27	145.20	148.88	147.04
	24	LO	*Lithapium anoectum*	1051A-16H-CC, 25-27	1051A-17X-2, 42-45	148.88	150.22	149.55
	25	FO	*Sethochytris triconiscus*	1051A-17X-5, 39-42	1051A-17X-CC, 16-19	154.69	156.82	155.76
	26	LO	*Theocotyle nigriniae* var. A	1051A-17X-CC, 16-19	1051A-18X-2, 43-45	156.82	160.13	158.48
	27	LO	*Thyrsocyrtis (Thyrsocyrtis) hirsuta*	1051A-17X-CC, 16-19	1051A-18X-2, 43-45	156.82	160.13	158.48
	28	FO	*Dictyoprora pirum*	1051A-18X-5, 42-45	1051A-18X-CC, 29-32	164.62	167.19	165.91
	29	FO	*Podocyrtis (Lampterium) trachodes*	1051A-18X-5, 42-45	1051A-18X-CC, 29-32	164.62	167.19	165.91
	30	LO	*Podocyrtis (Lampterium) sinuosa*	1051A-20X-5, 37-39	1051A-20X-CC, 27-29	183.77	187.16	185.47
	31	FO	*Thyrsocyrtis (Pentalacorys) krooni*	1051A-20X-5, 37-39	1051A-20X-CC, 27-29	183.77	187.16	185.47
	32	LO	*Lithapium plegmacantha*	1051A-21X-2, 41-43	1051A-22X-2, 29-31	188.91	198.39	193.65
	33	LO	*Podocyrtis (Lampterium) helenae*	1051A-21X-2, 41-43	1051A-22X-2, 29-31	188.91	198.39	193.65
	34	EVOL	*Podocyrtis (Lampterium) sinuosa - P. (L.) mitra*	1051A-21X-2, 41-43	1051A-22X-2, 29-31	188.91	198.39	193.65
RP13	35	LO	*Eusyringium lagena*	1051A-22X-2, 29-31	1051A-22X-CC, 25-27	198.39	202.65	200.52
	36	FO	*Lithapium mitra*	1051A-22X-2, 29-31	1051A-22X-CC, 25-27	198.39	202.65	200.52
	37	EVOL	*Lithapium anoectum - L. mitra*	1051A-22X-2, 29-31	1051A-22X-CC, 25-27	198.39	202.65	200.52
	38	LO	*Sethocyrtis austellus*	1051A-22X-2, 29-31	1051A-22X-CC, 25-27	198.39	202.65	200.52
	39	FO	*Sethocyrtis chrysallis*	1051A-22X-CC, 25-27	1051A-23X-5, 40-42	202.65	212.60	207.63
	40	EVOL	*Eusyringium lagena - E. fistuligerum*	1051A-22X-CC, 25-27	1051A-23X-CC, 36-38	202.65	215.98	209.32
	41	FO	*Podocyrtis (Lampterium) helenae*	1051A-24X-2, 39-42	1051A-24X-5, 40-42	217.69	222.20	219.95
	42	LO	*Buryella clinata*	1051A-24X-2, 39-42	1051A-24X-5, 40-42	217.69	222.20	219.95
	43	FO	*Podocyrtis (Lampterium) mitra*	1051A-24X-CC, 17-19	1051A-25X-2, 42-45	225.43	227.32	226.38
	44	LO	*Amphicraspedum prolixum* gr.	1051A-24X-5, 40-42	1051A-27X-2, 40-42	222.20	246.50	234.35
	45	FO	*Sethocyrtis austellus*	1051A-26X-2, 46-48	1051A-26X-5, 42-44	236.96	241.42	239.19
	46	LO	*Thyrsocyrtis (Pentalacorys) tensa*	1051A-27X-2, 40-42	1051A-27X-5, 39-41	246.50	250.99	248.75
	47	LO	*Podocyrtis (Podocyrtoges) dorus*	1051A-27X-2, 40-42	1051A-27X-5, 39-41	246.50	250.99	248.75
	48	FO	*Eusyringium fistuligerum*	1051A-27X-2, 40-42	1051A-27X-5, 39-41	246.50	250.99	248.75
	49	LO	*Stylosphaera coronata coronata*	1051A-28X-2, 42-44	1051A-28X-5, 35-37	256.12	260.55	258.34
	50	LO	*Spongodiscus phrix*	1051A-28X-5, 35-37	1051A-29X-2, 39-41	260.55	265.69	263.12
	51	LO	*Theocotyle venezuelensis*	1051A-28X-CC, 27-29	1051A-29X-2, 39-41	263.98	265.69	264.84
RP12	52	LO	*Thyrsocyrtis (Thyrsocyrtis) robusta*	1051A-28X-CC, 27-29	1051A-29X-2, 39-41	263.98	265.69	264.84
	53	LO	sporadic *Theocorys (?) phyzella* (=max range)	1051A-31X-5, 39-41	1051A-32X-2, 39-42	289.39	294.49	291.94
	54	LO	*Theocotyle conica*	1051A-31X-5, 39-41	1051A-32X-2, 39-42	289.39	294.49	291.94
	55	LO	*Theocotyle cryptocephala*	1051A-31X-5, 39-41	1051A-32X-2, 39-42	289.39	294.49	291.94
	56	FO	*Theocotyle nigriniae* var. A	1051A-32X-5, 36-39	1051A-32X-CC, 35-37	298.96	300.41	299.69
	57	LO	*Theocotyle nigriniae*	1051A-32X-CC, 35-37	1051A-33X-2, 34-36	300.41	304.04	302.23
	58	LO	*Podocyrtis (Podocyrtoges) diamesa*	1051A-35X-2, 47-49	1051A-35X-5, 36-39	323.47	327.86	325.67
	59	LO	*Lamptonium fabaeforme fabaeforme*	1051A-35X-5, 36-39	1051A-35X-CC, 33-35	327.86	330.50	329.18
	60	LO	*Theocorys anaclasta*	1051A-35X-CC, 33-35	1051A-36X-2, 39-41	330.50	332.99	331.75
	61	EVOL	*Thyrsocyrtis (Pentalacorys) tensa - T. (P.) triacantha*	1051A-36X-2, 39-41	1051A-36X-CC, 34-36	332.99	340.89	336.94
	62	FO	*Podocyrtis (Podocyrtoges) dorus*	1051A-36X-5, 36-38	1051A-36X-CC, 34-36	337.46	340.89	339.18
	63	LO	*Lamptonium fabaeforme constrictum*	1051A-36X-5, 36-38	1051A-36X-CC, 34-36	337.46	340.89	339.18
	64	LO	*Lamptonium fabaeforme chaunothorax*	1051A-36X-CC, 34-36	1051A-37X-2, 41-43	340.89	342.61	341.75
	65	FO	*Thyrsocyrtis (Pentalacorys) triacantha*	1051A-37X-2, 41-43	1051A-37X-CC, 13-16	342.61	347.06	344.84
	66	FO	*Eusyringium lagena*	1051A-37X-2, 41-43	1051A-37X-CC, 13-16	342.61	347.06	344.84

The events are listed according to their mean depth in the core and numbered sequentially from one to 200.

1051B. The ash layers vary in thickness from a few millimetres to 3 cm (Norris *et al.* 1998).

Biostratigraphy and biochronology

Excellent age control for Hole 1051A was provided by calcareous microfossils. Nannofossils and foraminifers are well-preserved in the upper and middle Eocene sequences, variable in the lower Eocene, and become poorly preserved near the P–E boundary, where foraminifers are recrystallized and infilled. Nannofossil preservation improves downhole in the upper Palaeocene sequence because of increased clay content and is moderate throughout the upper part of the lower Palaeocene sequence.

As radiolarian zonal schemes have not been directly correlated with the geomagnetic polarity time scale for the Eocene and Palaeocene time intervals, radiolarian events have to be calibrated to the time scale via calcareous nannofossil and foraminifer events or zones. Correlation of the 1051A radiolarian events with calcareous nannofossil datum levels (Table 1 (provided in the envelope at the back of this volume)) is determined using zonal assignments provided

Table 2. *Summary of first and last occurrences and evolutionary transitions recorded in Table 1*

RP11	67	LO	*Periphaena tripyramis tripyramis*	1051A-37X-2, 41-43	1051A-37X-CC, 13-16	342.61	347.06	344.84
	68	EVOL	*Theocotyle nigriniae - T. cryptocephala*	1051A-37X-2, 41-43	1051A-37X-CC, 13-16	342.61	347.06	344.84
	69	LO	*Calocycloma castum*	1051A-37X-CC, 13-16	1051A-38X-2, 55-57	347.06	352.45	349.76
	70	FO	*Theocorys acroria* var. A	1051A-38X-2, 55-57	1051A-38X-5, 48-50	352.45	356.88	354.67
	71	FO	*Calocycloma ampulla*	1051A-38X-2, 55-57	1051A-39X-2, 39-41	352.45	361.99	357.22
	72	FO	*Theocotyle conica*	1051A-38X-CC, 26-29	1051A-39X-2, 39-41	359.81	361.99	360.90
	73	LO	*Theocorys acroria*	1051A-39X-2, 39-41	1051A-39X-5, 47-49	361.99	366.57	364.28
	74	FO	*Lithapium anoectum*	1051A-39X-5, 47-49	1051A-39X-CC, 35-38	366.57	369.72	368.15
	75	FO	*Amphicraspedum murrayanum* var. A	1051A-39X-5, 47-49	1051A-39X-CC, 35-38	366.57	369.72	368.15
	76	LO	*Thyrsocyrtis (Thyrsocyrtis) tarsipes*	1051A-39X-CC, 35-38	1051A-40X-2, 99-101	369.72	372.19	370.96
	77	FO	*Dictyoprora mongolfieri*	1051A-40X-2, 99-101	1051A-40X-5, 49-51	372.19	376.19	374.19
RP10	78	FO	*Lithapium plegmacantha*	1051A-40X-2, 99-101	1051A-40X-5, 49-51	372.19	376.19	374.19
	79	LO	*Calocyclas hispida* var. A	1051A-40X-2, 99-101	1051A-40X-5, 49-51	372.19	376.19	374.19
	80	FO	*Periphaena tripyramis triangula*	1051A-40X-2, 99-101	1051A-41X-CC, 16-18	372.19	380.58	376.39
	81	FO	*Periphaena tripyramis tripyramis*	1051A-40X-2, 99-101	1051A-41X-CC, 16-18	372.19	380.58	376.39
	82	FO	*Spongatractus pachystylus*	1051A-40X-2, 99-101	1051A-41X-CC, 16-18	372.19	380.58	376.39
	83	FO	*Theocorys anapographa*	1051A-40X-2, 99-101	1051A-41X-CC, 16-18	372.19	380.58	376.39
	84	LO	*Podocyrtis (Podocyrtoges) aphorma*	1051A-40X-5, 49-51	1051A-41X-CC, 16-18	376.19	380.58	378.39
	85	FO	*Theocotyle venezuelensis*	1051A-40X-5, 49-51	1051A-41X-CC, 16-18	376.19	380.58	378.39
	86	LO	*Spongodiscus cruciferus*	1051A-40X-5, 49-51	1051A-41X-CC, 16-18	376.19	380.58	378.39
	87	LO	*Buryella tetradica s.s.*	1051A-40X-5, 49-51	1051A-41X-CC, 16-18	376.19	380.58	378.39
	88	LO	*Giraffospyris lata*	1051A-40X-5, 49-51	1051A-41X-CC, 16-18	376.19	380.58	378.39
	89	LO	*Spongodiscus quartus quartus*	1051A-40X-5, 49-51	1051A-41X-CC, 16-18	376.19	380.58	378.39
	90	LO	*Spongomelissa adunca*	1051A-40X-5, 49-51	1051A-41X-CC, 16-18	376.19	380.58	378.39
	91	LO	*Lamptonium pennatum*	1051A-40X-5, 49-51	1051A-41X-CC, 16-18	376.19	380.58	378.39
	92	LO	*Theocotylissa alpha*	1051A-40X-5, 49-51	1051A-41X-CC, 16-18	376.19	380.58	378.39
	93	LO	*Lithochytris archaea*	1051A-40X-5, 49-51	1051A-41X-CC, 16-18	376.19	380.58	378.39
	94	LO	*Phormocyrtis striata exquisita*	1051A-40X-5, 49-51	1051A-41X-CC, 16-18	376.19	380.58	378.39
	95	FO	*Lamptonium fabaeforme constrictum*	1051A-40X-5, 49-51	1051A-41X-CC, 16-18	376.19	380.58	378.39
	96	FO	*Dictyophimus craticula*	1051A-40X-5, 49-51	1051A-41X-CC, 16-18	376.19	380.58	378.39
	97	FO	*Lithochytris vespertilio*	1051A-40X-5, 49-51	1051A-41X-CC, 16-18	376.19	380.58	378.39
	98	FO	*Lithocyclia ocellus* gr.	1051A-40X-5, 49-51	1051A-41X-CC, 16-18	376.19	380.58	378.39
	99	FO	*Podocyrtis (Lampterium) sinuosa*	1051A-40X-5, 49-51	1051A-41X-CC, 16-18	376.19	380.58	378.39
	100	FO	*Thyrsocyrtis (Pentalacorys) tensa*	1051A-40X-5, 49-51	1051A-41X-CC, 16-18	376.19	380.58	378.39
	101	FO	*Thyrsocyrtis (Thyrsocyrtis) robusta*	1051A-40X-5, 49-51	1051A-41X-CC, 16-18	376.19	380.58	378.39
	102	FO	*Podocyrtis (Podocyrtoges) diamesa*	1051A-40X-5, 49-51	1051A-41X-CC, 16-18	376.19	380.58	378.39
	103	FO	*Rhopalocanium ornatum*	1051A-40X-5, 49-51	1051A-41X-CC, 16-18	376.19	380.58	378.39
	104	FO	*Spongodiscus phrix*	1051A-40X-5, 49-51	1051A-41X-CC, 16-18	376.19	380.58	378.39
	105	FO	*Dictyoprora amphora*	1051A-40X-5, 49-51	1051A-41X-CC, 16-18	376.19	380.58	378.39
Hiatus	106	EVOL	*Lithochytris archaea - L. vespertilio*	1051A-40X-5, 49-51	1051A-41X-CC, 16-18	376.19	380.58	378.39
RP9	107	LO	*Phormocyrtis cubensis*	1051A-41X-CC, 16-18	1051A-42X-2, 40-42	380.58	391.80	386.19
	108	LO	*Lamptonium sanfilippoae*	1051A-41X-CC, 16-18	1051A-42X-2, 40-42	380.58	391.80	386.19
	109	LO	*Amphicraspedum murrayanum*	1051A-41X-CC, 16-18	1051A-42X-2, 40-42	380.58	391.80	386.19
	110	FO	*Lychnocanoma bellum*	1051A-42X-2, 40-42	1051A-42X-5, 38-40	391.80	396.28	394.04
	111	LO	*Theocorys (?) phyzella* (consistant range)	1051A-42X-5, 38-40	1051A-42X-CC, 38-40	396.28	399.79	398.04
	112	EVOL	*Theocotylissa alpha - T. ficus*	1051A-42X-5, 38-40	1051A-43X-CC, 47-49	396.28	409.46	402.87
	113	FO	*Thyrsocyrtis (Thyrsocyrtis) rhizodon*	1051A-43X-2, 41-43	1051A-43X-5, 37-40	401.41	405.87	403.64
	114	FO	*Calocyclas hispida s.s.*	1051A-43X-2, 41-43	1051A-43X-5, 37-40	401.41	405.87	403.64
	115	LO	*Pterocodon (?) tenellus*	1051A-43X-5, 37-40	1051A-43X-CC, 47-49	405.87	409.46	407.67
	116	FO	*Dictyoprora pirum* var. A	1051A-43X-5, 37-40	1051A-43X-CC, 47-49	405.87	409.46	407.67
	117	FO	*Theocotylissa ficus*	1051A-43X-5, 37-40	1051A-43X-CC, 47-49	405.87	409.46	407.67
	118	FO	*Theocorys anaclasta*	1051A-43X-5, 37-40	1051A-43X-CC, 47-49	405.87	409.46	407.67
RP8	119	LO	*Phormocyrtis turgida*	1051A-43X-CC, 47-49	1051A-44X-2, 36-38	409.46	410.96	410.21
	120	LO	*Bekoma bidartensis*	1051A-44X-2, 36-38	1051A-44X-5, 41-44	410.69	415.51	413.10
	121	FO	*Podocyrtis (Podocyrtoges) aphorma*	1051A-44X-2, 36-38	1051A-44X-5, 41-44	410.69	415.51	413.10
	122	FO	*Theocotyle cryptocephala*	1051A-44X-5, 41-44	1051A-44X-CC, 42-44	415.51	418.97	417.24
	123	LO	*Pterocodon (?) anteclinata*	1051A-45X-2, 51-53	1051A-45X-5, 44-46	420.71	425.14	422.93
	124	FO	*Lamptonium sanfilippoae*	1051A-45X-2, 51-53	1051A-45X-5, 44-46	420.71	425.14	422.93
	125	LO	*Lychnocanoma auxilla*	1051A-45X-CC, 41-43	1051A-46X-5, 42-44	428.58	434.72	431.65
	126	FO	*Buryella clinata*	1051A-46X-CC, 33-35	1051A-47X-2, 31-33	438.16	439.71	438.94
	127	EVOL	*Pterocodon (?) anteclinata - Buryella clinata*	1051A-46X-CC, 33-35	1051A-47X-2, 31-33	438.16	439.71	438.94
RP7	128	FO	*Theocotyle nigriniae*	1051A-47X-2, 31-33	1051A-47X-5, 45-47	439.71	444.35	442.03
	129	FO	*Thyrsocyrtis (Thyrsocyrtis) hirsuta*	1051A-47X-5, 45-47	1051A-47X-CC, 52-54	444.35	447.76	446.06
	130	LO	*Pterocodon (?) ampla*	1051A-47X-CC, 52-54	1051A-48X-2, 47-49	447.76	449.47	448.62
	131	FO	*Lamptonium fabaeforme chaunothorax*	1051A-47X-CC, 52-54	1051A-48X-2, 47-49	447.76	449.47	448.62
	132	FO	*Spongatractus balbis*	1051A-47X-CC, 52-54	1051A-48X-5, 41-43	447.76	453.91	450.84
	133	LO	*Theocotylissa auctor*	1051A-48X-2, 47-49	1051A-48X-5, 41-43	449.47	453.91	451.69
	134	LO	*Stylotrochus alveatus*	1051A-48X-2, 47-49	1051A-48X-5, 41-43	449.47	453.91	451.69

The events are listed according to their mean depth in the core and numbered sequentially from one to 200.

by the Leg 171B Shipboard Scientific Party (1998). These Shipboard Scientific Party calibrations were estimated using the time scale of Berggren *et al.* (1995). The hachured intervals in Table 1 (provided in the envelope at the back of this volume) indicate uncertainty in limits of zonal assignment.

Radiolarian biostratigraphy

The radiolarian zones first used to date Eocene radiolarian-bearing sediments were proposed by Riedel & Sanfilippo (1970) based on sediments from Deep Sea Drilling Project (DSDP) Leg 4 in the Caribbean and from the exposure of the Oceanic Formation at Bath Cliff, Barbados. Subsequent studies of the biostratigraphic ranges of Palaeogene radiolarians from the tropical Pacific (Moore 1971; Dinkelman 1973) and the Gulf of Mexico (Foreman 1973; Sanfilippo & Riedel 1973) resulted in the revised Cenozoic radiolarian biozonations by Riedel & Sanfilippo (1978) and Sanfilippo *et al.* (1985). Radiolarian zones for the upper Palaeocene succession were also described by Nishimura (1987, 1992) on the basis of her work on North Atlantic sequences

Table 2. *Summary of first and last occurrences and evolutionary transitions recorded in Table 1*

	135	FO	*Lophocyrtis biaurita*	1051A-48X-5, 41-43	1051A-48X-CC, 52-54	453.91	457.28	455.60
	136	FO	*Calocycloma castum*	1051A-49X-CC, 33-36	1051A-50X-2, 41-43	460.45	462.11	461.28
	137	FO	*Theocotylissa alpha*	1051A-50X-2, 41-43	1051A-50X-5, 41-43	462.11	466.61	464.36
	138	LO	*Spongurus (?) irregularis*	1051A-51X-5, 36-38	1051A-52X-2, 39-41	473.16	478.39	475.78
	139	EVOL	*Phormocyrtis striata exquisita - P. striata striata*	1051A-52X-2, 39-41	1051A-53X-5, 36-39	478.39	492.46	485.43
	140	FO	*Lithochytris archaea*	1051A-52X-6, 82-84	1051A-52X-CC, 41-45	484.82	486.39	485.61
	141	LO	*Lychnocanoma (?) parma*	1051A-52X-CC, 41-45	1051A-53X-1, 80-82	486.39	486.90	486.65
	142	LO	*Velicucullus palaeocenica*	1051A-53X-1, 80-82	1051A-53X-3, 80-82	486.90	489.91	488.41
	143	LO	*Bekoma (?) oliva*	1051A-53X-2, 56-58	1051A-53X-5, 36-39	488.16	492.46	490.31
	144	LO	*Pterocodon (?) poculum*	1051A-53X-6, 80-82	1051A-53X-7, 30-32	494.41	495.41	494.91
	145	FO	*Lychnocanoma (?) parma*	1051A-54X-4, 103-105	1051A-54X-5, 40-42	501.23	502.10	501.67
	146	LO	*Stylosphaera goruna*	1051A-54X-5, 40-42	1051A-54X-CC, 43-46	502.10	504.15	503.13
	147	LO	*Theocorys (?) aff. phyzella*	1051A-54X-5, 40-42	1051A-54X-CC, 43-46	502.10	504.15	503.13
	148	FO	*Thyrsocyrtis (Thyrsocyrtis) tarsipes*	1051A-55X-1, 69-71	1051A-55X-2, 78-80	505.99	507.58	506.79
	149	FO	*Theocorys acroria*	1051A-55X-2, 78-80	1051A-55X-3, 93-95	507.58	509.23	508.41
	150	FO	*Theocotylissa auctor*	1051A-55X-2, 78-80	1051A-55X-3, 93-95	507.58	509.23	508.41
	151	FO	*Theocorys (?) phyzella*	1051A-55X-2, 78-80	1051A-55X-3, 93-95	507.58	509.23	508.41
RP7	152	FO	*Pterocodon (?) ampla*	1051A-55X-3, 93-95	1051A-55X-4, 55-57	509.23	510.35	509.79
	153	EVOL	*Lamptonium pennatum - L. fab. fabaeforme*	1051A-54X-5, 40-42	1051A-56X-5, 40-42	502.10	521.30	511.70
	154	FO	*Calocyclas hispida* var. A	1051A-54X-CC, 43-46	1051A-56X-5, 40-42	504.15	521.30	512.73
LPTM	155	FO	*Phormocyrtis turgida*	1051A-55X-4, 55-57	1051A-56X-1, 86-88	510.35	515.76	513.06
	156	FO	*Podocyrtis (Podocyrtis) papalis*	1051A-55X-4, 55-57	1051A-56X-1, 86-88	510.35	515.76	513.06
	157	FO	*Giraffospyris lata*	1051A-56X-3, 80-82	1051A-56X-4, 78-80	518.70	520.18	519.44
	158	LO	*Bekoma campechensis*	1051A-56X-4, 78-80	1051A-56X-5, 40-42	520.18	521.30	520.74
	159	FO	*Spongomelissa adunca*	1051A-56X-5, 40-42	1051A-56X-6, 80-82	521.30	523.20	522.25
	160	FO	*Lamptonium fabaeforme fabaeforme*	1051A-56X-6, 80-82	1051A-56X-CC, 52-54	523.20	524.72	523.96
	161	FO	*Phormocyrtis striata striata*	1051A-56X-6, 80-82	1051A-56X-CC, 52-54	523.20	524.72	523.96
	162	FO	*Ampicraspedum murrayanum*	1051A-56X-6, 80-82	1051A-56X-CC, 52-54	523.20	524.72	523.96
	163	FO	*Amphicraspedum prolixum* gr.	1051A-56X-6, 80-82	1051A-56X-CC, 52-54	523.20	524.72	523.96
	164	FO	*Phormocyrtis cubensis*	1051A-57X-2, 40-42	1051A-57X-3, 105-107	526.40	528.55	527.48
	165	LO	*Lamptonium (?) incohatum*	1051A-57X-2, 40-42	1051A-57X-5, 40-42	526.40	530.90	528.65
	166	FO	*Theocorys (?) aff. phyzella*	1051A-57X-5, 40-42	1051A-57X-CC, 48-50	530.90	534.38	532.64
	167	LO	*Buryella pentadica*	1051A-57X-CC, 48-50	1051A-58X-2, 40-42	534.38	536.00	535.19
	168	LO	*Clathrocycloma capitaneum*	1051A-57X-CC, 48-50	1051A-58X-2, 40-42	534.38	536.00	535.19
	169	FO	*Spongodiscus cruciferus*	1051A-58X-2, 40-42	1051A-58X-5, 40-42	536.00	540.50	538.25
	170	LO	*Lamptonium (?) colymbus*	1051A-58X-2, 40-42	1051A-58X-5, 40-42	536.00	540.50	538.25
	171	FO	*Bekoma bidartensis*	1051A-58X-2, 40-42	1051A-58X-5, 40-42	536.00	540.50	538.25
	172	FO	*Pterocodon (?) anteclinata*	1051A-58X-2, 40-42	1051A-58X-5, 40-42	536.00	540.50	538.25
	173	FO	*Lychnocanoma auxilla*	1051A-58X-2, 40-42	1051A-58X-5, 40-42	536.00	540.50	538.25
	174	FO	*Pterocodon (?) poculum*	1051A-58X-2, 40-42	1051A-58X-5, 40-42	536.00	540.50	538.25
	175	LO	*Bekoma (?) demissa robusta*	1051A-58X-2, 40-42	1051A-58X-6, 106-108	536.00	542.66	539.33
	176	LO	*Bekoma (?) demissa demissa*	1051A-58X-2, 40-42	1051A-59X-5, 30-32	536.00	550.00	543.00
	177	FO	*Stylotrochus nitidus*	1051A-58X-6, 106-108	1051A-58X-CC, 34-36	542.66	543.91	543.29
	178	FO	*Spongodiscus quartus quartus*	1051A-58X-6, 106-108	1051A-59X-2, 42-44	542.66	545.62	544.14
	179	FO	*Pterocodon (?) tenellus*	1051A-59X-2, 42-44	1051A-59X-5, 30-32	545.62	550.00	547.81
	180	LO	*Lychnocanoma (?) costata*	1051A-59X-2, 42-44	1051A-59X-5, 30-32	545.62	550.00	547.81
	181	LO	*Buryella tetradica* var. A	1051A-59X-2, 42-44	1051A-59X-5, 30-32	545.62	550.00	547.81
	182	LO	*Lychnocanoma (?) pileus*	1051A-60X-1, 80-82	1051A-60X-2, 40-42	554.10	555.20	554.65
	183	FO	*Buryella tetradica s.s.*	1051A-60X-2, 40-42	1051A-60X-CC, 34-36	555.20	556.16	555.68
	184	LO	*Phormocyrtis striata praexquisita*	1051A-60X-2, 40-42	1051A-60X-CC, 34-36	555.20	556.16	555.68
	185	FO	*Spongurus (?) irregularis*	1051A-60X-2, 40-42	1051A-60X-CC, 34-36	555.20	556.16	555.68
RP6	186	FO	*Dendrospyris fragoides*	1051A-61X-2, 41-43	1051A-61X-4, 50-52	558.31	561.40	559.86
	187	FO	*Dendrospyris golli*	1051A-61X-2, 41-43	1051A-61X-4, 50-52	558.31	561.40	559.86
	188	LO	*Thyrsocyrtis (?) annikae*	1051A-61X-2, 41-43	1051A-61X-4, 50-52	558.31	561.40	559.86
	189	FO	*Lamptonium (?) incohatum*	1051A-61X-4, 50-52	1051A-62X-2, 45-47	561.40	564.85	563.13
	190	FO	*Bekoma (?) demissa demissa*	1051A-61X-4, 50-52	1051A-62X-2, 45-47	561.40	564.85	563.13
	191	FO	*Lamptonium (?) colymbus*	1051A-61X-4, 50-52	1051A-62X-2, 45-47	561.40	564.85	563.13
	192	LO	*Anthocyrtis ? sp. aff. A. mespilus*	1051A-62X-2, 45-47	1051A-62X-3, 80-82	564.85	566.70	565.78
	193	FO	*Phormocyrtis striata exquisita*	1051A-62X-4, 38-40	1051A-62X-5, 39-41	567.78	569.29	568.54
	194	LO	*Theocorys ? sp. aff. T. acroria*	1051A-62X-CC, 60-62	1051A-63X-2, 35-37	572.63	574.35	573.49
	195	FO	*Phormocyrtis striata praexquisita*	1051A-62X-CC, 60-62	1051A-63X-2, 35-37	572.63	574.35	573.49
	196	FO	*Lychnocanoma babylonis* gr.	1051A-63X-2, 35-37	1051A-63X-5, 40-42	574.35	578.90	576.63
	197	FO	*Siphocampe (?) quadrata*	1051A-63X-5, 40-42	1051A-64X-2, 47-49	578.90	584.07	581.49
	198	FO	*Lychnocanoma (?) costata*	1051A-64X-5, 40-42	1051A-64X-CC, 62-66	588.50	592.15	590.33
	199	FO	*Thyrsocyrtis (?) annikae*	1051A-64X-5, 40-42	1051A-64X-CC, 62-66	588.50	592.15	590.33
	200	FO	*Lychnocanoma (?) pileus*	1051A-67X-5, 33-36	1051A-68X-2, 36-38	617.43	622.56	620.00

The events are listed according to their mean depth in the core and numbered sequentially from one to 200.

from DSDP Legs 43 and 93, and Hollis (1993) proposed five zones for the lower part of the Palaeocene succession based on land sections from New Zealand. The South Pacific zones have been further revised and extended into the middle Eocene succession (Strong *et al.* 1995; Hollis 1997; Hollis *et al.* 1997) as a result of a re-examination of radiolarian faunas from eastern Marlborough, New Zealand, and DSDP Sites 208 and 277 along with improved age control provided by foraminifera and calcareous nannofossils. Sanfilippo & Nigrini (1998*b*) revised the tropical zonation and assigned code numbers for zones from the Pacific, Indian and Atlantic Oceans. They also included a stratigraphically ordered list of radiolarian events that falls within each tropical zone and assigned mean numerical ages for many of the zonal boundary events. Comparison of events in Hole 1051A with those in the tropics is based on the Sanfilippo & Nigrinis' (1998*b*) list of radiolarian events and is discussed in detail in the Remarks section for each zone. We have applied the Sanfilippo & Nigrini (1998*b*) zonation and their standardized code numbers (RP16–RP6) throughout this study and define the top and bottom of each zone.

Although the middle Eocene to Palaeocene sediments from Hole 1051A contained over 200 discernible radiolarian events (first and last occurrences and evolutionary transitions), the tropical radiolarian zonation of Sanfilippo & Nigrini (1998*b*) was found to be marginally applicable for dating and correlating the Hole 1051A radiolarian faunas. Many first and last occurrences (FO and LO, respectively) of species that define low-latitude tropical zones were either missing or proved to have different ranges in Hole 1051A from those in the tropics. Correlation of radiolarian events in Hole 1051A with the early (top of radiolarian Zone RP8) to mid-Eocene (base of Zone RP12) tropical events are shown in Fig. 3. This time interval contains a number of taxa at Site 1051 that have different first and last occurrences from those in the tropics. Heavy lines in the correlation column in Fig. 3 indicate zonal markers. Accuracy of the correlation depends on many factors including core recovery, preservation, reworking, consistent recognition of taxa, geographical variability of taxa, absence from or presence in some regions and diachronism of datum levels. As the tropical and Atlantic datasets are considered robust, crossing lines in the correlation column in Fig. 3 indicate diachronism.

Moore *et al.* (1993) investigated the diachronism of late Neogene radiolarian events in the eastern equatorial Pacific and were able to relate the temporal pattern of first and last occurrences of these diachronous events to shifts in the boundaries of ocean currents. A comparison of the stratigraphic ranges of the species investigated from western North Atlantic Hole 1051A with those from the tropics suggests that a high proportion of them have diachronous first and/or last occurrences relative to tropical species. Preliminary evaluation of the temporal pattern of diachronous datum levels in Hole 1051A does not reveal a coherent trend to explain why some first appearances occur earlier in Hole 1051A than in the tropics whereas others occur later. The same can be said for last occurrences. The following species are examples of species having earlier FOs in Hole 1051A than in the tropics: *Calocyclas turris*, *Dictyoprora pirum s.s.*, *Lychnocanoma amphitrite*, *Phormocyrtis striata striata*, *Theocotyle cryptocephala*, *T. nigriniae* and *Thyrsocyrtis (Thyrsocyrtis) hirsuta*. Species having later FOs in Hole 1051A than in the tropics are *Buryella tetradica*, *Calocyclas hispida*, *Eusyringium fistuligerum*, *Lamptonium sanfilippoae*, *L. fabaeforme constrictum*, *Lithocyclia ocellus* group, *Podocyrtis (Lampterium) trachodes*, *P. (Podocyrtoges) diamesa*, *P. (P.) dorus*, *Pterocodon ampla*, *Thyrsocyrtis (Pentalacorys) tensa* and *T. (Thyrsocyrtis) tarsipes*. Examples of species having earlier LOs in Hole 1051A than in the tropics are: *Podocyrtis (Lampterium) trachodes*, *P. (L.) helenae*, *P. (Podocyrtoges) diamesa*, *Pterocodon ampla*, *Theocotylissa auctor* and *Sethochytris triconiscus*. Species having later LOs in Hole 1051A than in the tropics are *Buryella pentadica*, *B. tetradica*, *Calocycloma castum*, *Giraffospyris lata*, *Phormocyrtis cubensis*, *P. striata striata*, *Theocorys phyzella*, *Theocotyle cryptocephala*, *Theocotylissa alpha*, *Thyrsocyrtis (Pentalacorys) tensa*, *T. (Thyrsocyrtis) hirsuta* and *T. (T.) tarsipes*. The absence of absolute numerical ages for the Palaeogene events makes it impossible to determine the degree of diachronism between individual events in Hole 1051 and those from the tropics.

Correlation between the tropical radiolarian zones, planktonic foraminiferal zones, calcareous nannofossil zones, polarity chrons and the Eocene–Palaeocene time scale is shown in Fig. 4, and occurrence of stratigraphically important radiolarians and the correlation of radiolarian zones with foraminiferal and calcareous nannofossil zones in Hole 1051A are shown in Table 1. Table 2 summarizes information regarding the top and bottom interval of 200 radiolarian events recorded in Table 1. The radiolarian biochronology and problems associated with correlating tropical radiolarian zones to mid-latitude Atlantic sediments at Site 1051 are discussed below.

RP16, *Podocyrtis (Lampterium) goetheana* Interval Zone

Definition. Interval from the morphotypic lowest occurrence of *Cryptocarpium azyx* to the morphotypic lowest occurrence of *Podocyrtis (Lampterium) goetheana*.

Remarks. Podocyrtis (Lampterium) goetheana was not encountered in Hole 1051A. Another event was chosen to recognize the lower limit of the zone. In the tropics the evolutionary transition from the *Lithocyclia ocellus* group to *L. aristotelis* group is considered synchronous with the lower limit of Zone RP16 (Sanfilippo & Nigrini 1998*b*). However, although both taxa occur in our material, the evolutionary transition is not evident at Site 1051. For this reason we have used the FO of *L. aristotelis* group to recognize the lower limit of Zone RP16. Of the events reported from the tropics that occur within the zone, the LOs of *Spongatractus pachystylus*, *Sethochytris triconiscus* and *Lithochytris vespertilio* occur slightly above the lower limit of the zone. Also, the FOs of *Thyrsocyrtis (Thyrsocyrtis) bromia* and *Dictyoprora armadillo* have not been observed and the FO of

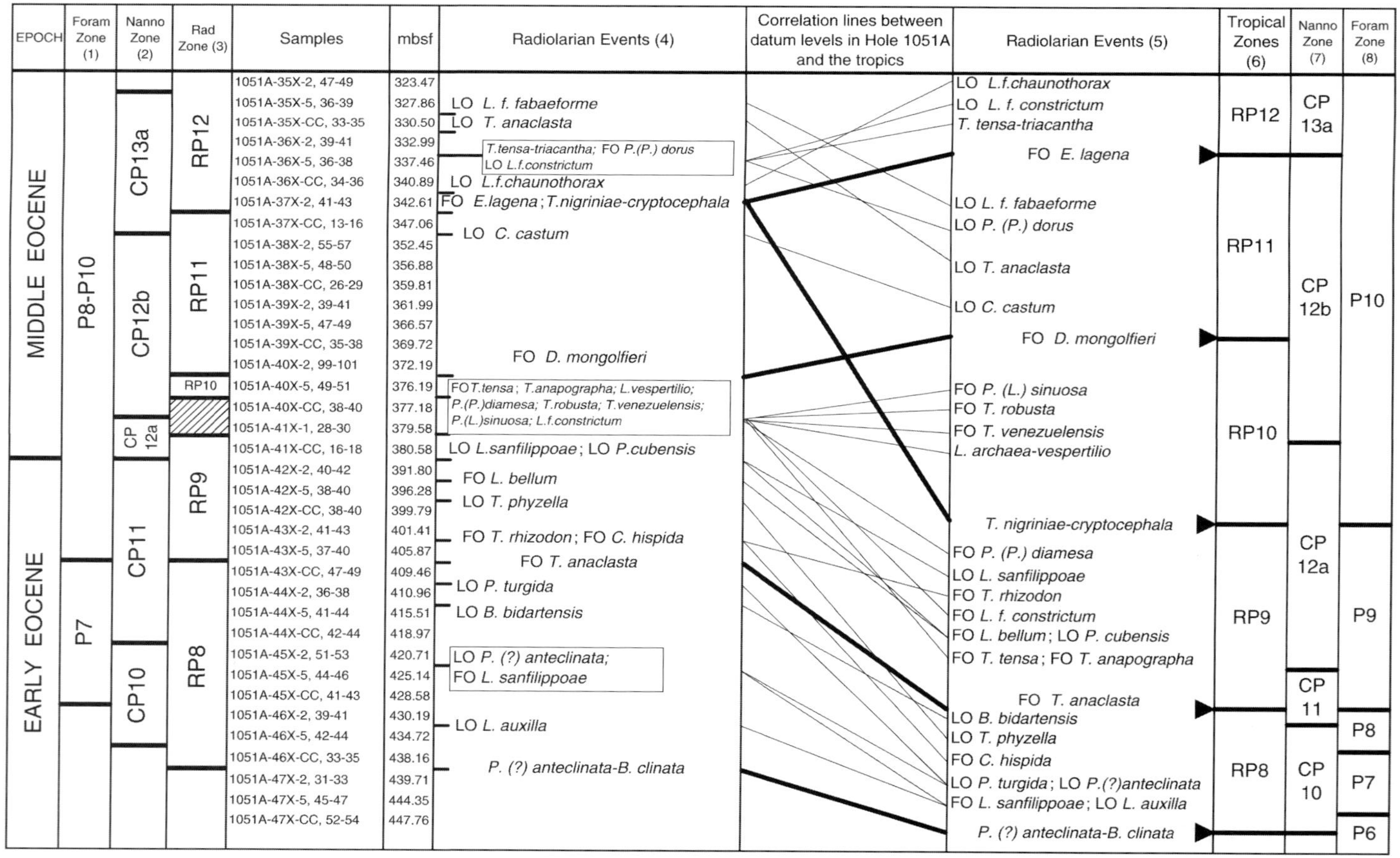

Fig. 3. Fence diagram showing correlation of tropical radiolarian events with those in Hole 1051A in the Early Eocene–Mid-Eocene time interval calibrated to foraminifera and calcareous nannofossil zones. Correlation lines between zonal marker taxa are marked in bold. Hachured interval indicate uncertainty in limits of zonal assignment. (1, 2) Leg 171B Shipboard Scientific Party (1998) foraminifer and calcareous nannofossil zones, respectively; (3) radiolarian zones, this study; (4) radiolarian events, this study; (5, 6) tropical radiolarian events and zones (Sanfilippo & Nigrini 1998*b*), respectively; (7) Okada & Bukry (1980); (8) Berggren *et al.* (1995).

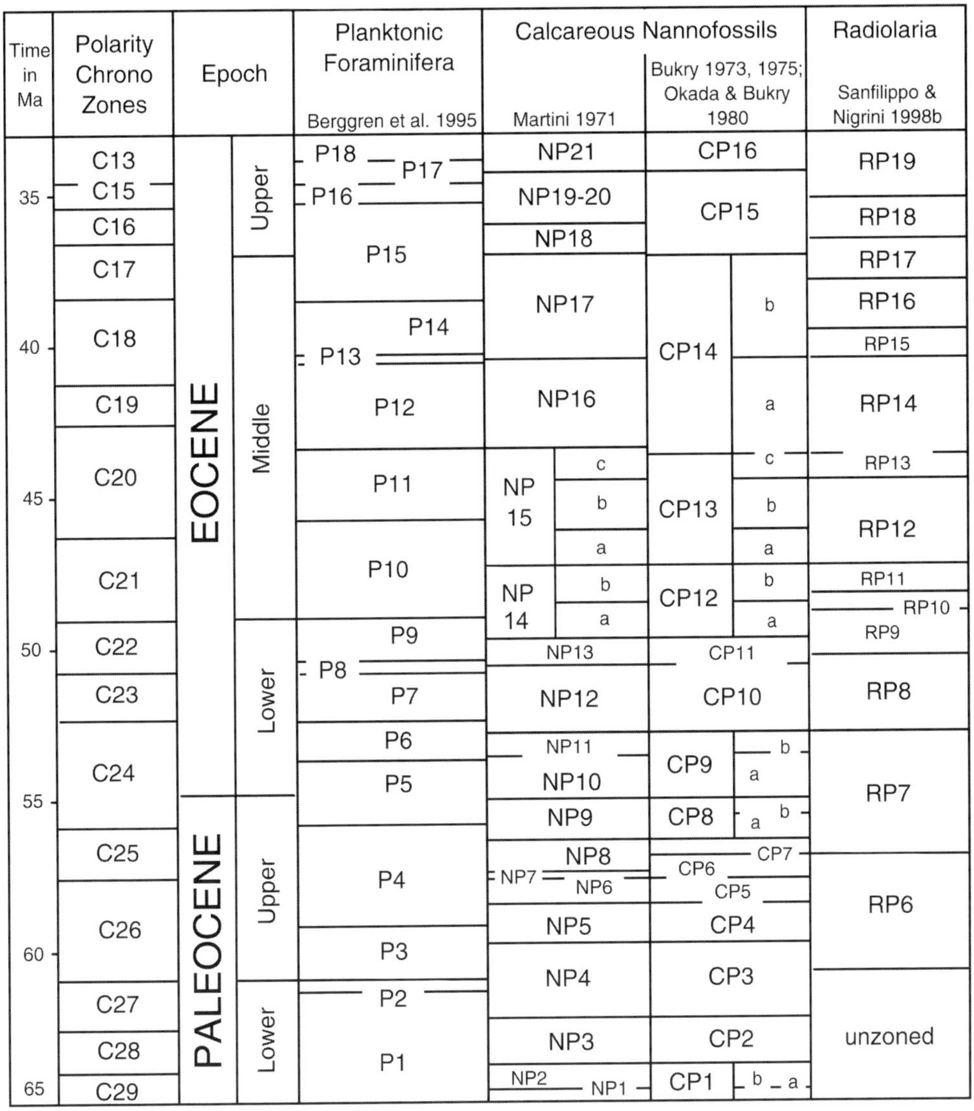

Time in Ma
Polarity Chrono Zones
Epoch
Planktonic Foraminifera
Berggren et al. 1995
Calcareous Nannofossils
Martini 1971
Bukry 1973, 1975; Okada & Bukry 1980
Radiolaria
Sanfilippo & Nigrini 1998b
EOCENE
PALEOCENE
Upper
Middle
Lower
unzoned

Fig. 4. Correlation of tropical radiolarian zones, planktonic foraminifera, calcareous nannofossil zones, polarity chrons and the Eocene–Palaeocene time scale (after Berggren *et al.* (1995); radiolarian zones from Sanfilippo & Nigrini (1998*b*) and correlation of radiolarian zones with the time scale from Hardenbol *et al.* (1998).

Dictyoprora pirum s.s. has a significantly earlier FO lower in the section in Zone RP14.

Age. Radiolarian Zone RP16 correlates with the upper part of calcareous nannofossil Zone CP14b.

RP15, *Podocyrtis (Lampterium) chalara* Lineage Zone

Definition. Interval from the morphotypic lowest occurrence of *Podocyrtis (Lampterium) goetheana* to the evolutionary transition from *Podocyrtis (Lampterium) mitra* to *Podocyrtis (Lampterium) chalara.*

Remarks. Of the events recorded within this zone at the tropics, the FO of *Podocyrtis (Podocyrtopsis) apeza* and the LO of *Lophocyrtis biaurita* are within the zone in Site 1051, the LO of *Podocyrtis (Lampterium) trachodes* is earlier in Zone RP14, and the LO of *Phormocyrtis striata striata*, usually synchronous with the lower limit of Zone RP15, is later above the top of the section. The FO of *Spongatractus klausi* n. sp. is in Zone RP15.

Age. Radiolarian Zone RP15 correlates with the lower half of calcareous nannofossil Zone CP14b and uppermost CP14a.

RP14, *Podocyrtis (Lampterium) mitra* Lineage Zone

Definition. Interval from the evolutionary transition from *Podocyrtis (Lampterium) mitra* to *Podocyrtis (Lampterium) chalara* to the evolutionary transition from *Podocyrtis (Lampterium) sinuosa* to *Podocyrtis (Lampterium) mitra.*

Remarks. The FO of *Sethochytris triconiscus* occurs in the zone, but other tropical radiolarian taxa that typically have their FOs or LOs in Zone RP14 are either missing (*Podocyrtis (Podocyrtoges) ampla s.s.* and *Podocyrtis (Lampterium) fasciolata*) or occur only sporadically (*Thyrsocyrtis (Pentalacorys) lochites*). The LO of *Podocyrtis (Lampterium) helenae* is synchronous with the lower boundary of Zone RP14, the LO of *Eusyringium lagena* occurs just below the RP14–RP13 boundary, which is somewhat earlier than in the tropics, and the FO of the relatively fragile *Artophormis barbadensis* is higher in the section, in Zone RP15, which is significantly later than in the tropics. The total range of *Podocyrtis (Lampterium) trachodes*, and the FOs of the newly described taxa *Calocyclas aphradia*, *Thyrsocyrtis (Pentalacorys) krooni* and *Thyrsocyrtis (Thyrsocyrtis) norrisi* all occur within the zone. *Calocyclas turris* and *Lychnocanoma amphitrite* also have their first occurrences in Zone RP14, which is significantly earlier than in the tropics, where both first occur at late Eocene Zone RP17.

Age. Radiolarian Zone RP14 correlates approximately with calcareous nannofossil Zone CP14a.

RP13, *Podocyrtis (Podocyrtoges) ampla* Lineage Zone

Definition. Interval from the evolutionary transition from *Podocyrtis (Lampterium) sinuosa* to *Podocyrtis (Lampterium) mitra* to the evolutionary transition from *Podocyrtis (Podocyrtoges) phyxis* to *Podocyrtis (Podocyrtoges) ampla.*

Remarks. Podocyrtis (Podocyrtoges) phyxis and *P. (P.) ampla* do not occur at this mid-latitude site and therefore cannot be used to define the lower zonal boundary. Instead, we have used the LO of *Theocotyle venezuelensis*, a tropical event considered approximately synchronous with the *P. (P.) phyxis* to *P. (P.) ampla* transition. Although many radiolarian events appear to be diachronous between the tropics and North Atlantic Site 1051, the LO of *T. venezuelensis* does fall within the zone in the expected order of events established from the tropics and can thus be considered a reliable event. Additional events that fall both within the zone at Site 1051 and in the tropics include the evolutionary transition from *Eusyringium lagena* to *E. fistuligerum*, the FO of *Podocyrtis (Lampterium) helenae*, and the LO of *Podocyrtis (Podocyrtoges) dorus*. Compared with the tropics, the FO of *Podocyrtis (Lampterium) trachodes* is much higher in the section in the overlying Zone RP14. The total range of the new taxon *Sethocyrtis austellus* and the FO of *S. chrysallis* n. sp. are in Zone RP13.

Age. Radiolarian Zone RP13 spans lowermost calcareous nannofossil Zone CP14a, Zone CP13c and uppermost Zone CP13b.

RP12, *Thyrsocyrtis (Pentalacorys) triacantha* Interval Zone

Definition. Interval from the evolutionary transition from *Podocyrtis (Podocyrtoges) phyxis* to *Podocyrtis (Podocyrtoges) ampla* to the lowest occurrence of *Eusyringium lagena.*

Remarks. We have used the LO of *Theocotyle venezuelensis* at this site to define the upper limit of the zone (see Remarks above). The lower limit of Zone RP12 is considered synchronous with the evolutionary transition of *Thyrsocyrtis (Pentalacorys) tensa* to *Thyrsocyrtis (Pentalacorys) triacantha* in the tropics. However, in Hole 1051A, this event occurs somewhat higher in the section but still within the zone. Although the radiolarian events that define the boundary between zones are critical for determining the biostratigraphy of marine sequences, events that fall within a zone can be equally valuable, especially when studying unknown sequences. A high number of events fall within this zone in the tropics and North Atlantic; these include the LOs of *Thyrsocyrtis (Thyrsocyrtis) robusta*, *Theocotyle conica*, *T. nigriniae*, *Theocorys anaclasta*, *Lamptonium fabaeforme chaunothorax* and *L. fabaeforme constrictum*. The FO of *Eusyringium fistuligerum* occurs higher in the section at Site 1051, in the overlying Zone RP13, than in the tropics, and the

LO of *Thyrsocyrtis hirsuta s.s.* is significantly higher, in Zone RP14. In the tropics the evolutionary transition from *Podocyrtis (Podocyrtoges) diamesa* to *P. (P.) phyxis* takes place within Zone RP12. At Site 1051, however, the descendant species has not been observed and the lineage terminates with the last occurrence of *P. (P.) diamesa*, in Zone RP12.

Age. Radiolarian Zone RP12 correlates with all but uppermost part of calcareous nannofossil Zone CP13b to lower Zone CP13a.

RP11, *Dictyoprora mongolfieri* Interval Zone

Definition. Interval from the lowest occurrence of *Eusyringium lagena* to the lowest occurrence of *Dictyoprora mongolfieri.*

Remarks. The FO of *Theocotyle conica* is within the zone, as in the tropics, but the evolutionary transition from *Theocotyle cryptocephala* to *T. conica*, known to take place within Zone RP11 in the tropics, could not be determined in Hole 1051A. In the tropics the lower limit of Zone RP11 is approximately synchronous with the LO of *Calocycloma castum*, but in Hole 1051A this event occurs higher in the section, near the top of the zone. Two other events, the FO of *Podocyrtis (Podocyrtoges) dorus* and LO of *Lamptonium fabaeforme fabaeforme*, that occur in Zone RP11 in the tropics occur higher in Hole 1051A, in the lower part of the overlying Zone RP12.

Age. Radiolarian Zone RP11 correlates with the lower calcareous nannofossil Zone CP13a to lower Zone CP12b.

RP10, *Theocotyle cryptocephala* Interval Zone

Definition. Interval from the lowest occurrence of *Dictyoprora mongolfieri* to the evolutionary transition from *Theocotyle nigriniae* to *Theocotyle cryptocephala.*

Remarks. Microfossil biostratigraphies indicate the presence of a *c.* 2 Ma hiatus and a radiolarian barren interval between 376.1 and 381.6 mbsf in Hole 1051A. Calcareous nannofossil Subzone CP12a is truncated, foraminiferal Zones P8–P10 spanning the lower to lower middle Eocene sequence could not be differentiated because of the absence of zonal markers for the base of Zones P10 and P9, and the large radiolarian faunal turnover (Table 2) from intervals bounding the hiatus all suggest that the uppermost part of radiolarian Zone RP9 and most of Zone RP10 are missing. Recognition of Zone RP10 is hampered by the fact that the evolutionary transition from *Theocotyle nigriniae* to *Theocotyle cryptocephala*, which marks the lower boundary of the zone in the tropics, occurs significantly higher in the 1051A section, at the RP12–RP11 boundary. The upper part of Zone RP10 is recognized only in sample 1051A-40X-5, 49–51 cm (376.19 mbsf), based on the presence of three taxa (*Podocyrtis (Lampterium) sinuosa*, *Theocotyle venezuelensis* and *Thyrsocyrtis (Thyrsocyrtis) robusta*) that have their first occurrences in Zone RP10 in the tropics. Assignment to Zone RP10 is further strengthened by the co-occurrence of a number of taxa that in the tropics and at Site 1051 range through the zone; some of these are *Lamptonium fabaeforme fabaeforme*, *L. fab. chaunothorax*, *L. fab. constrictum*, *Lychnocanoma bellum*, *Podocyrtis (Podocyrtis) papalis*, *Theocorys anaclasta*, *Theocotylissa ficus*, *Thyrsocyrtis (Pentalacorys) tensa*, *T. (Thyrsocyrtis) hirsuta* and *T. (T.) rhizodon* (Fig. 3).

Age. Radiolarian Zone RP10 correlates with the lower part of the calcareous nannofossil Zone CP12b.

RP9, *Phormocyrtis striata striata* Interval Zone

Definition. Interval from the evolutionary transition from *Theocotyle nigriniae* to *Theocotyle cryptocephala* to the lowest occurrence of *Theocorys anaclasta.*

Remarks. The upper part of this zone is truncated by the *c.* 2 Ma hiatus. Of the events that take place within Zone RP9 in the tropics, the LO of *Lamptonium sanfilippoae* and the FO of *Thyrsocyrtis (Thyrsocyrtis) rhizodon* occur in the zone at this site and the evolutionary transition from *Spongatractus balbis* to *S. pachystylus* has not been observed because the ancestral species has a much longer stratigraphic range. Also, the FO of *Podocyrtis (Podocyrtoges) diamesa* is above the hiatus in the overlying Zone RP10. The lower limit of the zone is approximately synchronous with a number of events in the tropics. Of these, the FO of *Lychnocanoma bellum* and the LO of *Phormocyrtis cubensis* fall within the lower part of the zone below the hiatus. The evolutionary transition from *Phormocyrtis striata exquisita* to

P. striata striata occurs earlier in Zone RP7, and the FO *Lamptonium fabaeforme constrictum* occurs just above the hiatus in the overlying Zone RP10 (Fig. 3). The taxon *Podocyrtis (Lampterium) acalles* is missing in Hole 1051A. Its absence in Hole 1051A is explained by the fact that its tropical biostratigraphic range is from the RP9–RP8 zonal boundary to Zone RP10 which approximately coincides with the hiatus in Site 1051A.

Age. Radiolarian Zone RP9 spans calcareous nannofossil Zone CP12a to the upper part of Zone CP11.

RP8, *Buryella clinata* Interval Zone

Definition. Interval from the lowest occurrence of *Theocorys anaclasta* to the evolutionary transition from *Pterocodon (?) anteclinata* to *Buryella clinata.*

Remarks. A high number of events occur within the zone in Hole 1051 and in the tropics, these include the LOs of *Lychnocanoma auxilla*, *Pterocodon (?) anteclinata* and *Phormocyrtis turgida* (Fig. 3). Of the events listed by Sanfilippo & Nigrini (1998*b*) to occur within the zone, only the LO of *Bekoma bidartensis* is within the zone at Site 1051, whereas the LOs of *Pterocodon ampla* and *Thyrsocyrtis (Thyrsocyrtis) tarsipes* are earlier, in the underlying Zone RP7. The FO of *Calocyclas hispida* and the evolutionary transition from *Theocotylissa alpha* to *T. ficus* also occur later, in Zone RP9. The LO of *Buryella tetradica s.s.* and the FOs of *Lithocyclia ocellus* and *Thyrsocyrtis (Pentalacorys) tensa* occur much later, in Zone RP10. The lower limit of the zone is usually considered synchronous with the FOs of *Lamptonium sanfilippoae*, *Theocotyle nigriniae* and *Thyrsocyrtis (Thyrsocyrtis) hirsuta*, but in Hole 1051A, the FO of *L. sanfilippoae* is slightly later but still within Zone RP8. The FOs of *T. nigriniae* and *T. (T.) hirsuta* occur earlier, in Zone RP7.

Age. Radiolarian Zone RP8 correlates with the lower half of calcareous nannofossil Zone CP11 to the base of Zone CP10.

RP7, *Bekoma bidartensis* Interval Zone

Definition. Interval from the evolutionary transition from *Pterocodon (?) anteclinata* to *Buryella clinata* to the lowest morphotypic occurrence of *Bekoma bidartensis.*

Remarks. Radiolarian events that occur within the zone both in the tropics and in Hole 1051A include the FOs of *Lamptonium fabaeforme chaunothorax*, *Calocycloma castum*, *Theocotylissa alpha* and *Podocyrtis (Podocyrtis) papalis*, the LO of *Bekoma campechensis*, and the evolutionary transition from *Lamptonium pennatum* to *L. fabaeforme fabaeforme.* The FO of *Pterocodon (?) anteclinata* is in Zone RP7 in the tropics, but appears lower in the section at this site in the topmost part of Zone RP6. *Lophocyrtis (Lophocyrtis) jacchia* occurs in moderate abundance in the tropics, but in Hole 1051A is too sporadic to be stratigraphically useful. The Palaeocene–Eocene boundary also falls within the zone and occurs in the interval between samples 1051A-54X-5, 40–42 cm and 1051A-54X-CC, 43–46 cm.

Age. Radiolarian Zone RP7 spans from the top of calcareous nannofossil Zone CP9 to mid Zone CP7-6.

RP6, *Bekoma campechensis* Interval Zone

Definition. Interval from the lowest morphotypic occurrence of *Bekoma bidartensis* to the lowest morphotypic occurrence of *Bekoma campechensis.*

Remarks. The presence of *Bekoma campechensis* in our lowermost sample indicates that the base of the zone is not present. Three events that occur in Zone RP6 in the tropics (the FO of *Phormocyrtis cubensis*, LO of *Buryella pentadica* and FO of *Pterocodon (?) ampla*), all occur higher in the Hole 1051A section, in the lowermost part of the overlying Zone RP7.
We were not able to apply Nishimura's (1992) Palaeocene Subzones to the Site 1051 *Bekoma campechensis* Zone, partly because one of her key markers, *Pterocodon (?) poculum*, has its first occurrence in the zone above, and her second marker, *Peritiviator (?) dumitricai*, did not occur in Site 1051 material. However, there are 30 FOs and LOs that fall within Zone RP6, which include the LO of *Anthocyrtis*? sp. aff. *A. mespilus*, LO of *Theocorys*? sp. aff. *T. acroria*, FO of *Lychnocanoma (?) costata*, FO of *Phormocyrtis striata exquisita*, FO of *Siphocampe (?) quadrata*, and the total ranges of *Lamptonium (?) colymbus*, *Lamptonium (?) incohatum*, *Lychnocanoma (?) pileus* and *Thyrsocyrtis (?) annikae.* Many of these taxa have distinct morphologies and/or robust test structures, and potentially could be used in the future to subdivide the *Bekoma campechensis* Zone into a number of North Atlantic subzones.

The presence of *B. campechensis*, in our lowermost sample (1051A-73X-CC, 20–22 cm, 643.95 mbsf), in foraminiferal Subzone P1c equivalent to nannofossil Zone CP3, is substantially older than previously reported from the tropics (Foreman 1973), North Atlantic (Nishimura 1987, 1992) and the Southwest Pacific (Strong *et al.* 1995; Hollis 1997; Hollis *et al.* 1997). The FO of *Buryella pentadica* also occurs below our lowermost sample, but is approximately equivalent in age to that reported by Nishimura (1992), and is older than its FO in the Southwest Pacific in nannofossil zone CP4–CP5 according to Hollis (1997). *Buryella tetradica s.s.* first appears in the upper part of Zone RP6 at the CP4–CP5 nannofossil zonal boundary at Site 1051, which is in sharp contrast to its earlier first occurrence in North Atlantic Sites 384 and 603 (Nishimura 1987, 1992). In Southwest Pacific localities reported by Hollis (1997), it appears before the FO of *B. campechensis* in middle nannofossil Zone CP3. These observations suggest that additional work is needed before the degree of diachronism in the lower Palaeocene sequences can be determined.

Age. Radiolarian Zone RP6 spans calcareous nannofossil Zone CP7–6 to Zone CP3.

Hiatuses: upper lower and lower middle Eocene radiolarian record

Calcareous evidence indicates the presence of two hiatuses at Site 1051, one in the lowermost middle Eocene sequence and a second in the upper Palaeocene sequence. A comprehensive study by Aubry (1995) shows that late early–early late Eocene boundary hiatuses are common in the Atlantic. Aubry (1995) also showed that, of the over 50 Atlantic sites recorded in the lower and middle Eocene sequences, continuous sedimentation in early Eocene time is well represented whereas the latest early and earliest mid-Eocene Epochs are characterized by hiatuses. Seismic evidence given by Norris *et al.* (this volume) also reveals the widespread occurrence of unconformities. Specifically, Norris *et al.* demonstrated that in the western North Atlantic, seismic Reflector A^c, which formed at the end of early Eocene time, is correlative with unconformities in every major ocean.

On the basis of nannofossil biostratigraphy from ODP Leg 171B, a 1–2 Ma hiatus is reported to occur near the early–mid-Eocene boundary between samples 1051A-40X-CC, 38–40 cm (377.18 mbsf), and 1051A-41X-CC, 16–18 cm (380.58 mbsf). Sample 1051A-40X-CC, 38–40 cm, can be correlated with the first appearance of the calcareous nannofossil *Rhabdosphaera inflata* marker species for the lower boundary of Zone CP12b (dated at 48.5 Ma) whereas sample 1051A-41CC, 16–18 cm, can be correlated with the first appearance of the nannofossil *Discoaster sublodoensis* marker for the lower boundary of Zone CP12a (dated at 49.7 Ma). Foraminiferal Zones P10–P8, which span the lower to lower middle Eocene interval, cannot be differentiated because of the absence of the zonal markers for the base of Zones P10 and P9. Presence of the middle Eocene hiatus is corroborated by an abnormally large radiolarian turnover. Twenty-six first and last occurrences are documented and show that most of radiolarian Zone RP10 and a substantial part of RP9 are missing.

Radiolarian recovery was remarkable throughout Hole 1051A except for a 2 m thick radiolarian barren interval occurring between Samples 171B-1051A-40X-CC, 38–40 cm (377.18 mbsf), and 1051A-41X-1, 28–30 cm (379.58 mbsf). The lithology at the top of this barren interval in Core 1051A-40X-CC is characterized by white limestone with greenish grey chert. Most of the underlying Core 1051A-41X-1 (except the very bottom of the core catcher) also contains limestone with nannofossils and a thin grey layer of clay at 75–78 cm with Fe oxide that may represent an altered ash layer. The sample immediately above the barren interval at 1051A-40X-5, 49–51 cm (376.19 mbsf), assigned to radiolarian Zone RP10 and described as siliceous nannofossil chalk, contains relatively well-preserved radiolarians and a large number of FOs which include *Dictyophimus craticula*, *Dictyoprora amphora*, *Lamptonium fabaeforme constrictum*, *Lithochytris vespertilio*, *Lithocyclia ocellus* group, *Periphaena tripyramis tripyramis*, *P. tripyramis triangula*, *Podocyrtis (Lampterium) sinuosa*, *Podocyrtis (Podocyrtoges) diamesa*, *Rhopalocanium ornatum*, *Spongatractus pachystylus*, *Spongodiscus phrix*, *Theocorys anapographa*, *Theocotylissa venezuelensis*, *Thyrsocyrtis (Pentalacorys) tensa* and *Thyrsocyrtis (Thyrsocyrtis) robusta*. The first sample, immediately below the barren interval at 1051A-41X-CC, 16–18 cm (380.58 mbsf), assigned to radiolarian Zone RP9, contains relatively well-preserved radiolarians and a number of LOs that include *Buryella tetradica s.s.*, *Giraffospyris lata*, *Lamptonium pennatum*, *Lithochytris archaea*, *Phormocyrtis striata exquisita*, *Podocyrtis (Podocyrtoges) aphorma*, *Spongodiscus cruciferus*, *Spongodiscus quartus quartus*, *Spongomelissa adunca* and *Theocotylissa alpha*. Figure 5 shows the number of first and last occurrences plotted

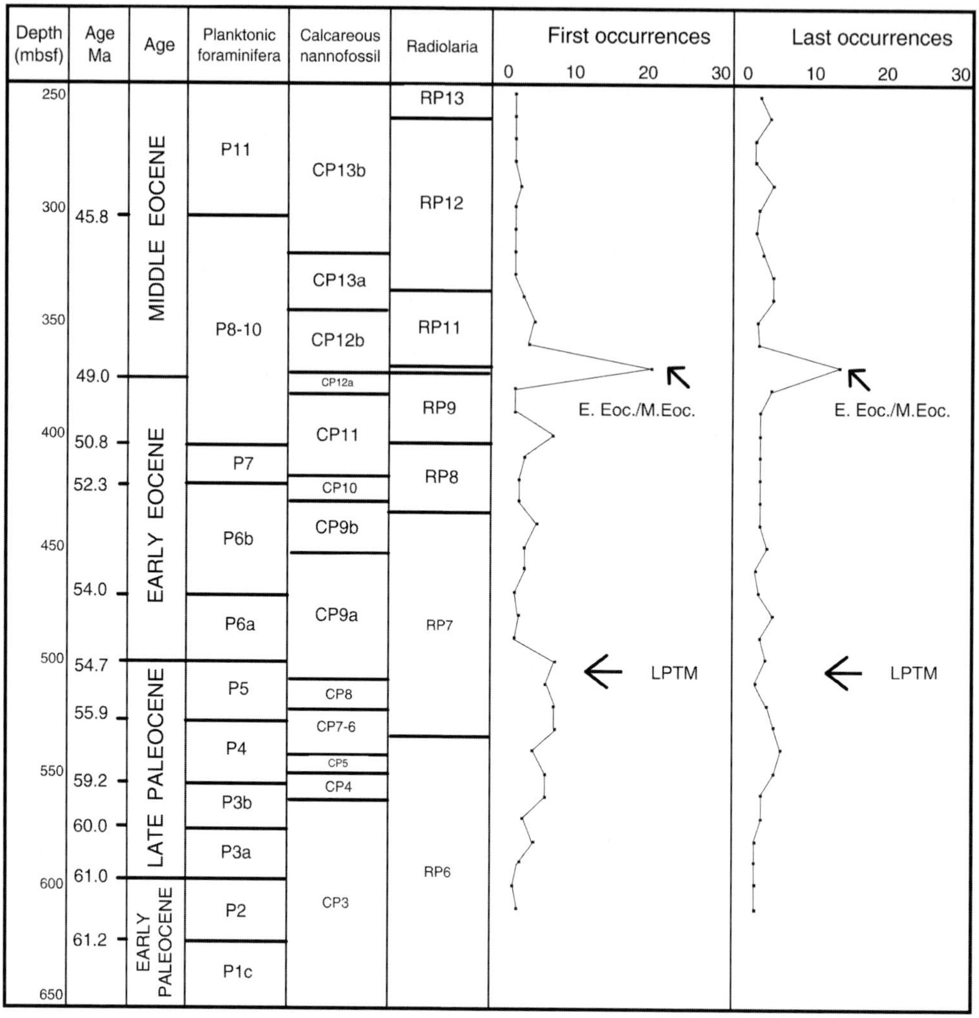

Fig. 5. Diagram showing the number of first and last radiolarian occurrences per 10 m intervals for the middle Eocene to lower Palaeocene sediments in Hole 1051A. Ages are from Berggren *et al.* (1995). Radiolarian zones, this study using Sanfilippo & Nigrini (1998*b*). Calcareous microfossil zones determined by the Shipboard Scientific Party (1998) using the zonal scheme of Berggren *et al.* (1995) for foraminifera, and Bukry (1973, 1975) and Okada & Bukry (1980) for calcareous nannofossils.

in 10 m increments through the middle Eocene to lower Palaeocene sequence including critical subepoch boundaries, the LPTM interval, and the early–mid-Eocene boundary.

A comparison of previously published data from DSDP Leg 10 shows that high radiolarian turnover also has been observed in the early–mid-Eocene interval reported from Site 94, where 25 radiolarian events occur between Cores 94-28-CC and 94-29-CC at the RP9–RP8 boundary (Foreman 1973, fig. 1; Sanfilippo & Riedel 1973, fig. 5). Events recorded from this site include the FOs of *Amphymenium splendiarmatum*, *Dendrospyris turriturcica turriturcica*, *D. turriturcica dasyotus*, *Dictyophimus craticula*, *Entapium chaenapium*, *Periphaena delta*, *P. tripyramis triangula*, *Podocyrtis aphorma*, *Podocyrtis diamesa*, *Podocyrtis platypus*, *Spongodiscus phrix*, *S. rhabdostylus*, *Stylosphaera coronata sabaca* and *Stylotrochus quadribrachiatus quadribrachiatus*. The LOs include those of *Amphicraspedum murrayanum*, *Dendrospyris acuta*, *Dictyospyris mellissium*, *Giraffospyris lata*, *Lithomespilus mendosa*, *Rhabdolithus ellida*, *Spongatractus balbis*,

Spongodiscus quartus bosoculus, *S. quartus quartus*, *Spongomelissa cucumella* and *S. euparyphus*. Although the absence of foraminifer Zone P9, a slight offset between the radiolarian *Phormocyrtis striata striata* (RP9) and *Buryella clinata* (RP8) Zones, and the lowest occurrence of calcareous nannofossil *Discoaster sublodoensis* between Cores 94-28 and 94-29 could indicate a stratigraphic gap at this level, the large coring gaps between 516 and 552 m precludes confirmation of a stratigraphic gap (Aubry 1995).

Although radiolarian Zones RP10 and RP9 could not be identified from DSDP Leg 95 Sites 612 and 613 in the western Atlantic because of poor silica preservation (Palmer 1987), calcareous nannofossils and foraminifera record an unconformable contact between the lower and middle Eocene units within Core 612-39 (347.3–357.0 mbsf). According to Valentine (1987), nannofossil Zone CP12b encompasses the unconformity, which is <2 Ma (Poag & Low 1987).

Seismic evidence by Norris *et al.* (this volume) shows that the presence of Reflector A^c in the Blake Nose sediments is correlative with worldwide latest early Eocene unconformities and that this interval is directly associated with major change in ocean circulation. Also, isotopic studies by Bralower *et al.* (1995), on the basis of planktonic and benthic foraminifer $\delta^{18}O$ stable isotopic records, indicate that at Site 865 (Allison Guyot, Mid-Pacific Mountains) in the equatorial Pacific, water temperatures rapidly cooled 3–6° C during late early Eocene time. Although additional work is needed to explain the palaeoceanographic mechanisms that produced the high radiolarian faunal turnover in the upper lower Eocene to lowermost middle Eocene sequence in Hole 1051A, we believe that the turnover can, in part, be attributed to changes in oceanic circulation and temperature gradient in the latest early Eocene time.

A second hiatus of *c.* 1 Ma occurs in the upper Palaeocene sequence between 1051A-58X-CC, 34–36 cm (543.9 mbsf), and 1051A-59X-1, 74–76 cm (544.4 mbsf), based on the absence of calcareous nannofossil Zone CP6 (Shipboard Scientific Party 1998). Abundant radiolarians assigned to the *Bekoma campechensis* (RP6) Zone cannot confirm this hiatus.

Radiolarians across the P–E boundary and LPTM interval

Palaeocene–Eocene boundary

Recent interest in the precise definition of Palaeogene boundaries has prompted the examination of radiolarian faunas from the Palaeocene–Eocene (P–E) Series boundary interval in deep-sea sediments. Twenty-four DSDP sites with radiolarian-bearing upper Palaeocene and/or lower Eocene sediments were re-examined by Sanfilippo & Nigrini (1998*a*). The results showed that none of the investigated sediment sequences contained the P–E boundary, and few of the deep-sea sites provided the opportunity to correlate upper Palaeocene–lower Eocene zonal schemes based on siliceous microfossils with those based on calcareous microfossils. Sanfilippo & Nigrini (1998*a*) identified an 'interval of non-radiolarian bearing sediments' corresponding to the upper part of the *Bekoma bidartensis* Zone (RP7), correlated it with the lower part of calcareous nannofossil Zone CP10 and all of CP9, and identified six first occurrences that approximate the P–E boundary (*Calocycloma castum*, *Theocotylissa auctor*, *Lamptonium fabaeforme fabaeforme*, *Podocyrtis (Podocyrtis) papalis*, *Giraffospyris lata* and *Phormocyrtis turgida*). In response to a growing need to define the P–E boundary and record events that immediately preceded and followed it within a 5 Ma span, Sanfilippo & Hull (1999) evaluated a section in western Cuba for its potential as a boundary stratotype. The P–E boundary in this succession falls within an unconformity and the material was too poorly preserved for detailed biostratigraphy. It was anticipated that the ODP Leg 165 Caribbean sediment sequences would resolve some of the biostratigraphic problems in the P–E interval and provide information on radiolarian evolution during the Late Palaeocene Thermal Maximum (LPTM). However, these sediments were mostly barren of radiolarians in the upper Palaeocene and lower Eocene sequences except for sporadic occurrences of abundant, poorly preserved forms (Nigrini & Sanfilippo 2000). Marked increase in radiolarian abundance and improved preservation relative to assemblages below and above the P–E boundary was observed (Nigrini & Sanfilippo 2000) in the LPTM interval at Site 1001 and Hole 999B. Biostratigraphy from calcareous microfossils was used to identify the LPTM interval (Sigurdsson *et al.* 1997). This clayey interval is characterized by reduced carbonate content and multiple ash layers.

The apparently complete Palaeocene–Eocene transition recovered on ODP Leg 171B from Hole 1051A provides a unique record of western North Atlantic radiolarian evolution. The boundary sediments are laminated siliceous nannofossil chalks in the lowermost Eocene sequence and parts of the upper Palaeocene sequence, with a 10 m thick, soft-sediment breccia occurring *c.* 10 m below the P–E boundary. This breccia occurs just above Chron C25n (55.9 Ma) and represents part of a small slump. The slump appears to be within or just below the interval in which Palaeocene benthic foraminifers become extinct. Benthic foraminifers are rare for more than 10 m within the extinction interval and the fauna becomes greatly reduced from about 40 taxa to seven. An impoverished benthic fauna remains for at least 50 m above the onset of the extinction interval (Shipboard Scientific Party 1998).

Placement of the Palaeocene–Eocene boundary in Hole 1051A in terms of foraminifers (P6a–P5 boundary), based on the LO of *Morozovella velascoensis* between sample 1051A-54X-5, 83–86 cm (502.53 mbsf), and 1051A-54X-CC, 43–46 cm (504.15 mbsf), was hampered by poor preservation including significant shell recrystallization and infilling of the foraminifer tests (Shipboard Scientific Party 1998). Similarly, determination of calcareous nannofossil biostratigraphic datum levels was difficult because of poor preservation, particularly near the base of the lower Eocene sequence (Shipboard Scientific Party 1998). In contrast, excellent radiolarian preservation above and below the boundary suggests that there are no major changes in the radiolarian fauna in the P–E boundary interval. Only three events are observed at the P–E boundary: the LO of *Theocorys (?)* aff. *phyzella* at the boundary, FO of *Lychnocanoma (?) parma* n. sp. just above the boundary and FO of *Thyrsocyrtis (Thyrsocyrtis) tarsipes* just below the boundary.

Late Palaeocene Thermal Maximum interval

Investigations of deep-sea sediments from high southern latitudes have demonstrated that there is a pattern of short-term (1–10 ka) fluctuations of temperature in the Palaeogene oceans (e.g. Kennett & Stott 1990; Stott & Kennett 1990; Zachos *et al.* 1993). These short-term excursions are superimposed on long-term (>1 Ma) climatic fluctuations (e.g. Kennett & Stott 1991; Miller 1992). The rapid climate change in latest Palaeocene time was one of the most dramatic warming events in the geological record, shown by the dramatic decrease in $\delta^{18}O$ values and negative excursion in $\delta^{13}C$ records of planktonic and benthic foraminifera. The event is believed to have been associated with a temporary change in deep-water sources from high to low latitudes (Kennett & Stott 1991; Pak & Miller 1992; Thomas 1992), which could have been the cause of the associated extinction of benthic foraminifera (Kennett & Stott 1990; Kennett & Stott 1991; Thomas 1992). The P–E transition at Site 1051 includes the carbon isotope event and the benthic foraminiferal extinction at 512.80 mbsf (Norris & Röhl 1999; Katz *et al.* 1999).

To examine radiolarian faunal change during the LPTM, we have sampled the uppermost Palaeocene and lowermost Eocene interval from 171B-1051A-53X to -57X (487–533 mbsf) using a higher sampling resolution than for the rest of Hole 1051A. The isotope peak denoting the LPTM occurs in the lower part of Core 1051A-55X at 509–513 mbsf (Norris & Röhl 1999). On the basis of one sample per core section, there is no notable change in the composition of the radiolarian assemblage except for a transient decrease in total abundance and preservation in the LPTM interval. There are only two first occurrences observed in the LPTM interval (*Podocyrtis (Podocyrtis) papalis* and *Phormocyrtis turgida*). If a broader 10 m interval is considered (1051A-55X-2 to -56X-2 (507.58–517.20 mbsf)), then there are six first occurrences, which include *Phormocyrtis turgida*, *Podocyrtis (Podocyrtis) papalis*, *Pterocodon (?) ampla*, *Theocorys acroria*, *Theocorys (?) phyzella* and *Theocotylissa auctor*. Six first occurrences would be significant if only the Eocene interval in Hole 1051A is considered, where the rate of first occurrences is commonly one to three per 10 m (Fig. 5). Through most of the upper Palaeocene sequence this rate is rather constant but slightly higher, varying from three to six per 10 m. A noteworthy exception to the low rate through the Eocene and Palaeocene interval is the increase to 18 first occurrences per 10 m at the lower–middle Eocene boundary (Fig. 5). Thus, the result from this and previous investigations (Sanfilippo & Nigrini 1998*a*; Nigrini & Sanfilippo 2000) suggests that radiolarian evolution was not noticeably affected, if at all, by the LPTM. It must be noted, however, that this information is somewhat biased, as it comes only from analyses of mostly biostratigraphic data and would be more useful, from a palaeoceanographic perspective, if quantitative data representing the total assemblage had been collected.

Comparison of Site 1051 data with those collected by Sanfilippo & Nigrini (1998*a*) indicates that, of the six first occurrences they reported to approximate the P–E boundary, two FOs denote the LPTM at Site 1051 (*Phormocyrtis turgida* and *Podocyrtis (Podocyrtis) papalis*). Of the remaining four, the FOs of *Calocycloma castum* and *Theocotylissa auctor* are well above the LPTM at Site 1051, and the FOs of *Lamptonium fabaeforme fabaeforme* and *Giraffospyris lata* are slightly below the LPTM (Table 1). Conversely, the FOs of *Theocorys acroria* and *Theocorys (?) phyzella*, which occur in a 10 m interval containing the LPTM at Site 1051, were also considered by Sanfilippo & Nigrini (1998*a*) to approximate to the P–E boundary. At Site 1051 the FO of *Pterocodon (?) ampla* is slightly lower below the P–E boundary but above the LPTM.

Conclusions

The radiolarian faunas from Site 1051 are remarkable in their abundance and preservation and, with the exception of a 2 m radiolarian-barren interval between samples 171B-1051A-40X-CC, 38–40 cm (377.18 mbsf), and 171B-1051A-41X-1, 28–30 cm (379.58 mbsf), a nearly complete radiolarian record for the upper middle Eocene to upper lower Palaeocene interval was recovered in Hole 1051A. Although the Palaeocene to middle Eocene sediments at Hole 1051A contain over 200 radiolarian events, the tropical radiolarian zonation of Sanfilippo & Nigrini (1998*b*) is only marginally applicable for dating and correlating the 1051A radiolarian faunas. Many first and last occurrences of species that define the low-latitude tropical zones are either missing or have different ranges in Hole 1051A compared with those in the tropics. It was not possible to subdivide the Palaeocene *Bekoma campechensis* (RP6) Zone into Nishimura's (1992) subzones, because one of the markers, *Peritiviator (?) dumitricai*, is missing from this site and the second marker, *Pterocodon (?) poculum*, has a diachronous first occurrence. However, because more than 30 taxa with distinct morphologies and robust skeletons have their FOs and LOs within Zone RP6, it should be possible to subdivide this zone in the future.

Comparison of the stratigraphic ranges in North Atlantic Hole 1051A with those from the tropics indicates that a high proportion of the Hole 1051A species have diachronous first and/or last occurrences relative to those from the tropics. A preliminary evaluation of the temporal patterns of these diachronous datum levels does not reveal a coherent trend to explain why some first occurrences occur earlier in Hole 1051A than in the tropics whereas others appear later. Of the investigated species, seven have FOs that are earlier in 1051A than in the tropics and 13 have FOs that are later than in the tropics. Also, six taxa in Hole 1051A have LOs that are earlier than in the tropics and 12 have LOs that are later. Absence of numerical ages for Palaeogene radiolarian events precludes determination of the degree of diachronism between individual events in Hole 1051 and those in the tropics.

Although placement of the Palaeocene–Eocene (P–E) boundary (1051A-54X-CC, 504.15 mbsf) was hampered by poor calcareous nannofossil and foraminifer preservation, the radiolarian faunas are excellently preserved across the P–E boundary interval in Hole 1051A. Using a sampling density of one sample per core section indicates that there is no gross change in the composition of the fauna across the boundary and only a minor increase in the number of first and last occurrences of species used for biostratigraphic analysis. The last occurrence of *Theocorys (?)* aff. *phyzella* is within the boundary interval, the FO of *Lychnocanoma (?) parma* n. sp. just above the boundary, and the FO of *Thyrsocyrtis (Thyrsocyrtis) tarsipes* just below the boundary interval.

Site 1051 is exceptional in containing the only known record of the Late Palaeocene Thermal Maximum (LPTM) radiolarian assemblage in the world's oceans. Radiolarian-bearing samples across the interval containing the LPTM isotope peak in 1051A-55X (509–513 mbsf) suggest little change in faunal composition across the interval. *Podocyrtis (Podocyrtis) papalis* and *Phormocyrtis turgida* have their first occurrences within this interval. If the LPTM interval is expanded to 10 m, only four additional species have their FOs in the expanded interval: (*Pterocodon (?) ampla*, *Theocorys acroria*, *Theocorys (?) phyzella* and *Theocotylissa auctor*.

Six first occurrences could be significant if one considered an average Eocene interval in Hole 1051A, where the rate of first occurrences is commonly one to three per 10 m, or through most of the upper Palaeocene sequence, where it varies from three to six. This information is somewhat biased, as it comes only from analyses of mostly biostratigraphic data and would be more useful, from a palaeoceanographic perspective, if quantitative data representing the total assemblage had been collected. Nevertheless, the result from this and previous investigations suggests that radiolarian evolution was not noticeably affected, if at all, by the LPTM.

The rapid change in lithology from siliceous nannofossil chalk in 171B-1051A-40X to limestone and chert in 171B-1051A-41X, combined with evidence from calcareous nannofossil datum levels indicative of 48.5–49.7 Ma confirm the presence of a 1–2 Ma hiatus near the lower–mid-Eocene boundary. Presence of the lowermost middle Eocene hiatus is corroborated by an abnormally large radiolarian turnover. Twenty-six first and last occurrences are documented and show that most of radiolarian Zone RP10 and a substantial part of Zone RP9 are missing. Thus, the radiolarian evidence from Hole 1051A suggests that the early Eocene Epoch is rather well represented whereas the latest early Eocene and the early mid-Eocene Epochs are not.

Hiatuses coeval with that in Site 1051 are known to be common in the lower–middle Eocene boundary interval in the Atlantic.

Systematic section

The continuously cored upper middle Eocene to upper lower Palaeocene sequence from Site 1051 provides a unique opportunity to study mid-latitude radiolarians from the Atlantic and to compare them with low-latitude species. We have encountered unusually high intraspecific variability in these faunas. As a result, it has been useful to tabulate taxa using a *sensu strictu* definition for a given taxon known from the tropics, and to informally describe variants of this taxon (e.g. 'var. A').

Superorder POLYCYSTINA Ehrenberg *emend.* Riedel 1967
Order SPUMELLARIA Ehrenberg 1875
Family ACTINOMMIDAE Haeckel 1862, *sensu* Riedel 1967

Genus *Spongatractus* Haeckel 1887

Spongatractus klausi Sanfilippo & Blome, new species
Fig. 6a–c

Type material. Holotype (Fig. 6b1, b2) from 171B-1051A-5H-2, 40–42 cm, in the middle Eocene interval that is equivalent to the low-latitude *Podocyrtis (Lampterium) goetheana* Zone (equivalent to calcareous nannofossil Zone CP14b).

Description. Spongy lanceolate-shaped disc with a pylome at one apex and a small spine at the other. The disc is thick in its middle part and is covered by spongy, dense meshwork with relatively small pores obscuring the internal structures. The margin is smooth. Most specimens have a distinct marginal pylome surrounded by spongy meshwork, occasionally a few short, smooth spines on either side of the pylome. In rare specimens, the spongy meshwork is thin enough to observe a lenticular cortical shell and a medullary shell. It has not been possible to observe the pore structure of the two shells or the number of medullary shells. Early in the range the shape is almost circular with a pointed apex and the pylome protruding at the opposite end, whereas later forms become more lanceolate or diamond shaped. The pylome is poreless or with a few, small, scattered subcircular pores.

Etymology. This species is named for Adam Klaus, Staff Scientist on ODP Leg 171B.

Dimensions. Measurements are based on 25 specimens throughout the range from 171B-1051A-2H-5, -5H-5, -8H-2 and -9H-5. Length of long axis 205–370 μm; length of short axis 145–275 μm; length of pylome protruding beyond the spongy margin 12–80 μm.

Distinguishing characters. This species is distinguished from other co-occurring spongy actinommids by its unusual shape, lack of concentric zonation of the spongy flange, and its distinct pylome.

Variability. The most variable character is the shape, from almost circular with a pointed apex to inflated lanceolate to rounded diamond-shape.

Distribution. This species occurs rarely in the upper middle Eocene *Podocyrtis (Lampterium) chalara* and *P. (L.) goetheana* Zones.

Phylogeny. Examination of broken, very rare transitional forms and forms from the lower part of the stratigraphic range of *Spongatractus klausi* n. sp. (Fig. 6c) suggest that this species evolved from *S. pachystylus* to terminate the lineage in the *Podocyrtis (Lampterium) goetheana* Zone.

Suborder CYRTIDA Haeckel 1862, *emend.* Petrushevskaya 1971
Family THEOPERIDAE Haeckel 1881, *emend.* Riedel 1967

Genus *Calocyclas* Ehrenberg 1847

Calocyclas aphradia Sanfilippo & Blome, new species
Fig. 6d–f

Unidentified theoperid in Sanfilippo & Riedel (1979, plate 1, fig. 12).

Type material. Holotype (Fig. 6d) from 171B-1051A-12H-5, 39–41 cm, in the middle Eocene interval that is equivalent to the lower part of the low-latitude *Podocyrtis chalara* Zone (equivalent to calcareous nannofossil Zone CP14b).

Description. Similar in general form to *Calocyclas hispida* (Fig. 6g), except that the thorax is poreless or with only a few small, subcircular scattered pores. Shell consists of three segments, with the collar and lumbar strictures expressed externally. Cephalis spherical, poreless or with a few small pores, bearing three conical horns of variable length. The apical spine is free in the cephalic cavity and continues externally as the apical horn. Auxiliary spines arising from the mitral arches close to the cephalic wall give rise

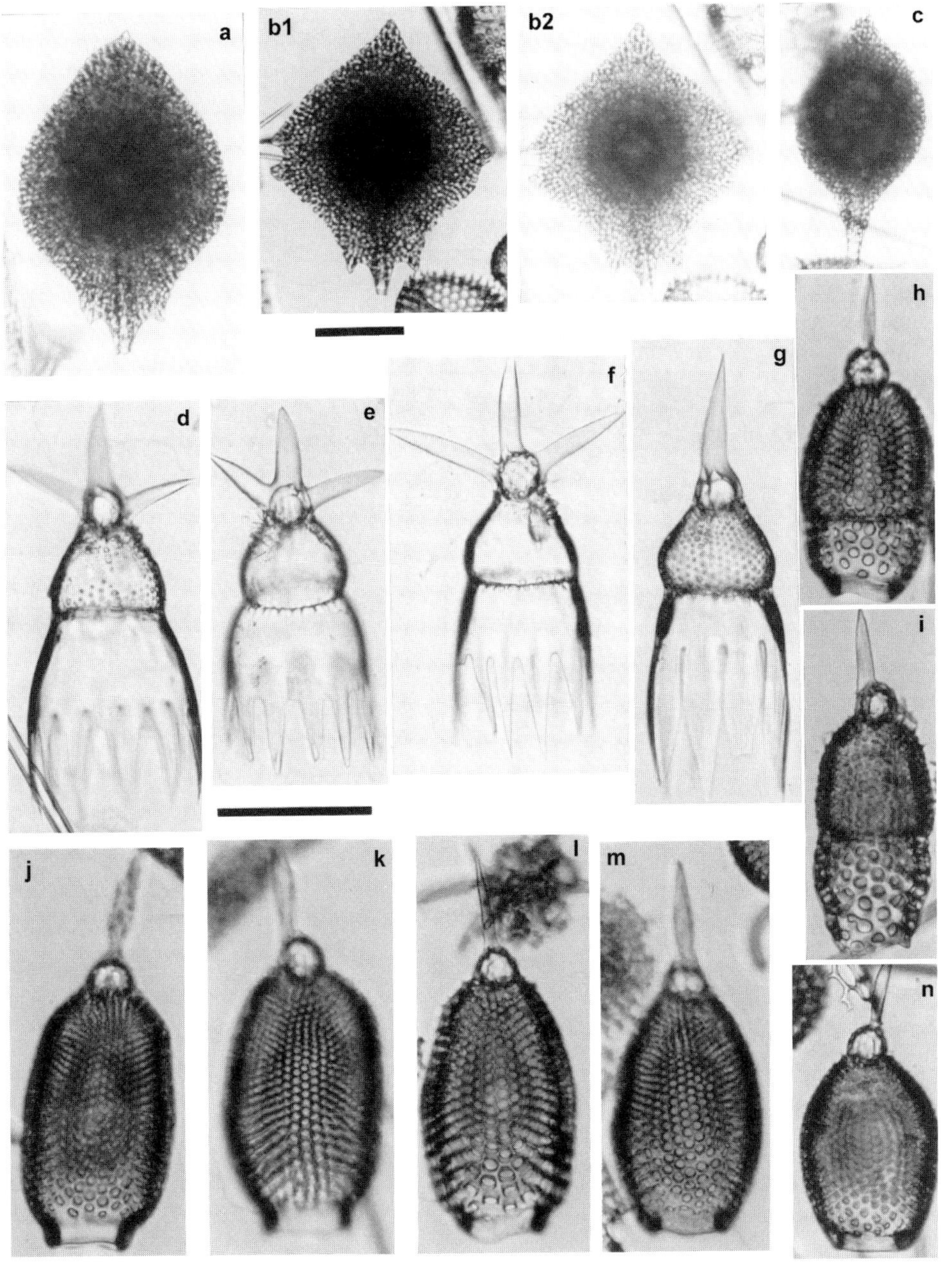

Fig. 6. In the figure explanations (also for Figs 7–11) the sample numbers and slide designations in the form 'Sl.1', 'Ph.2', and 'Cs.1' indicate preparations in our collection, and designations in the form 'K29/3' indicate England Finder positions (Riedel & Foreman 1961) of the illustrated specimens on the slides. Scale bars represent 100 μm. For (**a**)–(**c**) use scale bar below (*b*), for (**d**)–(**n**) use scale bar below (**e**). (**a–c**) *Spongatractus klausi* n. sp. (**a**) 1051A-2H-5, 39–41 Cs.1 O6/1. (**b1**, **b2**) Holotype, 1051A-5H-2, 40–42 Cs.1 K29/0. (**b2**) focused on internal structure. (**c**) 1051A-8H-2, 41–43 Cs.2 E40/0. (**d–f**) *Calocyclas aphradia* n. sp. (**d**) Holotype, 1051A-12H-5, 39–41 Cs.1 J12/4. (**e**) 1051A-10H-2, 40–42 Cs.1 K5/0. (**f**) 1051A-2H-5, 39–41 Cs.1 P40/2. (**g**) *Calocyclas hispida* (Ehrenberg) *s.s.* 1051A-16H-5, 40–42 Cs.2 W19/0. (**h**, **i**) *Sethocyrtis austellus* n. sp. (**h**) Holotype, 1051A-24X-5, 40–42 Cs.2 Q30/3. (**i**) 1051A-26X-2, 46–48 Sl.1 E40/0. (**j–n**) *Sethocyrtis chrysallis* n. sp. (**j**) Holotype, 1051A-8H-2, 41–43 Cs.2 E19/0. (**k**) 1051A-8H-2, 41–43 Cs.2 C40/0. (**l**) 1051A-22X-2, 39–41 Ph.1 X41/4. (**m**) 1051A-7H-5, 40–42 Ph.1 O8/4. (**n**) 1051A-16H-5, 40–43 Ph.1 N14/4.

to the two additional horns. Thorax campanulate, poreless, or rarely with a few small circular pores in the collar region. Early in the range small pores are scattered along ridges on the thorax caused by three spines (dorsal and two primary laterals) prolonged in the upper half of the thorax and protruding as small thorns. The vertical spine extends beyond the cephalic wall near the collar stricture. The proximal part of the cylindrical to truncate-conical abdomen is poreless, except for a row of very small circular pores just below the lumbar stricture. The remainder consists of 12–20 subparallel to tapering lamellar feet or teeth.

Etymology. The name is derived from the Greek noun *aphradia* (feminine), folly, fool.

Dimensions. Measurements based on 25 specimens throughout the stratigraphic range of the species from 171B-1051A-2H-5, -8H-5, -9H-5,-12H-2 and -12H-5. Length (excluding horn) 135–245 µm; length of cephalothorax 65–90 µm; length of hyaline part of abdomen 45–70 µm; length of teeth 30–90 µm; length of horn 45–80 µm; breadth of thorax 50–90 µm; breadth of hyaline abdomen 80–125 µm; distal breadth of teeth 90–125 µm.

Distinguishing characters. Calocyclas aphradia n. sp. is distinguished by its three horns and by having a shell that is hyaline except for a row of small pores just below the lumbar stricture.

Variability. Slight variation in the size of the three horns. Early forms may have a few scattered pores in the collar region.

Distribution. Calocyclas aphradia n. sp. occurs rarely in the middle Eocene *Podocyrtis (Podocyrtoges) ampla* Zone through the *Podocyrtis (Lampterium) goetheana* Zone. A similar form was previously recorded (Sanfilippo & Riedel 1974, plate 3, figs 5 and 6) from the Indian Ocean in the *Podocyrtis (Lampterium) mitra* Zone, and from the Pacific Ocean and the Caribbean (Sanfilippo & Riedel 1976, p. 155) in the *Thyrsocyrtis bromia* Zone (probably *Calocyclas bandyca* Zone).

Phylogeny. Calocyclas aphradia n. sp. is an offshoot of *Calocyclas hispida*.

Genus *Lychnocanoma* Haeckel 1887

Lychnocanoma (?) parma Sanfilippo & Blome, new species
Fig. 7n1, n2

Type material. Holotype (Fig. 7n1, n2) from 171B-1051A-53X-7, 30–32 cm, in the lower Eocene part of the *Bekoma bidartensis* Zone (equivalent to calcareous nannofossil Zone CP9a).

Description. A large two- or three-segmented shell with three shovel-shaped latticed feet. Cephalis spherical to subspherical, with a few circular pores. The apical horn has never been observed intact. Thorax inflated-conical to subhemispherical with a slightly constricted aperture, consisting of a smooth, unobtrusive rim. Three broad, coarsely latticed feet extend almost straight, or slightly tapering, from the lower margin of the thorax. These forms may be considered three-segmented when the feet are joined proximally by one or two rows of coarse lattice. Thoracic pores circular to subcircular, quincuncially arranged. The shell surface is roughened by minute thorns arising from the intervening pore bars.

Etymology. From the Latin noun *parma* (feminine), small light shield, an allusion to the shield-like feet.

Dimensions. Measurements based on 10 specimens from 171B-1051A-53X-7 and -53X-4. Total length (excluding horn) 195–250 µm; length of cephalothorax 120–170 µm; length of feet 65–110 µm; breadth of thorax 125–170 µm.

Distinguishing characters. This species is distinguished from *Lychnocanoma bellum* by not having three hollow feet, from *Lamptonium (?) colymbus* by not having three three-bladed wings extending from the upper part of the inflated thorax, and from *L. (?) incohatum* by having shovel-shaped latticed feet and thoracic pores that are not arranged in distinct longitudinal rows.

Variability. The shovel-shaped latticed feet are in some specimens connected proximally by one or two rows of coarse pored lattice attached to the aperture of the thorax.

Distribution. Lychnocanoma (?) parma n. sp. occurs only rarely and has a short stratigraphic range in the lower Eocene part of the *Bekoma bidartensis* Zone.

Phylogeny. Unknown.

Remarks. This lychnocaniid form is questionably assigned to *Lychnocanoma* Haeckel (*sensu* Foreman 1973). The relationship of

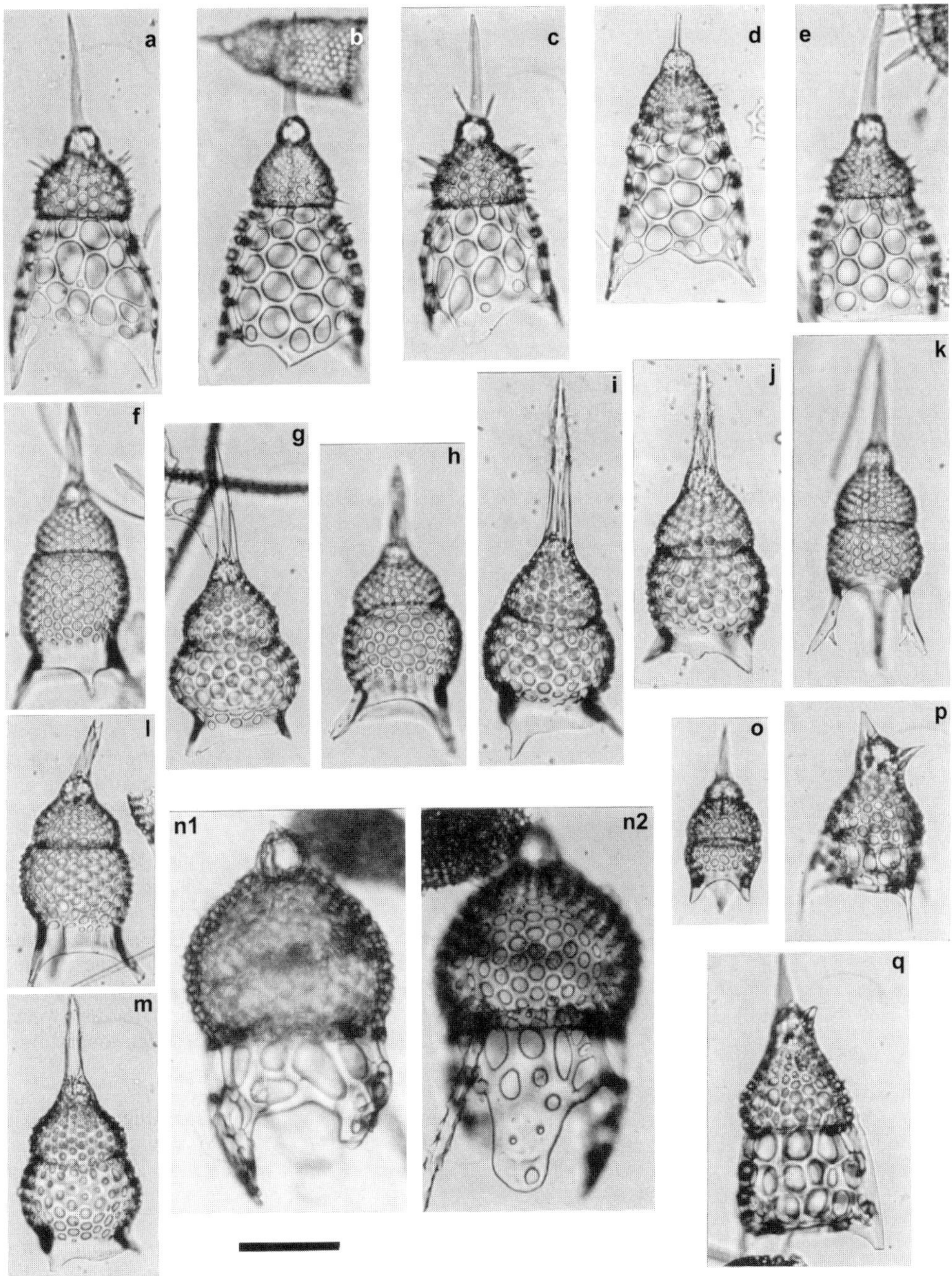

Fig. 7. Scale bar represents 100 μm. (**a–e**) *Thyrsocyrtis (Pentalacorys) krooni* n. sp. (**a**) 1051A-4H-2, 40–42 Cs.2 E31/4. (**b**) Holotype, 1051A-8H-5, 41–43 Cs.2 N44/0. (**c**) 1051A-4H-2, 40–42 Cs.2 H43/3. (**d**) 1051A-17X-2, 42–45 Cs.1 O38/0. (**e**) 1051A-2H-5, 39–41 Cs.1 H25/0. (**f–j, l, m**) *Thyrsocyrtis (Thyrsocyrtis) norrisi* n. sp. (**f**) 1051A-15H-2, 42–44 Cs.1 S10/2. (**g**) Holotype, 1051A-4H-2, 40–42 Cs.2 F28/0. (**h**) 1051A-15H-2, 42–44 Cs.1 H11/1. (**i**) 1051A-4H-2, 40–42 Cs.2 G17/2. (**j**) 1051A-5H-2, 40–42 Cs.2 N8/3. (**l**) 1051A-15H-2, 42–44 Cs.1 H10/2. (**m**) 1051A-15H-2, 42–44 Cs.1 S6/0. (**k, o**) *Thyrsocyrtis (Thyrsocyrtis) rhizodon* Ehrenberg. (**k**) 1051A-15H-2, 42–44 Cs.1 H47/0. (**o**) 1051A-44X-5, 41–44 Sl.1 H45/0. (**n1, n2**) *Lychnocanoma (?) parma* n. sp. Holotype, 1051A-53X-7, 30–32 Cs.1 F44/1. (**n2**) Opposite view. (**p, q**) *Thyrsocyrtis (?) annikae* Nishimura. (**p**) 1051A-62X-2, 45–47 Cs.1 U17/0. (**q**) 1051A-62X-2, 45–47 Cs.1 C15/0.

Lychnocanoma (?) parma n. sp. to the type species of the genus (designated by Campbell (1954, p. D124)) *Lychnocanoma clavigerum* Haeckel (1887, p. 1230, plate 61, fig. 4) is uncertain because of the presence of latticed rather than solid feet.

Genus *Sethocyrtis* Haeckel 1887

Sethocyrtis austellus Sanfilippo & Blome, new species
Fig. 6h, i

Type material. Holotype (Fig. 6h) from 171B-1051A-24X-5, 40–42 cm, in the middle Eocene interval that is equivalent to the low-latitude *Podocyrtis (Podocyrtoges) ampla* Zone (equivalent to calcareous nannofossil Zone CP13b).

Description. This form is similar to *Sethocyrtis chrysallis* n. sp. in most of its characters except for being three-segmented rather than two-segmented. The cephalis is poreless or with a few scattered pores and bears a slender, sometimes roughened, conical horn of variable length. Some specimens have a small pore at the base of the horn. The thoracic pores are small, closely packed, circular, uniform in size and arrangement in longitudinal rows. Collar and lumbar strictures distinct. The abdomen is somewhat narrower than the thorax, with pores that are 1.5–2 times larger than the thoracic ones, quincuncially arranged. Abdomen terminates in a differentiated hyaline peristome.

Etymology. The name is derived from the Latin noun *austellus*, meaning gentle south wind.

Dimensions. Measurements based on 20 specimens from 171B-1051A-24X-5, -25X-2, -25X-5, and -26X-2. Total length (excluding horn) 140–200 μm; length of cephalothorax 90–120 μm; length of abdomen 35–100 μm; length of horn 20–65 μm; breadth of thorax 75–100 μm; breadth of abdomen 75–100 μm. There are commonly 15 pores on half the circumference on the thorax, and 8–11 pores on half the circumference of the abdomen.

Distinguishing characters. It is distinguished from its descendant *Sethocyrtis chrysallis* n. sp. in having three rather than two segments, from the Antarctic lophocyrtid taxa *Aphetocyrtis rossi* and *A. gnomabax* in possessing a horn, a longer more porous thorax, and in a constricting abdominal peristome, and from *Lophocyrtis (Lophocyrtis?) semipolita* by its longer thorax and shorter abdomen with a differentiated termination.

Variability. The length of the abdomen and the development of the peristome are somewhat variable.

Distribution. This species occurs in rare abundance, and has a short stratigraphic range in the middle Eocene *Podocyrtis (Podocyrtoges) ampla* Zone.

Phylogeny. Origin unknown; gave rise to *S. chrysallis* n. sp. as described under that species.

Sethocyrtis chrysallis Sanfilippo & Blome, new species
Fig. 6j–n

Sethocyrtis sp. (Chen 1975, p. 459, plate 1, figs 4 and 5); Takemura (1992, p. 747, plate 7, figs 14 and 15); Strong *et al.* (1995 (in part), p. 209, fig. 11w); Takemura & Ling (1997, p. 114, plate 1, fig. 11). *Sethocyrtis* sp. A Hollis *et al.* (1997, p. 65, plate 6, fig. 7).

Type material. Holotype (Fig. 6j) from ODP 1051A-8H-2, 41–43 cm, in the middle Eocene *Podocyrtis (Lampterium) chalaka* Zone (equivalent to calcareous nannofossil Zone CP14b).

Description. Two-segmented shell bearing a slender conical horn. The acorn-shaped thorax varies slightly in length, and terminates in a slightly constricted aperture surrounded by a hyaline peristome. The cephalis is poreless and bears a slender, sometimes roughened, conical horn of variable length. The thoracic pores are small, closely packed, circular, uniform in size and arrangement in longitudinal rows. In the early part of its range the pores in the distal third of the thorax are somewhat larger than the proximal ones.

Etymology. The name is derived from the Greek noun *chrysallis* (feminine), chrysalis.

Dimensions. Measurements based on 30 specimens from 171B-1051A-8H-5, -16H-5 and -17H-2. Length (excluding horn) 125–190 μm; length of horn 50–100 μm; breadth of thorax 85–110 μm; breadth of peristome usually 12 μm.

Distinguishing characters. It differs from other co-occurring two-segmented forms in the arrangement of the thoracic pores and the somewhat constricted hyaline peristome.

Variability. The length of the thorax is somewhat variable. In the early part of its range the distal thoracic pores are slightly larger than the proximal ones.

Distribution. From its first appearance in the *Podocyrtis (Podocyrtoges) ampla* Zone, *S. chrysallis* n. sp. occurs in moderate abundance into late Eocene time.

Phylogeny. Although no transitional forms have been observed, it seems clear that *S. chrysallis* n. sp. evolved rapidly from *S. austellus* n. sp. in the uppermost part of the *Podocyrtis (Podocyrtoges) ampla* Zone by loss of the abdominal segment.

Genus *Thyrsocyrtis* Ehrenberg 1847

Thyrsocyrtis (Pentalacorys) krooni
Sanfilippo & Blome, new species
Fig. 7a–e

Thyrsocyrtis tetracantha (Ehrenberg) Riedel & Sanfilippo (1970, p. 527 (in part)); Riedel & Sanfilippo (1978, p. 81 (in part), plate 10, fig. 9). *Thyrsocyrtis (Pentalacorys) tetracantha* (Ehrenberg) Sanfilippo & Riedel (1982, p. 176 (in part), plate 1, fig. 11); Sanfilippo *et al.* (1985, p. 690 (in part), fig. 26.8 b).

Type material. Holotype (Fig. 7b) from 171B-1051A-8H-5, 41–43 cm, in the middle Eocene interval that is equivalent to the low-latitude *Podocyrtis (Lampterium) chalara* Zone (equivalent to calcareous nannofossil Zone CP14b).

Description. Three-segmented shell in which the large-pored abdomen forms the major part. Cephalis subspherical, poreless or with a few small pores, bearing a cylindrical to elongate-conical horn of variable length and thickness. Collar stricture distinct. Thorax considerably shorter than abdomen, broadly conical or slightly inflated, with small subcircular pores and a rough to very thorny surface. Lumbar stricture distinct. Abdomen thick-walled with large subcircular pores (3–5 along its length), and a narrow poreless distal rim, occasionally with a thorny surface. In some specimens the abdomen is very long, constricts slightly in the lower half, and expands to terminate in a narrow, barely differentiated margin. The termination is variable, commonly as broad as a pore-bar, but may be smooth, undulating or drawn out as 4–6 short flat, hyaline points.

Etymology. This species is named for Dick Kroon, Co-Chief Scientist on ODP Leg 171B.

Dimensions. Measurements based on 45 specimens from 171B-1051A-2H-5, -8H-5, -11H-2, -13H-5, -18X-2, -19X-2 and -19X-5. Total length (excluding horn) 185–295 μm; length of cephalothorax 75–95 μm; length of abdomen 100–200 μm; length of horn 50–145 μm; breadth of thorax 75–105 μm; breadth of abdomen 110–165 μm. There are 4–6 pores on half the abdominal circumference.

Distinguishing characters. Thyrsocyrtis (Pentalacorys) krooni n. sp. is very similar to *T. (P.) tetracantha* and is distinguished from it by the presence of foot-like porous structures rather than solid cylindrical feet, and from *T. (P.) triacantha* by the absence of a differentiated peristome.

Variability. The termination of the abdomen varies from a narrow barely differentiated margin to a slightly thickened, smooth pore-bar, which may be undulating or drawn out as 4–6 short flat, hyaline points. The presence or absence of thorns or spines on the shell surface are variable characters.

Distribution. This species occurs in moderate abundance from its earliest occurrence in the middle Eocene interval near the lower boundary of the *Podocyrtis (Lampterium) mitra* Zone (equivalent to calcareous nannofossil Zone CP14a) into the *P. (L.) goetheana* Zone.

Phylogeny. Forms from the early part of the stratigraphic range of *Thyrsocyrtis (Pentalacorys) krooni* n. sp. and rare transitional forms indicate that this species arose from *T. (P.) triacantha* by loss of the differentiated peristome and the three long, cylindrical feet (Fig. 7d).

Thyrsocyrtis (Thyrsocyrtis) norrisi
Sanfilippo & Blome, new species
Fig. 7f–j, l, m

Type material. Holotype (Fig. 7g) from 171B-1051A-4H-2, 40–42 cm, in the middle Eocene interval that is equivalent to the low-latitude *Podocyrtis (Lampterium) goetheana* Zone (equivalent to calcareous nannofossil Zone CP14b).

Description. Three-segmented skeleton with truncate-conical thorax and barrel-shaped abdomen. Cephalis subspherical, with rather few small pores, bearing a long, robust horn, which is bladed in its proximal half and conical in its distal half. There is a ring of small thorns or spinules at the junction from bladed to conical, and beyond this the horn is roughened or thorny. Collar stricture is moderately distinct. Thorax campanulate, with small subcircular pores, separated from the abdomen by a distinct lumbar stricture. Abdomen barrel shaped to inflated cylindrical, wider and usually longer than the thorax, with pores commonly 2–4 times as large as those on the thorax. Surface of thorax and abdomen rough. The porous part of the abdomen is constricted distally before terminating in a wide, flared hyaline peristome from which three feet arise. The feet are divergent prolongations of the peristome and vary in length.

Etymology. This species is named for Richard D. Norris, Co-Chief Scientist on ODP Leg 171B.

Dimensions. Measurements are based on 30 specimens throughout its range from 171B-1051A-4H-2, -12H-2 and -14H-5. Total length (excluding horn) 170–240 μm; length of cephalothorax 65–90 μm; length of abdomen 95–145 μm (including short feet); length of horn 55–150 μm; breadth of thorax 80–120 μm; breadth of abdomen 100–145 μm. There are usually 10 pores on half the abdominal circumference.

Distinguishing characters. *Thyrsocyrtis (Thyrsocyrtis) norrisi* n. sp. differs from its ancestor *T. (T.) rhizodon* by having larger pores on the abdomen, a longer more robust horn and a flared peristome from which the divergent feet arise. It differs from *T. (T.) bromia* by having smaller and fewer abdominal pores and a flared peristome with longer feet. *Thyrsocyrtis (T.) norrisi* n. sp. and *T. (T.?) pinguisicoides* O'Connor 1999 (plate 4, figs 28–32; plate 7, figs 28a–31) have many morphological characters in common such as a very long, stout bladed horn that is roughened and thorny with a corona of small spinules in the distal part, but *T. (T.) norrisi* n. sp. differs from *T. (T.?) pinguisicoides* by having a broad flared peristome and the presence of feet. It is possible that these two taxa are geographical or ecological variants of *T. (T.) bromia*. *Thyrsocyrtis (Thyrsocyrtis) norrisi* n. sp. has a longer stratigraphic range than both *T. (T.) bromia* and *T. (T.?) pinguisicoides*. *Thyrsocyrtis (Thyrsocyrtis) norrisi* n. sp. differs from *Thyrsocyrtis (Pentalacorys) tetracantha* and *T. (P.) krooni* n. sp. by its distinct peristome, which is wider than a pore-bar.

Variability. Early forms have a less inflated abdomen with smaller pores, and a slightly narrower peristome with longer more divergent feet.

Distribution. This species occurs in moderate abundance from its earliest occurrence in the middle Eocene interval from the upper part of the *Podocyrtis (Lampterium) mitra* Zone (equivalent to calcareous nannofossil Zone CP14a) to the *P. (L.) goetheana* Zone.

Phylogeny. Only rare transitional forms have been found; however, from observation of forms in the early part of the stratigraphic range of *Thyrsocyrtis (Thyrsocyrtis) norrisi* n. sp. it appears that *T. (T.) norrisi* n. sp. evolved from *T. (T.) rhizodon* (Fig. 7f, h).

Species list and taxonomic notes

Amphicraspedum murrayanum Haeckel
Amphicraspedum murrayanum Haeckel (1887, p. 523, plate 44, fig. 10); Sanfilippo & Riedel (1973, p. 524, plate 10, figs 3–6; plate 28, fig. 1).

Amphicraspedum murrayanum Haeckel var. A
Fig. 8a
Remarks. Under this name we have tabulated a form similar to *Amphicraspedum murrayanum* in all respects except for its smaller size, reduced median swelling and less pronounced longitudinal structure in the arms, especially in the expanded terminations of the arms. The length of the skeleton is approximately half that of *A. murrayanum*. This form co-occurs intermittently with *A. murrayanum* and extends well above the latest occurrence of typical *A. murrayanum*. The stratigraphic range can be determined in the middle Eocene sequence above the latest occurrence of *A. murrayanum* from the *Dictyoprora mongolfieri* Zone into the *Podocyrtis goetheana* Zone.

Amphicraspedum prolixum Sanfilippo & Riedel group, Sanfilippo & Riedel (1973, p. 524, plate 10, figs 7–11; plate 28, figs 3 and 4).

Anthocyrtis (?) sp. aff. *A. mespilus* Ehrenberg
Anthocyrtis mespilus Ehrenberg (1854, plate 36, fig. 13); Ehrenberg (1875, p. 66, plate 6, fig. 4); Haeckel (1887, p. 1269); Nishimura (1992, p. 331, plate 6, figs 12 and 13).

Artophormis barbadensis (Ehrenberg)
Calocyclas barbadensis Ehrenberg (1873, p. 217; 1875, plate 18, fig. 8).
Artophormis barbadensis (Ehrenberg), Haeckel (1887, p. 1459); Riedel & Sanfilippo (1970, p. 532, plate 13, fig. 5).

Bekoma bidartensis Riedel & Sanfilippo
Fig. 11q
Bekoma bidarfensis Riedel & Sanfilippo (1971, p. 1592, plate 7, figs 1–7); Foreman (1973, p. 432,

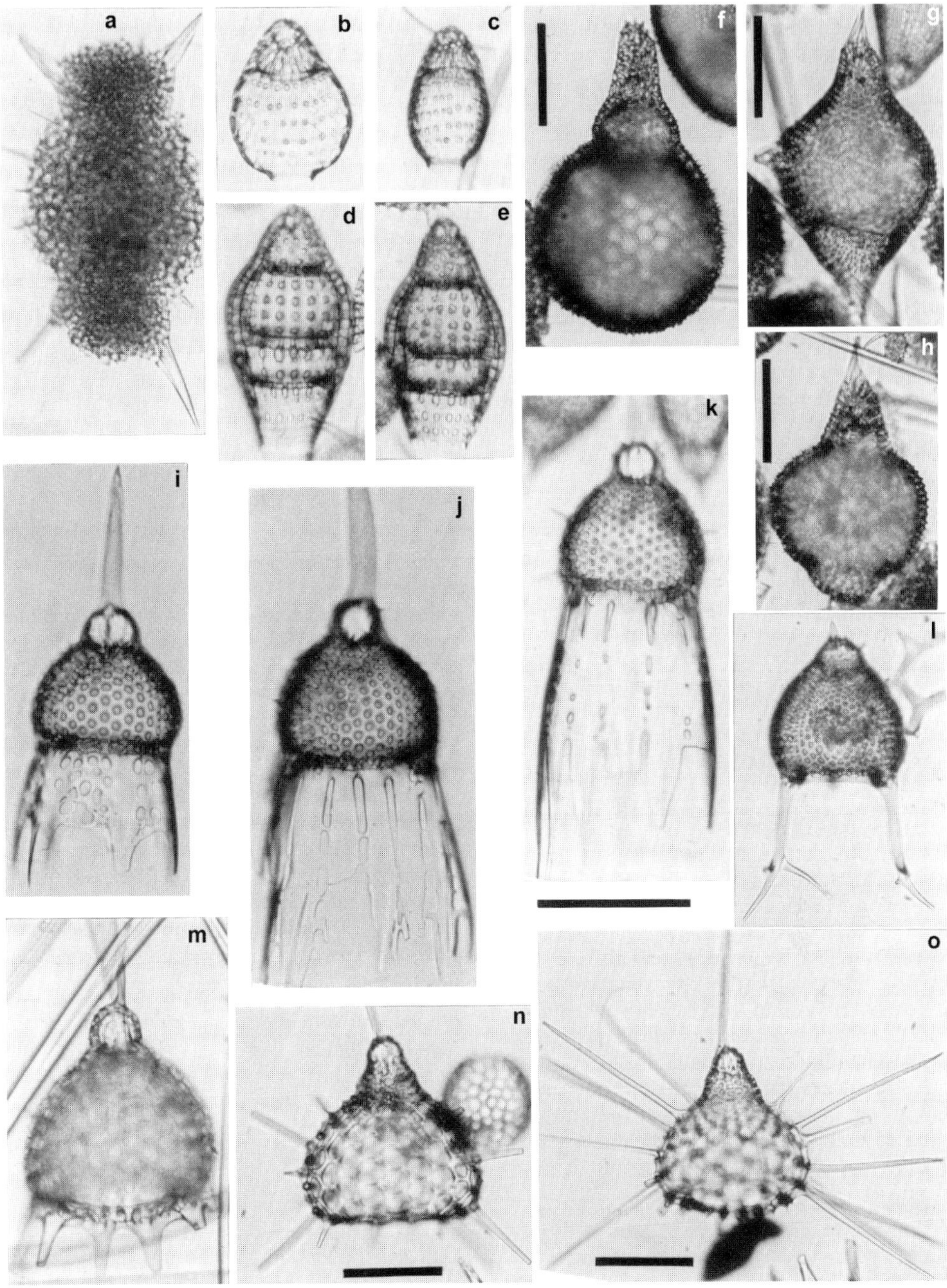

Fig. 8. Scale bars represent 100 μm. For (**a**)–(**e**) and (**i**)–(**m**) use scale bar below (**k**). (**a**) *Amphicraspedum murrayanum* Haeckel var. A. 1051A-14H-5, 40–42 Cs.1 R32/3. (**b**) *Dictyoprora pirum* (Ehrenberg) *s.s.* 1051A-6H-5, 40–42 Ph.1 V22/0. (**c**) *Dictyoprora pirum* (Ehrenberg) var. A. 1051A-17X-2, 42–45 Ph.1 D40/2. (**d**, **e**) *Buryella tetradica* Foreman var. A. (**d**) 1051A-67X-2, 47–49 Ph.1 H17/0. (**e**) 1051A-67X-5, 33–35 Ph.1 N9/4. (**f**–**h**) *Lamptonium sanfilippoae* Foreman. (**f**) 1051A-42X-2, 40–42 Cs.1 Q12/1. (**g**) 1051A-47X-5, 45–47 Cs.1 B27/4. (**h**) 1051A-45X-2, 51–53 Cs.1 P7/1. (**i**–**k**) *Calocyclas turris* Ehrenberg. (**i**) 1051A-2H-5, 39–41 Cs.1 Y15/1. (**j**) 1051A-11H-5, 40–42 Cs.2 Q6/0. (**k**) 1051A-8H-5, 41–43 Cs.1 O35/0. (**l**) *Bekoma (?) demissa demissa* Foreman. 1051A-62X-1, 105–107 Cs.2 G19/0. (**m**) *Calocyclas hispida* (Ehrenberg) var. A. 1051A-53X-5, 36–39 Cs.2 N20/0. (**n**, **o**) *Calocycloma castum* (Haeckel). (**n**) 1051A-2H-5, 39–41 Cs.1 E37/2. (**o**) 1051A-6H-2, 40–42 Cs.1 T27/0.

plate 3, figs 20 and 21; plate 10, fig. 6). (The original spelling of the specific name as *bidarfensis* was based on an error in copying the name of the type locality Ibbaritz–Bidart. This *lapsus calami* was corrected by Riedel & Sanfilippo (1978, p. 65).

Remarks. Included are forms varying slightly in the length and the width of the three feet and the size of the shell. Later forms have a tendency to have shorter, thicker feet and the size of the thorax is closer to the maximum for its size range.

Bekoma campechensis Foreman

Bekoma campechensis Foreman (1973, p. 432, plate 3, fig. 24; plate 10, figs 1 and 2 (not fig. 4)); Nishimura (1992, p. 332, plate 5, figs 1–4; plate 13, fig. 1).

Remarks. The original description by Foreman (1973, p. 432) included a reference to illustrations of *B. campechensis* (plate 3, figs 20 and 21; plate 10, figs 1, 2 and 4). Foreman's plate 10, fig. 4 is not *B. campechensis* but *B. divaricata.* Forms at Site 1051A are variable in the size of the shell, and the length and thickness of the three feet. Towards the end of its stratigraphic range the domed cephalis becomes shorter and not so markedly hyaline, the feet become shorter and thicker, and the lattice between the feet becomes more strongly developed.

Bekoma (?) demissa demissa Foreman

Fig. 8l

Bekoma (?) demissa Foreman (1973, p. 432, plate 3, fig. 22; plate 10, fig. 5).

Bekoma (?) demissa demissa Foreman, Nishimura (1992, p. 333, plate 6, fig. 1).

Remarks. Some of the observed forms are intermediate between *B. demissa demissa* and *B. demissa robusta* in that they have three rather thick feet that extend downward from the distal thoracic margin terminating in two or three short lateral branches. However, these branches have never been observed to join in the specimens included under this name.

Bekoma (?) demissa robusta Nishimura

Bekoma (?) demissa robusta Nishimura (1992, p. 333, plate 6, figs 2 and 3).

Remarks. Forms included under this name have thick thoracic shell walls and very robust feet that are always joined by lateral branches.

Bekoma (?) oliva Nishimura

Fig. 11j

Bekoma (?) oliva Nishimura (1992, p. 333, plate 5, figs 10 and 11; plate 13, fig. 6).

Remarks. Specimens of *Bekoma (?) oliva* encountered in the Site 1051 material agree well with those described by Nishimura (1992) from the northwest Atlantic DSDP Site 384, but their stratigraphic range is somewhat longer than that reported, extending from the lower Palaeocene *Bekoma campechensis* Zone well into the Upper Palaeocene *Bekoma bidartensis* Zone.

Buryella clinata Foreman

Buryella clinata Foreman (1973, p. 433, plate 8, figs 1–3; plate 9, fig. 19); Foreman (1975, p. 620, plate 9, figs 35 and 36).

Buryella pentadica Foreman

Buryella pentadica Foreman (1973, p. 433, plate 8, fig. 8; plate 9, figs 15 and 16).

Buryella tetradica Foreman *s.s.*

Buryella tetradica Foreman (1973, p. 433, plate 8, figs 4 and 5; plate 9, figs 13 and 14); Sanfilippo *et al.* (1985, p. 668, figs 14.3a and b); Hollis (1997, p. 81, plate 21, figs 16–19).

Remarks. We have followed Foreman's (1973) original description in its strictest sense, admitting only four-segmented forms with a subovate shell. The pores on the third segment are uniform, circular, and arranged in longitudinal and transverse rows. Forms with a transversely segmented fourth segment have been tabulated as *Buryella tetradica* var. A. In these forms the fourth, inverted truncate-conical, segment is usually more delicate. *Buryella tetradica s.s.* ranges from the Upper Palaeocene *Bekoma campechensis* Zone through the lower Eocene sequence, with its latest occurrence near the lower Eocene–middle Eocene boundary in the *Theocotyle cryptocephala* Zone. Suspected contaminants or reworked specimens have been found well into the Middle Eocene *Podocyrtis (Lampterium) mitra* Zone.

Buryella tetradica Foreman var. A

Fig. 8d and e

Remarks. Similar to *Buryella tetradica s.s.* in all respects except for having the fourth, inverted truncate-conical, segment transversely segmented. These forms commonly show an external expression at the lumbar stricture. *Buryella tetradica* var. A has an earlier stratigraphic range than that of *B. tetradica s.s.* in the lower to upper Palaeocene *Bekoma campechensis* Zone, with a slight overlap of their stratigraphic ranges in the upper part of the *B. campechensis* Zone.

Calocyclas hispida (Ehrenberg) *s.s.*

Fig. 6g

Anthocyrtis hispida Ehrenberg (1873, p. 216; 1875, plate 8, fig. 2).

Cycladophora hispida (Ehrenberg) Riedel & Sanfilippo (1970, p. 529, plate 10, fig. 9; 1971, p. 1593, plate 3B, figs 10 and 11); Moore (1971, p. 741, plate 4, figs 6 and 7).

Calocyclas hispida (Ehrenberg) Foreman (1973, p. 434, plate 1, figs 12–15; plate 9, fig. 18).
Remarks. The name is applied in a strict sense to forms with lamellar teeth that are closely spaced and rather broad and parallel sided, at least proximally. *Calocyclas hispida s.s.* ranges from the lower Eocene *Phormocyrtis striata striata* Zone through the upper Eocene interval.

Calocyclas hispida (Ehrenberg) var. A
Fig. 8m
Calocyclas sp. Foreman (1973, p. 434, plate 1, fig. 16).
Remarks. These forms are similar in all respects to *C. hispida s.s.* but have narrower, distally tapering and more widely spaced lamellar teeth extending from the thoracic apertural margin. Their earlier stratigraphic range is from the lower Eocene part of the *Bekoma bidartensis* Zone into the middle Eocene *Dictyoprora mongolfieri* Zone, overlapping in its later part with the stratigraphic range of *C. hispida s.s.* This form occurs rarely in these sediments.

Calocyclas turris Ehrenberg
Fig. 8i–k
Calocyclas turris Ehrenberg (1873, p. 218; 1875, plate 18, fig. 7); Foreman (1973, p. 434); Sanfilippo *et al.* (1985, p. 669, fig. 15.1a–c).
Remarks. In the tropics, this short-ranged species evolves from its ancestor *C. hispida* in late Eocene time. However, at the investigated site, this evolutionary transition takes place earlier, in the middle Eocene *Podocyrtis (Lampterium) mitra* Zone. The transitional forms are variable in thoracic size, development of thorns on the thorax and the manner in which the lamellar teeth are joined. The lamellar teeth are usually joined by bars to form an abdomen with longitudinal rows of pores. In these mid-latitude forms, the bars vary in thickness, and some specimens appear to have an almost hyaline abdomen with small pores arranged longitudinally at the junctions of the lamellar teeth.

Calocycloma ampulla (Ehrenberg)
Eucyrtidium ampulla Ehrenberg (1854, plate 36, fig. 15a–c; 1873, p. 22).
Calocycloma ampulla (Ehrenberg), Foreman (1973, p. 434, plate 1, figs 1–5; plate 9, fig. 20).

Calocycloma castum (Haeckel)
Fig. 8n and o
Calocyclas casta Haeckel (1887, p. 1384, plate 73, fig. 10); Cita *et al.* (1970, p. 404, plate 2, fig. H).
Calocycloma castum (Haeckel) Foreman (1973, p. 434, plate 1, figs 7, 9 and 10).
Remarks. This species is variable in size, and the typical distinct change in contour between the conical upper part of the thorax and the inflated part is not always as obvious as in tropical material. Commonly the specimens have very long slender spines arising from the junctions of the intervening pore bars.

Clathrocycloma capitaneum Foreman
Clathrocycloma capitaneum Foreman (1973, p. 434, plate 2, fig. 5; plate 11, fig. 11).

Cryptocarpium ornatum (Ehrenberg)
Cryptoprora ornata Ehrenberg (1873, p. 222; 1875, plate 5, fig. 8); Sanfilippo *et al.* (1985, p. 693, fig. 27.2a and b).
Cryptocarpium ornatum (Ehrenberg) Sanfilippo & Riedel (1992, pp. 6 and 36, plate 2, figs 18–20).

Dendrospyris fragoides Sanfilippo & Riedel
Dendrospyris fragoides Sanfilippo & Riedel (1973, p. 526, plate 15, figs 8–13; plate 31, figs 13 and 14).

Dendrospyris golli Nishimura
Dendrospyris golli Nishimura (1992, p. 330, plate 3, figs 1 and 2; plate 12, fig. 11).

Dictyophimus craticula Ehrenberg
Dictyophimus craticula Ehrenberg (1873, p. 223; 1875, plate 5, figs 4 and 5); Sanfilippo & Riedel (1973, p. 529, plate 19, fig. 1; plate 33, fig. 11).

Dictyoprora amphora (Haeckel) group
Dictyocephalus amphora in Haeckel (1887, p. 1305, plate 62, fig. 4).
Dictyoprora amphora (Haeckel) Nigrini (1977, p. 250, plate 4, figs 1 and 2).

Dictyoprora mongolfieri (Ehrenberg)
Eucyrtidium mongolfieri Ehrenberg (1854, plate 36, fig. 18, B lower; 1873, p. 230).
Dictyoprora mongolfieri (Ehrenberg) Nigrini (1977, p. 250, plate 4, fig. 7).

Dictyoprora pirum (Ehrenberg) *s.s.*
Fig. 8b
Eucyrtidium pirum Ehrenberg (1873, p. 232; 1875, plate 10, fig. 14).
Dictyoprora pirum (Ehrenberg) Nigrini (1977, p. 251, plate 4, fig. 8).
Remarks. The first appearance of *Dictyoprora pirum s.s.* occurs earlier at this site than previously recorded. In Hole 1051A the lowest occurrence is in the middle Eocene *Podocyrtis (Lampterium) mitra* Zone, whereas in low latitudes it occurs in the upper part of the *P. (L.) goetheana* Zone.

Dictyoprora pirum (Ehrenberg) var. A
Fig. 8c
Remarks. *Dictyoprora pirum* var. A is similar to *D. pirum s.s.* except for not being laterally compressed. Its abdominal breadth is narrower

(50–70 μm) when compared with *D. pirum s.s.* (65–85 μm). In the material from Hole 1051A, *D. pirum* var. A occurs in moderate abundance from the early Eocene *Phormocyrtis striata striata* Zone to overlap with the first occurrence of *D. pirum* in the middle Eocene *Podocyrtis (Lampterium) mitra* Zone.

Eusyringium fistuligerum (Ehrenberg)
Fig. 9a–d
Eucyrtidium fistuligerum Ehrenberg (1873, p. 229; 1875, plate 9, fig. 3).
Eusyringium fistuligerum (Ehrenberg) Riedel & Sanfilippo (1970, p. 527, plate 8, figs 8 and 9); Foreman (1973, p. 435, plate 11, fig. 6); Sanfilippo *et al.* (1985, p. 670, fig. 17.1a and b).
Remarks. Great variation has been observed in the shape of the thorax from extremely pyriform to very slender, in wall thickness, and in length and robustness of the horn. The different morphologies co-occur throughout the range of the species. Except for the earliest occurring representatives which have short, thin-walled tubular prolongations that do not preserve well, all specimens are well developed with long tubular prolongations.

Eusyringium lagena (Ehrenberg)
Lithopera lagena Ehrenberg (1873, p. 241; 1875, plate 3, fig. 4).
Eusyringium lagena (Ehrenberg) Riedel & Sanfilippo (1970, p. 527, plate 8, figs 5–7); Foreman (1973, p. 436, plate 11, figs 4 and 5); Sanfilippo *et al.* (1985, p. 672, fig. 17.2a–c).

Giraffospyris lata
Giraffospyris lata Goll (1969, p. 334, plate 58, figs 22, 24–26).

Lamptonium (?) colymbus Foreman
Lamptonium (?) colymbus Foreman (1973, pp. 435–436, plate 6, fig. 2; plate 11, figs 15 and 19).

Lamptonium fabaeforme chaunothorax Riedel & Sanfilippo
Lamptonium (?) fabaeforme (?) chaunothorax Riedel & Sanfilippo (1970, p. 524, plate 5, figs 8 and 9); Sanfilippo *et al.* (1985, p. 673, fig. 18.3).

Lamptonium fabaeforme constrictum Riedel & Sanfilippo
Lamptonium (?) fabaeforme (?) constrictum Riedel & Sanfilippo (1970, p. 523, plate 5, fig. 7); Sanfilippo *et al.* (1985, p. 674, fig. 18.4).

Lamptonium fabaeforme fabaeforme (Krasheninnikov)
[?] *Cyrtocalpis fabaeformis* Krasheninnikov (1960, p. 296, plate 3, fig. 11).
Lamptonium (?) fabaeforme fabaeforme (Krasheninnikov) Riedel & Sanfilippo (1970, p. 523, plate 5, fig. 6); Foreman (1973, p. 436, plate 6, figs 6–9); Sanfilippo *et al.* (1985, p. 674, fig. 18.2).

Lamptonium (?) incohatum Foreman
Lamptonium (?) incohatum Foreman (1973, p. 436, plate 6, fig. 1; plate 11, fig. 18).

Lamptonium pennatum Foreman
Lamptonium pennatum Foreman (1973, p. 436, plate 6, figs 3–5; plate 11, fig. 13).

Lamptonium sanfilippoae Foreman
Fig. 8f–h
Lamptonium sanfilippoae Foreman (1973, p. 436, plate 6, figs 15 and 16; plate 11, figs 16 and 17).
Remarks. Also included are two-segmented pyriform specimens generally similar to *L. sanfilippoae* except for lacking the massive layer of rough shell material that normally extends upward from the cephalis to cover the base of the bladed horn. These forms show great external morphological variation.

Lithapium anoectum Riedel & Sanfilippo
Lithapium anoectum Riedel & Sanfilippo (1973, p. 516, plate 24, figs 6 and 7).

Lithapium mitra (Ehrenberg)
Cornutella mitra Ehrenberg (1873, p. 221; 1875, plate 2, fig. 8).
Lithapium (?) mitra (Ehrenberg), Riedel & Sanfilippo (1970, p. 520, plate 4, figs 6 and 7).

Lithapium plegmacantha Riedel & Sanfilippo
Lithapium (?) plegmacantha Riedel & Sanfilippo (1970, p. 520, plate 4, figs 2 and 3).

Lithochytris archaea Riedel & Sanfilippo
Fig. 9e and f
Lithochytris archaea Riedel & Sanfilippo (1970, p. 528, plate 9, fig. 7; 1971, plate 7, fig. 13); Foreman (1973, p. 436, plate 2, figs 4 and 5).
Remarks. Specimens of *L. archaea* in this material are very robust with short abdominal segments, short feet and a cephalis that is enclosed in the base of the thick, broad-based horn.

Lithochytris vespertilio Ehrenberg
Lithochytris vespertilio Ehrenberg (1873, p. 239; 1875, plate 4, fig. 10); Riedel & Sanfilippo (1971, p. 528, plate 9, figs 8 and 9).

Lithocyclia aristotelis (Ehrenberg) group
Astromma aristotelis Ehrenberg (1847, p. 55, fig. 10).
Lithocyclia aristotelis (Ehrenberg) group, Riedel & Sanfilippo (1970, p. 522, plate 13, figs 1 and 2).

Lithocyclia ocellus Ehrenberg group

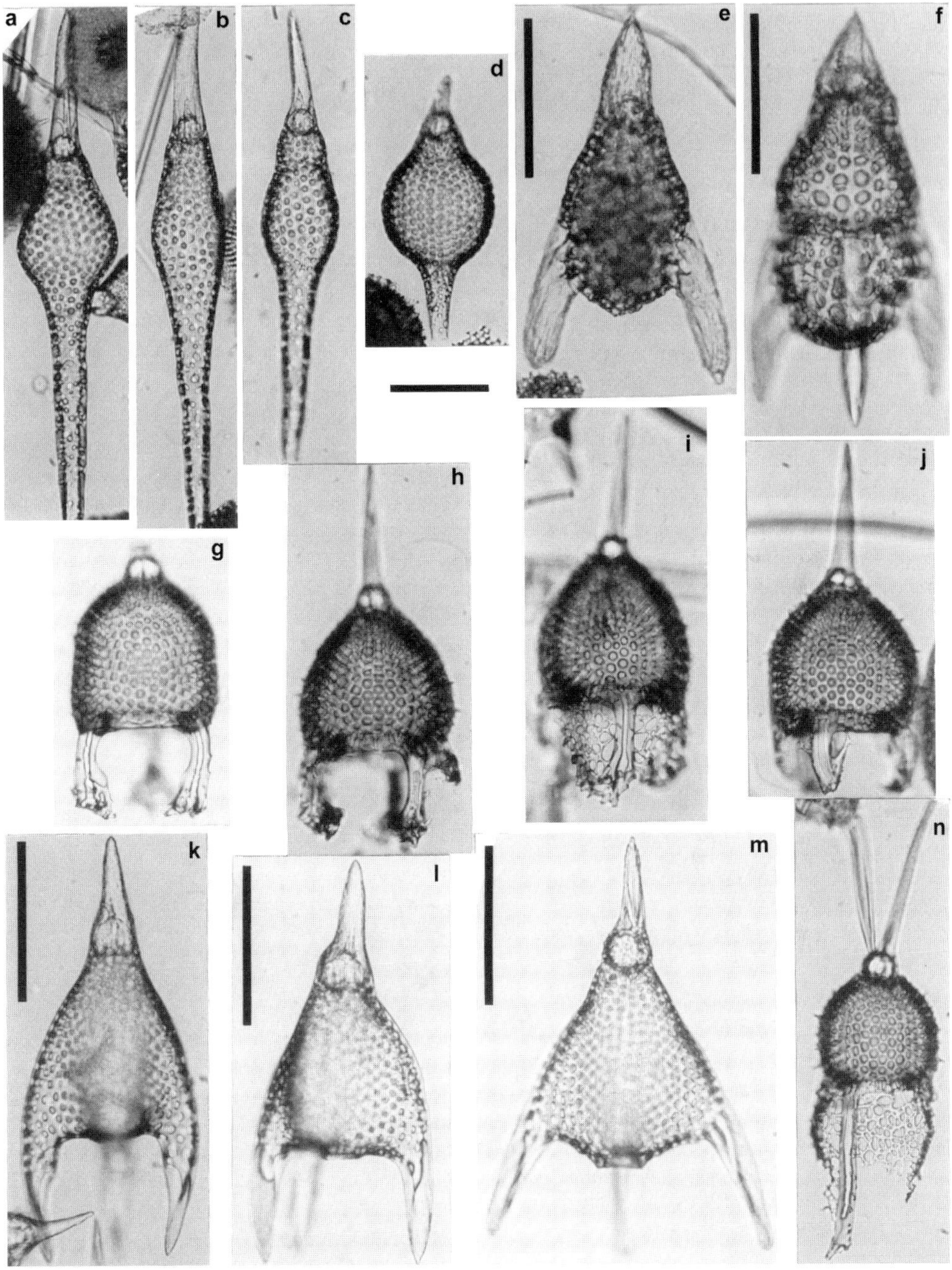

Fig. 9. Scale bars represent 100 μm. For (**a**)–(**d**), (**g**)–(**j**) and (**n**) use scale bar below (**d**). (**a**–**d**) *Eusyringium fistuligerum* (Ehrenberg). (**a**) 1051A-6H-2, 40–42 Cs.2 K11/2. (**b**) 1051A-6H-2, 40–42 Cs.2 K19/0. (**c**) 1051A-3H-5, 40–42 Cs.1 W20/3. (**d**) 1051A-3H-5, 40–42 Cs.2 Q23/4. (**e**, **f**) *Lithochytris archaea* Riedel & Sanfilippo. (**e**) 1051A-50X-2, 40–43 Sl.2 K40/0. (**f**) 1051A-46X-2, 39–41 Cs.2 L32/0. (**g**–**j**) *Lychnocanoma amphitrite* Foreman. (**g**) 1051A-15H-5, 41–43 Cs.1 J36/4. (**h**) 1051A-12H-5, 39–41 Cs.1 U10/0. (**i**) 1051A-15H-5, 41–43 Cs.2 K20/2. (**j**) 1051A-16H-5, 40–42 Cs.2 R12/2. (**k**–**m**) *Lychnocanoma lucerna* Ehrenberg. (**k**) 1051A-2H-5, 37–41 Cs.2 F18/0. (**l**) 1051A-5H-2, 40–42 Cs.2 J45/4. (**m**) 1051A-2H-5, 39–41 Cs.1 W20/1. (**n**) *Lychnocanoma bellum* (Clark & Campbell). 1051A-23X-5, 40–42 Cs.1 X22/2.

Lithocyclia ocellus Ehrenberg (1854, plate 36, fig. 30; 1873, p. 240)
Lithocyclia ocellus Ehrenberg, Riedel & Sanfilippo (1970, p. 522, plate 5, figs 1 and 2).

Lophocyrtis biaurita (Ehrenberg)
Eucyrtidium biaurita Ehrenberg (1873, p. 226; 1875, p. 70, plate 10, figs 7 and 8).
Lophocyrtis biaurita (Ehrenberg), Haeckel (1887, p. 1411); Cita *et al.* (1970, p. 404, plate 2, figs I–K); Foreman (1973, p. 442, plate 8, figs 23–26).

Lychnocanoma amphitrite Foreman
Fig. 9g–j
Lychnocanoma amphitrite Foreman (1973, p. 437, plate 11, fig. 10).
Remarks. In spite of the abdominal lattice being commonly missing, the large size, massive, thick-walled thorax and robust, short, distally stubby feet easily distinguish this species from its ancestor *L. bellum*. Its stratigraphic range is longer at this mid-latitude location than in the tropics, ranging from the Middle Eocene *Podocyrtis (Lampterium) mitra* Zone through the Upper Eocene sequence. Takemura (1992) similarly observed lower to middle Eocene occurrences of *L. amphitrite* from the Kerguelen Plateau in the Southern Ocean.

Lychnocanoma auxilla Foreman
Lychnocanoma auxilla Foreman (1973, p. 437, plate 2, fig. 6; plate 11, figs 1 and 2).

Lychnocanoma babylonis (Clark & Campbell) group
Dictyophimus babylonis Clark & Campbell (1942, p. 67, plate 9, figs 32 and 36).
Lychnocanoma babylonis (Clark & Campbell) group Foreman (1973, p. 437, plate 2, fig. 1)

Lychnocanoma bellum (Clark & Campbell)
Fig. 9n
Lychnocanium bellum Clark & Campbell (1942, p. 72, plate 9, figs 35 and 39); Riedel & Sanfilippo (1970, p. 529, plate 10, fig. 5); Riedel & Sanfilippo (1971, p. 1595).
Lychnocanoma bellum (Clark & Campbell) Foreman (1973, p. 437, plate 1, fig. 17; plate 11, fig. 9).

Lychnocanoma (?) costata Nishimura
Fig. 11b.
Lychnocanoma (?) costata Nishimura (1992, p. 342, plate 6, figs 4–6).

Lychnocanoma lucerna Ehrenberg
Fig. 9k–m.
Lychnocanium lucerna Ehrenberg (1847, p. 55, unnumbered plate, fig. 5).
Remarks. Similar to *Sethochytris triconiscus* in most respects. The shell is more rounded in outline than that of *S. triconiscus*, the feet are latticed, tubular only in their proximal part, terminating in long, solid conical spines. The solid spines are usually conical but in some specimens they are weakly bladed, or rarely, with a few small pores. The thoracic aperture is constricted, about the same diameter as the cephalis, and surrounded by a short hyaline peristome. *Lychnocanoma lucerna* co-occurs with *S. triconiscus* and persists beyond its latest occurrence into the upper Eocene *Podocyrtis (Lampterium) goetheana* Zone.

Lychnocanoma (?) pileus Nishimura
Fig. 11r, see below.
Lychnocanoma (?) pileus Nishimura (1992, p. 344, plate 6, figs 7 and 8; plate 13, fig. 5).

Periphaena tripyramis tripyramis (Haeckel)
Triactis tripyramis Haeckel (1887, p. 432, plate 33, fig. 6).
Triactis tripyramis tripyramis Haeckel, Riedel & Sanfilippo (1970, p. 521, plate 4, fig. 8).
Periphaena tripyramis tripyramis (Haeckel), Sanfilippo & Riedel (1973, p. 523, plate 9, figs 7–9).

Periphaena tripyramis triangula (Sutton)
Phacotriactus triangula Sutton (1896, p. 61).
Periphaena tripyramis triangula (Sutton) Riedel & Sanfilipo (1970, p. 521, plate 4, figs 9 and 10); Sanfilippo & Riedel (1973, p. 523, plate 9, figs 7–9).

Phormocyrtis cubensis (Riedel & Sanfilippo)
Eucyrtidium cubense Riedel & Sanfilippo (1971, p. 1594, plate 7, figs 10 and 11).
Phormocyrtis cubensis (Riedel & Sanfilippo) Foreman (1973, p. 438, plate 7, figs 11, 12 and 14).

Phormocyrtis striata exquisita (Kozlova)
Podocyrtis exquisita Kozlova in Kozlova & Gorbovets (1966, p. 106, plate 17, fig. 2).
Phormocyrtis striata exquisita (Kozlova) Foreman (1973, p. 438, plate 7, figs 1–4, 7–8; plate 12, fig. 5).

Phormocyrtis striata praexquisita Nishimura
Phormocyrtis striata praexquisita Nishimura (1992, p. 346, plate 9, figs 1–3).

Phormocyrtis striata striata Brandt
Phormocyrtis striata Brandt (1935, p. 55, plate 9, fig. 12); Riedel & Sanfilippo (1970, p. 532, plate 10, fig. 7).
Phormocyrtis striata striata Brandt, Foreman (1973, p. 438, plate 7, figs 5, 6 and 9).
Remarks. Whereas the evolutionary transition from *Phormocyrtis striata exquisita* to *P. striata striata* takes place at the lower boundary of the

P. striata striata Zone in the tropics, it occurs earlier at Site 1051, in the *Bekoma bidartensis* Zone. The transition is difficult to determine precisely because of the prolonged co-occurrence of ancestor and descendant. During their overlapping ranges the two taxa are rather common and occur in similar abundance.

Phormocyrtis turgida (Krasheninnikov)
Lithocampe turgida Krasheninnikov (1960, p. 301, plate 3, fig. 17).
Phormocyrtis turgida (Krasheninnikov) Foreman (1973, p. 438, plate 7, fig. 10; plate 12, fig. 6).

Podocyrtis (Lampterium) chalara Riedel & Sanfilippo
Podocyrtis (Lampterium) chalara Riedel & Sanfilippo (1970, p. 535, plate 12, figs 2 and 3); Riedel & Sanfilippo (1978, p. 71, plate 8, fig. 3, text-fig. 3); Sanfilippo *et al.* (1985, p. 697, fig. 30.11).

Podocyrtis (Lampterium) fasciolata Nigrini
Podocyrtis (Podocyrtis) ampla fasciolata Nigrini (1974, p. 1069, plate 1K, figs 1 and 2; plate 4, figs 2 and 3).
Podocyrtis (Lampterium) fasciolata Nigrini, Sanfilippo *et al.* (1985, p. 697, fig. 30.7).
Remarks. Only sporadic, isolated occurrences have been observed.

Podocyrtis (Lampterium) helenae Nigrini
Podocyrtis (Lampterium) helenae Nigrini (1974, p. 1070, plate lL, figs 9–11; plate 4, figs 4 and 5); Sanfilippo *et al.* (1985, p. 698, fig. 30.13).

Podocyrtis (Lampterium) mitra Ehrenberg
Fig. 10a and b
Podocyrtis mitra Ehrenberg (1854, plate 36, fig. B20; 1873, p. 251); *non* Ehrenberg (1875, plate 15, fig. 4); Riedel & Sanfilippo (1970, p. 534, plate 11, figs 5 and 6; 1978, text-fig. 3); Sanfilippo *et al.* (1985, p. 698, fig. 30.10).
Remarks. In the investigated material this species exhibits great variation in the length and breadth of the abdomen. Towards the later part of its range the distal abdominal pores tend to lose their longitudinal alignment.

Podocyrtis (Lampterium) sinuosa Ehrenberg
Podocyrtis sinuosa Ehrenberg (1873, p. 253; 1875, plate 15, fig. 5); Riedel & Sanfilippo (1970, p. 534, plate 11, figs 3 and 4; 1978, text-fig. 3); Sanfilippo *et al.* (1985, p. 698, fig. 30.9).

Podocyrtis (Lampterium) trachodes Riedel & Sanfilippo
Fig. 10c
Podocyrtis (Lampterium) trachodes Riedel & Sanfilippo (1970, p. 535, plate 11, fig. 7; plate 12, fig. 1); Sanfilippo *et al.* (1985, p. 699, fig. 30.14).
Remarks. The rough surface on the abdomen is restricted to the distal half in the forms encountered at Site 1051.

Podocyrtis (Podocyrtopsis) apeza Sanfilippo & Riedel
Fig. 10d
Podocyrtis (Podocyrtopsis) apeza Sanfilippo & Riedel (1992, p. 14, plate 3, figs 13–15).

Podocyrtis (Podocyrtopsis) apeza Sanfilippo & Riedel var. A
Fig. 10g–i
Remarks. The illustrated form precedes *P. (P.) apeza*, with which it has many characters in common.

Podocyrtis (Podocyrtis) papalis Ehrenberg
Fig. 10f
Podocyrtis papalis Ehrenberg (1847, p. 55, fig. 2); Riedel & Sanfilippo (1970, p. 533, plate 11, fig. 1); Sanfilippo & Riedel (1973, p. 531, plate 20, figs 11–14; plate 36, figs 2 and 3).
Remarks. Variable in size. Included are co-occurring small squat forms and forms with very short abdomens.

Podocyrtis (Podocyrtoges) ampla Ehrenberg
Podocyrtis (?) ampla Ehrenberg (1873, p. 248; 1875, plate 16, fig. 7).
Podocyrtis (Podocyrtis) ampla Ehrenberg, Riedel & Sanfilippo (1970, p. 533, plate 12, figs 7 and 8).
Podocyrtis (Podocyrtoges) ampla Ehrenberg, Sanfilippo & Riedel (1992, p. 14, plate 5, fig. 4).
Remarks. Only very rare isolated occurrences have been found.

Podocyrtis (Podocyrtoges) aphorma Riedel & Sanfilippo
Podocyrtis (Lampterium) aphorma Riedel & Sanfilippo (1970, p. 534, plate 11, fig. 2); Sanfilippo & Riedel (1973, p. 532, plate 20, figs 7 and 8).
Podocyrtis (Podocyrtoges) aphorma Riedel & Sanfilippo, Sanfilippo & Riedel (1992, p. 14).

Podocyrtis (Podocyrtoges) diamesa Riedel & Sanfilippo
Podocyrtis (Podocyrtis) diamesa Riedel & Sanfilippo (1970, p. 533 (*pars*), plate 12, fig. 4, *non* figs 5 and 6); Sanfilippo & Riedel (1973, p. 531, plate 20, figs 9 and 10; plate 35, figs 10 and 11).
Podocyrtis (Podocyrtoges) diamesa Sanfilippo & Riedel (1992, p. 14).

Podocyrtis (Podocyrtoges) dorus Sanfilippo & Riedel

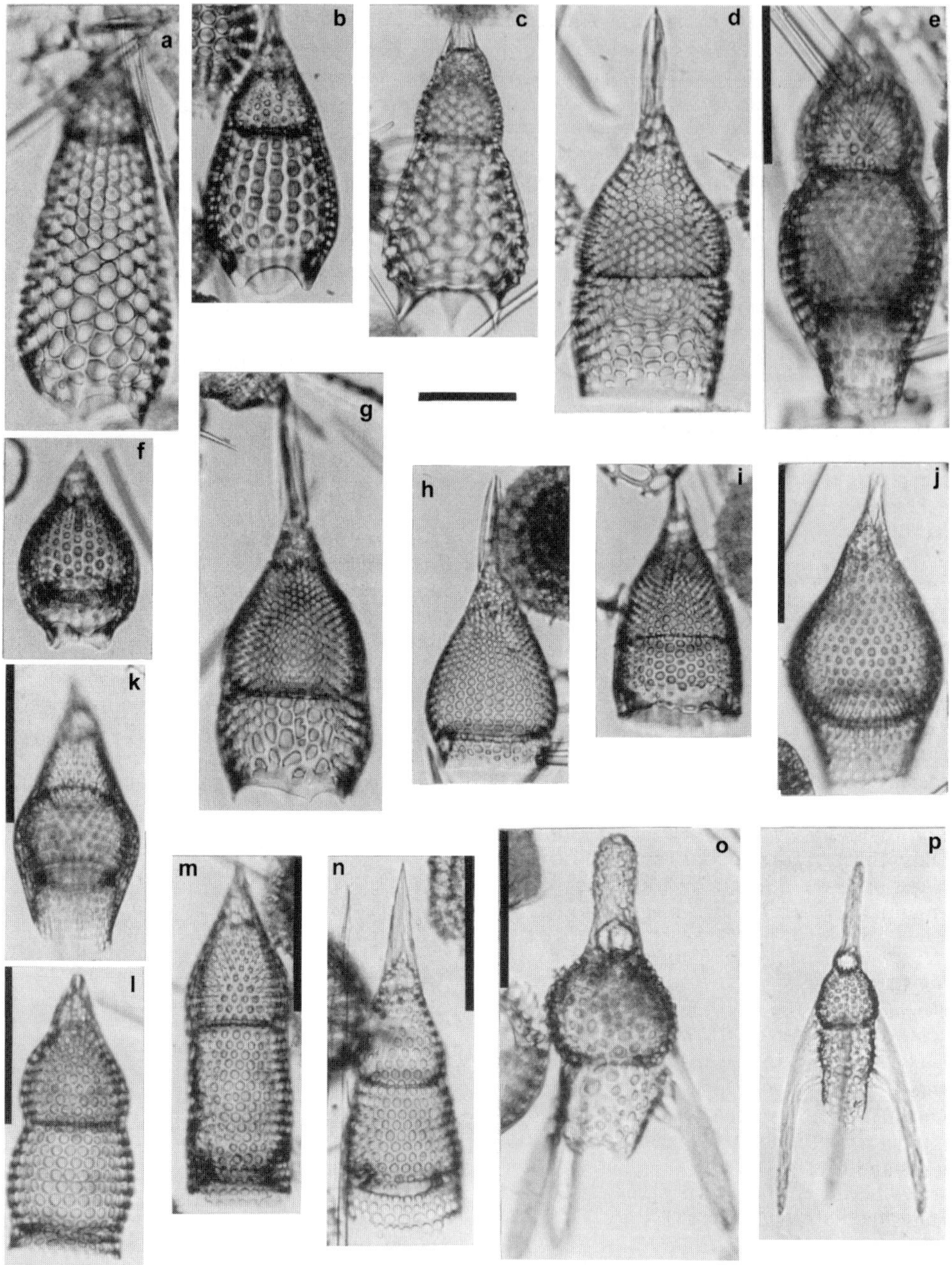

Fig. 10. Scale bars represent 100 μm. For (**a**)–(**d**), (**f**)–(**i**) and (**p**) use scale bar below (**c**). (**a**, **b**) *Podocyrtis (Lampterium) mitra* Ehrenberg. (**a**) 1051A-2H-5, 39–41 Cs.1 M7/0. (**b**) 1051A-25X-5, 37–39 Cs.2 N22/2. (**c**) *Podocyrtis (Lampterium) trachodes* Riedel & Sanfilippo. 1051A-16H-5, 40–42 Cs.2 D26/0. (**d**) *Podocyrtis (Podocyrtopsis) apeza* Sanfilippo & Riedel. 1051A-4H-2, 40–42 Cs.1 G27/0. (**f**) *Podocyrtis (Podocyrtis) papalis* Ehrenberg. 1051A-38X-2, 55–57 Ph.1 O10/4. (**g**–**i**) *Podocyrtis (Podocyrtopsis) apeza* Sanfilippo & Riedel var. A. (**g**) 1051A-10H-2, 40–42 Cs.1 F36/0. (**h**) 1051A-15H-2, 42–44 Cs.2 O29/3. (**i**) 1051A-9H-5, 40–42 Cs.1 T26/1. (**e**, **j**–**n**) *Pterocodon (?) tenellus* Foreman. (**e**) 1051A-53X-5, 36–39 Ph.2 V24/3. (**j**) 1051A-46X-5, 42–44 Sl.2a K21/0. (**k**) 1051A-46X-2, 39–41 Cs.2 Q11/0. (**l**) 1051A-47X-2, 31–33 Sl.2 N6/4. (**m**) 1051A-46X-5, 42–44 Sl.2a C33/2. (**n**) 1051A-45X-2, 51–53 Cs.1 V41/0. (**o**, **p**) *Rhopalocanium ornatum* Ehrenberg. (**o**) 1051A-26X-2, 46–48 Cs.1 G29/4. (**p**) 1051A-5H-5, 39–41 Cs.2 A9/0.

Podocyrtis (Podocyrtis) dorus Sanfilippo & Riedel (1973, p. 531, plate 35, figs 12–14).
Podocyrtis (Podocyrtoges) dorus Sanfilippo & Riedel (1992, p. 14, plate 5, fig. 3).

Pterocodon (?) ampla (Brandt)
Theocyrtis ampla Brandt (1935, p. 56, plate 9, figs 13–15).
Pterocodon (?) ampla (Brandt) Foreman (1973, p. 438, plate 5, figs 3–5).

Pterocodon (?) anteclinata Foreman
Pterocodon (?) anteclinata Foreman (1975, p. 621, plate 9, figs 32–34).

Pterocodon (?) poculum Nishimura
Pterocodon (?) poculum Nishimura (1992, p. 350, plate 8, figs 1–3; plate 13, fig. 13).

Pterocodon (?) tenellus Foreman
Fig. 10e, j–n
Pterocodon (?) tenellus Foreman (1973, p. 439, plate 5, fig. 7; plate 12, fig. 4).

Remarks. Great variation has been observed in this species. Thorax varies from subhemispherical, inflated subhemispherical to conical, and the abdomen from barrel-shaped to inverted, truncate conical. In some specimens the uppermost three segments form a smooth-sided cone, whereas in others, the strictures between the segments on the contour show a curvy outline. A consistent feature is the presence of very uniform, small, circular pores, set in upwardly directed frames arranged quincuncially in transverse and longitudinal rows. The size is variable.

Rhopalocanium ornatum Ehrenberg
Fig. 10o and p
Rhopalocanium ornatum Ehrenberg (1847, fig. 3; 1854, plate 36, fig. 9; 1873, p. 256; 1875, plate 17, fig. 8); Foreman (1973, p. 439, plate 2, figs 8–10; plate 12, fig. 3).
Remarks. The forms encountered herein are unusually well developed with robust, long horns and feet.

Sethochytris triconiscus Haeckel
Fig. 11a
[?] *Sethochytris triconiscus* Haeckel (1887, p. 1239, plate 57, fig. 13); Riedel & Sanfilippo (1970, p. 528, plate 9, figs 5 and 6); Sanfilippo *et al.* (1985, p. 680, fig. 22.1a–d).
Remarks. Observed more frequently and in greater abundance than in low-latitude material. Commonly the tubular feet are very long, terminally closed with a short thorn.

Siphocampe (?) quadrata (Petrushevskaya & Kozlova)
Lithamphora sacculifera quadrata Petrushevskaya & Kozlova (1972, p. 539, plate 30, figs 4–6; plate 24, fig. 7).
Lithomitra docilis Foreman (1973, p. 431, plate 8, fig. 20–22; plate 9, figs 3–5).
Siphocampe (?) quadrata (Petrushevskaya & Kozlova) Nigrini (1977, p. 257, plate 3, fig. 12); Takemura (1992, p. 743, plate 7, fig. 7); Hollis *et al.* (1997, p. 55, plate 4, fig. 27).

Spongatractus balbis Sanfilippo & Riedel
Spongatractus balbis Sanfilippo & Riedel (1973, p. 518, plate 2, figs 1–3; plate 25, figs 1 and 2).
Remarks. At the time of the evolutionary transition of *Spongatractus balbis* to *S. pachystylus*, it is difficult to distinguish the two taxa from each other, because of a combination of heavy development of the spongy meshwork on the ancestor and the effects of dissolution.

Spongatractus pachystylus (Ehrenberg)
Spongosphaera pachystyla Ehrenberg (1873, p. 256; 1875, plate 26, fig. 3).
Spongatractus pachystylus (Ehrenberg) Sanfilippo & Riedel (1973, p. 519, plate 2, figs 4–6; plate 25, fig. 3).

Spongodiscus cruciferus (Clark & Campbell)
Spongasteriscus cruciferus Clark & Campbell (1942, p. 50, plate 1, figs 1–6, 8, 10, 11 and 16–18.
Spongodiscus cruciferus (Clark & Campbell) Sanfilippo & Riedel (1973, p. 524, plate 11, figs 14–17; plate 28, figs 10 and 11).

Spongodiscus phrix Sanfilippo & Riedel
Spongodiscus phrix Sanfilippo & Riedel (1973, p. 525, plate 12, figs 1 and 2; plate 29, fig. 2).

Spongodiscus quartus quartus (Borisenko)
Staurodictya quartus Borisenko (1958, plate 2, fig. 5)
Spongodiscus quartus quartus (Borisenko) Sanfilippo & Riedel (1973, p. 525, plate 12, figs 6 and 7; plate 29, figs 5 and 6).

Spongomelissa adunca Sanfilippo & Riedel
Spongomelissa adunca Sanfilippo & Riedel (1973, p. 529, plate 19, figs 3 and 4; plate 34, figs 1–6).

Spongurus (?) irregularis Nishimura
Spongodiscid gen. et sp. indet in Sanfilippo & Riedel (1973, plate 27, fig. 10).
Spongurus (?) irregularis Nishimura (1992, p. 327, plate 2, figs 7–9; plate 12, figs 3 and 7); Blome (1992, p. 645–646, plate 3, fig. 19).

Stylosphaera coronata coronata Ehrenberg
Stylosphaera coronata Ehrenberg (1873, p. 258; 1875, plate 25, fig. 4).
Stylosphaera coronata coronata Ehrenberg, Sanfilippo & Riedel (1973, p. 520, plate 1,

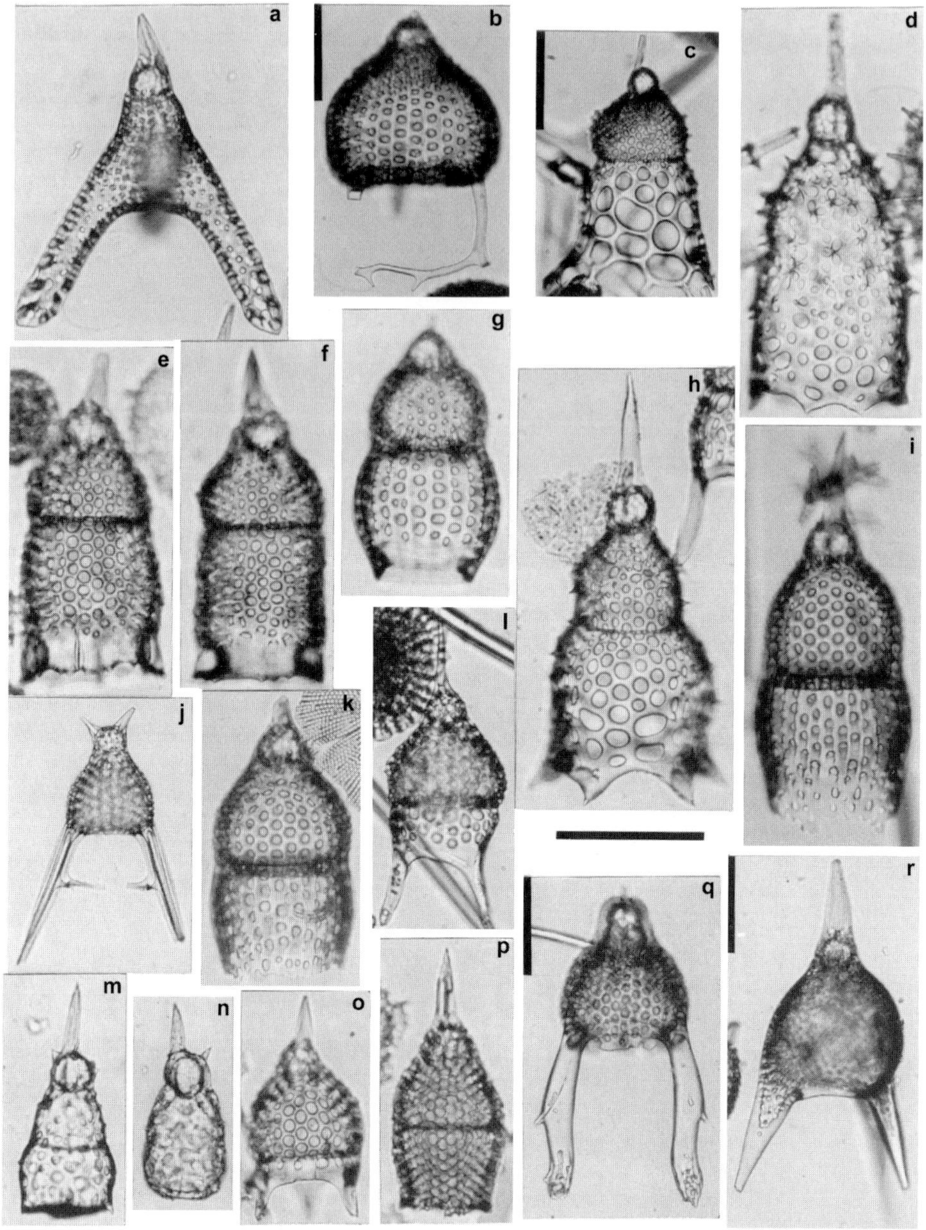

Fig. 11. Scale bars represent 100 μm. For (**a**), (**d**) and (**e**)–(**p**) use scale bar below (**h**). (**a**) *Sethochytris triconiscus* Haeckel. 1051A-14H-5, 40–42 Cs.1 Q34/1. (**b**) *Lychnocanoma (?) costata* Nishimura. 1051A-62X-1, 105-107 Cs.2 F13/1. (**c**) *Theocorys acroria* Foreman. 1051A-42X-2, 40–42 Cs.1 W20/3. (**d, h**) *Theocorys acroria* Foreman var. A. (**d**) 1051A-2H-5, 39–41 Cs.1 P22/4. (**h**) 1051A-35X-5, 36–39 Cs.1 H43/0. (**e, f**) *Theocotyle nigriniae* Riedel & Sanfilippo. (**e**) 1051A-42X-2, 40–42 Sl.2 M15/4. (**f**) 1051A-42X-2, 40–42 Cs.1 F21/4. (**g**) *Theocorys (?) phyzella* Foreman. 1051A-44X-5, 41–44 Sl.1 R36/4. (**i, k**) *Theocorys* (?) aff. *phyzella*. (**i**) 1051A-56X-2, 80–82 Cs.2 U22/0. (**k**) 1051A-56X-2, 80–82 Cs.2 K27/0. (**j**) *Bekoma (?) oliva* Nishimura. 1051A-59X-5, 30–32 Cs.2 C11/1. (**l, o**) *Thyrsocyrtis (Thyrsocyrtis) tarsipes* Foreman. (**l**) 1051A-53X-5, 36–37 Ph.2 D45/3. (**o**) 1051A-48X-2, 47–49 Ph.2 S33/3. (**m, n**) *Theocorys anapographa* Riedel & Sanfilippo var. A. (**m**) 1051A-5H-2, 40–42 Ph.2 D44/0. (**n**) 1051A-5H-2, 40–42 Ph.2 D44/2. (**p**) *Theocotyle nigriniae* Riedel & Sanfilippo var. A. 1051A-24X-5, 40–42 Ph.1 O29/0. (**q**) *Bekoma bidartensis* Riedel & Sanfilippo. 1051A-47X-5, 45–47 Cs.1 G13/0. (**r**) *Lychnocanoma (?) pileus* Nishimura. 1051A-62X-5, 39–41 Cs.2 W28/2.

figs 13–17; plate 25, fig. 4); Nishimura (1992, p. 325, plate 1, fig. 2; plate 11, fig. 9).

Stylosphaera goruna Sanfilippo & Riedel
Stylosphaera goruna Sanfilippo & Riedel (1973, p. 521, plate 1, figs 20–22; plate 25, figs 9 and 10); Nishimura (1992, p. 326, plate 1, fig. 1; plate 11, fig. 8).

Stylotrochus alveatus Sanfilippo & Riedel
Stylotrochus alveatus Sanfilippo & Riedel (1973, p. 525, plate 13, figs 4; plate 30, figs 3 and 4).

Stylotrochus nitidus Sanfilippo & Riedel
Stylotrochus nitidus Sanfilippo & Riedel (1973, p. 525, plate 13, figs 9–14; plate 30, figs 7–10).

Theocorys acroria Foreman
Fig. 11c
Theocorys acroria Foreman (1973, p. 439, plate 5, figs 11–13; plate 12, fig. 2).

Theocorys acroria Foreman var. A
Fig. 11d and h
Remarks. Two- or three-segmented shell similar to *Theocorys acroria* with a conical to campanulate thorax. Cephalis poreless, or with a few small pores, bearing a stout conical horn. The abdomen is broader than the thorax, cylindrical to bell-shaped and widening gradually to terminate in a slightly differentiated pore-bar, which is pulled out into 6–10 flat, triangular points. Thoracic pores are small, subcircular, quincuncially arranged. Collar and lumbar strictures distinct. Abdominal pores increase in size distally to become very large, and are commonly regularly arranged. This species exhibits great variation in length of the segments, presence of thorns on the surface, arrangement of pores on the abdomen and the distal termination. Later forms become two-segmented with a thorny surface, and the pores become irregularly arranged. The stratigraphic range of this species post-dates that of *T. acroria* in the middle Eocene sequence from the lower *Thyrsocyrtis (Pentalacorys) triacantha* Zone into the *Podocyrtis (Lampterium) goetheana* Zone.

Theocorys (?) sp. aff. *T. acroria* Foreman
Theocorys (?) sp. aff. *T. acroria* Foreman, Nishimura (1992, plate 4, fig. 7).
Remarks. Three-segmented shell similar to *Theocorys acroria* with a conical to campanulate thorax with shoulders that do not have thorns. Abdomen widens gradually to become flared, with an undifferentiated margin that is smooth, undulating or scalloped. The large, spherical cephalis is porous and bears an apical and a vertical short horn of variable length. Collar and lumbar stricture moderately distinct. Thorax with small subcircular pores. The abdominal pores are rounded, irregular in size and shape. In the proximal part of the abdomen, just below the lumbar stricture is a row of closely spaced, small pores separating the distal part of the abdomen in which the pores tend to be very large.

Theocorys anaclasta Riedel & Sanfilippo
Theocorys anaclasta Riedel & Sanfilippo (1970, p. 530, plate 10, figs 2 and 3); Riedel & Sanfilippo (1978, p. 76, plate 1, figs 6–8); Sanfilippo *et al.* (1985, p. 683, fig. 24.1a–d).

Theocorys anapographa Riedel & Sanfilippo
Theocorys anapographa Riedel & Sanfilippo (1970, p. 530, plate 10, fig. 4).

Theocorys anapographa Riedel & Sanfilippo var. A
Fig. 11m and n
Remarks. This small hyaline form co-occurs with *Theocorys anapographa* in the later part of its stratigraphic range. The cephalis appears relatively large compared with the small skeleton. The shell is two- or three-segmented, with a few subcircular pores, varying in size, scattered over the surface. Rare completely hyaline forms have been observed. Termination closed or with a very small aperture.

Theocorys (?) phyzella Foreman
Fig. 11g
Theocorys (?) phyzella Foreman (1973, p. 440, plate 5, fig. 8; plate 12, fig. 1).

Theocorys (?) aff. *phyzella*
Fig. 11i and k
Remarks. This species is similar to *Theocorys (?) phyzella* in all respects except for lacking the distal peristome. The pores on the abdomen are less regularly aligned than those in *T. (?) phyzella* and the ridges between the pore rows less distinct. *Theocorys* (?) aff. *phyzella* precedes *T. (?) phyzella*.

Theocotyle conica Foreman
Theocotyle (Theocotyle) cryptocephala (?) conica Foreman (1973, p. 440, plate 4, fig. 11; plate 12, figs 19 and 20).
Theocotyle conica Sanfilippo & Riedel (1982, p. 177, plate 2, fig. 1).

Theocotyle cryptocephala (Ehrenberg)
[?] Eucyrtidium cryptocephalum Ehrenberg (1873, p. 227; 1875, plate 11, fig. 11).
Theocotyle cryptocephala (Ehrenberg) Sanfilippo & Riedel (1982, p. 178, plate 2, figs 4–7).

Theocotyle nigriniae Riedel & Sanfilippo
Fig. 11e and f

Theocotyle cryptocephala (?) nigriniae Riedel & Sanfilippo (1970, p. 525, plate 6, fig. 5 (*non* 6)).
Theocotyle nigriniae Riedel & Sanfilippo, Sanfilippo & Riedel (1982, p. 178, plate 2, figs 1–3).

Theocotyle nigriniae Riedel & Sanfilippo var. A
Fig. 11p
Theocotyle nigriniae Riedel & Sanfilippo, Sanfilippo & Riedel (1982, p. 178 (in part), plate 2, fig. 1); Sanfilippo *et al.* (1985, p. 685 (in part), fig. 25.1b).
Remarks. Previous investigations show that *Theocotyle nigriniae* is accompanied by a markedly smaller form (length (excluding horn) 85–150 μm; length of cephalothorax 65–75 μm; length of abdomen 25–70 μm; length of horn usually 50 μm; breadth of thorax 75–90 μm; breadth of abdomen 60–100 μm). The horn is weakly bladed with a few small thorns at the base. Three small wings have been observed near the lumbar stricture in a few specimens. The abdomen is commonly slightly constricted in its lower half, and often closed terminally by a transverse lattice-plate. This form has been tabulated separately herein.

Theocotyle venezuelensis Riedel & Sanfilippo
Theocotyle venezuelensis Riedel & Sanfilippo (1970, p. 525, plate 6, figs 9 and 10; plate 7, figs 1 and 2); Sanfilippo & Riedel (1982, p. 179, plate 2, figs 8–12); Sanfilippo *et al.* (1985, p. 685, fig. 25.4a–c).

Theocotylissa alpha Foreman
Theocotyle (Theocotylissa) alpha Foreman (1973, p. 441, plate 4, figs 13–15 (*non* 14); plate 12, fig. 16); Foreman (1975, p. 621).
Theocotylissa alpha Foreman, Sanfilippo & Riedel (1982, p. 179, plate 2, figs 16 and 17).
Remarks. Early forms are at the smaller end of the size range for this species.

Theocotylissa auctor (Foreman)
Theocotyle (Theocotylissa) auctor Foreman (1973, p. 441, plate 4, figs 8–10; plate 12, fig. 13).
Theocotylissa auctor (Foreman) Sanfilippo & Riedel (1982, p. 180, plate 2, figs 14 and 15).

Theocotylissa ficus (Ehrenberg)
Eucyrtidium ficus Ehrenberg (1873, p. 228; 1875, plate 11, fig. 19).
Theocotylissa ficus (Ehrenberg), Sanfilippo & Riedel (1982, p. 180, plate 2, figs 19 and 20); Sanfilippo & Nigrini (1998*a*, p. 273, plate 13.2, fig. 21).

Thyrsocyrtis (?) annikae Nishimura
Fig. 7p and q
Thyrsocyrtis (?) annikae Nishimura (1992, p. 356, plate 7, figs 4–6).
Remarks. Size is variable.

Thyrsocyrtis (Thyrsocyrtis) hirsuta (Krasheninnikov)
Podocyrtis hirsutus Krasheninnikov (1960, p. 300, plate 3, fig. 16).
Thyrsocyrtis (Thyrsocyrtis) hirsuta (Krasheninnikov) Sanfilippo & Riedel (1982, p. 173, plate 1, figs 3 and 4).
Remarks. Size is variable, especially the length and robustness of the horn and feet, the length and breadth of the abdomen, and the size of the abdominal pores.

Thyrsocyrtis (Thyrsocyrtis) rhizodon Ehrenberg
Fig. 7k and o
Thyrsocyrtis rhizodon Ehrenberg (1873, p. 262; 1875, p. 94, plate 12, fig. 1); Sanfilippo & Riedel (1982, p. 173, plate 1, figs 14–16; plate 3, figs 12–17).
Remarks. Earlier forms are somewhat smaller than later ones.

Thyrsocyrtis (Thyrsocyrtis) robusta Riedel & Sanfilippo
Thyrsocyrtis hirsuta robusta Riedel & Sanfilippo (1970, p. 526, plate 8, fig. 1).
Thyrsocyrtis (Thyrsocyrtis) robusta Riedel & Sanfilippo, Sanfilippo & Riedel (1982, p. 174, plate 1, fig. 5).

Thyrsocyrtis (Thyrsocyrtis) tarsipes Foreman
Fig. 11l and o
Thyrsocyrtis tarsipes Foreman (1973, p. 442, plate 3, fig. 9; plate 12, fig. 14).
Thyrsocyrtis (Thyrsocyrtis) tarsipes Foreman, Sanfilippo & Riedel (1982, p. 174, plate 1, figs 1 and 2).
Remarks. Rare specimens have been observed in which the abdomen is only one pore row wide.

Thyrsocyrtis (Pentalacorys) tensa Foreman
Thyrsocyrtis hirsuta tensa Foreman (1973, p. 442, plate 3, figs 13–16; plate 12, fig. 8).
Thyrsocyrtis (Pentalacorys) tensa Foreman, Sanfilippo & Riedel (1982, p. 176, plate 1, figs 6 and 7; plate 3, figs 1 and 2).

Thyrsocyrtis (Pentalacorys) triacantha (Ehrenberg)
Podocyrtis triacantha Ehrenberg (1873, p. 254; 1875, plate 13, fig. 4).
Thyrsocyrtis (Pentalacorys) triacantha (Ehrenberg) Sanfilippo & Riedel (1982, p. 176, plate 1, figs 8–10; plate 3, figs 3 and 4).

Velicucullus (?) palaeocenica Nishimura
Velicucullus spp. Sanfilippo & Riedel (1973, p. 530, plate 20, figs 5 and 6; plate 34, fig. 14).
Velicucullus (?) palaeocenica Nishimura (1992, p. 331, plate 3, figs 7 and 9).

We thank the crew, co-chiefs and Scientific Shipboard Party of ODP Leg 171B for their generous collaboration at sea, and ODP for providing the samples. This study greatly benefited from discussions with R. D. Norris, D. Kroon and D. Bukry at several stages. The junior author thanks L. Snee and R. Thompson for their support in travelling to Scripps Institution of Oceanography for the final writing stages of the manuscript. The authors thank J. Barron and J. Self-Trail for their constructive reviews of an earlier version of the manuscript. We also wish to thank J. P. Caulet and C. J. Hollis for their reviews and valuable comments on the manuscript.

References

AUBRY, M.-P. 1995. From chronology to stratigraphy: interpreting the lower and middle Eocene stratigraphic record in the Atlantic Ocean. *In*: BERGGREN, W. A., KENT, D. V., AUBRY, M.-P. & HARDENBOL, J. (eds) *Geochronology, Time Scales and Global Stratigraphic Correlation. A Unified Temporal Framework for a Historical Geology*. SEPM, Special Publications, **54**, 213–274.

BERGGREN, W. A., KENT, D. V., SWISHER, C. C., III & AUBRY, M.-P. 1995. A revised Cenozoic geochronology and chronostratigraphy. *In*: BERGGREN, W. A., KENT, D. V., AUBRY, M.-P. & HARDENBOL, J. (eds) *Geochronology, Time Scales, and Global Stratigraphic Correlation. A Unified Temporal Framework for a Historical Geology*. SEPM, Special Publications, **54**, 129–212.

BLOME, C. D. 1992. Radiolarians from Leg 122, Exmouth and Wombat Plateaus, Indian Ocean. *In*: VON RAD, U., HAQ, B. U. *et al.* (eds) *Proceedings of the Ocean Drilling Program, Scientific Results*, 122. Ocean Drilling Program, College Station, TX, 633–652.

BORISENKO, N. N. 1958. Palaeocene Radiolaria of Western Kubanj. *In*: KRYLOV, A. P. (ed.) *Problems in Geology, Drilling and Exploitation of Wells*. Trudy Vsesoyuznyi Neftegazovyi Nauchno-Issledovalelskii Institut (VNII), Krasnodarskii Filial, 81–100.

BRALOWER, T. J., ZACHOS, J. C., THOMAS, E. *et al.* 1995. Late Palaeocene to Eocene palaeoceanography of the equatorial Pacific Ocean: stable isotopes recorded at Ocean Drilling Program Site 865, Allison Guyot. *Paleoceanography*, **10**(4), 841–865.

BRANDT, R. 1935. Die Mikropalaeontologie des Heiligenhafener, Kieseltones (Ober-Eozan) Radiolarien; Systematik. *In*: WETZEL, E. O. (ed.) *Jahresbericht des Niedersächsischen geologischen Vereins*, **27**, 48–59.

BUKRY, D. 1973. Low latitude coccolith biostratigraphic zonation. *In*: EDGAR, N. T., SAUNDERS, J. B. *et al.* (eds) *Initial Reports of the Deep Sea Drilling Project*, **15**. US Government Printing Office, Washington, DC, 685–703.

BUKRY, D. 1975. Coccolith and silicoflagellate stratigraphy, northwestern Pacific Ocean, Ocean Drilling Project Leg 32. *In*: LARSON, R. L., MOBERLY, R. *et al.* (eds) *Initial Reports of the Deep Sea Drilling Project*, **32**. US Government Printing Office, Washington, DC, 685–677.

CAMPBELL, A. S. 1954. Radiolaria. *In*: MOORE, R. C. (ed.) *Treatise in Invertebrate Paleontology*, Part D, Protista 3. Geological Society of America, Boulder, CO; University of Kansas Press, Lawrence, D11–D195.

CHEN, P. H. 1975. Antarctic Radiolaria. *In*: HAYES, D. E., FRAKES, L. A. *et al.* (eds) *Initial Reports of the Deep Sea Drilling Project*, **28**. US Government Printing Office, Washington, DC, 437–513.

CITA, M. B., NIGRINI, C. A. & GARTNER, S. 1970. Biostratigraphy, Leg 2. *In*: PETERSON, M. N. A., EDGAR, N. T. *et al.* (eds) *Initial Reports of the Deep Sea Drilling Project*, **2**. US Government Printing Office, Washington DC, 391–411.

CLARK, B. L. & CAMPBELL, A. S. 1942. *Eocene Radiolarian Faunas from the Monte Diablo Area, California*. Geological Society of America, Special Papers, **39**, 1–112.

DINKELMAN, M. G. 1973. Radiolarian stratigraphy: Leg 16, Deep Sea Drilling Project. *In*: VAN ANDEL, T. H., HEATH, G. R. *et al.* (eds) *Initial Reports of the Deep Sea Drilling Project*, **16**. US Government Printing Office, Washington, DC, 747–813.

EHRENBERG, C. G.EHRENBERG, C. G. 1847. Über die mikroskopischen kieselschaligen Polycystinen als mächtige Gebirgsmasse von Barbados und über das Verhältniss deraus mehr als 300 neuen Arten bestehenden ganz eigenthümlichen Formengruppe jener Felsmasse zu den jetzt lebenden Thieren und zur Kreidebildung. Eine neue Anregung zur Erforschung des Erdlebens. *Königliche Preussische Akademie der Wissenschaften zu Berlin, Bericht, Jahre 1847*, 40–60.

EHRENBERG, C. G. 1854. *Mikrogeologie. Das Erden und Felsen schaffende Wirken des unsichtbar kleinen selbststandigen Lebens auf der Erde*. Leopold Voss, Leipzig, Germany.

EHRENBERG, C. G. 1873. Grössere Felsproben des Polycystinen-Mergels von Barbados mit weiteren Erläuterungen. *Königliche Preussische Akademie der Wissenschaften zu Berlin, Monatsberichte, Jahre 1873*, 213–263.

EHRENBERG, C. G. 1875. Fortsetzung der mikrogeologischen Studien als Gesammt-Uebersicht der mikroskopischen Paläontologie gleichartig analysirter Gebirgsarten der Erde, mit specieller Rücksicht auf den Polycystinen-Mergel von Barbados. *Königliche Akademie der Wissenschaften zu Berlin, Abhandlungen, Jahre 1875*, 1–225.

FOREMAN, H. P. 1973. Radiolaria of Leg 10 with systematics and ranges for the families Amphipyndacidae, Artostrobiidae, and Theoperidae. *In*: WORZEL, J. L., BRYANT, W. *et al.* (eds) *Initial Reports of the Deep Sea Drilling Project*, **10**. US Government Printing Office, Washington, DC, 407–474.

FOREMAN, H. P. 1975. Radiolaria from the North Pacific, Deep Sea Drilling Project, Leg 32. *In*: LARSON, R. L., MOBERLY, R. *et al.* (eds) *Initial Reports of the Deep Sea Drilling Project*, **32**. US

Government Printing Office, Washington, DC, 579–676.

GOLL, R. M. 1969. Classification and phylogeny of Cenozoic Trissocyclidae (Radiolaria) in the Pacific and Caribbean basins. Part II. *Journal of Paleontology*, **43**, 322–339.

HAECKEL, E. 1862. *Die Radiolarien (Rhizopoda Radiaria). Eine Monographie.* Reimer, Berlin, Germany.

HAECKEL, E. 1881. Entwurf eines Radiolarien-Systems auf Grund von Studien der *Challenger*-Radiolarien (Basis for a radiolarian classification from the study of Radiolaria of the *Challenger* collection). *Jenaische Zeitschrift für Naturwissenschaft*, **15**, 418–472.

HAECKEL, E. 1887. *Report on the Radiolaria collected by H.M.S.* Challenger *during the years 1873–1876.* Report on the Scientific Results of the Voyage of the H.M.S. *Challenger*, Zoology, **18**.

HARDENBOL, J., THIERRY, J., FARLEY, M. B., JACQUIN, TH., DE GRACIANSKY, P.-C. & VAIL, P. R. 1998. Mesozoic and Cenozoic sequence chronostratigraphic framework of European basins. *In*: DE GRACIANSKY, P.-C., HARDENBOL, J., JACQUIN, TH. & VAIL, P. R. (eds) *Mesozoic and Cenozoic Sequence Stratigraphy of European Basins*, SEPM, Special Publications, **60**, charts.

HOLLIS, C. J. 1993. Latest Cretaceous to Late Palaeocene radiolarian biostratigraphy: a new zonation from the New Zealand region. *Marine Micropaleontology*, **21**, 295–327.

HOLLIS, C. J. 1997. *Cretaceous–Palaeocene Radiolaria from Eastern Marlborough, New Zealand.* Institute of Geological and Nuclear Sciences, Monograph, **17**.

HOLLIS, C. J., WAGHORN, D. B., STRONG, C. P. & CROUCH, E. M. 1997. *Integrated Palaeogene biostratigraphy of DSDP Site 277 (Leg 29); foraminifera, calcareous nannofossils, Radiolaria and palynomorphs.* Institute of Geological and Nuclear Sciences, Science Report, **7**.

KATZ, M. E., PAK, D. K., DICKENS, G. R. & MILLER, K. G. 1999. The source and fate of massive carbon input during the Latest Palaeocene Thermal Maximum: new evidence from the North Atlantic Ocean. *Science*, **236**, 1531–1533.

KENNETT, J. P. & STOTT, L. D. 1990. Proteus and Proto-Oceanus: ancestral Palaeogene oceans as revealed from Antarctic stable isotopic results; ODP Leg 113. *In*: BARKER, P. F., KENNETT, J. P. *et al.* (eds) *Proceedings of the Ocean Drilling Program, Scientific Results*, **113**. Ocean Drilling Program, College Station, TX, 865–880.

KENNETT, J. P. & STOTT, L. D. 1991. Abrupt deep-sea warming, palaeoceanographic changes and benthic extinctions at the end of the Palaeocene. *Nature*, **353**, 225–229.

KOZLOVA, G. E. & GORBOVETZ, A. N. 1966. *Radiolarians of the Upper Cretaceous and Upper Eocene deposits of the West Siberian Lowland.* Trudy vsesoyuznogo neftyanogo nauchno-issledovatelskogo geologorazvedochnogo instituta (VNIGRI), **248**, 1–159 [in Russian].

KRASHENINNIKOV, V. A. 1960. Some radiolarians of the Lower and Middle Eocene of the Western Caucasus. *Mineralogicko-Geologicka i Okhrana Nedr SSSR Vsesoyuznogo Nauchno-Issledovatelskogo Geologorazved Neftyanogo Instituta*, **16**, 271–308 [in Russian].

MARTINI, E. 1971. Standard Tertiary and Quaternary calcareous nannoplankton zonation. *In*: FARINACCI, A. (ed.). *Proceedings of the II Planktonic Conference.* Rome, 39–785.

MILLER, K. G. 1992. Middle Eocene to Oligocene stable isotopes, climate and deep water history; the terminal Eocene event? *In*: PROTHERO, D. R. & BERGGREN, W. A. (eds) *Eocene–Oligocene Climatic and Biotic Evolution.* Princeton University Press, Princeton, NJ, 160–177.

MOORE, T. C. 1971. Radiolaria. *In*: TRACEY, J. I. JR, SUTTON, G. H. *et al.* (eds) *Initial Reports of the Deep Sea Drilling Project*, **8**. US Government Printing Office, Washington, DC, 727–775.

MOORE, T. C., SHACKLETON, N. J. & PISIAS, N. G. 1993. Palaeoceanography and the diachrony of radiolarian events in the eastern equatorial Pacific. *Paleoceanography*, **8**(5), 567–586.

NIGRINI, C. 1974. Cenozoic Radiolaria from the Arabian Sea, DSDP Leg 23. *In*: DAVIES, T. A., LUYENDYK, B. P. *et al.* (eds) *Initial Reports of the Deep Sea Drilling Project*, **23**. US Government Printing Office, Washington, DC, 1051–1121.

NIGRINI, C. 1977. Tropical Cenozoic Artostrobiidae (Radiolaria). *Micropaleontology*, **23**, 241–269.

NIGRINI, C. & SANFILIPPO, A. 2000. Palaeogene Radiolaria from Sites 998, 999 and 1001 in the Caribbean. ODP Leg 165. *In*: LECKIE, R. M., SIGURDSSON, M., ACTON, G. D. & DRAPER, G. (eds) *Proceedings of the Ocean Drilling Program, Scientific Results*, **165**. Ocean Drilling Program, College Station, TX, 57–81.

NISHIMURA, A. 1987. Cenozoic Radiolaria in the western North Atlantic, Site 603, Leg 93 of the Deep Sea Drilling Project. *In*: VAN HINTE, J. E., WISE, S. W. JR *et al.* (eds) *Initial Reports of the Deep Sea Drilling Project*, **93**. US Government Printing Office, Washington, DC, 713–737.

NISHIMURA, A. 1992. Palaeocene radiolarian biostratigraphy in the northwest Atlantic at Site 384, Leg 43, of the Deep Sea Drilling Project. *Micropaleontology*, **38**, 317–362.

NORRIS, R. D. & RÖHL, U. 1999. Carbon cycling and chronology of climate warming during the Palaeocene–Eocene transition. *Nature*, **401**, 775–778.

NORRIS, R. D., KLAUS, A. & KROON, D. 2001. Mid-Eocene deep water, the Late Palaeocene Thermal Maximum and continental slope mass wasting during the Cretaceous–Palaeogene impact. *This volume.*

NORRIS, R. D., KROON, D., KLAUS, A. *et al.* (eds) 1998. *Proceedings of the Ocean Drilling Program, Initial Reports*, **171B**. Ocean Drilling Program, College Station, TX.

O'CONNOR, B. 1999. Radiolaria from the late Eocene Oamaru diatomite, South Island, New Zealand. *Micropaleontology*, **45**, 1–55.

OKADA, H. & BUKRY, D. 1980. Supplementary modification and introduction of code numbers to the low latitude coccolith biostratigraphic zonation (Bukry 1973; 1975). *Marine Micropaleontology*, **5**, 321–325.

PAK, D. K. & MILLER, K. G. 1992. Palaeocene to Eocene benthic foraminiferal isotopes and assemblages: implications for deep water circulation. *Paleoceanography*, **7**, 405–422.

PALMER, A. A. 1987. Cenozoic radiolarians from Deep Sea Drilling Project sites 612 and 613 (Leg 95, New Jersey Transect) and Atlantic Slope Project Site ASP 15. *In*: POAG, C. W., WATTS, A. B. *et al.* (eds) *Initial Reports of the Deep Sea Drilling Project*, **95**. US Government Printing Office, Washington, DC, 339–357.

PETRUSHEVSKAYA, M. G. 1971. On the natural system of polycystine Radiolaria (Class Sarcodina). *In*: FARINACCI, A. (ed.) *Proceedings of the II Planktonic Conference, Roma, 1970*. Edizioni Tecnoscienza, Rome, 981–992.

PETRUSHEVSKAYA, M. G. & KOZLOVA, G. E. 1972. Radiolaria: Leg 14, Deep Sea Drilling Project. *In*: HAYES, D. E., PIMM, A. C. *et al.* (eds) *Initial Reports of the Deep Sea Drilling Project*, **14**. US Government Printing Office, Washington, DC, 495–648.

POAG, C. W. & LOW, D. 1987. Unconformable sequence boundaries at Deep Sea Drilling Project Site 612, New Jersey Transect: their characteristics and stratigraphic significance. *In*: POAG, C. W. & WATTS, A. B. (eds) *Initial Reports of the Deep Sea Drilling Project*, **16**. US Government Printing Office, Washington, DC, 453–498.

RIEDEL, W. R. 1967. Subclass Radiolaria. *In*: HARLAND, W. B., HOLLAND, C. H., HOUSE, M. R. *et al.* (eds) *The Fossil Record. A Symposium with Documentation*. Geological Society, London, 291–298.

RIEDEL, W. R. & FOREMAN, H. P. 1961. Type specimens of North American Paleozoic Radiolaria. *Journal of Paleontology*, **35**(3), 628–632.

RIEDEL, W. R. & SANFILIPPO, A. 1970. Radiolaria, Leg 4, Deep Sea Drilling Project. *In*: BADER, R. G., GERARD, R. D. *et al.* (eds). *Initial Reports of the Deep Sea Drilling Project*, **4**. US Government Printing Office, Washington, DC, 503–575.

RIEDEL, W. R. & SANFILIPPO, A. 1971. Cenozoic Radiolaria from the western tropical Pacific, Leg 7. *In*: WINTERER, E. L., RIEDEL, W. R. *et al.* (eds) *Initial Reports of the Deep Sea Drilling Project*, **7**. US Government Printing Office, Washington, DC, 1529–1672.

RIEDEL, W. R. & SANFILIPPO, A. 1973. Cenozoic Radiolaria from the Caribbean, Deep Sea Drilling Project, Leg 15. *In*: EDGAR, N. T., SAUNDERS, J. B. *et al.* (eds) *Initial Reports of the Deep Sea Drilling Project*, **15**. US Government Printing Office, Washington, DC, 705–751.

RIEDEL, W. R. & SANFILIPPO, A. 1978. Stratigraphy and evolution of tropical Cenozoic radiolarians. *Micropaleontology*, **24**, 61–96.

SANFILIPPO, A. & HULL, D. M. 1999. Upper Palaeocene–lower Eocene radiolarian biostratigraphy of the San Francisco de Paula Section, western Cuba: regional and global comparisons. *Micropaleontology*, **45**, supplement 2, 57–82.

SANFILIPPO, A. & NIGRINI, C. 1998*a*. Upper Palaeocene–Lower Eocene deep-sea radiolarian stratigraphy and the Palaeocene–Eocene series boundary. *In*: AUBRY, M.-P., LUCAS, S. G. & BERGGREN, W. A. (eds) *Late Palaeocene–Early Eocene Climatic and Biotic Events in the Marine and Terrestrial Records*. Columbia University Press, New York, 244–276.

SANFILIPPO, A. & NIGRINI, C. 1998*b*. Code numbers for Cenozoic low latitude radiolarian biostratigraphic zones and GPTS conversion tables. *Marine Micropaleontology*, **33**, 109–156.

SANFILIPPO, A. & RIEDEL, W. R. 1973. Cenozoic Radiolaria (exclusive of theoperids, artostrobiids and amphipyndacids) from the Gulf of Mexico, DSDP Leg 10. *In*: WORZEL, J. L., BRYANT, W. *et al.* (eds) *Initial Reports of the Deep Sea Drilling Project*, **10**. US Government Printing Office, Washington, DC, 475–611.

SANFILIPPO, A. & RIEDEL, W. R. 1974. Radiolaria from the west–central Indian Ocean and Gulf of Aden, DSDP Leg 24. *In*: FISHER, R. L., BUNCE, E. T. *et al.* (eds) *Initial Reports of the Deep Sea Drilling Project*, **24**. US Government Printing Office, Washington, DC, 997–1035.

SANFILIPPO, A. & RIEDEL, W. R. 1976. Radiolarian occurrences in the Caribbean Region. *Publication de la VII conference géologique des Caraibes du 30 juin au 12 juillet 1974 (VII Conference Géologique des Caraibes, Cayenne, Département Français de la Guyane*), 145–168.

SANFILIPPO, A. & RIEDEL, W. R. 1979. Radiolaria from the northeastern Atlantic Ocean, DSDP Leg 48. *In*: MONTADERT, L., ROBERTS, D. G. *et al.* (eds) *Initial Reports of the Deep Sea Drilling Project*, **48**. US Government Printing Office, Washington, DC, 493–511.

SANFILIPPO, A. & RIEDEL, W. R. 1982. Revision of the radiolarian genera *Theocotyle, Theocotylissa* and *Thyrsocyrtis*. *Micropaleontology*, **28**, 170–188.

SANFILIPPO, A. & RIEDEL, W. R. 1992. The origin and evolution of Pterocorythidae (Radiolaria): a Cenozoic phylogenetic study. *Micropaleontology*, **38**, 1–36.

SANFILIPPO, A., WESTBERG-SMITH, M. J. & RIEDEL, W. R. 1985. Cenozoic Radiolaria. *In*: BOLLI, H. M., SAUNDERS, J. B. & PERCH-NIELSEN, K. (eds) *Plankton Stratigraphy*. Cambridge University Press, Cambridge, 631–712.

SHIPBOARD SCIENTIFIC PARTY 1998. Site 1051. *In*: NORRIS, R. D., KROON, D., KLAUS, A. *et al.* (eds) *Proceedings of the Ocean Drilling Program, Initial Reports*, **171B**. Ocean Drilling Program, College Station, TX, 171–239.

SIGURDSSON, H., LECKIE, R. M., ACTON, G. D. *et al.* (eds) 1997. *Proceedings of the Ocean Drilling Program, Initial Reports*, **165**. Ocean Drilling Program, College Station, TX.

STOTT, L. D. & KENNETT, J. P. 1990. Antarctic Palaeogene planktonic foraminifer biostratigraphy: ODP Leg 113, Sites 689 and 690.

In: BARKER, P. F., KENNETT, J. P. *et al.* (eds) *Proceedings of the Ocean Drilling Program, Scientific Results*, **113**. Ocean Drilling Program, College Station, TX, 549–569.

STRONG, C. P., HOLLIS, C. J. & WILSON, G. J. 1995. Foraminiferal, radiolarian and dinoflagellate biostratigraphy of Late Cretaceous to Middle Eocene pelagic sediments (Muzzle Group), Mead Stream, Marlborough, New Zealand. *New Zealand Journal of Geology and Geophysics*, **38**, 171–212.

SUTTON, H. J., 1896. Radiolaria: a new genus from Barbados. *American Monthly Microscopical Journal*, **17**(194), 61–62.

TAKEMURA, A. 1992. Radiolarian Palaeogene biostratigraphy in the southern Indian Ocean, Leg 120. *In*: WISE, S. W. JR, SCHLICH, R. *et al.* (eds) *Proceedings of the Ocean Drilling Program, Scientific Results*, **120**. Ocean Drilling Program, College Station, TX, 735–756.

TAKEMURA, A. & LING, H.-Y. 1997. Eocene and Oligocene radiolarian biostratigraphy from the Southern Ocean: correlation of ODP Legs 114 (Atlantic Ocean) and 120 (Indian Ocean). *Marine Micropaleontology*, **30**, 97–116.

THOMAS, E. 1992. Cenozoic deep-sea circulation: evidence from deep-sea benthic foraminifera. *In*: KENNETT, J. P. & WARNKE, D. A. (eds) *Antarctic Palaeoenvironment: a Perspective on Global Change, Part One*, Antarctic Research Series, American Geophysical Union, **56**, 141–165.

VALENTINE, P. C. 1987. Lower Eocene calcareous nannofossil biostratigraphy beneath the Atlantic Slope and Upper Rise off New Jersey—new zonation based on Deep Sea Drilling Project Sites 612 and 613. *In*: POAG, C. W., WATTS, A. B. *et al.* (eds) *Initial Reports of the Deep Sea Drilling Project*, **95**. US Government Printing Office, Washington, DC, 359–394.

ZACHOS, J. C., LOHMANN, K. C., WALKER, J. C. & WISE, S. W. 1993. Abrupt climate change and transient climates in the Palaeogene: a marine perspective. *Journal of Geology*, **100**, 191–213.

Mid- to Late Eocene organic-walled dinoflagellate cysts from ODP Leg 171B, offshore Florida

CAROLINE A. VAN MOURIK[1], HENK BRINKHUIS[2] & GRAHAM L. WILLIAMS[3]

[1]*Department of Geology and Geochemistry, Stockholm University, S-106 91 Stockholm, Sweden (e-mail: Caroline.vanMourik@geo.su.se)*

[2]*Laboratory of Palaeobotany and Palynology, Utrecht University, Budapestlaan 4, NL-3584, CD Utrecht, The Netherlands*

[3]*Geological Survey of Canada (Atlantic), Bedford Institute of Oceanography, P.O. Box 1006, Dartmouth, N.S., Canada B2Y 4A2*

Abstract: The well-calibrated mid- to late Eocene sediment record of ODP Leg 171B (Site 1053A, Blake Nose) allows a detailed stratigraphic and palaeoenvironmental analysis of the dinoflagellate cyst (dinocyst) content. The recovered assemblages are a mixture of inner neritic, outer neritic and oceanic species. The autochthonous dinoflagellates, principally those of the *Impagidinium* group, indicate an oceanic milieu, with possibly some shallowing of water depth towards the top of the section. This trend is also indicated by a corresponding increase of inner neritic dinocysts. The close agreement in the abundance peaks of inner neritic dinocysts and terrestrial palynomorphs indicats that both are allochthonous. This is confirmed by the much higher number of neritic species found in JOIDES Holes 1 and 2, on the continental shelf of eastern Florida, immediately to the west of the Blake Nose. Lower-latitude species found in Hole 1053A, but not occurring at higher latitudes during late Eocene time, are *Diphyes colligerum* and *Thalassiphora delicata*. The presence of these, and other lower-latitude species, confirms that warmer-water conditions persisted during mid- to early late Eocene time in the vicinity of Site 1053. Eighteen new taxa are described, two of them formally: *Charlesdowniea proserpina* sp. nov. and *Oligosphaeridium anapetum* sp. nov.

Ocean Drilling Program (ODP) Leg 171B, Site 1053 is situated in the western North Atlantic, in 1629.5 m water depth on the Blake Nose, offshore of eastern Florida, USA (29°59′ N, 76°31′ W; Fig. 1). Two holes were drilled, 1053A and 1053B. Our samples were taken from 2.74 m below sea floor (mbsf) to 180.22 mbsf in Hole 1053A. A detailed description of Site 1053 has been provided by the Shipboard Scientific Party (1998*a*).

The primary ODP objective at Site 1053 was to increase knowledge of deeper-water conditions in Eocene time. The middle to upper Eocene sediments, identified by multichannel seismic survey at this site, are known to have been deposited in water depths close to those of today, that is, at about 1630 m (Shipboard Scientific Party 1998*a*). As this is equivalent to the depth of modern intermediate waters in the North Atlantic, the site should provide some insight into deeper-water circulation in that part of the Atlantic in Eocene time. Another major advantage was the thickness of the upper Eocene section, as identified on the seismic profile. The thick Eocene section drilled at Site 1053 provides a detailed record of the evolution of mid- to late Eocene palaeoclimates (Shipboard Scientific Party 1998*a*). The deep–water sediments of Site 1053 and their excellent calibration offer a unique opportunity to analyse the organic-walled dinoflagellate cyst (dinocyst) distribution and to assess if they are useful for age determination in oceanic environments.

JOIDES Holes 1 and 2 drilled in 1965 on the continental shelf of offshore eastern Florida, due west of the Blake Plateau (Schlee & Gerard 1965; Fig. 1), also encountered Eocene sediments, which contain rich dinoflagellate assemblages. One of the authors (G.L.W.) and the late W. Drugg studied dinocysts from this material but did not publish any data. We include some of these JOIDES data in this paper, as several of the species also occur in Hole 1053A.

From: KROON, D., NORRIS, R. D. & KLAUS, A. (eds). 2001 *Western North Atlantic Palaeogene and Cretaceous Palaeoceanography*. Geological Society, London, Special Publications, **183**, 225–251.

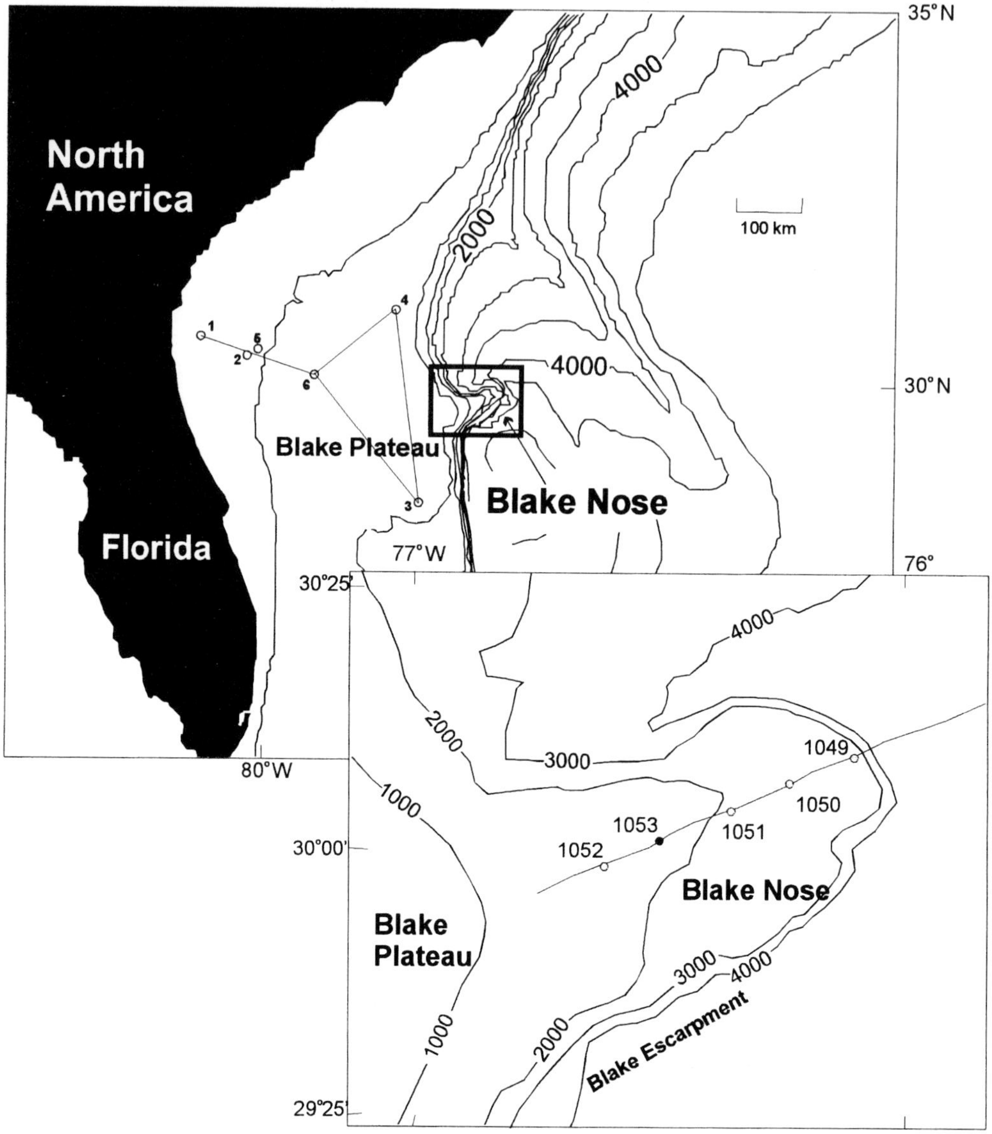

Fig. 1. Location map showing proximity of Hole 1053A to the other Sites of Leg 171B (Shipboard Scientific Party 1998*b*) and the JOIDES Holes (Schlee & Gerard 1965).

Material and methods

Material

The 183.2 m thick section retrieved from Hole 1053A consists of upper middle to upper Eocene deposits. The Shipboard Scientific Party (1998*a*) described one lithological unit with four subunits (Fig. 2). The uppermost, Subunit 1A, is a pale brown foraminiferal ooze with manganese oxide nodules. This unit is only 5 cm thick. Subunit 1B is a pale yellow nannofossil ooze with siliceous microfossils, with a total thickness of 13.1 m. Subunit 1C is likewise a siliceous nannofossil ooze, but is greenish grey, and is the thickest of the four subunits at 125.81 m. The hole bottoms in Subunit 1D, a siliceous nannofossil chalk that is 44.2 m thick.

JOIDES Hole 1 penetrated 73.7 m of upper Eocene sediments between 108.2 and 189.7 m. Hole 2 penetrated 149.4 m of upper Eocene sediments between 149.4 and 298.8 m (Schlee & Gerard 1965).

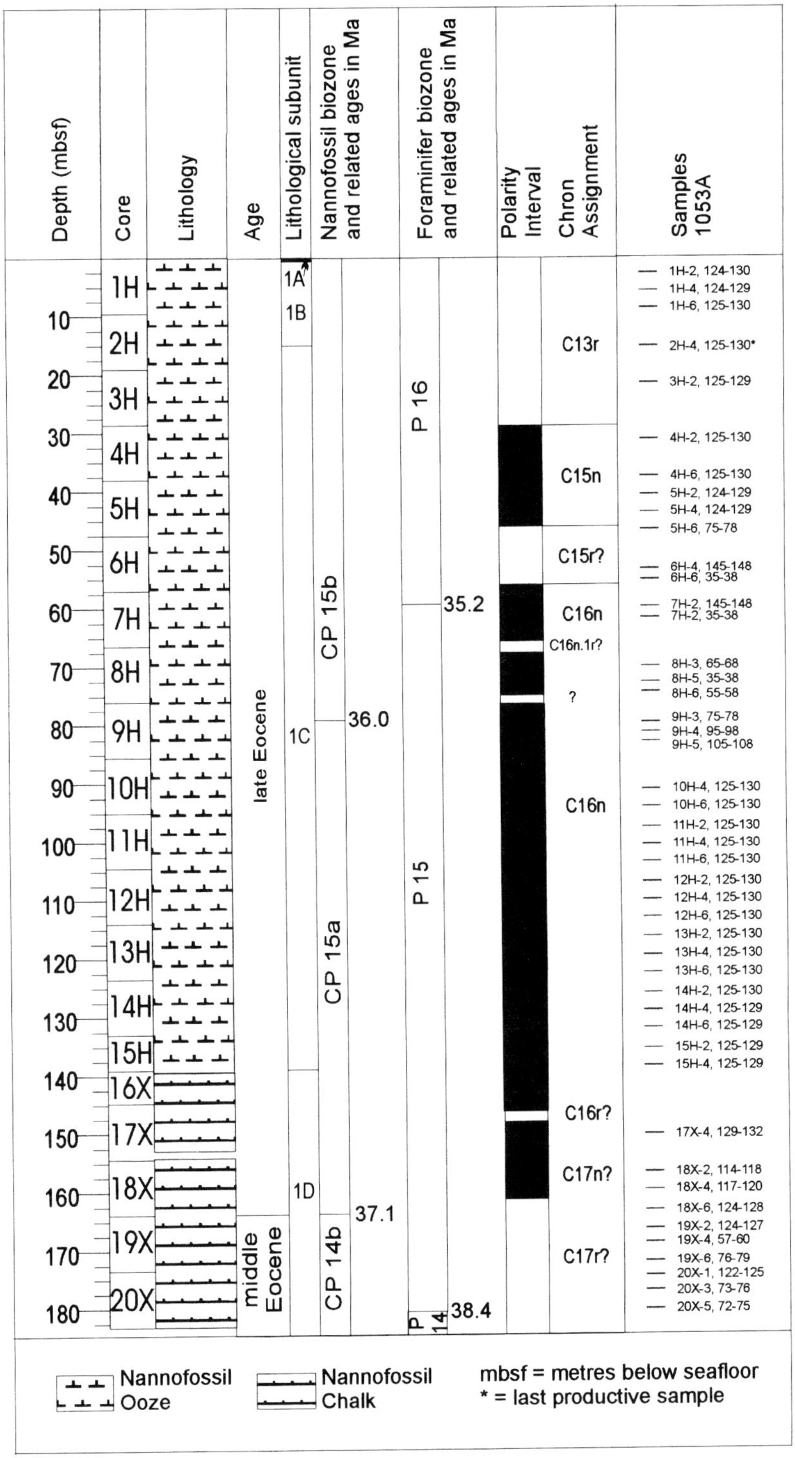

Fig. 2. Stratigraphy of Hole 1053A, showing lithology, lithological subunits, planktonic foraminiferal and nannofossil zonation, polarity and palaeomagnetic subdivisions and samples used in this study. From Shipboard Scientific Party (1998*a*).

Age assessment

Fossil groups previously studied from Hole 1053A include planktonic and benthic foraminifera and calcareous nannofossils (Fig. 2; Shipboard Scientific Party 1998*a*). According to the nannofossil data, the upper 163 m of sediments are of late Eocene age: the top 76 m are assigned to Subzone CP15b (NP19–20) and 86–164 mbsf are assigned to Subzone CP15a (NP18). The lowermost 20 m are placed in the middle Eocene Zone CP14b (NP17). This determination is qualified with the statement that it is difficult to define the mid–upper Eocene boundary because of reworking of the middle Eocene nannofossil marker species *Chiasmolithus grandis*. The planktonic foraminifera data confirm that the sequence spans the middle and upper Eocene zones. The following subdivisions are recognized: 0–59.20 mbsf, upper Eocene Zone P16; 62.2–181.2 mbsf, upper Eocene Zone P15; section 6 of the bottom core, 20, is included in the middle Eocene Zone P14. According to the Shipboard Scientific Party (1998*a*), the top of Zone P14 was barely penetrated in Holes 1053A and 1053B. Except for the absence of *P. semivoluta*, the assemblage from the interval is very similar to the assemblage from lowermost Zone P15. The ages of the zones, based on Berggren *et al.* (1995), are given in Fig. 2.

Shipboard palaeomagnetic results were not unambiguous in terms of assignment of polarity zones but were partially improved post cruise by analysing discrete samples. The polarity chron assignments are based on the polarity zone patterns, the nannofossil biostratigraphy and the chronostratigraphy of Berggren *et al.* (1995). Hole 1053A is interpreted to span Chrons C13r–C17r, with a possible condensation near the middle to upper Eocene boundary (Chron C17n; Shipboard Scientific Party 1998*a*).

Processing

To remove the carbonates, we placed each sample (about 5 g) in 30 % hydrochloric acid (HCl) for about 7 h. After decanting the acid and washing in distilled water, we added hydrofluoric acid (HF) and agitated the sample for 2 h to remove the silicates. After another 7 h, we decanted the HF and repeated the HCl and HF treatment, followed by a wash in HCl. We sieved the residues through a 15 μm mesh sieve and then gave 10–20 s in an ultrasonic bath. We stained the samples for 10 s with a 10 g/l Safranin-O solution 1 : 20 ethanol to distilled water and rewashed. The stained residue was transferred to a glass tube and centrifuged for 5 m at 2500 r.p.m. After transferring the residue to a glass vial we prepared two slides, using glycerine jelly as the mounting medium. Only one slide was made from samples 1053A-8H-3, 65–68 cm, and 1053A-10H-4, 125–130 cm, because of insufficient residue.

Counting procedures

For the purpose of this study, we counted 46 samples, with a minimum of 100 palynomorphs from each. The palynomorphs were divided into the seven categories: bisaccate pollen, other pollen and spores, organic inner walls of foraminifera (foram linings), prasinophyte algae, acritarchs, indeterminable dinocysts and determinable dinocysts. Within the determinable dinocysts we identified individual species. The only exceptions were most of the species of *Operculodinium* and *Spiniferites*, which we grouped as genera, and all the species of *Nematosphaeropsis* and *Cannosphaeropsis*, which we grouped together. We counted broken but identifiable parts as a half, quarter or eighth of a specimen. If possible, we counted 200 dinocysts in each sample. The remainder of the slides was scanned for taxa not included in the counts. Additional taxa are marked in the distribution chart by an asterisk.

Dinocyst taxonomy follows Williams *et al.* (1998*a*). We formally describe two new species and provide brief notes on 16 informal taxa.

Images

Light microscope photographs were made using a JVC TK-C1380E digital colour video camera attached to a Leica DMLP polarized light microscope and connected to a Leica Q500 Imaging Workstation. The Leica QWIN image analysis program was used to digitally capture the images. Using this program, we could enhance brightness and contrast of the images and add a scale bar before the images were saved. We saved the images as JPEG files. In the caption of the images, we state the sample, slide and England Finder coordinates.

All material is curated at the Department of Geology and Geochemistry, Stockholm University, Sweden.

Previous Late Eocene deep-water dinocyst studies

There are few published papers on Eocene dinocyst assemblages from deeper-water palaeoenvironments. Studies based on lower-latitude pelagic sections are principally available from Italy (Biffi & Manum 1988; Brinkhuis 1992, 1994; Brinkhuis & Biffi 1993).

Brinkhuis & Biffi (1993) proposed nine dinoflagellate zones for the pelagic Eocene–Oligocene strata of central Italy. At the Massignano section (the Eocene–Oligocene Global Stratotype and Stratigraphic Point (GSSP) section), they recognized four zones within the upper Eocene (Priabonian) sequence. These span the planktonic foraminiferal zones P15 (in part), P16, P17 and P18 (the lowermost part). In ascending order the zones were: the *Melitasphaeridium pseudorecurvatum* Zone, the *Schematophora speciosa* Zone, the *Cordosphaeridium funiculatum* Zone and the *Achomosphaera alcicornu* Zone. The top of the *Achomosphaera alcicornu* Zone is defined by the first occurrence (FO) of *Glaphyrocysta semitecta* in the central Tethys.

Within earliest Oligocene time, as established at the GSSP site, Brinkhuis & Biffi (1993) recognized, in ascending order, the *Glaphyrocysta semitecta* Zone, the *Areosphaeridium diktyoplokum* Zone, the

Reticulatosphaera actinocoronata Zone and the *Corrudinium incompositum* Zone.

Unfortunately, studies from mid- to high-latitude Eocene to Oligocene deep-water settings are rather scarce (Manum 1976; Costa & Downie 1979; Goodman 1983; Head & Norris 1989; Damassa *et al.* 1990), and poorly calibrated. Therefore, although Eocene–Oligocene taxa have been documented from such sections, their chronostratigraphic usefulness is poorly known.

Palaeoecology

Palaeoecological data on deep-sea fossil dinocysts are sparse. Brinkhuis (1992 1994) analysed the Eocene–Oligocene boundary beds of the Priabonian type area, northern Italy. He grouped the dinocysts into four groups: the *Spiniferites*, *Operculodinium*, *Homotryblium* and *Deflandrea* groups. The first three are gonyaulacaceans, nearly all of which are today autotrophic; the *Deflandrea* group includes peridiniaceans. The *Spiniferites* group seemed to be characteristic of open-marine neritic water masses. Davey & Rogers (1975) and Wall *et al.* (1977) have shown that the concentrations of this group in Recent sediments can reach high concentrations offshore, as a result of transportation. Extant taxa assigned to the *Operculodinium* group occur in oceanic to restricted marine environments. Brinkhuis (1994) considered this group to typify restricted marine to open-marine neritic water masses.

Homotryblium is an extinct genus with a tabulation similar to the extant genus *Pyrodinium*. The cyst equivalent of *Pyrodinium*, *Polysphaeridium*, is abundant in lower latitudes and predominantly in restricted marine to inner-marine environments. At present *Homotryblium* is generally assumed to be most abundant in restricted marine environments such as lagoons. Williams & Bujak (1977) were the first to propose that high relative abundances of *Homotryblium* indicate hypersaline, lagoonal conditions. Their findings were supported by the studies of Köthe (1990) and Brinkhuis (1992, 1994). As shown by Brinkhuis, the distribution patterns of *Homotryblium* at Priabona and Bressana (NE Italy), when compared with the palaeoenvironmental interpretations of Setiawan (1983), support the model of an inner neritic, probably lagoonal habitat. The *Deflandrea* group, although extinct, may be heterotrophic, as are the living protoperidinians, which are also marine. High concentrations of protoperidinioids are characteristic of areas of enriched nutrient supply such as zones of upwelling. Brinkhuis (1994) tentatively linked high concentrations of the *Deflandrea* group to such environments.

Besides the four groups, Brinkhuis also grouped other taxa that he considered to occupy predictable ecological niches. The *Areoligera* complex, taken to also include *Glaphyrocysta*, seems to be characteristic of marginal marine to inner neritic water masses in low to mid-latitudes, sometimes where there are carbonate build-ups. *Areosphaeridium* species, especially *Areosphaeridium diktyoplokum*, are interpreted as indicating outer neritic environments, especially in higher latitudes.

In the proposed model for dinocyst distribution, Brinkhuis (1994) showed high concentrations of *Impagidinium*, *Nematosphaeropsis* and *Cannosphaeropsis* in oceanic environments. This is in accordance with findings based on modern sediments by Wall *et al.* (1977) and Neogene assemblages by Head *et al.* (1989). Indeed, Wall *et al.* (1977, p. 151) noted that all five species of *Impagidinium* (as *Leptodinium*) were restricted to the outermost continental shelf, slope, rise and abyssal sediments. The minimum water depth in which the genus occurs is 100 m.

Recent studies of Oligocene dinocysts from central Italy by two of us (H.B. and G.L.W.) suggest that several species of *Hystrichokolpoma* occur most commonly in deeper-water environments, such as in slope or abyssal depths.

Temperature control of dinocyst assemblages is crucial to understanding fluctuations in species occurrences. Brinkhuis & Biffi (1993), in their study of the Massignano and Monte Cagnero Eocene–Oligocene boundary sections, identified both low- and high-latitude dinocyst species. The latter included *Glaphyrocysta semitecta*, *Corrudinium incompositum*, *Achomosphaera alcicornu*, *Impagidinium velorum* and *Rottnestia borussica*. Lower-latitude taxa include *Deflandrea granulata*, *Hemiplacophora semilunifera*, *Homotryblium* spp., *Impagidinium brevisulcatum*, *I. dispertitum*, *I. maculatum*, *Schematophora speciosa* and *Systematophora ancyrea*.

Results

Throughout Hole 1053A dinocysts are more numerous than terrestrial palynomorphs. In 75% of the 46 samples up to 100 palynomorphs (spores, pollen, acritarchs and dinocysts) could be counted (Table 1). Five samples contained less than ten palynomorphs (Table 1). About 200 dinocysts could be counted in two-thirds of the samples (Fig. 3). In the remaining, numbers vary between six and 179, except for the uppermost Core, 1053A-1H, which is barren. This interval coincides with lithological Subunits 1A and 1B, the pale brown foraminiferal ooze and the pale yellow nannofossil ooze. The most productive samples are from Subunit 1C, the light greenish grey, siliceous nannofossil ooze. In the lower part of Hole 1053A, Subunit 1D, the nannofossil chalk, palynomorph abundances are lower and we were not always able to generate counts of 100 palynomorphs or 200 dinocysts (Table 1).

Diatom and radiolarian fragments occur in all samples, except in Core 1053A-1H. These fragments presumably survived because they are pyritized, which protects them from the hydrochloric and hydrofluoric acids used to concentrate the palynomorphs (Batten 1996).

Biostratigraphy

The distribution of dinocysts in Hole 1053A is shown in Fig. 3. The samples contain several taxa

Table 1. *Palynomorph counts*

Samples 1053A	Section	Interval from	to	Depth (mbsf)	Total counted polynomorphys*	Dinocysts (%)	Indeterminable dinocysts (%)	Foram linings (%)	Acritarchs (%)	Prasinophyte algae (%)	Bisaccate pollen (%)	Pollen and spores (%)
1H	2	124	130	2.74	1							100
1H	4	124	129	5.74	4	75					25	
1H	6	125	130	8.75	1							100
2H	4	125	130	15.25	227	75	1	3	3			18
3H	2	125	129	21.75	165	56	4	9	2	2	26	
4H	2	125	130	31.25	154	65	1	1	2	2	1	28
4H	6	125	130	37.25	152	74	5	1	1	2		17
5H	2	124	129	40.74	137	60	2	2	4	1	29	
5H	4	124	129	43.74	149	60	3	3	3	5		27
5H	6	75	78	46.25	156	87	4		1		1	8
6H	4	145	148	53.45	128	84	6	1	1	1		7
6H	6	35	38	55.35	135	64	4		1	4		27
7H	2	145	148	59.95	188	71	1	5	1	2		20
7H	4	35	38	61.85	154	75	6		5	2	1	11
8H	3	65	68	70.15	132	83	4	2	2	1		8
8H	5	35	38	72.85	178	53	1	2	5			39
8H	6	55	58	74.55	164	73	4	2	2	5		13
9H	3	75	78	79.75	4	75						25
9H	4	95	98	81.45	144	83	2	1	3	4		7
9H	5	105	108	83.05	180	87		2	2	1		9
10H	4	125	130	91.25	156	76	5	1	2	1		14
10H	6	125	130	94.25	132	58	2	4	3	2		31
11H	2	125	130	97.75	142	68	3	1	3	6	1	19
11H	4	125	130	100.75	157	74	2		1	3		21

11H	6	125	130	103.75	151	81	3	1	2	2	1	9
12H	2	125	130	107.25	61	72	2		2	5		20
12H	4	125	130	110.25	182	88	2		1	2	1	7
12H	6	125	130	aa3.25	151	66	3	2	3	6	1	21
13H	2	125	130	116.75	133	74	4	2	2	6		13
13H	4	125	130	119.75	161	86	2	1	1	1		9
13H	6	125	130	122.75	32	73	3		3	6		13
14H	2	125	130	126.25	134	48	3	3	1	27		18
14H	4	125	129	129.25	157	90	3		1	1		5
14H	6	125	129	132.25	54	76	2		2	11		9
15H	2	125	129	135.75	148	55	3	1	4	11	1	25
15H	4	125	129	138.75	187	83	2		1			14
17X	4	129	132	150.49	47	89	4					6
18X	2	114	118	156.94	136	63	4		1	14		18
18X	4	117	120	159.97	47	81	6					13
18X	6	124.5	128	163.45	138	72	3	1	2	9	1	13
19X	2	124	127	166.64	41	90			2		5	2
19X	4	47	60	168.97	90	91	3				1	4
19X	6	76	79	172.16	134	46	2	1	2	40		7
20X	1	122	125	174.72	17	76	6			12		6
20X	3	73	76	17723	7	71	14					14
20X	5	72	75	180.22	122	64	4		2	22		8

*Absolute palynomorph counts with the objective counting at least 100 specimens. The other columns give the relative abundances of the seven palynomorph groups.

Samples: 1053A Core	1H	1H	1H	2H	3H	4H	4H	5H	5H	5H	6H	6H	7H	7H	8H	8H	8H	9H	9H	9H	10H
Section	2	4	6	4	2	2	6	2	4	6	4	6	2	4	3	5	6	3	4	5	4
Interval: from	124	124	125	125	125	125	125	124	124	75	145	35	145	35	65	35	55	75	95	105	125
to	130	129	130	130	129	130	130	129	129	78	148	38	148	38	68	38	58	78	98	108	130
Depth mbsf	2.74	5.74	8.75	15.25	21.75	31.25	37.25	40.74	43.74	46.25	53.45	55.35	59.95	61.85	70.15	72.85	74.55	79.75	81.45	83.05	91.25
Total dino's counted	0	3	0	196	261	298	248	275	280	226	252	236	286	247	242	265	259	3	247	276	237
Spiniferites sp. A																				3	
Samlandia chlamydophora sensu Stover and Hardenbol, 1994									*		*										
Nematosphaeropsis / Cannosphaeropsis spp.						2	2			2		5	2		2		1		9		
Oligosphaeridium anapetum sp. nov.					1																
Hystrichokolpoma cinctum					4	2	1			4	1		2	2		6				2	2
Hystrichosphaeropsis sp. of Biffi and Brinkhuis, 1993					4	3		2	1								*				2
Tectatodinium pellitum					1		5	3	4			5		2			1		*		
Operculodinium spp. (pars)				6	19	18	16	15	23	22	14	17	27	11	25	33	24		21	19	20
Pentadinium laticinctum				3	1	2	2	6	1	2	2	3	1	1	9	2	3	1	3	3	2
Impagidinium dispertitum				1	1		*	*	3		1	1	*	4	1	5	2		1	*	
Impagidinium sp. of Biffi and Brinkhuis, 1993				2		2		1	*	1	1				1	2	2		1		
Hystrichokolpoma rigaudiae				2	7	4	2	3	2	3	8	2	4		9	2	4		2	4	23
Homotryblium floripes				42	39	7		3	6	37	2	3	7			1	2		2	2	1
Spiniferites spp.(pars)		3		102	147	195	182	203	183	96	196	169	197	189	160	149	186	2	185	219	167
Areosphaeridium diktyoplokum									2												
Impagidinium cf. *I. aculeatum*				3		5		1	2	2	1						4		2	3	1
Systematophora ancyrea					1	2			2	2	2	2	7	2	9	3	2			6	
Melitasphaeridium pseudorecurvatum														1			1		1		
Deflandrea phosphoritica						4		1	1							2	*				
Glaphyrocysta intricata					1		5		1				2		1	1	*		*		
Dinopterygium cladoides sensu Morgenroth, 1966				1	1			*	1	1			1	*			1				
Dapsilidinium pastielsii				7	3	7	2	3	6	8	1	6	3	1	3	6	*		1	2	3
Cordosphaeridium minimum				6	1			1	3	1	*	1	1								
Impagidinium velorum											1										
Impagidinium cf. *I. velorum*																					
Cribroperidinium cf. *C. tenuitabulatum*						2	1	*	*	2				1		2					1
Thalassiphora pelagica					1		*	*	2					18			1		1		3
Hemiplacophora sp. A				4	3	11	10	10	1	8			3						2	*	
Charlesdowniea cf. *C. coleothrypta*					*		2		*		1		1			1					
Impagidinium maculatum																					
Lingulodinium machaerophorum				1		12	4	3	15	2	3	4	3				4		2	3	4
Impagidinium spp. (pars)													1								
Enneadocysta multicornuta										1			2	1							
Corrudinium incompositum								3	1		1	4	1		1						
Areoligera? sp. B						1	*	*	1		2	8	1	3	4	3	*		3	3	1
Xenicodinium sp. A						4		4		3	1			2	2	4	1			1	3
Dapsilidinium simplex					2			1	3			1					2			1	1
Systematophora placacantha				1	1	4	3	3		5	3	2	4	1	4	23			3		
Diphyes colligerum				1		2			*				*	1					1		
Phthanoperidinium comatum					*		*	1	3		2		3	1	2	3	*		2		3
Schematophora speciosa													3				*				
Cerebrocysta bartonensis						1			1							1	*			1	
Hystrichokolpoma truncatum																					
Lentinia serrata																					
Deflandrea granulata				1	1	1		1	1	3				*							
Thalassiphora patula					2		*	*	1		*			1							
Phthanoperidinium spp (pars)						*			*	3											
Cordosphaeridium gracile					2		1				1									1	
Batiacasphaera compta					8	2	1			3						2	1				
Phthanoperidinium sp. A				1			*												*		
Pentadinium lophophorum									1								1		1		
Impagidinium brevisulcatum						1		*	*			1	2				1				
Forma B				4			1	1	1		3						6				
Thalassiphora delicata					*	2				3				*		1				*	
Homotryblium tenuispinosum					3				1	10	1		2				*				
Forma A				5	2		7				4				3	9	8			1	
Wetzeliella symmetrica																					
Charlesdowniea proserpina sp. nov.					*	2		1	1				3			1	*				
Areoligera? sp. A									2	1		1	1	3							
Distatodinium cf. *D. craterum*				1					2	1				1	1				1		
Hemiplacophora semilunifera								4	2					1	2	1			2		
Samlandia chlamydophora				1	1			1				1	1	*	3	1	1		1	1	
Turbiosphaera filosa																				1	
Wetzeliella gochtii																1					
Cassidium fragile				1									1								
Cordosphaeridium fibrospinosum									*												
Achilleodinium sp. A									*												
Glaphyrocysta sp. A					1		1		*												
Rottnestia borussica					1																
Wetzeliella articulata					1																
Stoveracysta? sp. A					*																
Operculodinium divergens					1																

* present but not included in the counts.

Fig. 3. See opposite page for caption.

with known stratigraphic ranges from well-dated surface sections in Europe. The source of the ages for these taxa is taken from Williams *et al.* (1998*b*), with some modifications from Brinkhuis & Weegink (pers. comm.). Several species have their last occurrences in late Eocene time (Fig. 4). These include *Areosphaeridium diktyoplokum* (range 51.73–33.5 Ma), *Batiacasphaera compta* (38.0–33.7 Ma), *Melitasphaeridium pseudorecurvatum* (54.2–33.3 Ma), *Glaphyrocysta intricata* (43.4–33.73 Ma), *Enneadocysta multicornuta* (42.2–35.1 Ma), *Schematophora speciosa* (36.7–35.2 Ma) and *Cordosphaeridium gracile* (77.4–35.1 Ma). Thus, the presence of these species confirms that the uppermost productive sample in Section 1053A-2H-4 cannot be younger than late Eocene time. Five species have a first occurrence in 1053A (Fig. 4). One species provides constraints on the oldest age, namely *Schematophora speciosa*, which does not occur in the middle Eocene sequence. The presence of *Batiacasphaera compta*, whose first occurrence is close to the top of the middle Eocene sequence, supports the assignment to late Eocene time. *Schematophora speciosa*, which has a first occurrence in 1053A-15H-4, has a known maximum age of 36.7 Ma. Below 1053A-15H-4 there are no dinocyst species with definitive ranges that would

10H	11H	11H	11H	12H	12H	12H	13H	13H	13H	14H	14H	14H	15H	15H	17X	18X	18X	18X	19X	19X	19X	20X	20X	20X
6	2	4	6	2	4	6	2	4	6	2	4	6	2	4	4	2	4	6	2	4	6	1	3	5
125	125	125	125	125	125	125	125	125	125	125	125	125	125	125	129	114	117	124.5	124	57	76	122	73	72
130	130	130	130	130	130	130	130	130	130	130	129	129	129	129	132	118	120	128	127	60	79	125	76	75
94.25	97.75	100.75	103.75	107.25	110.25	113.25	116.75	119.75	122.75	126.25	129.25	132.25	135.75	138.75	150.49	156.94	159.97	163.45	166.64	168.97	172.16	174.72	177.23	180.22
234	237	136	221	50	275	236	260	144	26	252	255	42	197	154	39	221	29	224	37	82	179	13	6	101
1					5						1			1										1
				1							1		2					1						1
3	3		4	1		2	3			1	1	2	1	1			2	6	3		3			3
																			1		6	1		3
7																					1			1
			1				1						1											1
1	1			1			*	1										2						1
17	22	10	46	2	29	18	24	17	8	29	30	8	10	30	3	12	3	12	3	7	2			4
1	3	9	3	4	*	5	3	3		3	3		5											1
1	3	1			3	1	1	2	2	1	*		11			3		1		1	4	1		4
2	*		1		5		1	5			4		3	4		6		1			6	1		4
4	14		*			1	4	1		3		2					1				3			1
3	1	3			7	13	11	2		3	6	5	1	1			1			1	2	1	1	6
155	159	88	153	38	207	181	193	104	10	186	192	22	138	92	32	179	16	191	27	69	132	9	2	70
					1								1				1						1	
1			2		*	1			1	2	1	1		3	1	1	2	2			5		2	
*		4	1	1	2	3		1		1	3		4	2		2		2			2	*		
		1	1					1		2	1									1	2			
	*	2			1		1			1				1				1	1		2			
*																					2			
								1	1		1			1							2			
4	1	3	2				1			2	1		1	4	1			1			3			
1							1														2			
																1		1		2				
	*																	2		1				
		1				1													1					
11		1		1		2											1		1					
4	1																	1						
										*								*						
							*										1							
1	2	*	3				4			1		1				2	1							
4																1								
3	3	5			2			3		1					1	1								
1	1		1		1	1	1		1	*			1			2								
			1		2	2	3		2	3	5		4	8		4								
			1		3					1	1					1								
	1						2									1								
	2	1			2	3	5	1		5			8			4								
											1			1		1								
3	1	1								*			1		1									
1														4										
		1			1									1										
													2											
								1					2											
1	3	3			2					2			1											
					1					*		1												
											1													
											1													
											1													
	*				1						*													
						2	*		1	4														
										*														
	11	1	1							1														
								1																
	1						*																	
		1					1																	
						*																		
						*																		
3	2			1																				
*	1																							
	1																							
1	*																							

* present but not included in the counts

Fig. 3. Range chart ordered according to first occurrences of the dinocyst species. The numbers refer to absolute counts, with total number of dinocysts counted for each sample given at the top. * Denotes presence of a species in that sample but not included in total count. Species citations given in the Taxonomic Appendix.

allow us to determine whether the section is of mid-Eocene age.

Of the four zones Brinkhuis & Biffi (1993) established in the upper Eocene sequence of central Italy the *Melitasphaeridium pseudorecurvatum* Zone and the *Schematophora speciosa* Zone could be recognized. The zonal boundary at Site 1053 is, as in Italy, situated in mid-Chron C16N. The younger zones cannot be recognized, because of the absence of indicative taxa, or the (local) early disappearances of taxa. For example, the last occurrence of *S. speciosa* is recorded in Chron C13R2 in Italy, and in Chron C16N at Site 1053.

The last occurrence of *Enneadocysta multicornuta* was observed in Section 1053A-5H-6. The extinction of this species is estimated at 35.1 Ma, which is in close agreement with the

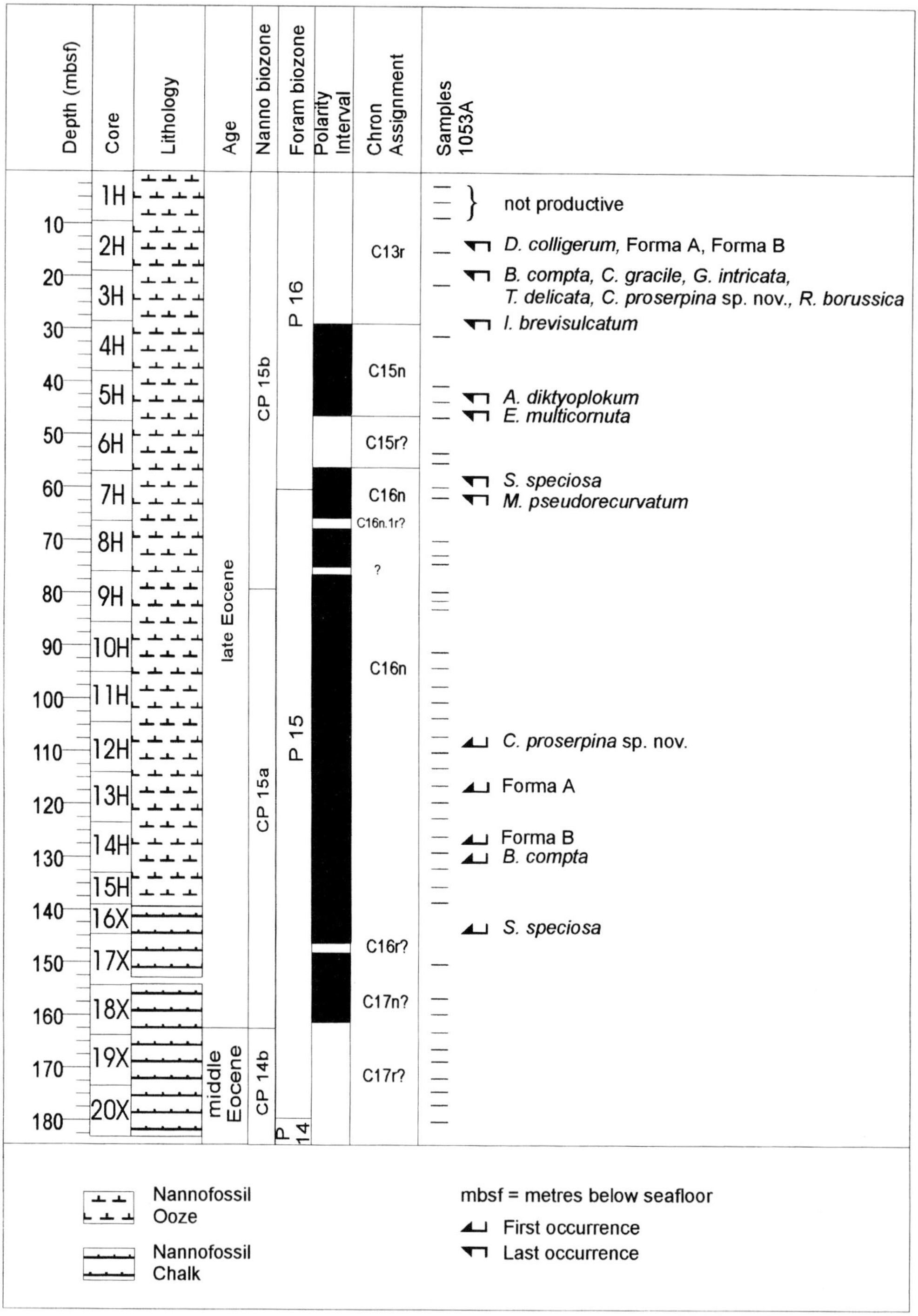

Fig. 4. Selective first and last occurrences of dinocysts in relation to the stratigraphy of Hole 1053A (Shipboard Scientific Party 1998*a*).

Table 2. *Absolute ages*

Sample 1053A-	Depth (mbsf)	Age (Ma) based on Nannofossils	Planktonic foraminifers	Dinoflagellate species
2H 1 66–67	10.16	34.3		
3H-CC	28.6	35.0		
5H-6 75–78	46.25			35.1 LO *E. multivornuta*
7H-2 70–71	59.2		35.2	
7H-2 145–148	59.95			35.2 LO *S. speciosa*
8H-CC	76.11	36.0		
15H-4 125–129	138.75			36.7 FO *S. speciosa*
18X-6 91–92	162.71	37.0		
18X-CC	163.76	37.10		
20X-6 24–28	181.24		38.4	

The absolute ages at Hole 1053A based on the nannofossils and planktonic forminifers (Shipboard Scientific Party 1998*a*) and dinoflagellates. All ages are based on the time scale of Berggren *et al.* (1995).

last occurrence of *Schematophora speciosa* in Section 1053A-7H-2 (35.2 Ma; Table 2).

One species, *Charlesdowniea proserpina* sp. nov., originally recorded from JOIDES Holes 1 and 2, appears to have a restricted stratigraphic range within late Eocene time. The informally described species Forma A is also identical to a taxon of late Eocene age in JOIDES Hole 1.

Descriptions of the new species *Charlesdowniea proserpina* and *Oligosphaeridium anapetum* are included in the Systematic Palynology section. A listing of all species and their full citations is given in the Taxonomic Appendix, including brief notes on 16 informal taxa.

Palaeoecology

The palaeoecological data are presented in two plots (Figs 5 and 6). The relative percentages of marine and non-marine palynomorphs is shown in Fig. 5. The relative percentages of restricted neritic, inner neritic, outer neritic, neritic to oceanic, and oceanic species is presented in Fig. 6. The relative percentages of marine and non-marine palynomorphs can be interpreted in terms of approximate distance from shore (Stover *et al.* 1996).

The plot of the relative percentages of marine and non-marine palynomorphs (Fig. 5) shows a dominance of marine palynomorphs, notably the dinocysts, throughout Hole 1053A. Pollen and spore percentages peak at 39 % in 1053A-8H-5, with only one other core, 1053A-10H-6, exceeding 30 %. Bisaccate pollen are included in the pollen and spores percentages of Fig. 5, but with their scattered occurrence (which hardly ever exceeds 1 %) they can be neglected.

The relative percentages of neritic, inner neritic, outer neritic, neritic to oceanic, and oceanic species, plotted in Fig. 6, include the species listed in Table 3. Only *Homotryblium* is considered to be a restricted neritic taxon. The inner neritic group includes all the peridinialeans and the gonyaulacalean genera *Achilleodinium, Areoligera, Cribroperidinium, Dapsilidinium, Glaphyrocysta, Lingulodinium, Melitasphaeridium, Systematophora* and the species *Operculodinium divergens*. The outer neritic group includes *Areosphaeridium, Cordosphaeridium, Dinopterygium, Diphyes, Enneadocysta, Hemiplacophora, Muratodinium, Pentadinium, Samlandia, Tectatodinium, Thalassiphora* and *Turbiosphaera*. The category neritic to oceanic is dominated by *Spiniferites* and includes *Operculodinium*. The oceanic group includes *Hystrichokolpoma, Impagidinium, Nematosphaeropsis* and *Cannosphaeropsis* (Table 3). Gonyaulacaceans consistently dominate the assemblages.

Neritic species show a peak of about 30 % in 1053A-5H-6 and minor peaks in Cores 1053A-8H-5 and 1053A-11H-4. Oceanic species are present in all the cores and reach a maximum percentage of about 30 % in 1053A-18X-4 and minor peaks in Cores 1053A-11H-6 and 1053A-15H-4. Abundances of the outer neritic species are low throughout the sequence from Hole 1053A.

All recorded species with a known or inferred sea surface temperature preference fit in the warm and/or temperate taxa group. This would indicate warm to warm temperate conditions throughout the analysed interval.

Discussion

The abundances of palynomorphs in most samples in Hole 1053A are surprisingly high, considering the lithologies and presumed distance from the shore (Fig. 2 and Table 1) The low numbers in Subunit 1B are probably due

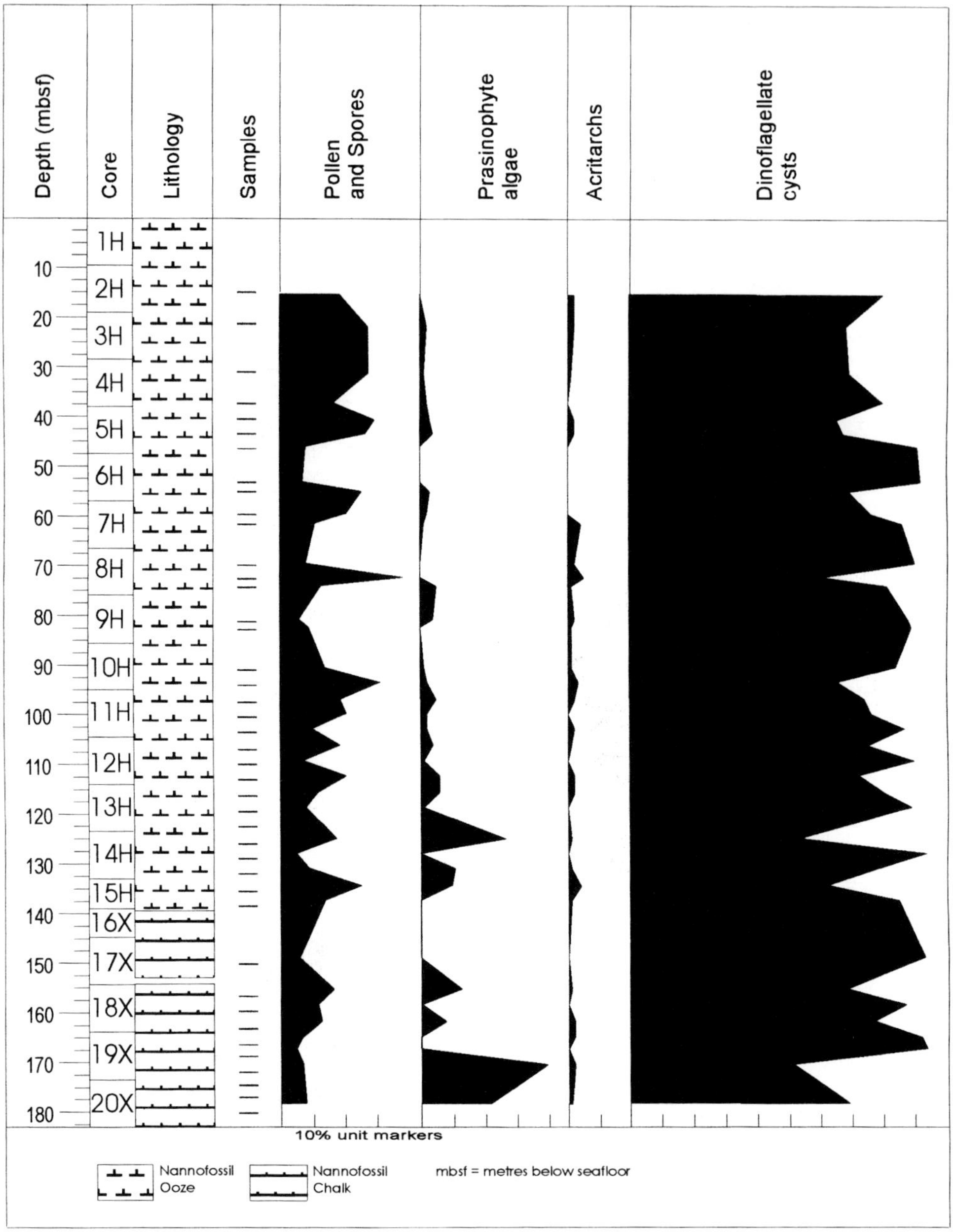

Fig. 5. Plot of relative percentages of non-marine (pollen and spores) and marine (dinocysts and acritarchs and prasinophyte algae) palynomorphs for Hole 1053A.

to most of the organic material being oxidized. This is supported by the occurrence of pyritized diatom and radiolarian fragments in all samples but Subunit 1B. Previous studies where recovery from similar lithologies has been poor, as in the samples processed from JOIDES Holes 3, 4 and 6, confirms the practicality of our processing technique, in which none of the samples were oxidized.

The higher abundances of terrestrial palynomorphs coincide generally with high relative abundances of inner neritic dinocyst group (Figs 5 and 6). As we know that this is a deep-water site, the transport mechanisms for the

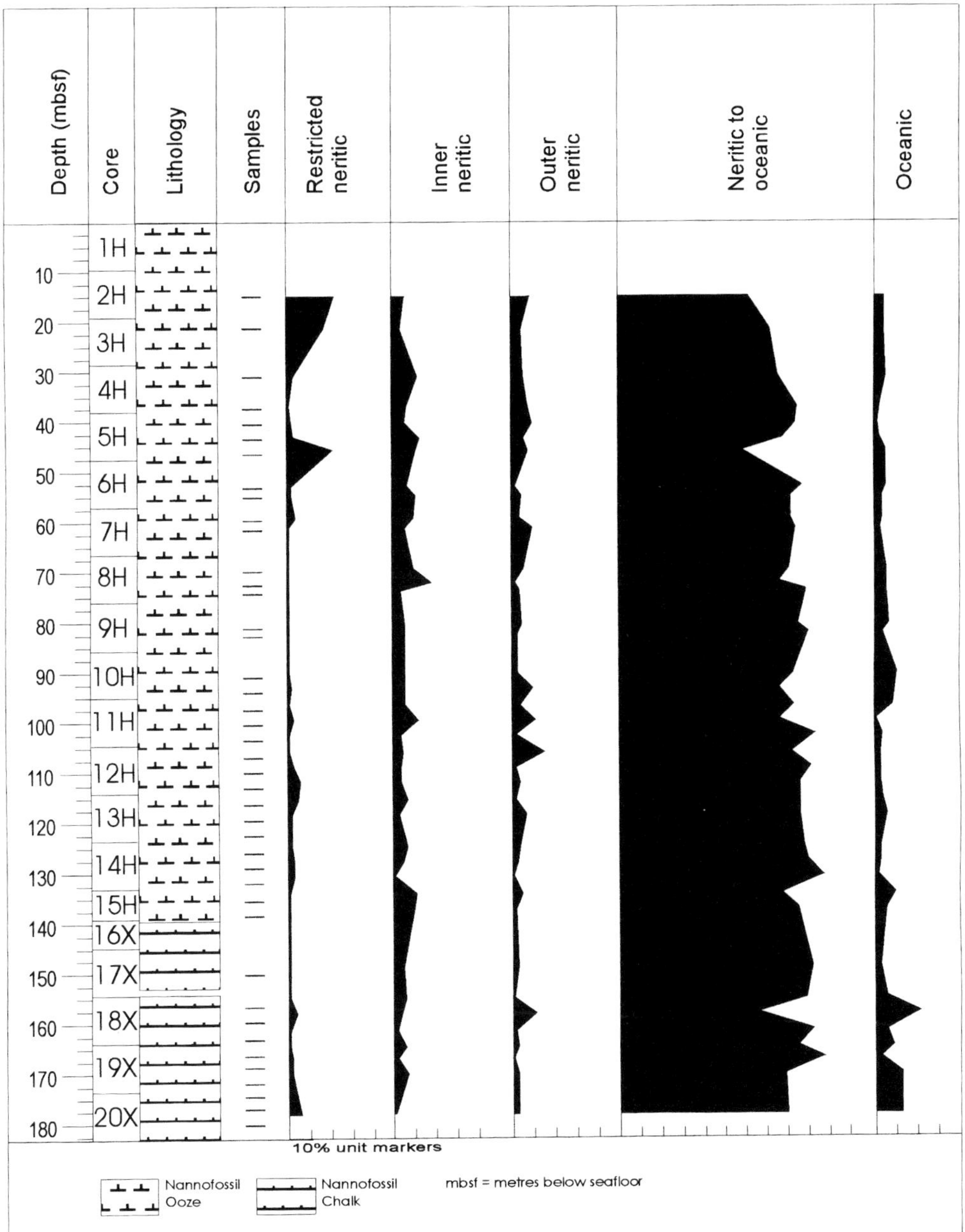

Fig. 6. Plot of relative percentages of restricted neritic, inner neritic, outer neritic, neritic to oceanic and oceanic species ratios for Hole 1053A. The species included in the five categories are listed in Table 2.

terrestrial palynomorphs and inner neritic dinocysts must be linked. The increased transport might be due to increased runoff and/or sea-level variations that changed the shelf configurations and possible its proximity to Site 1053A.

The plot of the restricted neritic ratio (*Homotryblium* spp.) shows maximum values towards the top of Hole 1053A in Cores 2H-4, 3H-2 and 5H-6 (Fig. 6). On the basis of the model of Brinkhuis (1992, 1994), our data suggest that sea level was falling during deposition of Cores 5 to 2,

Table 3. *Listing of species for the five categories from restricted neritic to oceanic*

Restricted neritic	Outer neritic	Neritic to oceanic	Oceanic
Homotryblium spp.	*Areosphaeridium* spp.	*Batiacasphaera* spp.	*Hystrichokolpoma* spp.
Inner neritic	*Cordosphaeridium* spp.	*Cassidium* spp.	*Impagidinium* spp.
Achilleodinium spp.	*Dinopterygium* spp.	*Cerebrocysta* spp.	*Nematosphaeropsis–Cannospaeropsis* spp.
Areoligera spp.	*Diphyes* spp.	*Corrudinium* spp.	
Charlesdowniea spp.	*Enneadocysta* spp.	*Distatodiniuum* spp.	
Cribroperidinium spp.	*Hemiplacophora* spp.	Forma A	
Dapsilidinium spp.	*Pentadinium* spp.	Forma B	
Deflandrea spp.	*Samlandia* spp.	*Hystrichosphaeropsis* spp.	
Glaphyrocysta spp.	*Tectatodinium* spp.	*Lithodinia* spp.	
Lentinia spp.	*Thalassiphora* spp.	*Oligosphaeridium* spp.	
Lingulodinium spp.	*Turbiosphaera* spp.	*Operculodinium* spp. (*pars*)	
Melitasphaeridium spp.		*Rottnestia* spp.	
Operculodinium divergens		*Schematophora* spp.	
Phthanoperidinium spp.		*Spiniferites* spp.	
Systematophora spp.		*Xenicodinium* spp.	
Wetzeliella spp.			

creating lagoonal environments near the coast, explaining an increased production of *Homotryblium*. A higher abundance of *Homotryblium* at Site 1053A may thus reflect an increased production in near-shore neritic environments resulting in an increased transport of these forms to Site 1053A.

The presence of one or more of the genera considered to be oceanic (*Hystrichokolpoma, Impagidinium*, *Nematosphaeropsis* and *Cannosphaeropsis*; see Fig. 6) confirms the deeper-water setting of Hole 1053A. Dale (1996) stated that even a few *Nematosphaeropsis–Cannosphaeropsis* or *Impagidinium* cysts can be taken as a definite signal of oceanic water influence.

The absolute ages in Hole 1053A, based on nannoplankton, planktonic foraminiferal and dinoflagellate data, are given in Table 2. The foraminiferal- and nannofossil-related ages are from the Shipboard Scientific Party (1998*a*). The dinoflagellate-related ages are from Brinkhuis & Biffi (1993) and Williams *et al.* (1998*b*). The absolute ages based on the three groups show general agreement, especially the identical foraminiferal and dinocyst age of 35.2 Ma. However, we cannot comment on the 38.4 Ma age for 1053A-20X-6 based on the planktonic foraminifera, as the oldest dinocyst age for that sample is 36.7 Ma.

No typically cold-water (or higher-latitude) species have been recorded in this study. However, the presence of several species extending into the upper Eocene sequence in Hole 1053A, but not found in the upper Eocene sequences from higher latitudes, suggests that they are warmer-water indicators. These include: *Rottnestia borussica*, *Diphyes colligerum* and *Thalassiphora delicata* (Fig. 4). All three species also occur in JOIDES Hole 1. *Impagidinium brevisulcatum*, and the undescribed species (*Impagidinium* sp. and *Hystrichosphaeropsis* sp. A) also recorded by Brinkhuis & Biffi (1993) from Italy, are assumed to be lower-latitude species.

The latests occurrences for *Areosphaeridium diktyoplokum* and *Schematophora speciosa* are generally taken to be at 33.5 Ma and 35.2 Ma, respectively, based on the Italian Eocene sections. At Site 1053A, these two species do not range as high in the succession as in Italy. This might be a local phenomenon, which indicates that further study in these areas is required. Interestingly, it coincides with the suggested sea-level fall in the upper part of the interval.

Another diachronous occurrence is *Wetzeliella symmetrica*, previously recorded with a first occurrence of 33.5 Ma in early Oligocene time at mid-latitudes. In 1053A its only occurrence is in 1053A-12H-6. The closest reliably dated section is 1053A-15H-4, with an age of 36.7 Ma. Thus, *Wetzeliella symmetrica* has a possible first occurrence just after 36.7 Ma. This species has also been recorded in the upper Eocene sequence of JOIDES Hole 1.

Conclusions

The dinocyst assemblages in Hole 1053A clearly indicate upper Eocene sediments from 15.25 to 138.75 mbsf. From 138.75 to 180.22 mbsf, there are no dinocyst species with definitive ranges that would allow us to determine if the section is of midd-Eocene age. There is good agreement with the planktonic foraminiferal and nannofossil data age estimates (Table 2). This is encouraging, as the use of dinocysts has been minimal for determining the age of deep-sea sediments and the information on lower-latitude assemblages is almost solely from the Tethyan area.

Most of the taxa could be placed in existing species but there are several new forms, including the new species *Oligosphaeridium anapetum* and *Charlesdowniea proserpina*, that are probably useful as warmer-water indicators.

The consistent presence of the oceanic taxa *Impagidinium* and *Nematosphaeropsis–Cannosphaeropsis* confirms the oceanic setting for Site 1053A. We consider the upward decrease in relative abundances of the *Impagidinium* group and the corresponding increase in the inner neritic dinocysts to indicate an increase of transport of neritic species.

The known oceanic setting and the much higher numbers of neritic dinocysts recovered from JOIDES Holes 1 and 2, on the continental shelf of eastern Florida due west of the Blake Plateau, confirm that the neritic dinocysts in our assemblages are allochthonous. Throughout the section similarity between the peaks of relative abundances of the inner neritic dinocysts and the terrestrial palynomorphs is observed. The coincidence of variations of these elements transported from the west implies there were changes in the configuration of the shelf environment. This might be increased runoff and/or sea-level variations.

The presence of many warm- and warm- to temperate-water species indicates that the surficial waters of the Blake Nose region were warmer in late Eocene time. This is further substantiated by the absence of known cold-water species.

Systematic Palynology

Division *Pyrrophyta* Pascher 1914
Class *Dinophyceae* Pascher 1914
Subclass *Peridiniphycidae* Fensome *et al.* 1993
Order *Peridiniales* Haeckel 1894
Family *Peridiniaceae* Ehrenberg 1831
Subfamily *Wetzelielloidideae* (Vozzhennikova 1961) Bujak & Davies 1983
Genus *Charlesdowniea* Lentin & Vozzhennikova 1989

Charlesdowniea proserpina sp. nov.
Fig. 7a–e

Etymology

Named for Proserpina, wife of the god Pluto in Greek mythology.

Holotype

Slide I, 1053A-8H-5, 35–38 (D45), Fig. 7b and c.

Diagnosis

Rhomboidal to pentagonal pericyst with short unequal antapical horns, soleiform archeopyle and reduced process complexes, which we term paraplate-centred process complexes, that are located towards the centre of the paraplates and are distally united by perforate membrane or trabeculae. Some of the processes are distally free.

Description

Shape

Pericyst. Ambitus rhomboidal to pentagonal (peridinioid), strongly compressed dorsoventrally, with either a short or no apical horn, short broad cingular horns or no cingular horns and unequal antapical horns, the left always the longer.

Endocyst. Ambitus rhomboidal, may be appressed to pericyst other than in vicinity of horns.

Pericoels. Cornucavate with a maximum of one apical, two cingular and one or two antapical pericoels. Alternatively, there is a single ambital pericoel uniting the pericoel areas beneath the horns.

Phragma. Periphragm of uniform thickness, about 1 μm, with very broad pandasutural zones devoid of ornamentation and paraplate-centred process complexes, although on individual paraplates these can migrate towards the apex on the epicyst and the antapex on the hypocyst. Processes of individual complex united distally by perforate membrane or trabeculae. Six to eight processes in most complexes. Some processes are free.

Endophragm. Less than 1 μm thick, smooth.

Paratabulation. Tabulation on pericyst expressed by process complexes and is as for genus. Paraplate complex 1′ rhomboidal and largest of the apicals. 2′ and 4′ represented by linear process complexes with processes often free. Process complex on paraplate 3′ of similar shape to 1′ but smaller. All three anterior intercalaries with small paraplate-centred process complexes and with 2a being quadrate. Of the precingulars, process complexes of 1″ and 7″ are same elongate triangular outline, 2″ and 6″ are linear complexes aligned longitudinally with processes commonly free. Paraplates 3″, 4″ and 5″ have linear process complexes aligned latitudinally. The five postcingulars include identical triangular, latitudinally elongate complexes on 1‴ and 5‴, identical longitudinally elongate complexes on 2‴ and 4‴ and a latitudinally elongate 3‴ complex. There are two, more or less equal antapical complexes, representing 1⁗ and 2⁗.

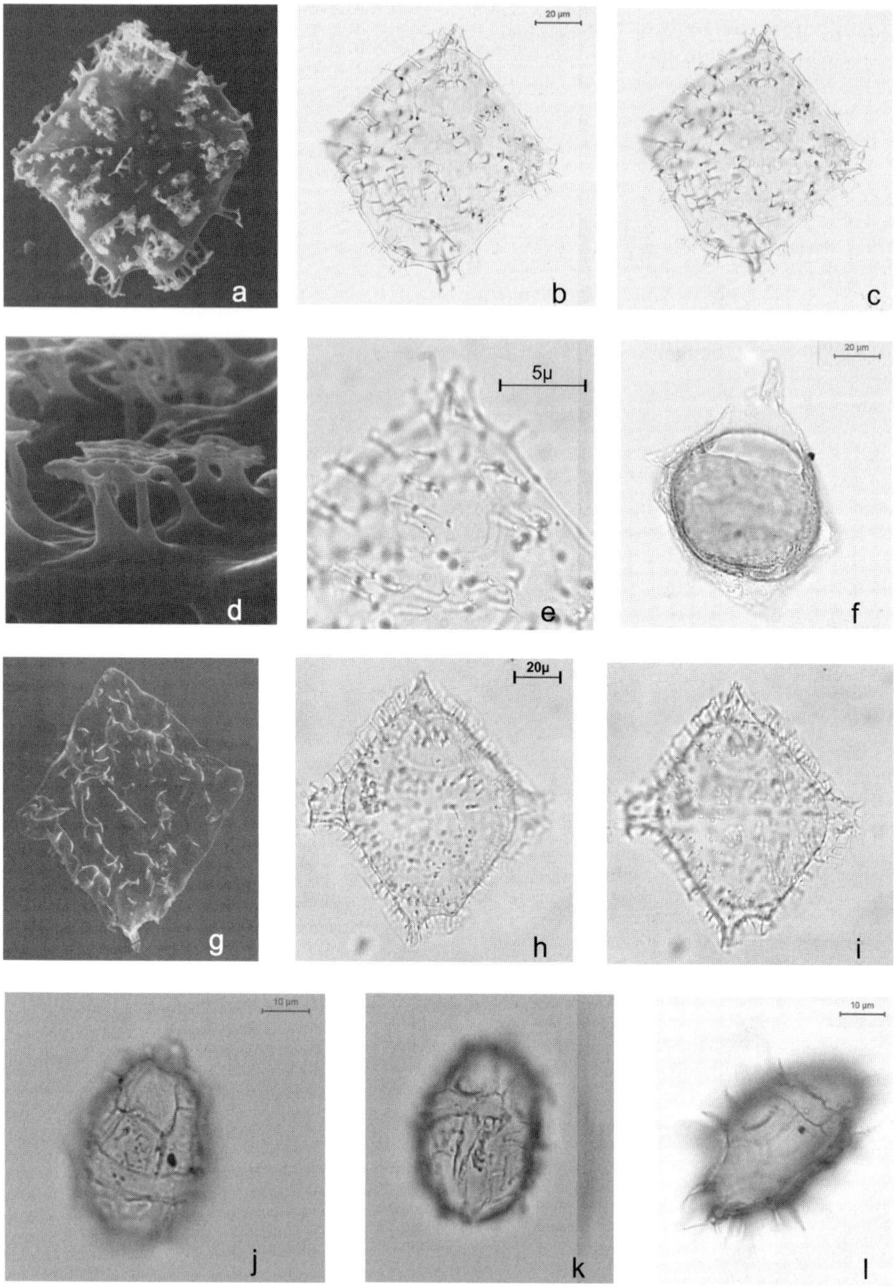

Fig. 7. In Figs 7–13, sample, slide number and England finder coordinates of all light microscope hotographys are included. View and/or focus levels are mentioned; D, dorsal; V, ventral; A, apical; AA, antapical; LL, left lateral; RL, right lateral; E, external; I, interior; hf, high focus; mf, mid-focus; lf, low focus. (**a**, **b**) (D, E, hf), (**c**) (D, mf), (**d**, **e**) *Charlesdowniea proserpina* sp. nov. sample/slide, JOIDES, Hole 1 (**a**, **d**), 8H-5, 35–38/I (**b**, **c**, **e**, D45). (**f**) (D, mf), *Deflandrea granulata* sample/slide, 2H-4, 125–130/I (Z41). (**g**, **h**) (D, I, lf), (**i**) (mf), *Charlesdowniea* cf. *C. coleothrypta* sample/slide, JOIDES, Hole 1 (**g**), 14H-2, 125–130/I (**h**, I, O35). (**j**) (D, E), (**k**) (V, I), *Phthanoperidinium* sp. A sample/slide, 19X-4, 057–060/I (A20). (**l**) (AA), *Phthanoperidinium comatum* sample/slide 8H-5, 035–038/I (F47).

Paracingulum. On the pericyst, cingulars marked by single row of processes, sometimes connected distally, sometimes free; slightly helicoidal.
Parasulcus. On the pericyst extending longitudinally from below 1′ to the antapical horn(s). Four process complexes, the largest, the posterior sulcal, sometimes being separated into two complexes and located in part on the antapical horns.
Archeopyle. Periarcheopyle: quadra, intercalary, soleiform, operculum remaining attached along anterior parasuture Q1. Sometimes there can be tearing along this parasuture to give the illusion of a detached operculum. Endoarcheopyle: appears to be same as periarcheopyle.

Dimensions

Pericyst. Length 113–116 μm, breadth 85–107 μm (n = 6).
Endocyst. Length 72–92 μm, breadth 74–89 μm.
Horns. Apical, 7–8 μm, antapical, 8–15 μm; process length, 4–12 μm.

Type locality

ODP Site 1053A, 31.25 mbsf, 40.74 mbsf, 43.74 mbsf, and 59.95 msbf, Blake Nose, offshore from Florida.

Other localities

JOIDES Hole 1 537′ (16 368 mbsf), Blake Plateau.

Stratigraphic range

Upper Eocene sequence.

Comparison

There is only one described species of *Charlesdowniea*, *C. limitata* (Stover & Hardenbol 1994) with a soleiform archeopyle. In *C. limitata*, the processes just in from the paraplate boundaries are united by distal trabeculae, so that they form simulate or penitabulate complexes, whose borders run parallel to the outline of the paraplates. A species from the middle Eocene succession of England that superficially resembles *C. proserpina* is *Charlesdowniea variabilis*. However, the archeopyle of *Charlesdowniea variabilis* (Bujak, in Bujak *et al*. 1980) Lentin & Vozzhennikova (1989) has a free operculum and the process complexes fill most of the paraplate.

Division *Pyrrophyta* Pascher 1914
Class *Dinophyceae* Pascher 1914
Subclass *Peridiniphycidae* Fensome *et al*. 1993
Order *Gonyaulacales* Taylor 1980
Family *Gonyaulacaceae* Lindemann 1923
Subfamily *Leptodinioideae* Fensome *et al*. 1993
Genus *Oligosphaeridium* Davey & Williams 1966

Oligosphaeridium anapetum sp. nov.
Fig. 8a–e

Etymology

Greek 'anapetos', meaning wide open, expanded.

Holotype

Slide I, 1053A-20X-5, 72–75/ I (P44). Fig. 8a–c.

Diagnosis

A species of *Oligosphaeridium* with tubiform processes whose walls may be perforate or incomplete towards the cingulum and which proximally form an annulate to arcuate contact with the cyst body.

Description

Shape. Body subspherical.
Wall relationships. Endophragm and periphragm in contact except where periphragm forms the processes.
Wall features. Processes intratabular, tubiform to cylindrical, striate, one per paraplate, absent on cingulum and some of sulcus. Processes open distally, expanded, with undulose to slightly serrate margins. Some processes perforated along their length or may be incomplete towards the cingulum, so that proximally the process contact with the cyst body is arcuate rather than annulate. Process width reflects location, with 2″ and 4′ being the broadest. Occasionally there is an adventitious process on the hypocyst. Periphragm between processes smooth to scabrate.

Paratabulation. Indicated by intratabular processes, gonyaulacacean, process formula: 6″, 5–6‴, 4″, 1p, 1⁗, 0–1s.
Archeopyle. Apical, principal archeopyle suture zigzag, presumably reflecting four apical paraplates. Operculum free, with four processes 1′ and 3′ of approximately equal size, 4′ and 1′ slender.
Paracingulum. Indicated by absence of processes and orientation of arcuate process bases.
Parasulcus. Indicated by one process, the posterior sulcal or devoid of processes.

Size

Main body. Length 80 μm; width, 52–64 μm.

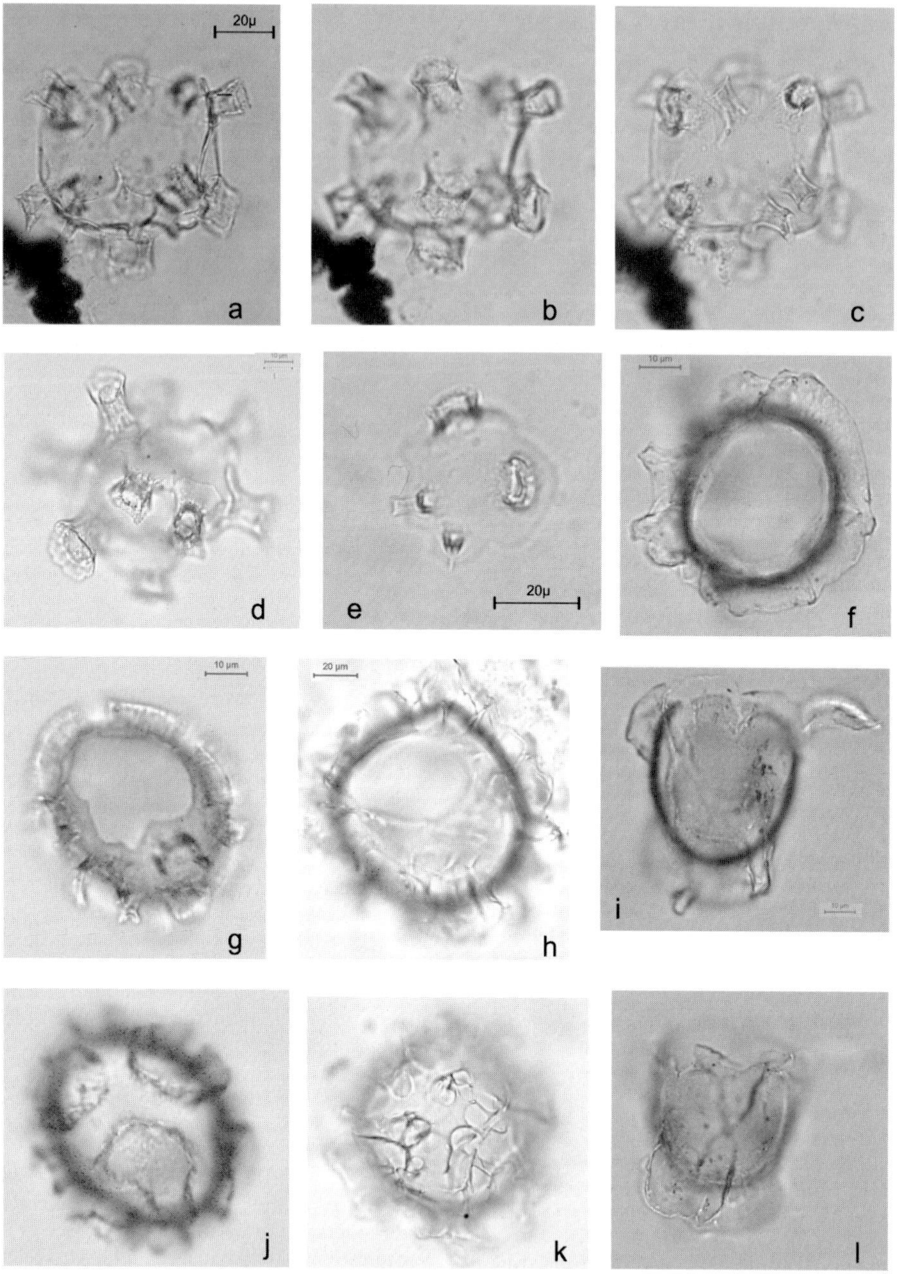

Fig. 8. (**a**) (hf), (**b**) (mf), (**c**) (lf), (**d**) (AA), (**e**) (operculum), *Oligosphaeridium anapetium* sp. nov. sample/slide 20X-5, 72–78 (**a**–**c**), E39; **e**, T47); 19X-6, 76–79/I (**d**, E39). (**f**) (A), (**i**) (D, hf), (**l**) (V, lf), *Hemiplacophora* sp. A sample/slide, 2H-4, 125–130/I (**f**, H36; **i**, **l**, J46; **d**, B27). (**g**) (A, I), (**j**) (AA, E), *Hemiplacophora semilunifera* sample/slide, 18H-4, 117–120/I (T51). (**h**) (V, I), (**k**) (D, E), *Samlandia sensu* Stover & Hardenbol sample/slide, 14H-4, 125–129/I (P26).

Processes. Length, 9–24 μm, width, 3–19 μm ($n = 7$).

Type locality

ODP Site 1053A, 166.64 mbsf, 172.16 msbf, and 174.72 msbf, Blake Nose, offshore from Florida.

Stratigraphic range

Upper middle to upper Eocene sequence.

Comparison

The most distinctive feature of *Oligosphaeridium anapetum* is the structure of the processes. These are striate, whereas all other species of the genus have processes with smooth walls. Some of the precingular and postcingular processes are also unusual in not being truly tubiform; rather that part of the process towards the cingulum is open. This is reminiscent of the type or ornamentation found in *Hemiplacophora semilunifera*. The variability in the number of postcingular, from five to six, and sulcal processes, none to one, plus the occasional presence of an adventitious process on the hypocyst also differentiate this species from other species of *Oligosphaeridium*.

Taxonomic appendix

The dinocyst taxonomy follows Williams *et al.* (1998*a*). Brief notes on 16 informal taxa are provided. Two new species are described in the Systematic Palynology section. When photographs are included in this paper, the figure number and part are given in the alphabetical list. In the figure captions the sample, slide and England Finder coordinates are stated.

Achilleodinium sp. A. (Fig. 11, and c)
Remarks. This species is included in *Achilleodinium* in having a precingular archeopyle and large precingular, postcingular and antpical processes. The single specimen bears a strong resemblance to *Florentinia buspina* (Davey & Verdier 1976) Duxbury 1981.
Dimensions: central body width 35 μm, length 38 μm; processes length 5–13 μm ($n = 10$).
Areoligera? sp. A (Fig. 11j–l).
Remarks. The overall morphology of this taxon resembles that of species of *Areoligera*. Notably the apical archeopyle, the dorsoventral flattening and the presence of dorsal penitabular processes supports attribution to the genus *Areoligera*. The wall surface of the central body is finely granular. However, the ventral processes are distally connected, a feature that would indicate attribution to the morphologically related genus *Glaphyrocysta*.
Dimensions: central body width 55–60 μm, length 45–50 μm; processes 35–40 μm ($n = 2$).
Areoligera? sp. B (Fig. 11f and i).
Remarks. Mainly broken specimens are found. The presence of quasi-penitabular processes on the dorsal surface supports attribution to *Areoligera*. The processes are solid, distally fenestrate and sometimes distally connected.
Dimensions: central body width 40–65 μm, length 40–50 μm; processes 25–40 μm ($n = 5$).
Areosphaeridium diktyoplokum (Klumpp 1953) emend. Stover & Williams 1995.

Batiacasphaera compta Drugg 1970.(Fig. 12j).

Cassidium fragile (Harris 1965) Drugg 1967 (Fig. 9f).
Cerebrocysta bartonensis Bujak in Bujak *et al.* 1980. (Fig. 13m).
Charlesdowniea cf. *C. coleothrypta* (Fig. 7g–i).
Remarks. This taxon differs from *C. coleothrypta* in having a soleiform archeopyle. It is an unusual soleiform archeopyle in having similar width and length.
Dimensions: pericyst width 110 μm, length 116 μm; endocyst width 80 μm, length 81 μm.
Charlesdowniea proserpina sp. nov. (Fig. 7a–e)
Holotype: Fig. 7b and c.
Cordosphaeridium fibrospinosum Davey & Williams 1966.
Cordosphaeridium gracile (Eisenack 1954) emend. Davey & Williams 1966.
Cordosphaeridium minimum (Morgenroth 1966) Benedek 1972 (Fig. 13a).
Corrudinium incompositum (Drugg 1970) Stover & Evitt 1978.
Cribroperidinium cf. *C. tenuitabulatum* (Fig. 13d and f).
Remarks. Differs from *C. tenuitabulatum* in the general absence of secondary sutures; surface scabrate.
Dimensions: central body width 59 μm, length 60 μm.
Dapsilidinium pastielsii (Davey & Williams 1966) Bujak *et al.* 1980.
Dapsilidinium simplex (White 1842) Bujak *et al.* 1980.
Deflandrea granulata Menéndez 1965 (Fig. 7f).
Deflandrea phosphoritica Eisenack 1938.
Dinopterygium cladoides sensu Morgenroth 1966.
Diphyes colligerum (Deflandre & Cookson 1955); emend. Goodman & Witmer 1985.
Distatodinium cf. *D. craterum* (Fig. 11a).

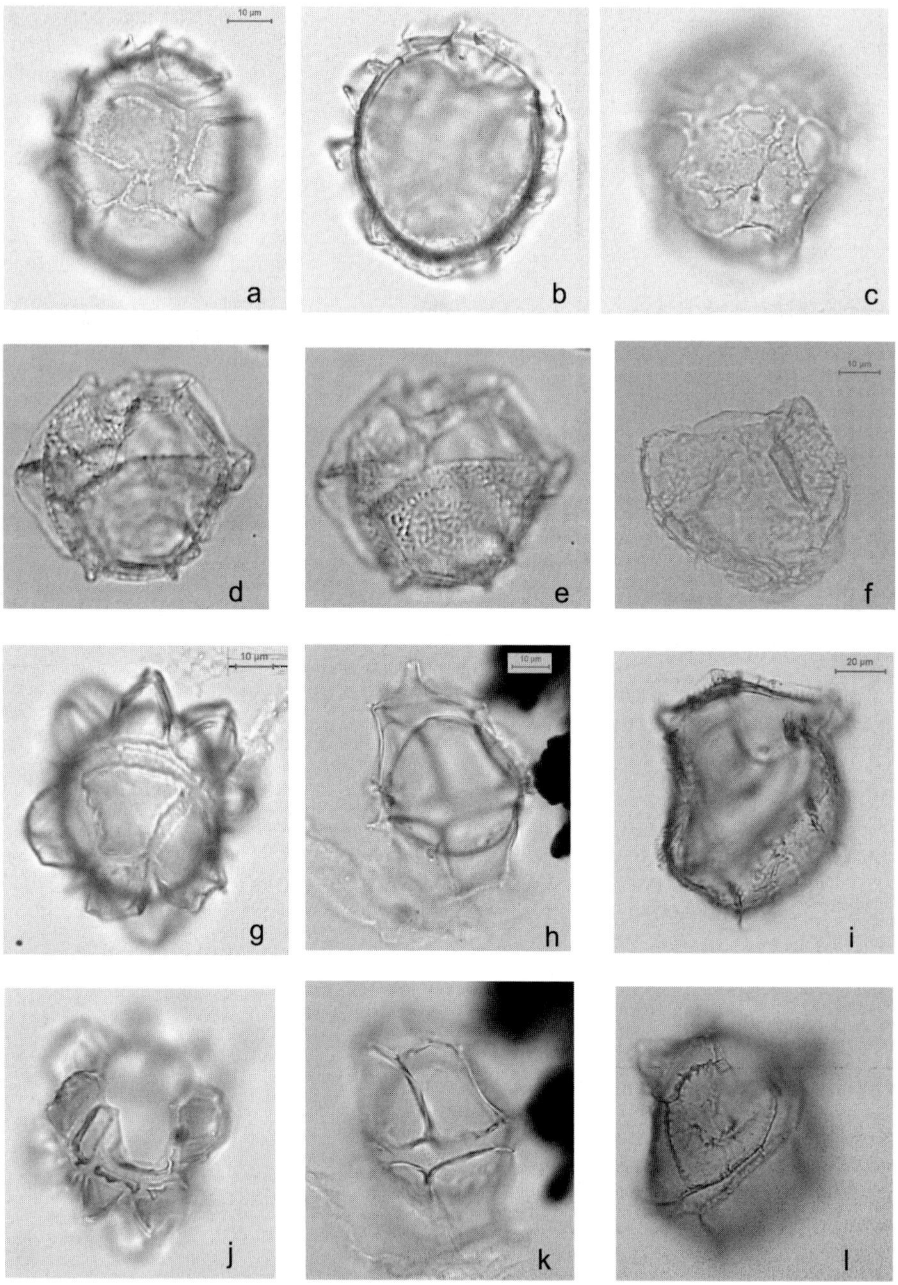

Fig. 9. (**a**) (D, hf), (**b**) (mf), (**c**) (V, lf), *Schematophora speciosa* sample/slide, 15H-4, 125–130/I (H37). (**d**) (V, mf), (**e**) (V, hf), *Pentadinium lophophorum* sample/slide, 8H-6, 55–58/I (S34). (**f**) *Cassidium fragile* sample/slide, 2H-4, 125–130/I (G17). (**g**) (AA, I), (**j**) (A, E), *Hystrichokolpoma truncata* sample/slide, 15H-2, 125–129/I (O42). (**h**) (mf), (**k**) (LL), *Hystrichosphaeropsis* sp. of Brinkhuis & Viffi (1993) sample/slide, 4H-2, 125–130/I ()27). (**i**) (mf), (**l**) (LL), *Impagidinium maculatum* sample/slide, 17X-4, 129–132/I (V19).

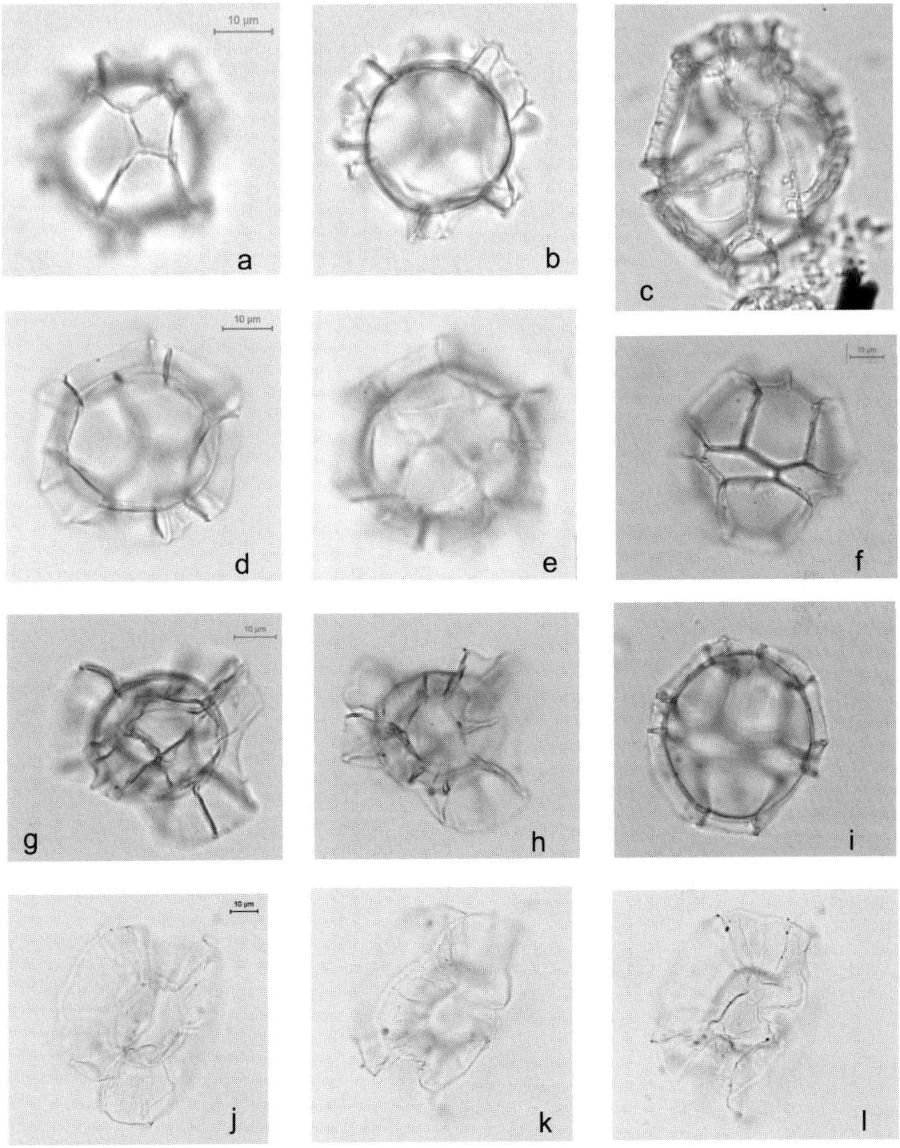

Fig. 10. (**a**) (AA, E), (**b**) (mf), *Impagidinium* cf. *I. aculeatum* sample/slide, 2H-4, 125–130/I (R29). (**c**) (V, E), *Impagidinium brevisulcatum* sample/slide, 7H-2, 145–148/II (E39). (**d**) (mf), (**e**) (AA), *Impagidinium* sp. of Brinkhuis & Biffi (1993) sample/slide, 13H–4, 125–130/I (F26). (**f**) (D, E), (**i**) (mf), *Impagidinium dispertitum* sample/slide, 2H-4, 125–130/I (F40). (**g**) (hf), (**h**) (mf), *Impagidinium* cf. *I. velorum* sample/slide, 19X-4, 057–060/I (N52). (**j**) (lf), (**k**) (mf), (**l**) (hf), *Impagidinium velorum* sample/slide, 4H-2, 125–130/I (F37).

Remarks. Differs from *D. craterum* in having more processes including cingular processes.

Enneadocysta multicornuta (Eaton 1971) Stover & Williams 1995.

Glaphyrocysta intricata (Eaton 1971) Stover & Evitt 1978. (Fig. 11g and h).

Glaphyrocysta sp. A (Fig. 12a and b).

Remarks. This taxon resembles *Cyclonephelium* sp. A of Williams & Brideaux 1975. The taxon has process complexes on all the precingulars apart from 6″ and process complexes on some of the postcingular and the antapical plates. Distally the processes within each complex

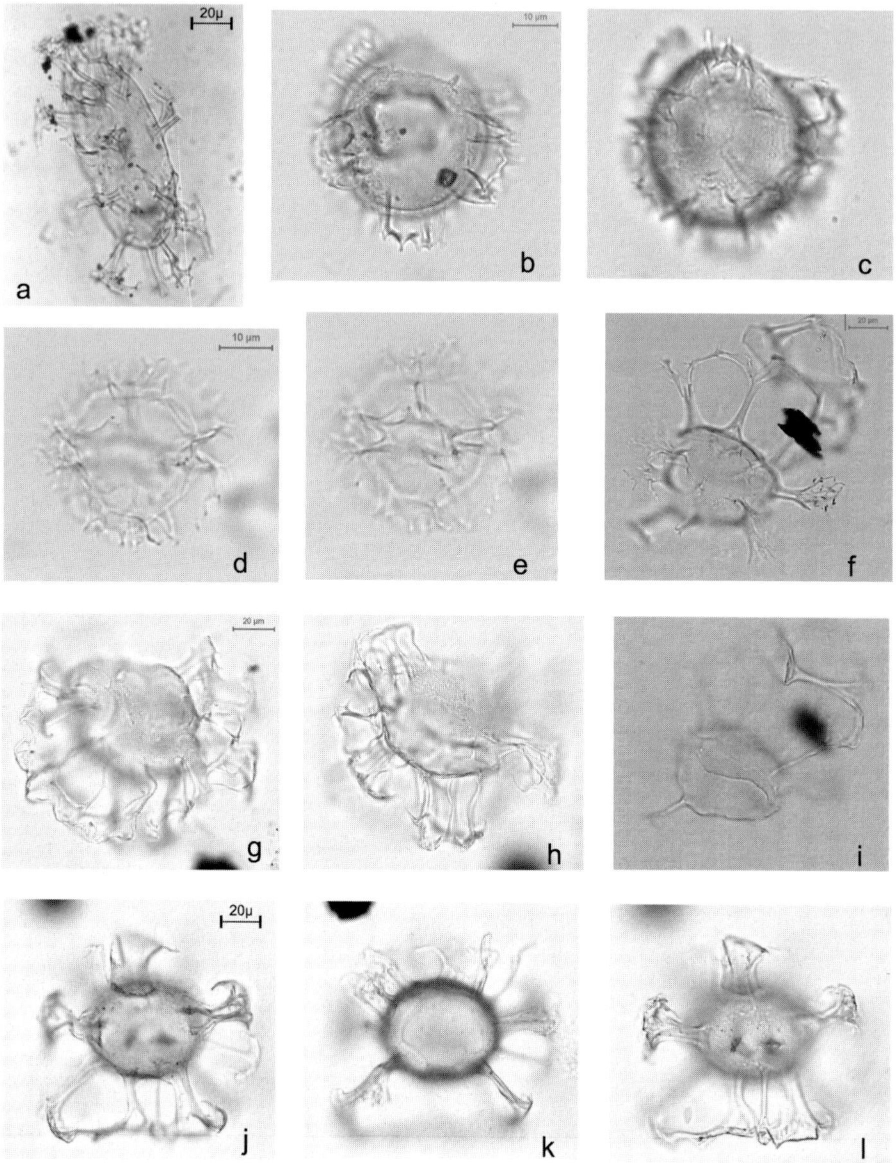

Fig. 11. (**a**) *Distatodinium* cf. *D. cratenum* sample/slide, 8H-3, 65–68/I (M48). (**b**) (D, E), (**c**) (V, I), *Achilleodinium* sp. A, sample/slide 5H-4, 124–129/II (T43). (**d**) (mf), (**e**) (D), *Spiniferites* sp. A. sample/slide, 14H-4, 125–129/I (C43). (**f, i**) *Areoligera*? sp. B sample/slide, 11H-4, 125–130/I (A35). (**g**) (V, E), (**h**) *Glaphyrocysta intricata* sample/slide, 8H-5, 035–038/I (C36). (**j, k**) (A, I), (**l**) (AA, E), *Areoligera*? sp. A sample/slide, 5H-6, 124–129/I (Q27).

can be united by trabeculae or perforate membranes. Complexes may be free or united by trabeculae.

Hemiplacophora semilunifera Cookson & Eisenack 1965 (Fig. 8g and j).
Hemiplacophora sp. A (Fig. 8f, i and l).

Remarks. The apical archeopyle (type tA), the incomplete penitabular septa, and absence of indication of the cingulum supports attribution to the genus *Hemiplacophora*. The endophragm (closely appressed to periphragm except below processes) is smooth. The periphragm is finely granulate. The septa, which can be perforate, are

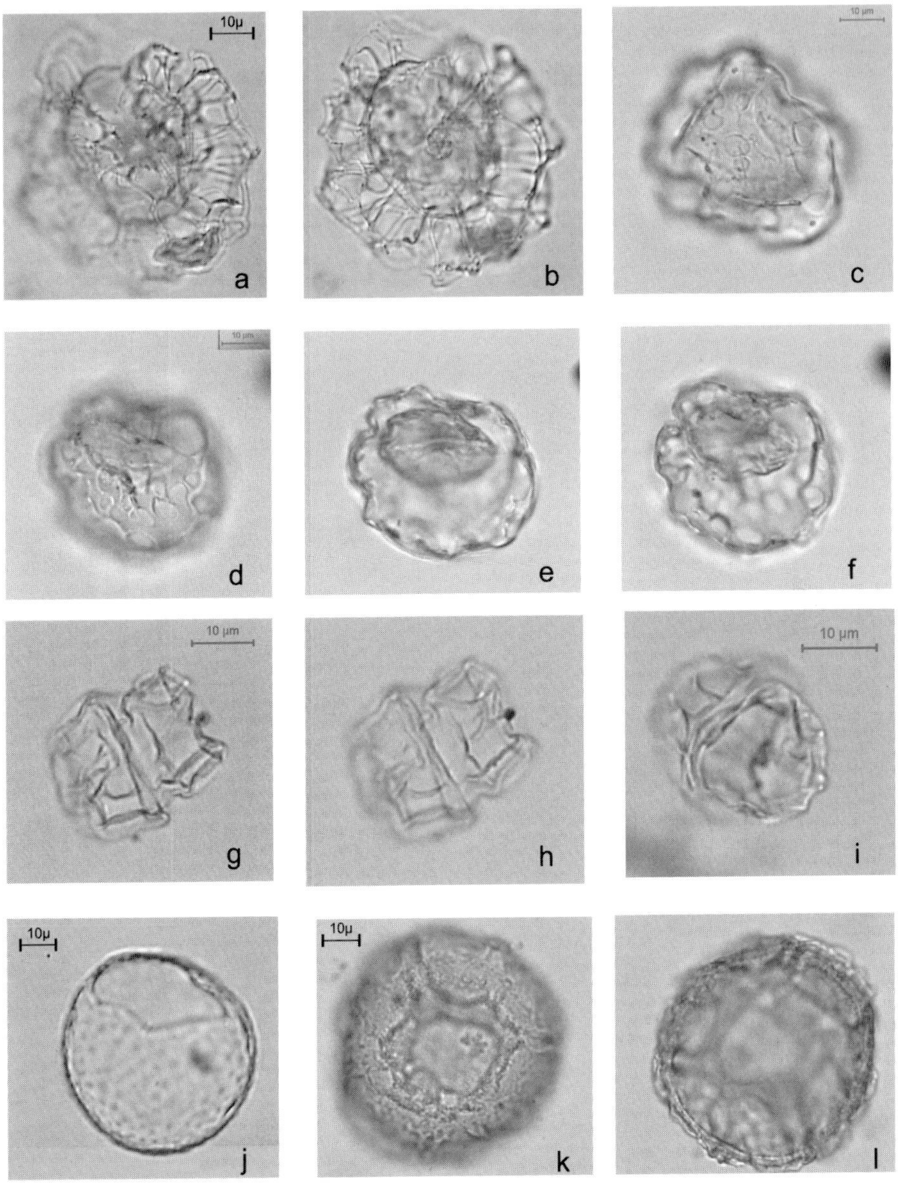

Fig. 12. (**a**) (hf), (**b**) (mf), *Glaphyrocysta* sp. A sample/slide 4H-6, 125–130/I (O40). (**c**) (hf), (**e**) (mf), (**f**) (lf), Forma A sample/slide, 2H-4, 125–130/I (**b**–**d**, L38; **f**, S46). (**g**) (hf), (**h**) (mf), (**i**) (hf), Forma B sample/slide, 2H-4, 125–130/I, (**g**, **h**, H34; **i**, B37). (**j**) (A, E), *Batiacasphaera compta* sample/slide, 3H-2, 125–129/I (O44), (**k**) (D, hf), (**l**) (mf) RL, I, *Stoveracysta*? sp. A sample/slide, 3H-2, 125–129/I (U26).

sometimes linear, sometimes arcuate, or occasionally, as with the antapical, mirror the outline of the paraplate. One specimen has a linear membrane encircling the archeopyle and on the periphragm plates. There is a break, however, in the sulcal area. When arcuatea, the septa are open towards the cingulum, which is devoid of ornamentation. This taxon differs from *H. semilunifera* by having larger and more irregular processes or septa.

Dimensions: central body 35–45 μm, septa 10–20 μm ($n = 10$).

Homotryblium floripes (Deflandre & Cookson 1955) emend. Stover 1975.

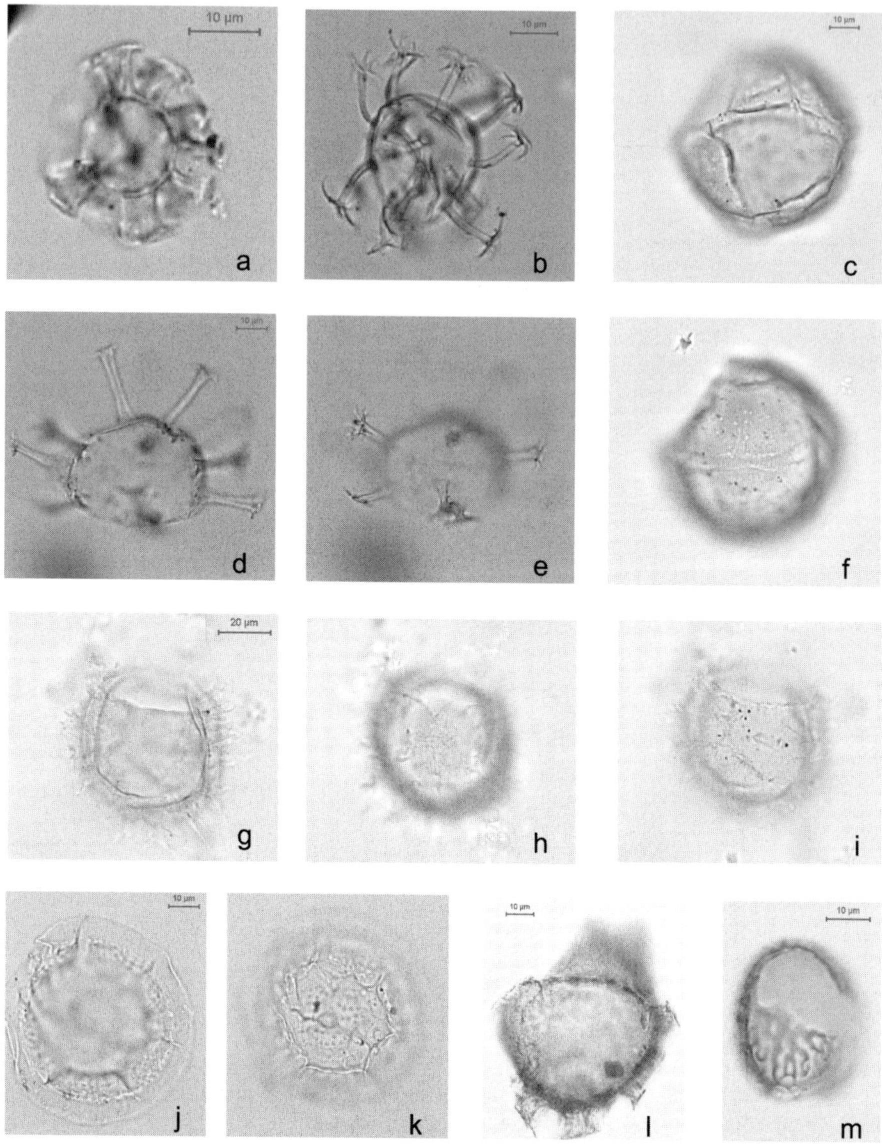

Fig. 13. (**a**) *Cordosphaeridium minimum* sample/slide, 2H-4, 125–130/I (J35). (**b**) *Melitasphaeridium pseudorecurvatum* sample/slide, 19X-4, 057–060/I (F30). (**c**) (RL, I), (**f**) (LL, E), *Cribroperidinium* cf. *C. tenuitabulatum* sample/slide, 11H-4, 125–130/I (F32). (**d**) (A), (**e**) (AA), *Homotryblium tenuispinosum* sample/slide, 2H-4, 125–130/I (J47). (**g**) (mf), (**h**) (lf), (**i**) (hf), *Xenicodinium* sp. A sample/slide, 14H-4, 125–129/I (D25). (**j**) (A, hf), (**k**) (mf), *Dinopterygium cladoides sensu* Morgenroth (1966) sample/slide, 2H-4, 125–130/I (R47). (**l**) (LL), *Turbiosphaera filosa* sample/slide, 9H-5, 105–108/I (Y29). (**m**) (V), *Cerebrocysta bartonensis sample/slide, 9H-5, 105–108/I (B35).*

Homotryblium tenuispinosum Davey and Williams 1966 (Fig. 13d and e).

Hystrichokolpoma cinctum Klumpp 1953

Hystrichokolpoma rigaudiae Deflandre & Cookson 1955.

Hystrichokolpoma truncata Biffi & Manum 1988. (Fig. 9g and j).

Hystrichosphaeropsis sp. of Brinkhuis & Biffi 1993 (Fig. 9h and k).

Impagidinium brevisulcatum Michoux 1985 (Fig. 10c).

Impagidinium cf. *I. aculeatum* (Wall 1967) Lentin & Williams 1981 (Fig. 10a and b).

Remarks. This species differs from *Impagidinium aculeatum* in having distally furcate processes or septa.
Impagidinium cf. *I. velorum* (Fig. 10g and h).
Remarks. This species differs from *Impagidinium velorum* in having incomplete septa.
Dimensions: central body 30 μm and septa 15 μm ($n = 1$).
Impagidinium dispertitum (Cookson & Eisenack 1965) Stover & Evitt 1978. (Fig. 10f and i).
Impagidinium maculatum (Cookson & Eisenack 1961) Stover and Evitt 1978. (Fig. 10i and l).
Impagidinium sp. of Brinkhuis & Biffi 1993 (Fig. 10d and e).
Impagidinium spp.
Impagidinium velorum Bujak 1984 (Fig. 10j, k and l).

Lentinia serrata Bujak in Bujak *et al.* 1980.
Lingulodinium machaerophorum. (Deflandre & Cookson 1955) Wall 1967.

Melitasphaeridium pseudorecurvatum (Morgenroth 1966) Bujak *et al.* 1980 (Fig. 13b).

Nematosphaeropsis–Cannosphaeropsis spp.

Oligosphaeridium anapetum sp. nov. (Fig. 8a–e)
Holotype: Fig. 8a–c.
Operculodinium divergens (Eisenack 1954) Stover & Evitt 1978.
Operculodinium spp. (pars.) Wall 1967, emend. Matsuoka *et al.* 1997.

Pentadinium laticinctum Gerlach 1961; emend. Benedek *et al.* 1982.
Pentadinium lophophorum Benedek 1972, emend. Benedek *et al.* 1982. (Fig. 9d and e).
Phthanoperidinium comatum (Morgenroth 1966) Eisenack & Kjellström 1971 (Fig. 7l).
Phthanoperidinium sp. A (Fig. 7j and k).
Remarks. The short apical horn and the indication of two antapical horns, an intercalary (2a) archeopyle, the clear peridinioid paratabulation as well as the clear cingulum are typical features for *Phthanoperidinium.* The surface is finely granular.
Dimensions: central body width 25 μm, length 35–40 μm, apical horn 4 μm ($n = 2$).
Phthanoperidinium spp. (pars).

Rottnestia borussica (Eisenack 1954) Cookson & Eisenack 1961.

Samlandia chlamydophora sensu Stover & Hardenbol 1994 (Fig. 8h and k).
Samlandia chlamydophora Eisenack 1954.
Schematophora speciosa Deflandre & Cookson 1955, emend. Stover 1975 (Fig. 9a–c).
Spiniferites sp. A (Fig. 11d and e).
Remarks: The precingular archeopyle, the sutural septa with distally bi- and trifurcate gonal processes supports the placement of this species in *Spiniferites.* It differs from other *Spiniferites* species by having relatively high sutural crests with typically denticulate margins.
Dimensions: central body 22–30 μm, septa 5–8 μm ($n = 4$).

Spiniferites spp. (pars).
Stoveracysta? sp. A (Fig. 12k and l).
Remarks. Very thick (up from 6 μm) spongeous wall, showing definite sutures, cingulum can be recognized. The archeopyle is apical. It is included in the genus Stoveracysta because of the apical archeopyle and the visible tabulation.
Dimensions: central body width 54–56 μm, length 64 μm.
Systematophora ancyrea Cookson & Eisenack 1965.
Systematophora placacantha (Deflandre & Cookson 1955) Davey *et al.* 1966; emend. May 1980.

Tectatodinium pellitum Wall 1967 emend. Head 1994.
Thalassiphora delicata Williams & Downie 1966; emend. Eaton 1976.
Thalassiphora patula (Williams & Downie 1966) Stover & Evitt 1978.
Thalassiphora pelagica (Eisenack 1954) Eisenack & Gocht 1960; emend. Benedek & Gocht 1981.
Turbiosphaera filosa (Wilson 1967) Archangelsky 1969. (Fig. 13l).

Wetzeliella articulata Eisenack 1938.
Wetzeliella gochtii Costa & Downie 1976.
Wetzeliella symmetrica Weiler 1956.

Xenicodinium sp. A (Fig. 13g–i).
Remarks. Although these forms have some similarity to *Operculodinium* spp., the wall structure is scabrate rather than reticulate. The archeopyle is precingular. We follow Stover & Hardenbol (1994), who assigned similar forms to *Xenicodinium.*
Dimensions: central body width 40–45 μm, length 45–50 μm, processes 10 μm ($n = 4$).

Incertae sedis.

Forma A (Fig. 12c–f).
Remarks. This is characterized as having a subspherical shape, and two wall layers. The inner body is smooth to granulate and subspherical. The outer wall is distinctly coarsely perforate and smooth. No indications of paratabulation or archeopyle have been observed. This form resembles Dinocyst sp. B of Head & Norris (1989).

Dimensions: overall size 45–50 μm, inner body 25–30 μm, average diameter of the perforations 5 μm ($n = 6$).
Affinity uncertain.

Forma B (Fig. 12g–i).
Remarks. This is characterized by its small size. There is a well-marked cingulum and an indication of peridinioid paratabulation. Most of the observed specimens have a short apical horn. No archeopyle was observed. The surface is smooth. Dimensions: central body width 20–25 μm and length 25–30 μm, apical horn 2 μm ($n = 4$).
Affinity uncertain.

We are grateful to J. Backman for reviewing an early version of the manuscript. The constructive remarks of our reviewers S. Damassa and J. Lucas-Clark were greatly appreciated. Financial support was provided by the Swedish Natural Research Council, the LPP-Foundation (Utrecht) and the 'Pieter Langerhuizen Lambertuszoon' Fund. This is NSG Publication 20000202 and Geological Survey of Canada Contribution 1999284.

References

BATTEN, D. J. 1996. Palynofacies. *In*: JANSONIUS, J. & MCGREGOR, D. C. (eds) *Palynology: Principles and Applications*, **3**. American Association of Stratigraphic Palynologists Foundation, Dallas, TX, 1011–1084.

BERGGREN, W. A., KENT, D. V., SWISHER, C. C. & AUBRY, M. P. 1995. A revised Cenozoic geochronology and chronostratigraphy. *In*: BERGGREN, W. A., KENT, D. V., AUBREY, M.-P. & HARDENBOL, J. (eds) *Geochronology Time Scales and Global Stratigraphic Correlation: a Unified Temporal Framework for a Historical Geology*: SEPM, *Special Publications*, **54**, 129–212.

BIFFI, U. & MANUM, S. B. 1988. Late Eocene–Early Miocene dinoflagellate cyst stratigraphy from the Marche Region (central Italy). *Bollettino della Società Palaeontologica Italiana*, **27**, 163–212.

BRINKHUIS, H. 1992. Late Palaeogene dinoflagellate cysts with special reference to the Eocene/Oligocene boundary. *In*: PROTHERO, D. R. & BERGGREN, W. A. (eds) *Eocene–Oligocene Climatic and Biotic Evolution*. Princeton Series in Geology and Paleontology, 327–340.

BRINKHUIS, H. 1994. Late Eocene to Early Oligocene dinoflagellate cysts from the Priabonian type-area (northeast Italy): biostratigraphy and palaeoenvironmental interpretation. *Palaeogeography, Palaeoclimatology, Palaeoecology*, **107**, 121–163.

BRINKHUIS, H. & BIFFI, U. 1993. Dinoflagellate cyst stratigraphy of the Eocene/Oligocene transition in central Italy. *Marine Micropalaeontology*, **22**, 131–183.

COSTA, L. I. & DOWNIE, C. 1979. The Wetzeliellaceae; Palaeogene dinoflagellates. *In*: *Proceedings of the 4th International Palynological Conference, Lucknow 1976–77*, **2**, 34–46.

DALE, B. 1996. Dinoflagellate cyst ecology: modelling and geological applications. *In*: JANSONIUS, J. & MCGREGOR, D. C. (eds) *Palynology: Principles and Applications*, **3**. American Association of Stratigraphic Palynologists Foundation, Dallas, TX, 1249–1275.

DAMASSA, S. P., GOODMAN, D. K., KIDSON, E. J. & WILLIAMS, G. L. 1990. Correlation of Palaeogene dinoflagellate assemblages to standard nannofossil zonation in North Atlantic DSDP sites. *Review of Palaeobotany and Palynology*, **65**, 331–339.

DAVEY, R. J. & ROGERS, J. 1975. Palynomorph distribution in Recent offshore sediments along two traverses off South West Africa. *Marine Geology*, **18**, 213–225.

GOODMAN, D. K. 1983. *Morphology, taxonomy and palaeoecology of Cretaceous and Tertiary organic-walled dinoflagellate cysts*. PhD thesis, Stanford University, CA.

HEAD, M. J. & NORRIS, G. 1989. Palynology and dinocyst stratigraphy of the Eocene and Oligocene in ODP Leg 105, Hole 647A, Labrador Sea. *In*: SRIVASTAVA, S. P. *et al*., (eds) *Proceedings of the Ocean Drilling Program, Scientific Results*, **105**, Ocean Drilling Program, College Station, TX, 515–550.

HEAD, M. J., NORRIS, G. & MUDIE, P. J. 1989. 27. Palynology and dinocyst stratigraphy of the Miocene in ODP Leg 105, Hole 645E, Baffin Bay. *In*: SRIVASTAVA, S. P. *et al*. (eds) *Proceedings of the Ocean Drilling Program, Scientific Results*, **105**, Ocean Drilling Program, College Station, TX, 467–514.

KÖTHE, A. 1990. Palaeogene dinoflagellates from northwest Germany—biostratigraphy and palaeoenvironment. *Geologisches Jahrbuch, Reihe A*, **118**, 3–111.

MANUM, S. B. 1976. Dinocysts in Tertiary Norwegian–Greenland Sea sediments (Deep Sea Drilling Project Leg 38), with observations on palynomorphs and palynodebris in relation to environment. *In*: TALWANI, M. *et al*. (eds) *Deep Sea Drilling Project, Initial Reports*, **38**. US Government Printing Office, Washington, DC, 897–919.

SCHLEE, J. & GERARD, R. 1965. *Cruise report and preliminary core log*, M/V Caldrill I—*17 April to 17 May 1965*. JOIDES Blake Panel report.

SETIAWAN, J. R. 1983. Foraminifera and microfacies of the type Priabonian. *Utrecht Micropalaeontological Bulletin*, **29**, 1–161.

SHIPBOARD SCIENTIFIC PARTY 1998*a*. Site 1053. *In*: NORRIS, R. D., KROON, D., KLAUS, A. *et al*. (eds) *Proceedings of the Ocean Drilling Program, Initial Reports*, **171B**. Ocean Drilling Program, Colege Station, TX, 321–347.

SHIPBOARD SCIENTIFIC PARTY 1998*b*. Introduction. *In*: NORRIS, R. D., KROON, D., KLAUS, A. *et al*. (eds) *Proceedings of the Ocean Drilling Program, Initial Reports*, **171B**. Ocean Drilling Program, College Station, TX, 5–10.

STOVER, L. E. & HARDENBOL, J. 1994. Dinoflagellates and depositional sequences in the Lower Oligocene (Rupelian) Boom Clay Formation, Belgium. *Bulletin de la Société Belge de Géologie*, **102**(1–2), 5–77 (cover date 1993, issue date 1994).

STOVER, L. E., BRINKHUIS, H., DAMASSA, S. P. *et al.* 1996. Mesozoic–Tertiary dinoflagellates, acritarchs and prasinophytes. *In*: JANSONIUS, J. & MCGREGOR, D. C. (eds) *Palynology: Principles and Applications*, **2**. American Association of Stratigraphic Palynologists Foundation, Dallas, TX, 641–750.

WALL, D., DALE, B., LOHMANN, G. P. & SMITH, W. K. 1977. The environmental and climatic distribution of dinoflagellate cysts in modern marine sediments from regions in the North and South Atlantic Oceans and adjacent areas. *Marine Micropalaeontology*, **2**, 121–200.

WILLIAMS, G. L. & BRIDEAUX, W. W. 1975. *Palynologic analyses of upper Mesozoic and Cenozoic rocks of the Grand Banks, Atlantic Continental Margin*. Geological Survey of Canada, Bulletin, **236**.

WILLIAMS, G. L. & BUJAK, J. P. 1977. Distribution patterns of some North Atlantic Cenozoic dinoflagellate cysts. *Marine Micropalaeontology*, **2**, 223–233.

WILLIAMS, G. L., LENTIN, J. K. & FENSOME, R. A. 1998*a*. *The Lentin and Williams Index of Fossil Dinoflagellates*, 1998 Edn. American Association of Stratigraphic Palynologists, Contributions Series, **34**.

WILLIAMS, G. L., BRINKHUIS, H., BUJAK, J. P. *et al.* 1998*b*. Dinoflagellates. *In*: HARDENBOL, J., THIERRY, J., FARLEY, M. B., JACQUIN, T., DE GRACIANSKY, P. C. & VAIL, P. R. (eds) *Mesozoic and Cenozoic Sequence Chronostratigraphic Framework of European Basins*. Society of Sedimentary Geology, Special Publication, **60**, 764–765.

North Atlantic climate variability in early Palaeogene time: a climate modelling sensitivity study

LISA CIRBUS SLOAN & MATTHEW HUBER

Department of Earth Sciences, University of California, Santa Cruz, Santa Cruz, CA 95064, USA (e-mail: lcsloan@earthsci.ucsc.edu)

Abstract: Understanding the nature and causes of the variability associated with past warm, high pCO_2 climates presents a significant challenge to palaeoclimate research. In this paper we investigate the early Eocene climatic response in the North Atlantic region to forcing from an indirect effect of atmospheric methane (via polar stratospheric clouds (PSCs)), and we investigate the response of the climate system to forcing from a combination of orbital insolation changes and high atmospheric pCO_2 concentration. We find that sea surface temperatures (SSTs), sea ice extent, net surface moisture, continental runoff and upwelling in the North Atlantic Ocean are all sensitive to those forcing factors, and that the degree of sensitivity is a function of location and season. Our results suggest that high-latitude SST values can vary by as much as 20 °C during the winter season in response to precessional and polar cloud forcing, whereas in contrast summer temperature varies by 4 °C or less. Model predictions of net surface moisture balance also vary substantially with our prescribed forcing. There is a large difference in variability between the localized net surface moisture results and the mean North Atlantic Ocean results, which suggests that large-scale assumptions about past surface ocean salinities and seawater $\delta^{18}O$ may need to be reassessed. According to model results, the influx of terrigenous material via continental runoff to the North Atlantic Ocean should be highly seasonal, with greatest runoff occurring in spring. Our model results also indicate that changes in wind-driven upwelling and in continental runoff on a precessional time scale should be seen in regions of the central North Atlantic.

Much work has gone into characterizing the warm mean climate states of late Cretaceous and early Cenozoic time (e.g. see summaries by Frakes *et al.* (1992) and Crowley & North (1991)). It has long been thought that greenhouse gases, most notably CO_2, were the dominant cause of past warm climates (e.g. Freeman & Hayes 1992; Crowley 1993; Berner 1994; Crowley & Kim 1995). In addition, early work on palaeoclimate records seemed to indicate a remarkable lack of climatic variability (e.g. Savin *et al.* 1975), which led to a characterization of past warm climates as being homogeneous, stable and not subject to the kind of variability common in Plio-Pleistocene time. In part, this idea arose from the relatively low temporal resolution of early palaeoclimate records. In contrast, more recent palaeoceanographic studies document high variability (on time scales less than 1 Ma) in deep-sea sediment characteristics, and presumably climate, throughout Cretaceous and early Palaeogene time (Herbert & D'Hondt 1990; Zachos *et al.* 1993, 1997; Bains *et al.* 1999; Norris *et al.* 1999). With increasing numbers of high-resolution records, these data suggest that past warm climates were no more stable than their cold, more recent, counterparts.

This high degree of climate variability suggests that factors other than tectonically driven greenhouse gases played roles in influencing climate. Furthermore, there is intriguing new evidence that Cenozoic climate variability has not been driven by atmospheric CO_2 variability (Pagani *et al.* 1999; Pearson & Palmer 1999). This idea implies either that greenhouse gases other than CO_2 have influenced the mean climate state during past climates, or that factors other than the direct effects of greenhouse gases played greater roles in driving climate change than has previously been explored. For climate variability on time scales of less than 1 Ma, forcing factors besides greenhouse gases must be considered to explain the variability. These ideas suggest that, especially for past warm climates, we need to (1) understand the climatic response to forcing from greenhouse gases other than CO_2, (2) more fully identify feedbacks between greenhouse gases and the rest of the climate system, and (3) explore the

From: KROON, D., NORRIS, R. D. & KLAUS, A. (eds). 2001 *Western North Atlantic Palaeogene and Cretaceous Palaeoceanography*. Geological Society, London, Special Publications, **183**, 253–272.
0305-8719/01/$15.00

response to forcing from combinations of factors that include but are not limited to the direct impacts of greenhouse gas changes (e.g. Bains *et al.* 1999).

It is not clear how oceanic and atmospheric processes operated under conditions of extreme warmth and high atmospheric pCO_2, such as those that existed in early Palaeogene time. Equally important, the sensitivity of these processes to changing climatic conditions is not well known. Thus, the early Palaeogene period provides a challenge to our understanding of how climate systems respond to and operate under conditions of extreme high-latitude warmth and relatively high atmospheric pCO_2.

The marine realm provides a more complete and extensive record of past climate processes than the terrestrial record. This is because there is greater ocean area than land area, and because global stratigraphic frameworks have been more completely established for marine than for continental sedimentary sequences. In addition, efforts by researchers associated with the Deep Sea Drilling Program and the Ocean Drilling Program have resulted in abundant records of deep-sea sediments that provide key information for reconstructing past climates, environments, and atmospheric and oceanic processes.

By combining scientific drilling and sedimentary and biogeochemical analyses of deep-sea sediments with modelling studies we can begin to more thoroughly document and understand past climate dynamics than has been accomplished previously. Currently, reconstructed linkages between hypothetical past climate processes and sedimentary responses are often qualitative or very rudimentary, especially for pre-Pleistocene time (e.g. Arthur & Garrison 1986). In some cases, the high-frequency variability recorded by sediments is not clearly linked to climate forcing, or the phasing is uncertain because of nonlinear processes (e.g. Morley & Huesser 1997; Zachos *et al.* 1997). These uncertainties occur as a function of the 'filtering' of climate signals by sedimentation processes. Climate models can be used to provide linkages between climate forcing factors and oceanic and atmospheric responses because they directly predict responses of the atmosphere and oceans to climate forcing (e.g. COHMAP Members, 1988; Oglesby & Park 1989; Crowley *et al.* 1992; Prell & Kutzbach 1992; Brickman *et al.* 1999).

In this study we use an atmospheric general circulation model (AGCM) to predict the sensitivity of North Atlantic climate to two forcing factors. We explore the indirect effects of high atmospheric methane concentrations (via the inclusion of polar stratospheric clouds (PSCs), explained below). We also investigate the influence of orbital forcing upon early Palaeogene climate. These two forcing factors may have influenced early Palaeogene climate and may have been responsible for some of the climatic variability recorded by marine sediments for this time. We focus upon model results in the North Atlantic region, particularly upon climatic factors that may most probably be manifested in the deep-sea sediment record.

Background

Atmospheric methane and polar stratospheric clouds

CO_2 has long been studied as an important greenhouse gas for past climates (e.g. Verbitsky & Oglesby 1992; Barron *et al.* 1993; Crowley & Kim 1995; Sloan & Rea 1995; Bush 1997) However, there are other greenhouse gases, most notably CH_4 and water vapour, that also may have had important impacts on climate at different times in Earth history. Several factors may have acted to increase tropospheric and stratospheric CH_4 concentrations and stratospheric water vapour concentrations during Palaeogene time, especially for late Palaeocene to early Eocene time.

It has been estimated that during Eocene time, there were at least three times the area of wetlands that existed today (Sloan *et al.* 1992). Given the overall warmer climate of late Palaeocene and early Eocene time (supported by diverse and independent fossil and isotopic evidence, e.g. Kennett & Stott 1991; Zachos *et al.* 1993, 1994; Greenwood & Wing 1995; Fricke *et al.* 1998; Wilf & Labandeira 1999; Wing *et al.* 2000), an extended methane emission season may have existed for wetlands. In addition, wetlands were likely to have been temperate in character (rather than polar). Temperate wetlands are estimated to produce at least three times the CH_4 fluxes that polar wetlands produce (Cao *et al.* 1998). Thus, to first order, we might expect a ten-fold increase in the flux of CH_4 into the troposphere during late Palaeocene and early Eocene time.

Increased tropospheric CH_4 fluxes may have affected conditions within both the troposphere and the stratosphere during past greenhouse climate intervals such as early Eocene time (Fig. 1). Researchers have suggested that the rate of CH_4 oxidation to water vapour may increase if the concentration of tropospheric CH_4 was substantially greater than at present. The amount of CH_4 that diffuses to the stratosphere may also increase in response to an increase in

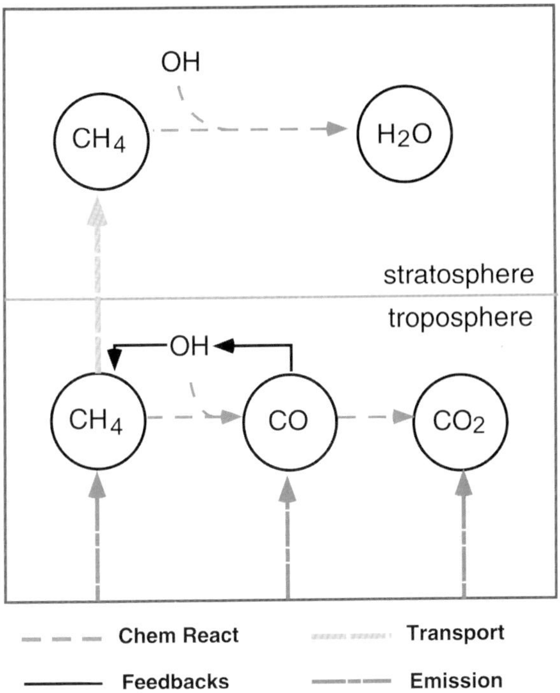

Fig. 1. Schematic diagram of methane oxidation in the atmosphere, from the perspective of climate effects. After Lelieveld *et al.* (1993).

tropospheric CH_4 concentration (Lelieveld *et al.* 1993). Additionally, an increase in CH_4 production may lead to an increase in the atmospheric residence time (Lelieveld *et al.* 1993). Thus we expect that the flux of CH_4 into the stratosphere and the standing concentration of CH_4 would be higher (Fig. 1).

The concentrations of tropospheric CH_4 also would be linked to tropospheric OH concentrations (Fig. 1). An increase in tropospheric CH_4 may have led to a decrease in tropospheric OH, amplifying the CH_4 increase (Chameides *et al.* 1977; Lelieveld *et al.* 1993). However, the lack of certainty in this proposed scenario is due in part to uncertainties in the amounts of OH, NO_x and water vapour that would have been present in the greenhouse atmosphere of early Eocene time.

PSCs are clouds of frozen water vapour that form in the lower stratosphere of the present atmosphere, in regions with temperatures at or below 195–187 K (Brasseur & Verstraete 1989). The temperature threshold limits PSCs to form in high-latitude regions during polar winter. PSCs are observed to form today in trace amounts (McCormick *et al.* 1989; David *et al.* 1997). These clouds warm the troposphere and the surface of the Earth by trapping outgoing longwave radiation. The albedo effect of PSCs is minimal because the clouds form only in polar night regions, and the net effect of these clouds is seasonally skewed towards winter warming.

There are two relationships that could increase the impact of PSCs during early Palaeogene time, over their present level. First, if tropospheric CH_4 was more abundant, stratospheric water vapour would also have been more abundant and PSCs would have been thicker. Second, if the troposphere was warmer than at present, the stratosphere would have been cooler. Lower stratospheric temperatures would have permitted the formation of more extensive PSCs than would occur today. For additional information on this topic we direct readers to Sloan *et al.* (1999).

With increased tropospheric CH_4 concentrations, stratospheric water vapour concentrations would increase and PSCs would be more extensive (Sloan & Pollard 1998; Sloan *et al.* 1999). Such conditions would be most likely to occur during greenhouse climates such as those in late Cretaceous or early Eocene time, and especially at the Late Palaeocene Thermal Maximum (LPTM). The peak warming of the LPTM has been hypothesized to have resulted from the catastrophic release of methane gas hydrates (Dickens *et al.* 1995, 1997; Bralower *et al.* 1997). Therefore, a condition of extensive or extreme PSCs, as a result of high atmospheric CH_4, might most apply to the LPTM, and to other periods similar to the LPTM that have not yet been fully identified in the geological record (e.g. Thomas & Zachos 1999).

In this study we investigate an extreme example of the effect of PSCs upon early Eocene climate. We prescribe PSCs because climate models do not yet have the capability of linking tropospheric methane to PSC dynamics (for more information, see Sloan *et al.* (1999) and Sloan & Pollard (1998)). This is an approximation of the clouds that we estimate would result in the presence of high (e.g. approximately ten-fold increase) atmospheric CH_4 concentrations. There are many other uncertainties in both the possible composition and structure of stratospheric clouds in Palaeogene time, and in the parameterization that we use to represent them. Despite these uncertainties, we believe that interesting possible physical mechanisms can be explored with this parameterization. These conditions are combined with a relatively high atmospheric CO_2 concentration (560 ppm) and ice-free land conditions in our model.

Orbital forcing

Several sedimentary records of Eocene age have been found to contain orbital-scale variability. Most notable is the record from the Green River Lake system of Eocene North America. These lake deposits exhibit varved sediments and sediment cycles with a frequency of *c.* 21 ka (Fischer & Roberts 1991; Roehler 1993). In addition, Eocene sediments recently recovered from the Blake Nose region in the western Atlantic Ocean contain high-frequency cycles that include a period of *c.* 22 ka (Kroon *et al.* 1999; Norris *et al.* 1999). Milankovitch-type variability has also been noted in lacustrine and marine sediments of other early Cenozoic and Mesozoic ages (e.g. Olsen 1986; Herbert & D'Hondt 1990; Park *et al.* 1993; Olsen & Kent 1996; Zachos *et al.* 1997).

We investigate the influence of the precessional cycle upon early Eocene climate, as that signal has been suggested in several different Palaeogene records (Fischer & Roberts 1991; Kroon *et al.* 1999; Norris *et al.* 1999). In our model we examine precessional forcing in conjunction with relative high atmospheric CO_2 concentration (560 ppm) and an ice-free Earth.

Model and methods

Climate model description

The climate model used in this study is Version 2 of the Genesis GCM (Thompson & Pollard 1997; Pollard *et al.* 1998). Version 2 of Genesis has been substantially modified from Version 1, and details of the modification have been provided by Thompson & Pollard (1997). Both versions of this model have been used extensively for palaeoclimate modelling studies (e.g. Sloan & Rea 1995; Kutzbach *et al.* 1996; Otto-Bleisner & Upchurch 1997; Pollard *et al.* 1998; Sloan & Pollard 1998). The model has been shown to perform on par with other GCMs in the simulation of present-day climate, and Version 2 of Genesis shows great improvement over Version 1 (Thompson & Pollard 1997).

The model contains both a diurnal cycle and a full seasonal solar cycle. The surface resolution of the model is 2° latitude by 2° longitude, and the horizontal resolution of the 18 atmospheric levels is spectral resolution T31 (*c.* 3.75° latitude × 3.75° longitude). Non-PSC clouds are predicted using prognostic 3D water cloud amounts. Individual greenhouse gas mixing ratios can be prescribed in the model for CO_2, CH_4 and N_2O. The effects of greenhouse gases are explicitly modelled in the IR radiation component of the model.

Model boundary conditions

Boundary conditions in our study were defined to represent early Eocene surface conditions. The land–sea distribution, terrestrial elevations and vegetation represent conditions during early Eocene time and are the same as those of Sloan & Rea (1995), interpolated to 2° × 2° resolution. Land surfaces were specified as ice-free. Soil composition was defined uniformly over land surfaces, with a value of 43% sand–39% silt–18% clay for all soil layers. Atmospheric pCO_2 was set at 560 ppm (2 × preindustrial value) for all of the experiments described below, atmospheric CH_4 and N_2O concentrations were set to preindustrial levels (0.700 and 0.285 ppm, respectively), and ozone was kept at modern values (Thompson & Pollard 1997). The solar constant was also specified at its modern value (1365 W m^{-2}).

In each case, sea surface temperatures (SSTs) were calculated by the model via the inclusion of a 50 m deep mixed layer (diffusive) slab ocean. As described by Thompson & Pollard (1997), meridional oceanic heat transport is calculated via a diffusion coefficient that depends upon latitude and the zonal fraction of land v. ocean at each latitude. Upward-directed heat flow under sea ice is set to 2 W m^{-2} in the Northern Hemisphere and 10 W m^{-2} in the Southern Hemisphere. The Norwegian Sea oceanic heat flux used by Thompson & Pollard (1997) and discussed in some palaeoclimate modelling studies was not used in our experiments, as it was designed for present-day conditions.

Model experiments

In this paper we present results from four experiments. The first case is our early Eocene control case (hereafter referred to as ECONTROL). This experiment includes the early Eocene boundary conditions described above. Orbital parameters were set at modern values. This case was run for 16 model years.

The second model experiment had the same boundary conditions as the ECONTROL case, with one addition. PSCs were prescribed to exist in the lower stratosphere (height of *c.*12 km) at latitudes within polar night. The clouds form a zonally symmetrical polar cap that expands and contracts seasonally. The clouds exist only within the winter half-year and extend to 66.5° latitude at maximum extent at winter solstice (Sloan & Pollard 1998). The prescribed fractional cover of the PSCs is 100 %, and the IR emissivity is 70 %. This cloud parameterization is slightly less extreme than that specified by Sloan & Pollard (1998); more details about this parameterization and the basis for it have been given by Sloan *et al.* (1999). This case (hereafter referred to as the PCLOUD case) was started from the 15th year of the ECONTROL case, and run for an additional 17 years.

The third experiment had the same boundary conditions as the ECONTROL case with the exception of altered orbital parameters. This case uses a minimum precession value with corresponding eccentricity and obliquity values taken from an orbital parameter time-series of the last 5 Ma (Berger 1978). The orbital configuration produces Northern Hemisphere perihelion during Northern Hemisphere winter solstice and aphelion during Northern Hemisphere summer solstice. This results in a relatively reduced Northern Hemisphere seasonal insolation cycle. This case (referred to as PMIN) was started from the 15th year of the control case, and run for an additional 10 years.

The last case had the same boundary conditions as the ECONTROL case with the exception of altered orbital parameters. This case represents the point in the same precessional cycle as above, but with precession at a maximum value, approximately 11.5 ka later than the PMIN case (with corresponding changes in the obliquity and eccentricity parameters). The orbital configuration results in Northern Hemisphere perihelion during the summer solstice and aphelion during Northern Hemisphere winter solstice. These conditions produce an amplified seasonal cycle of solar insolation for the Northern Hemisphere, with most of the impact occurring in Northern Hemisphere summer. This case (PMAX) was started from the 15th year of the ECONTROL case, and run for an additional 10 years.

In all cases the results were averaged over the final 3 years of model output. We focus here upon results in the North Atlantic region (Fig. 2). We consider the climatic factors that may have most influenced processes potentially recorded in the palaeoceanographic record.

Results

Surface temperature

North Atlantic SSTs show substantial responses to the polar cloud and orbital forcing, especially at high latitudes. For the period of December–February (DJF), in the North Atlantic at any given latitude the SST variability between the four cases increases with increasing latitude, to a maximum difference of 20 °C. Equatorward of 58° N latitude all four cases produce Atlantic SSTs that are within 1–2 °C of each other (Fig. 3a). At latitudes lower than 30° the PMAX case produces the lowest SST values, and the PMIN and PCLOUD cases produce the highest values. Poleward of latitude 58° N, SSTs in the four cases diverge more widely, especially poleward of 60° N latitude (the muted SST response to forcing at latitudes 58–60° N is a function of the palaeogeography and averaged ocean area at those latitudes, which includes western Tethys (Fig. 2, inset). At latitude 60° N the SST values differ by up to 6 °C, depending upon the forcing factor, and at 70° N, SSTs vary by up to 20 °C between the four cases. Across all latitudes poleward of 58° latitude the highest SST values are produced in the PCLOUD case, and the lowest values occur in the PMIN case. The PMAX case produces higher SST values than the ECONTROL case.

For the period of June–August (JJA), SSTs from the four cases show a uniform pattern at all latitudes: SSTs are highest in the PMAX case, followed by values from the PCLOUD case and then values from the ECONTROL case (Fig. 3b). SST values are lowest in the PMIN case. Unlike the DJF results, the difference between SST

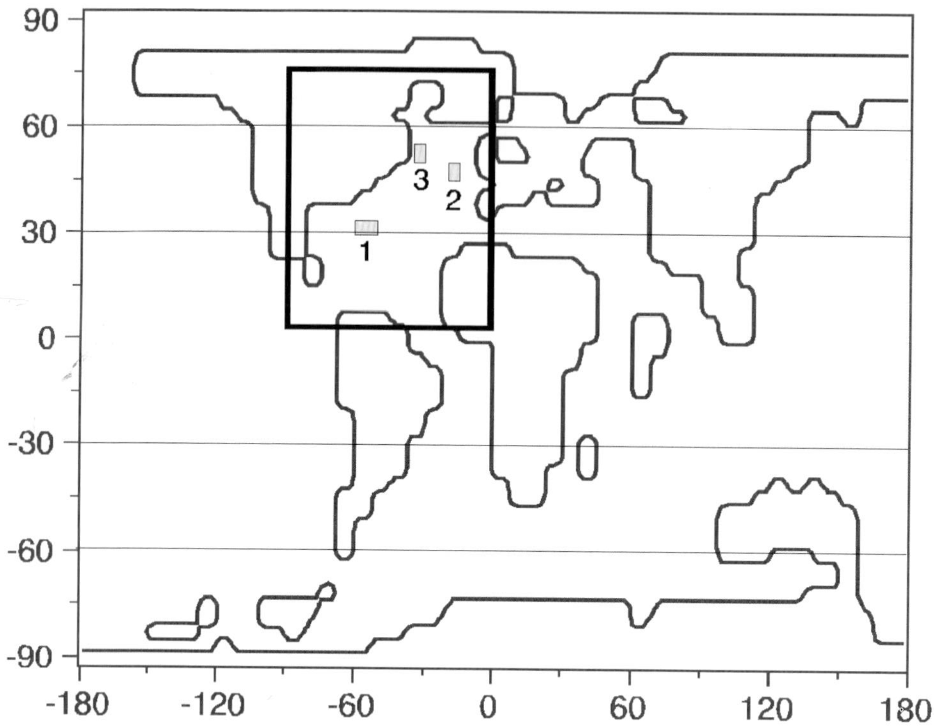

Fig. 2. Early Eocene palaeogeography used in the model experiments. Rectangle indicates North Atlantic region, which is the focus of our results and the domain of results presented in Figs 3–7 and 9. Labelled squares are sites for results shown in Figs 4 and 7 and discussed in the text.

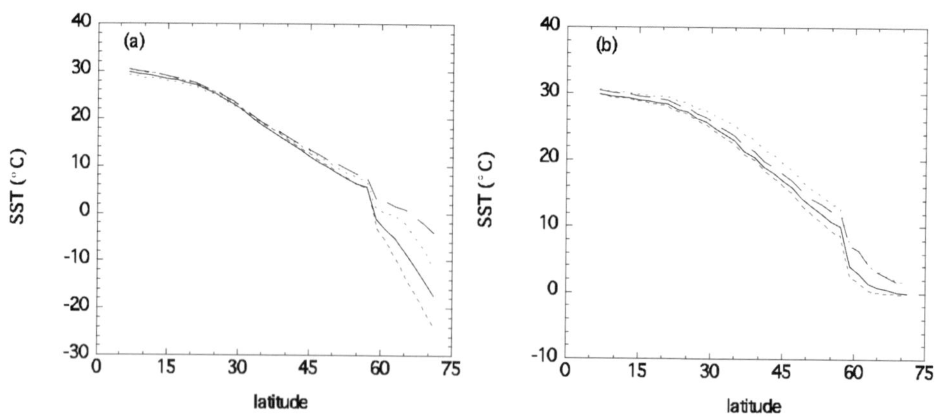

Fig. 3. North Atlantic SST results for the four model cases. Continuous line, ECONTROL case; long dashed line, PCLOUD case; short dashed line, PMIN case; dotted line, PMAX case. Left, results for December–February (DJF); right, results for June–August (JJA).

values in the four cases at any given latitude does not regularly increase with increasing latitude. The SST temperature variability between the four cases is 1–2 °C between the equator and 20° N latitude, increasing to a peak variability of *c.* 4 °C between 45 and 62° N latitude. Temperature variability between the four cases at highest latitudes of the North Atlantic is only *c.* 2 °C.

To illustrate the impact of the different forcing factors upon the annual cycle of SSTs we present the annual cycle of SSTs for three locations in the

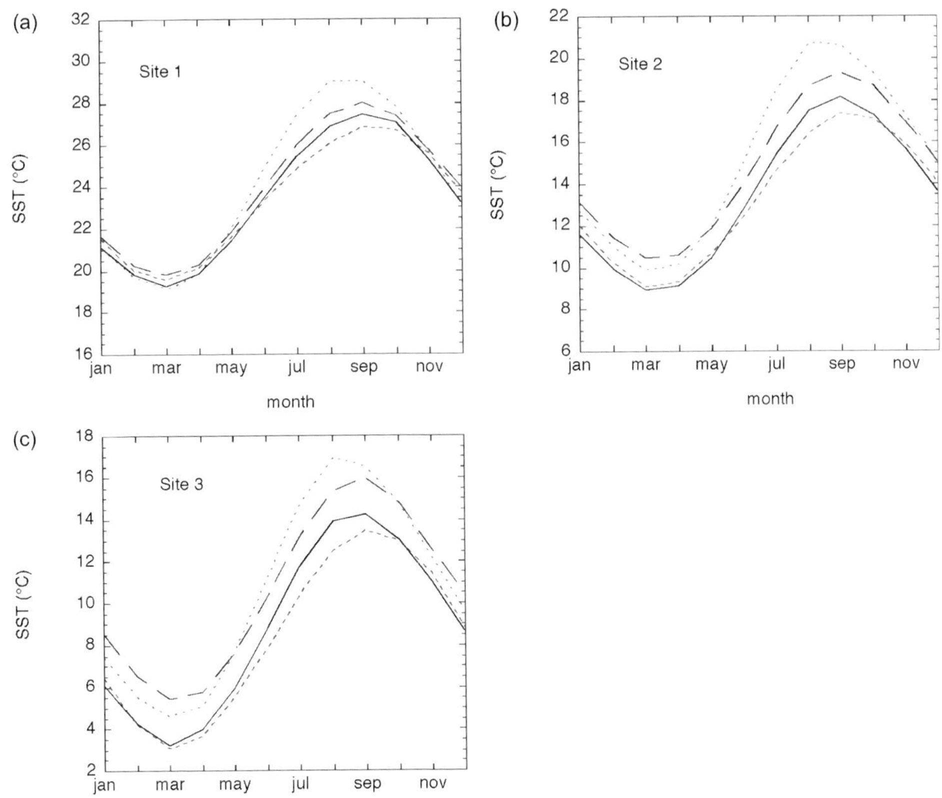

Fig. 4. Annual cycle of SSTs for three locations in the northwestern Atlantic Ocean (locations shown in Fig. 2): (**a**) Site 1, 29–22° N, 60–50° W; (**b**) Site 2, 45–50° N, 20–15° W; (**c**) Site 3, 50–55° N, 35–30° W. Line patterns the same as in Fig. 3.

western North Atlantic (Fig. 2). The southernmost location (Site 1, 29–33° N, 60–50° W) is near the early Eocene Blake Nose location (Norris *et al.* 1999). Temperature variability between the four cases is greatest in the period of July–October, and is smallest in November–February (Fig. 4a). In the peak summer months the SSTs vary by up to 3 °C between cases, depending upon the forcing factor. In the winter (February–April) SSTs values differ by less than 1 °C. At Site 1 the PMAX case produces the highest summer temperatures, followed by the PCLOUD case and the ECONTROL case. The PMIN case produces the lowest summer SST values. The PCLOUD case produces the highest winter SST values, followed by the PMIN case, and the ECONTROL and PMAX cases.

The second site (Site 2, 45–50° N, 20–25° W) shows a slight increase in the variability of mean monthly temperature between the four cases over the lower-latitude Site 1 results (Fig. 4b). As at Site 1, temperature variability is largest in summer (July–October) and smallest in winter (November–February). In the summer months the PMAX case produces the highest SST values, followed by the PCLOUD case, the ECONTROL case, and the PMIN case. This is the same hierarchy of temperature responses as at Site 1, and the temperature range between the cases is approximately the same (*c.* 3 °C). In the winter months, the highest SST values are produced in the PCLOUD case, as at Site 1. In contrast to Site 2 SSTs, the two coldest results occur in the ECONTROL and the PMIN cases. SST variability in the winter months is *c.* 2 °C. There is an increase in temperature variability between the cases for the period of November–June over results at Site 1.

The northernmost site (Site 3, 50–55° N, 35–30° W) shows a slight increase in the mean monthly SST variability between the four cases over results at Site 2 and a larger increase over results at Site 1 (Fig. 4c). For all months of the year the SST variability is at least 1.5 °C between the four cases. Summer SST variability is *c.* 4° and winter SST variability is *c.* 2.5 °C. The

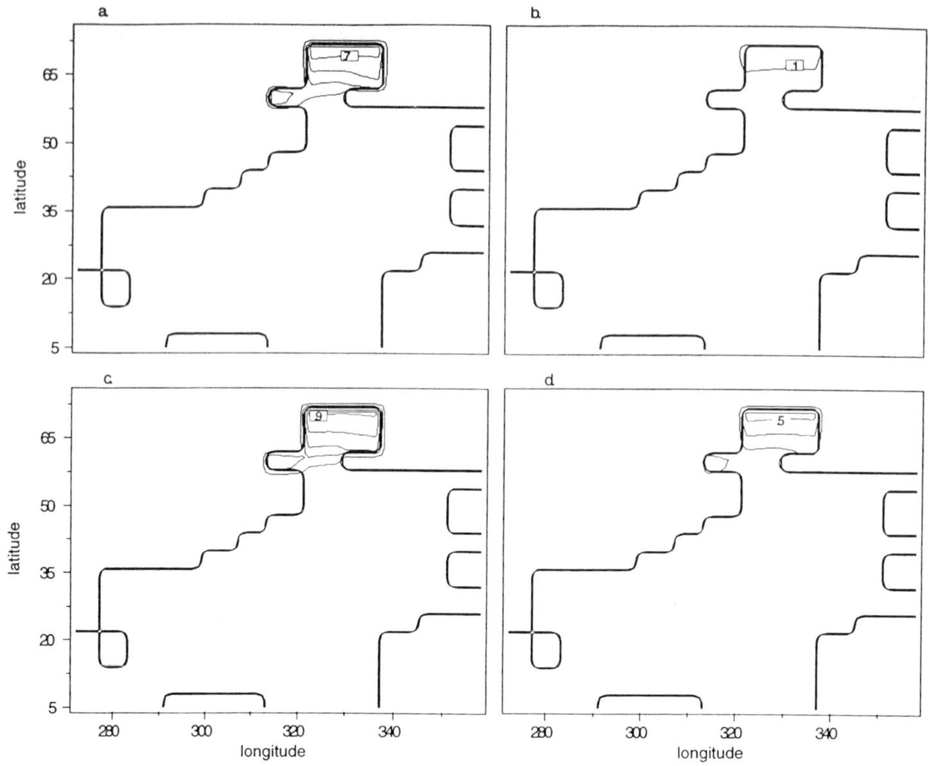

Fig. 5. Sea ice fraction (fraction of ocean surface covered by any thickness of sea ice at a given latitude) for DJF in the North Atlantic Ocean: (**a**) ECONTROL; (**b**) PCLOUD; (**c**) PMIN; (**d**) PMAX. Contour interval 0.2 m.

hierarchy of SST values is the same in the summer months as for the other two sites. In the winter months the highest and second-highest SST values are produced again by the PCLOUD and PMAX cases, respectively, and the lowest SST values are produced by the PMIN case.

Sea ice

The fraction of North Atlantic Ocean covered by sea ice during Northern Hemisphere winter (DJF) shows a high degree of variability between the four cases (Fig. 5). The PMIN case produces the greatest fraction of sea ice at all latitudes, and that case has sea ice extending over the largest area, to 54° N latitude. The ECONTROL case has the next greatest fraction of sea ice, with a southern extent of sea ice to the same latitude as in the PMIN case. The PMAX case has *c.* 10% less sea ice at any given latitude than results from the ECONTROL case, and most sea ice in the PMAX case is found poleward of 64° N latitude. In the PCLOUD case the smallest fraction of sea ice is produced and virtually no sea ice occurs equatorward of 65° N latitude in DJF.

Net moisture balance

We calculate the surface ocean net moisture balance as the difference between precipitation and evaporation in the model. For the North Atlantic region during DJF this net moisture balance differs by up to 1 mm day^{-1} between the four cases at any given latitude (Fig. 6a). There is net evaporation between the equator and *c.* 37° N latitude in all cases, and net precipitation poleward of this point (except for the PCLOUD case, discussed below). Between the equator and 35° N the PMAX case produces the lowest rate of net evaporation over the ocean surface, and the PMIN case produces the highest evaporation rate. The highest net precipitation rates are produced by the PCLOUD and the PMIN cases. Notably, at highest latitudes in the North Atlantic the PCLOUD case produces net evaporation of surface ocean waters whereas the other three cases produce net precipitation (Fig. 6a).

In JJA the ECONTROL, PMIN, and PCLOUD cases produce net evaporation conditions between 12 and 33° N latitude, and net

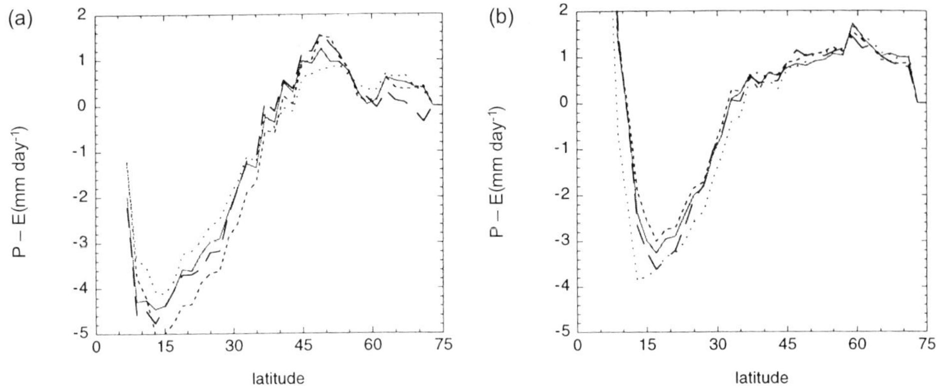

Fig. 6. Net surface ocean moisture balance, calculated as precipitation minus evaporation, in mm day^{-1} for the North Atlantic Ocean: (**a**) DJF; (*b*) JJA. Line patterns as in Fig. 3.

precipitation conditions at all other latitudes (Fig. 6). The PMAX case produces net evaporative conditions over a slightly larger area than the other cases, from 8–36° N latitude. The variability in net moisture between the cases is largest in the tropics, but is nowhere greater than *c.* 1 mm day^{-1}.

As with the SST values, we can examine the annual cycle of net moisture for the four cases at specific sites. For any given site, the variability in net moisture values is much greater than for the latitudinal averages of the results for the entire North Atlantic Ocean (compare Figs 6 and 7). At Site 1, the lowest-latitude site, net moisture values have their highest variability during the June–October period, up to 2.5 mm day^{-1}, and their lowest variability in March and April (Fig. 7a). Results also show that in all cases the net moisture balance is negative (net evaporation occurring) in December–February, and is positive (net precipitation) in all four cases only during September and October. All other months have values ranging from positive to negative between the cases. Values are typically most positive in the ECONTROL and PMIN cases and most negative in the PCLOUD and PMAX cases.

At Site 2, net moisture values are positive year-round in all four cases (Fig. 7b). As at Site 1, the values are highly variable between the cases, by up to 1.5 mm day^{-1} in any given month. Differences between the cases are greatest for results in December–March and smallest in the period of May–August. The largest net moisture values also occur in December through March. In general, the PMIN case produces the most positive values throughout the year.

At Site 3, net moisture values are also positive year-round in all four cases (Fig. 7c). The variability between the results for any given month is greater than 1 mm day^{-1} and is greatest for the period of October–February. Unlike results at Sites 1 and 2, there is no consistent seasonal signal seen in the net moisture values across the annual cycle for the four cases.

Continental runoff

In the Genesis climate model (like most AGCMs), runoff from continents does not route back into the model ocean. However, the continental runoff is calculated in the model, based upon a balance of precipitation and ground saturation (Thompson & Pollard 1997). We have identified continental-scale drainage basins for our palaeogeography and palaeotopography, and we examine the annual cycle of continental runoff that would be shed to the northwestern Atlantic from these basins. We consider this process because continental runoff can affect many biological and geochemical processes that can be recorded by deep-sea sediments (e.g. Stallard & Edmond 1983; Raymo *et al.* 1988; Hodell *et al.* 1990; Froelich *et al.* 1992; Bluth & Kump 1994). We calculated the mean monthly total runoff into the North Atlantic from (1) east–central and southeast North America, (2) northeast North America, and (3) the southern half of Greenland (Fig. 8).

The runoff from east–central and southeast north America into the Northwestern Atlantic shows a maximum volume of runoff occurring in April in all four cases, but the maximum runoff varies from *c.* 420 to 550 mm day^{-1} (i.e. 1260–1650 cm per month) between the cases (Fig. 8a). All cases show the same annual cycle, that of a single largest pulse of continental runoff

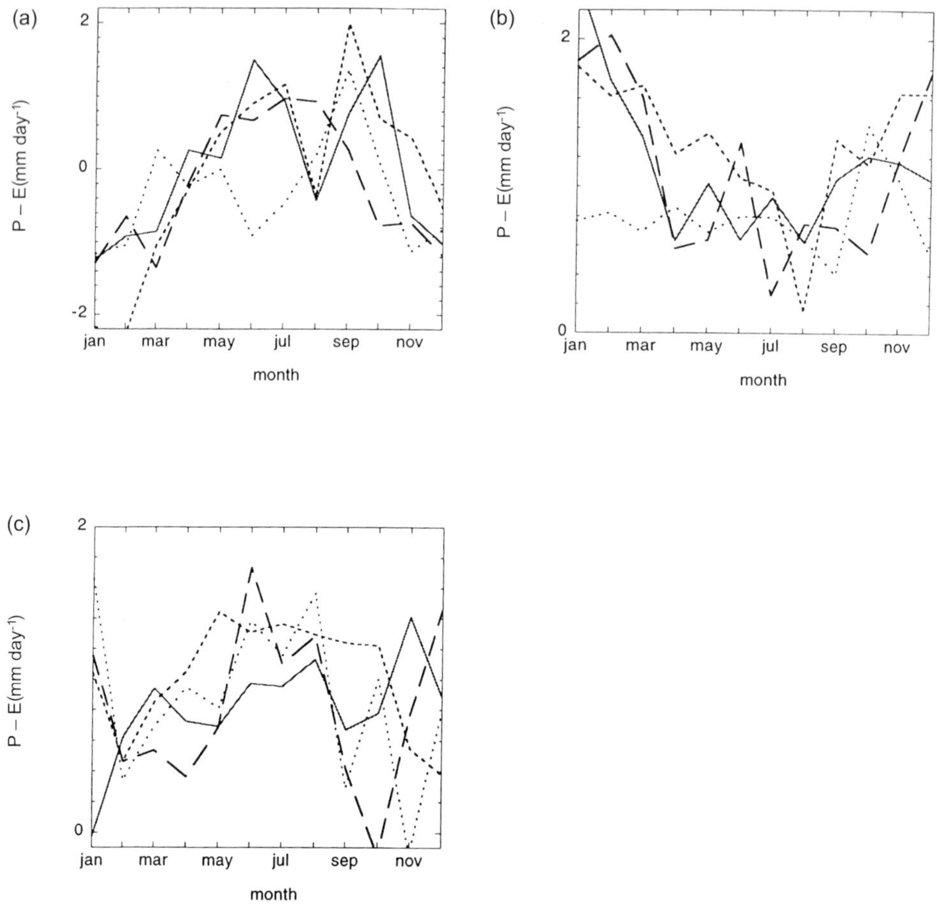

Fig. 7. Annual cycle of net surface moisture for three locations in the northwestern Atlantic Ocean (locations shown in Fig. 2): (**a**) Site 1, 29–22° N, 60–50° W; (**b**) Site 2, 45–50° N, 20–15° W; (**c**) Site 3, 50–55° N, 35–30° W. Line patterns the same as in Fig. 3.

occurring in the spring (March–May), and relatively constant lower runoff (*c.* 100 cm per month) occurring over the rest of the year. The duration of peak runoff (i.e. >200 mm day^{-1}) time varies slightly between the cases.

For the northeast region of North America, all cases show a maximum pulse of spring runoff (April–June) from the continent, but the actual timing of that pulse varies between the cases (Fig. 8b). In the ECONTROL case the maximum runoff occurs during a period spanning April–June, whereas the PCLOUD and PMAX cases have the maximum pulse in April and May. In the PMIN case the maximum pulse occurs in May and June. In May, the maximum runoff value in each case varies from *c.* 350 to 530 mm day^{-1} (i.e. 1050–1590 cm per month). All cases show little to no runoff occurring in the period from July to March. However, for this period runoff is slightly greater in the PCLOUD and PMAX cases than in the other two cases, especially for July–September (Fig. 8b).

Runoff from southern Greenland shows peak runoff also occurring in spring (April–June) for all four cases, but the timing and magnitude of the runoff varies between the cases (Fig. 8c). In May, peak runoff values range from a minimum of *c.* 230 mm day^{-1} (690 cm per month) in the PMIN case, to a high value of *c.* 500 mm day^{-1} (1500 cm per month) in the PMAX case in May. The ECONTROL case has maximum runoff of 400 mm day^{-1}, and the PCLOUD case has a maximum value of *c.* 330 mm day^{-1}. All cases show some runoff occurring year-round from southern Greenland, with the greatest year-round amount occurring in the PCLOUD case (Fig. 8c).

Upwelling

We calculated the wind-driven upwelling that our model winds would produce if we had a dynamic ocean component in our model. In general, upwelling is driven by wind stress curl. The relationship between winds and upwelling can be expressed as the divergence of the Ekman transport,

$$\nabla \cdot \boldsymbol{v}_{\epsilon} \frac{1}{\rho_{\mathrm{W}}} \boldsymbol{k} \cdot \nabla \times \left(\frac{\tau}{f}\right) \qquad (1)$$

calculated from curl of the wind stress, τ, which has been calculated from AGCM lowest atmospheric level wind fields, as given by Trenberth *et al.* (1990). In this expression, f is the Coriolis parameter, and ρ_{W} is the density of seawater (1027 kg m^{-3}).

The Ekman transport divergence provides a reliable estimate of wind-driven upwelling rates (Hellerman & Rosenstein 1983; Pedlosky 1996) at the present day. Explicitly linking upwelling to primary productivity and thence to proxy data records (e.g. mean accumulation rates, barium isotopes) is beyond the scope of this paper. We consider this a preliminary investigation of these issues, and we leave making explicit linkages with the proxy data record for later work.

In DJF, surface winds produce fairly strong upwelling (1×10^{-4} cm s^{-1} peak value, 0.5×10^{-4} cm s^{-1} overall) throughout much of the northwest Atlantic poleward of 45° N latitude. This upwelling is stronger (by a factor of two) and more extensive than that produced by the same models for the present day (not shown). In the ECONTROL and PCLOUD cases (Fig. 9) the upwelling extends across the North Atlantic to the eastern margins of the ocean (PCLOUD results not shown in Fig. 9 because they are the same as the ECONTROL result). Upwelling is much stronger (2–3 times) in the North Atlantic in the PMIN case relative to the PMAX case (not shown), although upwelling occurs in the same areas in both cases (Fig. 9). Upwelling in all cases extends southward along the east coast of North America, nearly to the Mississippi Embayment, and there is upwelling along the northern coast of South America and the northwestern margin of Africa. Upwelling along the margins of western Tethys shows slight differences between the model cases (Fig. 9).

Upwelling in JJA in all cases shows a reduction in the areas experiencing upwelling in the North Atlantic poleward of 45° latitude, relative to DJF conditions (Fig. 9). Upwelling is most restricted in that area in the PMAX case, with no upwelling occurring in the open ocean. In the ECONTROL and PCLOUD cases there is slightly more upwelling in the eastern North Atlantic around 45° N. In the PMIN case upwelling in the central North Atlantic is most extensive of all cases in JJA. Upwelling along the margins of western Tethys also shows some variability between the cases. Overall, there is greater variability in upwelling between the four cases in JJA than in DJF.

Discussion

Model results indicate that many oceanic and atmospheric processes are sensitive to changes in PSC forcing and to changes in orbital forcing. The primary responses by the atmosphere occur because of changes in incoming solar radiation (a controlling forcing in all seasons but minimized in winter), changes in outgoing longwave radiation (most important in the winter), and changes in the amount of sea ice (important in summer for changes in total receipt of solar radiation, and in winter for heat loss from the surface ocean to the atmosphere). Other processes, including precipitation, evaporation, winds and upwelling, respond to these factors. The variability of the climate responses to the two forcing factors examined in this study provides a first-order estimate of the uncertainties associated with assumptions about past climatic conditions, and of the processes that are included in palaeoceanographic interpretations.

Implications for variability of responses to forcing

Sea surface temperatures. The large variation in DJF SST values found poleward of 60° latitude in our results is due to the varying amounts of sea ice (in part itself influenced by palaeogeography) and insolation distributions. The difference in JJA SST results is due mostly to insolation differences, with small differences introduced by spring sea ice extents between cases (not shown). The presence of PSCs results in higher winter temperatures, with greatest warming occurring in high latitudes (Figs 3 and 4). This warming is due to reduced loss of longwave radiation during the winter via a direct response to the presence of the clouds (Sloan & Pollard 1998). There is also an indirect summer warming that occurs in response to the PSCs, caused by the reduction of winter sea ice in that case (Fig. 5), and the subsequently increased absorption of solar radiation by the newly ice-free ocean regions. As a result, there is a large winter SST response to PSCs and a smaller response in the summer (Fig. 3).

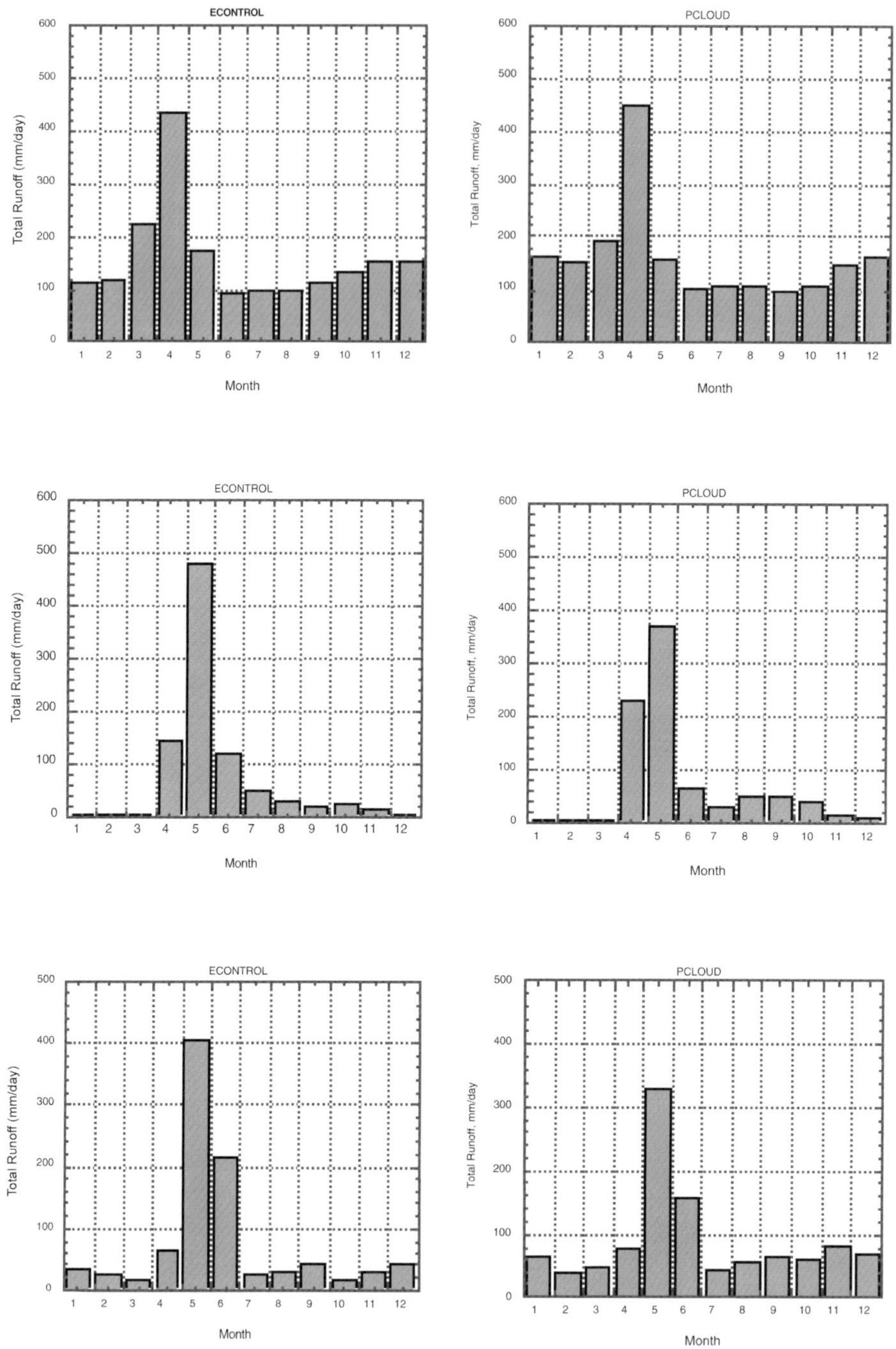

Fig. 8. See opposite page for caption.

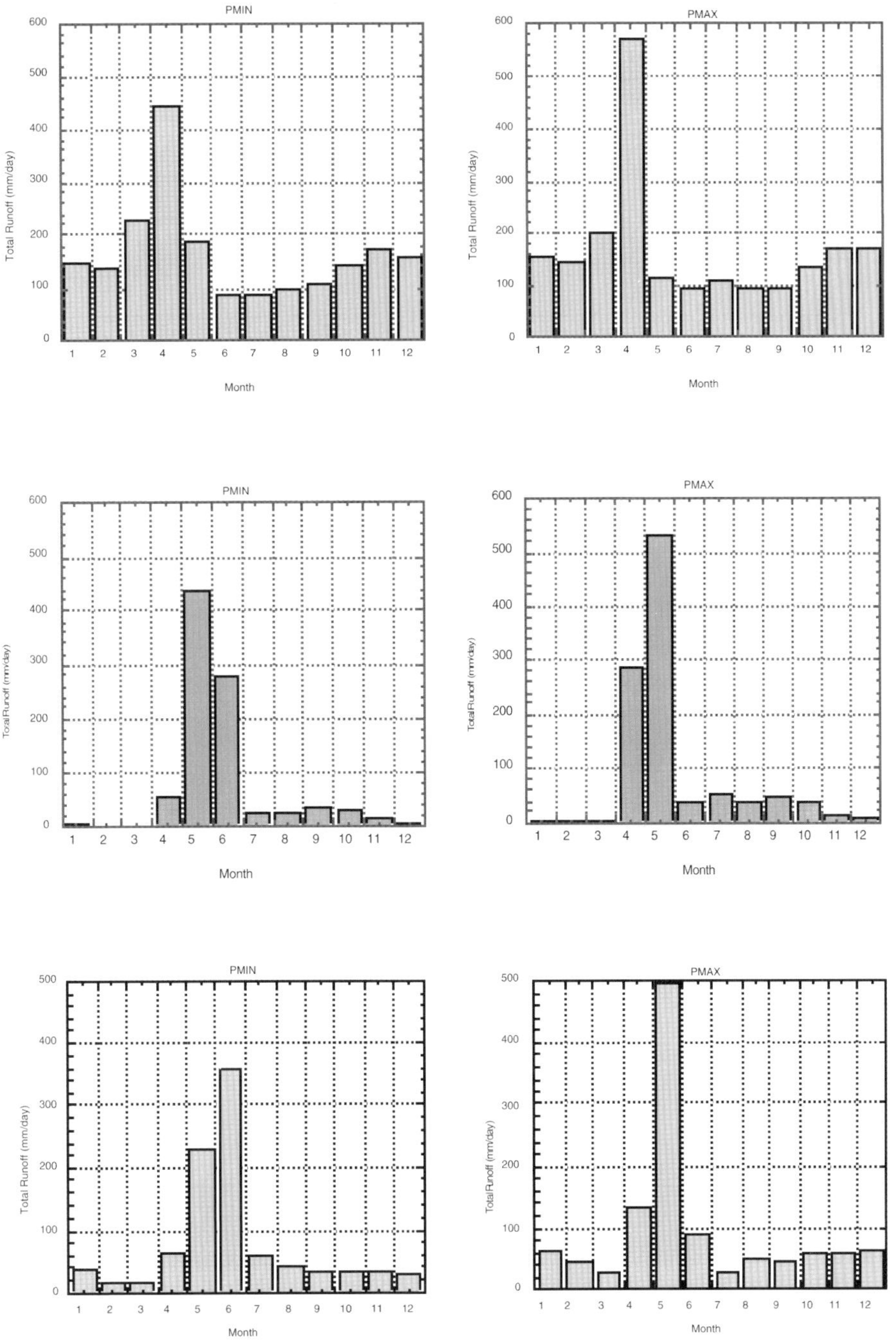

Fig. 8. Mean monthly total runoff (mm day^{-1}), from (top) east–central and southeast North America, (middle) northeast North America and (bottom) southern Greenland, for ECONTROL, PCLOUD, PMIN and PMAX cases. (Note different scales on *y*-axes between locations.)

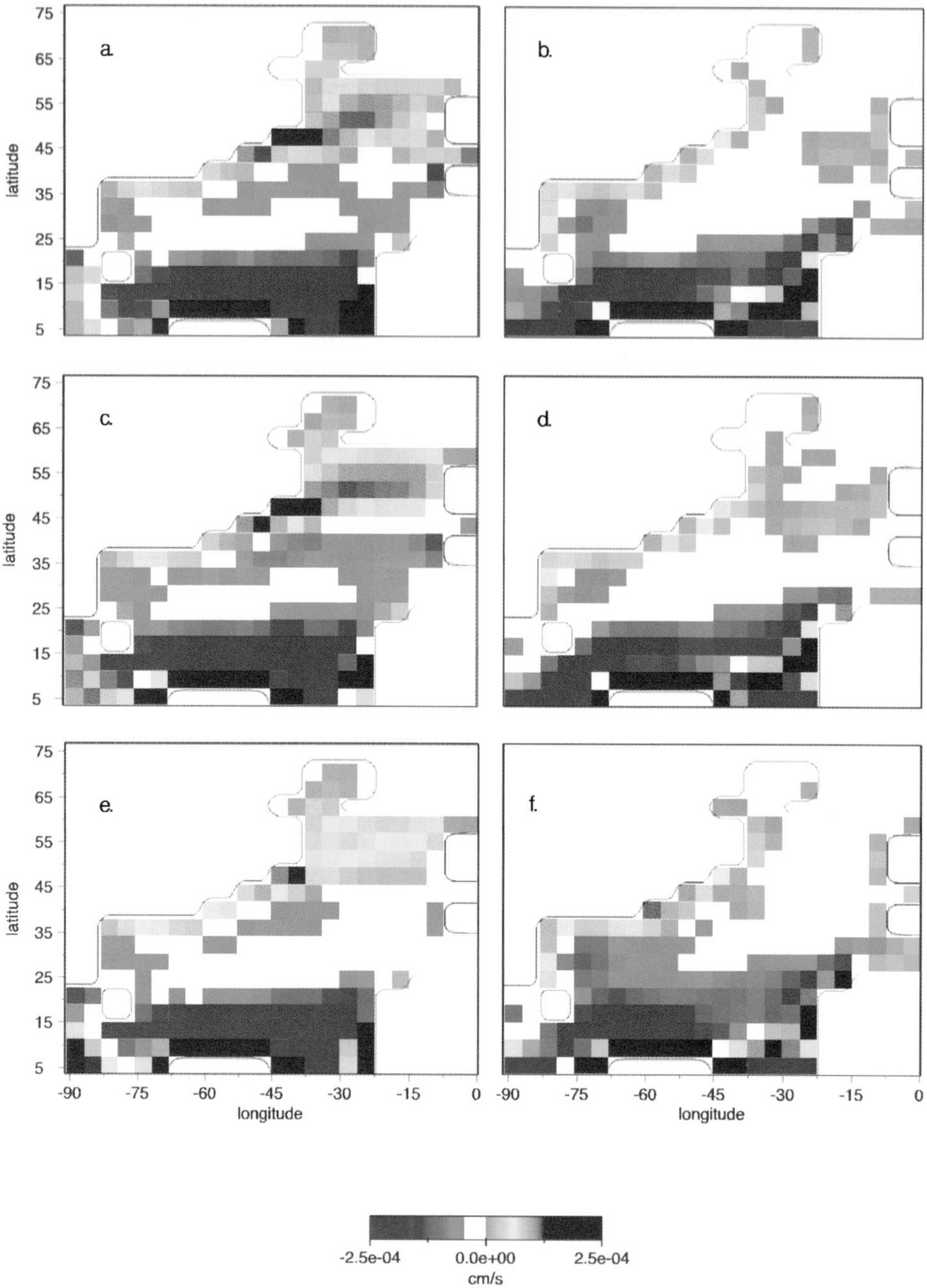

Fig. 9. Wind-driven upwelling for the North Atlantic region, with coloured areas denoting regions of upwelling (green to red) and downwelling (blue to purple): (a) ECONTROL DJF; (b) ECONTROL JJA; (c) PMIN DJF; (d) PMIN JJA; (e) PMAX DJF; (f) PMAX JJA. PCLOUD results not shown because they are the same as the ECONTROL results.

Response to this forcing factor is greatest at high latitudes and minimal in the subtropics and tropics. This result is most clearly seen by comparing the annual cycle of temperature responses in the four cases at the three sites. The PCLOUD case produces the highest winter SSTs at all sites (Fig. 4), but not the highest summer temperatures. In addition, the difference between the winter SSTs in the PCLOUD case vs the other three cases increases with increasing latitude (compare Fig. 4a and 4c).

The precessional forcing results in smaller overall SST responses than occur in the PCLOUD case (Fig. 3), but the forcing is important across a larger latitudinal range than in the PCLOUD case. The response is driven primarily by changes in incoming solar radiation, with a small indirect temperature perturbation caused by sea ice changes. The PMAX case, with the amplified seasonal insolation cycle, produces the highest summer SST values at every site. This is a function of perihelion occurring in Northern Hemisphere summer in this case; incoming solar radiation is much higher than in the other cases. In contrast, in the winter, the PMAX case results in the lowest SST value only at Site 1, the lowest latitude site, even though the PMAX case receives less winter insolation than the other cases. At Sites 2 and 3 the ECONTROL and PMIN cases produce lower winter SST values than the PMAX case. That the PMAX case receives less winter insolation than the other cases is not as important as the summer insolation component. The difference in DJF insolation between the PMAX and PMIN cases at 30° N latitude is *c.* 45 W m^{-2}, whereas at 50° N the difference is *c.* 20 W m^{-2}. The small difference in DJF insolation and the overall low amount of incoming solar radiation mean that other factors, such as sea ice, can introduce seasonal thermal lags and feedbacks and have important influences upon SSTs at higher latitudes (Crowley & North 1991; Short *et al.* 1991).

Overall, the variation in SST values that occurs in our cases suggests that localized responses to forcing such as precession and high-latitude warming (the PSCs) should be taken into account in SST interpretations (e.g. Zachos *et al.* 1994). The importance of localized effects upon temperature increases with increasing latitude.

Sea ice. Areas of the North Atlantic Ocean covered by sea ice show substantial response to the PSC and precessional forcing (Fig. 5). The greatest extent of sea ice is produced in our ECONTROL and PMIN cases, and these two cases also have the lowest SST values at high latitudes (Fig. 3). The PCLOUD case produces higher SSTs and less sea ice in response to the reduced winter longwave radiation caused by the PSCs (Sloan & Pollard 1998). The reduced longwave radiation results in reduced cooling of the winter polar region. In the PMAX case the SST values are higher and the sea ice extent less than those of the ECONTROL and PMIN cases for a different reason. In this case the higher winter temperatures and reduced sea ice occur because of residual heat from the increased summer warming.

Net surface moisture balance. Net moisture balance over the surface ocean has a broad pattern that is expected from global precipitation and evaporation patterns, and from the strong influence of temperature upon evaporation. In DJF and JJA net evaporation occurs in the tropics and subtropics. The greatest amount of net evaporation occurs in the PMIN case, where SSTs are relatively high (Fig. 3) and so is evaporation (not shown). Temperatures in the PMAX case are slightly lower (Fig. 3) and as a result evaporation is reduced. At high latitudes in DJF, the high SSTs in the PCLOUD case, combined with the increased area of open ocean (as a result reduced sea ice), results in more evaporation relative to the other cases. This produces a greater variability in the net surface moisture response. In contrast, in JJA at high latitudes the surface temperatures, evaporation and precipitation rates are similar between the four cases, resulting in less variable surface moisture conditions.

Viewed by location, the net surface moisture results are most variable between the cases when SST values are highest (compare Site 1 and Site 3 of Figs 4 and 7). This is expected, given the relationship between saturation vapour pressure and temperature (Clausius–Clapeyron equation; Wallace & Hobbs 1977). At the lowest latitudes we find the highest absolute temperatures, and so have the greatest evaporation response to a small temperature change. At higher latitudes the influence of temperature upon evaporation rates is less; there is a relative increase in the influence of precipitation and a relative decrease in the influence of evaporation upon the net surface moisture balance.

The large difference in variability between the localized net surface moisture results (Fig. 7) and the mean North Atlantic Ocean results (Fig. 6) suggests that local contributions to the surface ocean seawater $\delta^{18}O$ may be large enough to raise doubts about using global-scale 'corrections' derived from zonally averaged, large-scale representations to account for heterogeneity of seawater $\delta^{18}O$ (e.g. Zachos *et al.* 1994). For the area within Site 1, the range of annually averaged net moisture values is from −0.8 mm day^{-1} to 0.2 mm day^{-1}. On the basis of this range, SST values interpreted for this area might vary by as much as 4 °C (Crowley & Zachos 2000). In terms of the latitudinal gradients of net surface moisture (e.g. Zachos *et al.* 1994) the zonal seasonal average results found in this study are similar to present-day values, with net positive moisture values (net precipitation) found poleward of 40° N.

Continental runoff. Runoff was not included in the coupled atmosphere–ocean aspect of our model, but this result provides a first-order approximation of the influence of climate forcing upon continental runoff. If we assume that the runoff would ultimately be shed into the North Atlantic, there are implications for records of continental denudation and clay accumulation, and the relative contribution of terrigenous v. biogenic sediment components. Runoff reflects the precipitation–evaporation balance over land, plus the saturation of moisture by a simple soil parameterization in the model (Thompson & Pollard 1997).

All four of our model cases show highly seasonal runoff distributions, with peak runoff occurring in the spring (April–June) in all cases. This result is consistent with some interpretations of Palaeogene climate from deep-sea clay records (e.g. Robert & Chamley 1991; Gibson *et al.* 1993). For all three drainage basins there is slightly less seasonality to the runoff (that is, there is relatively more runoff occurring at off-peak months) in the PCLOUD case (Fig. 8). Some clay studies have suggested increased high-latitude warmth and more year-round precipitation at the LPTM (Robert & Chamley 1991; Gibson *et al.* 1993; Robert & Kennett 1994). Of our four cases presented here, the PCLOUD case best approximates the LPTM as a climate with extreme warming at high latitudes. The results from our PCLOUD case provide some support for the idea of more year-round precipitation at the LPTM. However, as we are examining runoff results for only limited areas of the globe this is a highly speculative conclusion.

The variability of runoff values to the North Atlantic between our four cases for all three drainage basins is extreme for some months. For example, in April for the east–central and southeast North America drainage basin, results vary by up to 3.9 m per month, with the greatest difference occurring between results from the ECONTROL and PMAX cases (Fig. 8). In May for the northeast North America drainage basin, variability between runoff in the cases is as great as 5.4 m per month (maximum difference between the PCLOUD and PMAX results). For the southern Greenland drainage basin in May a maximum runoff difference of 8.1 m per month occurs between the PMIN and PMAX cases (Fig. 8). These results suggest that precessional forcing and high-latitude warmth such as generated by the PSCs should produce noticeable changes to influx of terrigenous material to the North Atlantic Ocean, and that the influx should be highly seasonal, with the greatest runoff occurring in Northern Hemisphere spring.

Wind-driven upwelling. The upwelling calculated in this study shows great sensitivity to the precessional forcing, and less sensitivity to the PSC forcing. The PSC forcing has the greatest impact upon high-latitude regions (discussed above), and so the response at mid-latitudes by upwelling is minor (relative to the ECONTROL, PMIN and PMAX case results). In our results precession modulates the presence or absence of open-ocean upwelling in the North Atlantic and governs the vigour of near-coastal upwelling in the North Atlantic and western Tethys (Fig. 9). There also is a highly seasonal response of upwelling to precessional forcing for most regions of the North Atlantic. One implication of our results is that palaeoproductivity, as it is linked to upwelling strength, may be modulated by precessional forcing. There is some support for this idea in the results of Norris & Rohl (1999) and Wade & Kroon (1999), but more comparisons of these results with palaeoceanographic data must be made in the future. In addition, further investigation into other combinations of orbital forcing and other boundary conditions will be useful to pursue this issue.

Implications for palaeoceanographic interpretations

Results of modelling studies such as shown here may be valuable for defining the links (and temporal leads or lags) between climate forcing and sedimentary responses. For example, the direct response of the wind-driven ocean circulation to precessional forcing should be immediate (on geological time scales), with no lags. The response of continental runoff to precipitation, and of ocean sedimentation to continental runoff, would more probably contain significant time lags. However, in both examples our results are of sufficiently large-magnitude difference between the cases (Figs 8 and 9) that changes in upwelling or in terrigenous input on a precessional time scale should be seen in regions of the central North Atlantic (and, in the case of upwelling processes, along the margins of western Tethys). In contrast, we do not find any appreciable changes in upwelling results in the tropical Atlantic, which suggests that orbital variability in these records, if present, may be due to influences other than precessional forcing of upwelling.

The predictions of regions that are likely to contain processes that are highly sensitive to particular types of climate forcing could be used to design future scientific drilling efforts. We can gauge our understanding of these past warm climate systems by predicting aspects of the palaeoceanographic system with models, such as shown in this study, and then testing the hypotheses with strategic scientific drilling and sedimentary analyses. For example, as shown here, our modelling results predict that climatic processes in the northwest region of the North Atlantic had a high sensitivity to the presence of PSCs via temperature and runoff processes. Our results also predict that in the central North Atlantic there was strong upwelling sensitivity to precessional forcing in late Palaeocene and early Eocene time. This is somewhat consistent with the results of Kroon *et al.* (1999). However, further drilling of sediments in these regions could more completely support or refute such hypotheses. If our hypotheses are correct, analyses of sediments from these regions may provide more valuable information than sediments from other, less climatically sensitive, ocean regions for this time period.

Given the variation in climate characteristics produced for a single warm climate under different forcing conditions as shown here, it is very likely that when such variations in climate forcing are applied across different warm time periods we will find an even greater range of responses. In addition, the response of the climate system to forcing may also be a function of the pCO_2 'basic state'. That is, the slowly varying pCO_2 condition that changes on geological time scales may set up the basic environment in which variability occurs (e.g. Zachos *et al.* 1993). Therefore the nature of variability (i.e. frequencies and mechanisms) may be different in, for example, late Palaeocene compared with late Cretaceous time, or even between late Palaeocene and early Eocene time.

Future work

Currently we cannot directly model many important components of the past climate system, including soil chemistry, palaeoproductivity, deep ocean circulation and ocean chemistry. It is likely that the inclusion of such elements into our modelled system will significantly affect the results shown here, but the nature of such impacts is unknown. It would be most useful to study these variability questions with climate models that contain fully coupled atmospheres and oceans, and we also need to actively couple biological and geochemical processes to these systems to gain a full assessment of climate variability for a given time period. We also need to consider other combinations of forcing factors, and other types of models for such studies. In current work, we plan to carry out a detailed comparison of ocean sediment records with model results from these and other sensitivity studies. Such comparisons will allow us to evaluate the modelling sensitivity study results from the context of marine observations.

There are also some caveats about the methods and assumptions used in this study that provide inspiration for future work. First, the mean annual SSTs produced by the ECONTROL case are significantly different (5–15 °C colder) from those interpreted from SST proxies. Although we have identified possible flaws in assumptions regarding SST interpretations, we must also consider flaws in our modelling methods that relate to this temperature difference. As this is a sensitivity study of the climate response to forcings about a mean case, the difference in SST values between our control case and proxy data may not be important (i.e. we are assuming that the variability is a linear response to forcing and not sensitive to biases in our simulation of an early Palaeogene mean climate). However, this discrepancy in the SST values deserves to be a major focus in future studies of past warm climates. As another consideration, the results of this study are sensitive to the oceanic heat transport strength that is specified in the model. In the absence of any constraints on ancient heat transport we have opted to use the ocean heat transport strength that is based upon present-day simulations (Thompson & Pollard 1997). However, if ocean heat transport were changed, all of our results might change somewhat, especially the seasonal SST cycles, mean SST gradients, net surface moisture balance, and upwelling distributions. This presents unique opportunities, as it may be possible to constrain ocean heat transport by comparing model results with proxy data interpretations of seasonal SST cycles or by comparing model-predicted upwelling distributions and proxy-interpreted primary productivity patterns.

Conclusions

Model results indicate that SSTs, wind-driven upwelling, continental runoff, sea ice distributions and net surface moisture balance are sensitive to changes in PSC forcing and to changes in precessional forcing. The model predictions of regions containing high sensitivity of these processes to particular types of climate forcing may be useful for designing future scientific drilling efforts.

In the four cases presented here SST values vary between the cases by up to 20 °C at high latitudes in winter, and by only *c.* 2 °C at low latitudes in summer. The extent of sea ice in the North Atlantic Ocean is also sensitive to the imposed forcing and leads to additional climate feedbacks. Net surface moisture balance results demonstrate large sensitivity to both forcing factors at all latitudes and seasons, and our results suggest that large-scale or general assumptions about past surface ocean salinities and seawater $\delta^{18}O$ may need to be reassessed.

Our continental runoff results demonstrate that precessional forcing and high-latitude warmth such as generated by the PSCs should produce noticeable changes in the influx of terrigenous material to the North Atlantic Ocean, and that the influx should be highly seasonal with peak values in Northern Hemisphere spring. Wind-driven upwelling responses show greatest variability to the forcing factors in the North Atlantic poleward of 45° N, and there are large seasonal variations in upwelling strength and location between the cases.

We thank the National Science Foundation (NSF) for supporting this research (EAR9814883 and ATM9810799), and we thank the D. and L. Packard Foundation for additional support of this work. Computing was carried out at the National Center for Atmospheric Research, which is supported by the NSF. We thank D. Kroon and D. Norris for inviting this submission, and we thank A. Weaver and G. Schmidt for helpful reviews, which improved the manuscript. L.C.S. thanks members of the Warm Climates Program Planning Group of the Ocean Drilling Program for stimulating discussions that helped to shape this work, and L.C.S. especially thanks D. Kroon for encouragement of this work.

References

ARTHUR, M. & GARRISON, R. 1986. Cyclicity in the Milankovitch band through geologic time: an introduction. *Paleoceanography*, **1**, 369–372.

BAINS, S., CORFIELD, R. & NORRIS, R. 1999. Mechanisms of climate warming at the end of the Palaeocene. *Science*, **285**, 724–727.

BARRON, E. J., FAWCETT, P. J., POLLARD, D. & THOMPSON, S. L. 1993. Model simulations of Cretaceous climates: the role of geography and carbon dioxide. *Philosophical Transactions of the Royal Society of London, Series. B*, **341**, 307–315.

BERGER, A. 1978. Long-term variations of daily insolation and Quaternary climatic changes. *Journal of Atmospheric Science*, **35**, 2362–2367.

BERNER, R. 1994. GEOCARB II: A revised model of atmospheric CO_2 over Phanerozoic time. *American Journal of Science*, **294**, 56–91.

BLUTH, G. & KUMP, L. 1994. Lithologic and climatologic controls of river chemistry. *Geochimica Cosmochimica Acta*, **58**, 2341–2359.

BRALOWER, T., THOMAS, D. J., ZACHOS, J. C. *et al.* 1997. High-resolution records of the late Palaeocene thermal maximum and circum-Caribbean volcanism: is there a causal link? *Geology*, **25**, 963–966.

BRASSEUR, G. & VERSTRAETE, M. M. 1989. Atmospheric chemistry–climate interactions. *In*: BERGER, A. *et al.* (eds) *Climate and Geo-Sciences*. Kluwer Academic, Dordrect, 279–302.

BRICKMAN, D., HYDE, W. & WRIGHT, D. 1999. Filtering of Milankovitch cycles by the thermohaline circulation. *Journal of Climate*, **12**, 1644–1658.

BUSH, A. 1997. Numerical simulation of the Cretaceous Tethys circumglobal current. *Science*, **275**, 807–810.

CAO, M., GREGSON, K. & MARSHALL, S. 1998. Global methane emission from wetlands and its sensitivity to climate change. *Atmospheric Environment*, **32**, 3293–3299.

CHAMEIDES, W. L., LIU, S. C. & CICERONE, R. J. 1977. Possible variations in atmospheric methane. *Journal Geophysical Research*, **82**, 1795–1798.

COHMAP Members 1988. Climatic changes of the last 18,000 years: observations and model simulations. *Science*, **241**, 1043–1052.

CROWLEY, T. J. 1993, Geological assessment of the Greenhouse Effect. *Bulletin of the American Meteorological Society*, **74**, 2363–2373.

CROWLEY, T. J. & KIM, K.-Y. 1995. Comparison of longterm greenhouse projections with the geologic record. *Geophysical Research Letters*, **22**, 933–936.

CROWLEY, T. J. & NORTH, G. R. 1991. *Paleoclimatology*. Oxford University Press, New York.

CROWLEY, T. J. & ZACHOS, J. 2000. Comparison of zonal temperature profiles for past warm time periods. *In*: HUBER, B. T., MACLEOD, K. & WING, S. L. (eds), *Warm Climates in Earth History*. Cambridge University Press, Cambridge, 50–76.

CROWLEY, T. J., KIM, K.-Y., MENGEL, J. & SHORT, D. 1992. Modeling 100,000-year climate fluctuations in pre-Pleistocene time series. *Science*, **255**, 705–707.

DAVID, C., GODIN, S., MEGIE, G., EMERY, Y. & FLESIA, C. 1997. Physical state and composition of polar stratospheric clouds inferred from airborne lidar measurements during SESAME. *Journal of Atmospheric Chemistry*, **27**, 1–16.

DICKENS, G., CASTILLO, M. M. & WALKER, J. C. G. 1997. A blast of gas in the latest Palaeocene: simulating first-order effects of massive dissociation of oceanic methane hydrate. *Geology*, **25**, 259–262.

DICKENS, G., O'NEIL, J., REA, K. & OWEN, R. 1995. Dissociation of oceanic methane hydrate as a cause of the carbon isotope excursion at the end of the Palaeocene. *Paleoceanography*, **10**, 965–971.

FISCHER, A. G. & ROBERTS, L. T. 1991. Cyclicity in the Green River Formation (lacustrine Eocene) of Wyoming. *Journal of Sedimentary Petrology*, **61**, 1146–1154.

FRAKES, L., FRANCIS, J. & SYKTUS, J. 1992. *Climate Modes of the Phanerozoic*. Cambridge University Press, Cambridge.

FREEMAN, H. & HAYES, J. 1992. Fractionation of carbon isotopes by phytoplankton and estimates of ancient CO_2 levels. *Global Biogeochemical Cycles*, **6**, 185–198.

FRICKE, H., CLYDE, W., O'NEIL, J. & GINGERICH, P. 1998. Evidence for rapid climate change in North America during the Latest Palaeocene thermal maximum: oxygen isotope compositions of biogenic phosphate from the Bighorn Basin (Wyoming). *Earth and Planetary Science Letters*, **160**, 193–208.

FROELICH, P., BLANC, V., MORTLOCK, R., CHILLRUND, S., DUNSTAN, W., UDOMKIT, A. & PENG, T.-H. 1992. River fluxes of dissolved silica to the ocean were higher during glacials: Ge/Si in diatoms, rivers, and ocean. *Paleoceanography*, **7**, 739–767.

GIBSON, T., BYBELL, L. & OWENS, J. 1993. Latest Palaeocene lithologic and biotic events in neritic deposits of southwestern New Jersey. *Paleoceanography*, **8**, 495–514.

GREENWOOD, D. R. & WING, S. L. 1995. Eocene continental climates and latitudinal temperature gradients. *Geology*, **23**, 1044–1048.

HELLERMAN, S. & ROSENSTEIN, M. 1983. Normal monthly wind stress over the world ocean with error estimates. *Journal of Physical Oceanography*, **17**, 1093–1104.

HERBERT, T. D. & D'HONDT, S. L. 1990. Precessional climate cyclicity in Late Cretaceous–Early Tertiary marine sediments: a high resolution chronometer of Cretaceous–Tertiary boundary events. *Earth and Planetary Science Letters*, **99**, 263–275.

HODELL, D., MEAD, G. & MUELLER, P. 1990. Variation in the strontium isotopic composition of seawater (8 Ma to present): implications for chemical weathering rates and dissolved fluxes to the oceans. *Chemical Geology*, **80**, 291–307.

KENNETT, J. & STOTT, L. 1991. Abrupt deep-sea warming, palaeoceanographic changes and abenthic extinctions at the end of the Palaeocene. *Nature*, **353**, 225–229.

KROON, D., NORRIS, R., KLAUS, A. & ODP Leg 171B Scientific Party 1999. Variability of extreme Cretaceous–Palaeogene climates: evidence from Blake Nose (ODP Leg 171B). *In*: ABRANTES, F. & MIX, A. (eds) *Reconstructing Ocean History: a Window into the Future*. Kluwer Academic, Dordrect; Plenum, New York.

KUTZBACH, J., BONAN, G., FOLEY, J. & HARRISON, S. P. 1996. Vegetation and soil feedbacks on the response of the African monsoon to orbital forcing in the early to middle Holocene. *Nature*, **384**, 623–626.

LELIEVELD, J., CRUTZEN, P. J. & BRUHL, C. 1993. Climate effects of atmospheric methane, *Chemosphere*, **26**, 739–768.

MCCORMICK, M., TREPTE, C. & PITTS, M. 1989. Persistence of polar stratospheric clouds in the Southern Polar Region. *Journal of Geophysical Research*, **94**, 11241–11251.

MORLEY, J. & HUESSER, L. 1997. Role of orbital forcing in east Asian monsoon climates during the last 350 kyr: evidence from terrestrial and marine climate proxies from core RC14-99. *Paleoceanography*, **12**, 483-493.

NORRIS, R., ROHL, U. & BAINS, S. 1999. Astronomically-tuned chronology for the Palaeocene–Eocene transition and the structure of the $\delta^{13}C$ anomaly. *In*: ANDREASSON, F., SCHMITZ, B. & THOMPSON, E. (eds) *Early Palaeogene Warm Climates and Biosphere Dynamics, Abstract Volume*. GFF, Geologiska Fóreningen, Goteborg, 1999.

NORRIS, R. D. & ROHL, U. 1999. Carbon cycling and chronology of climate warming during the Palaeocene/Eocene transition. *Nature*, **401**, 775–778.

OGLESBY, R. & PARK, J. 1989. The effect of precessional insolation changes on Cretaceous climate and cyclic sedimentation. *Journal of Geophysical Research*, **94**, 14793–14816.

OLSEN, P. 1986. A 40-million year lake record of early Mesozoic orbital climate forcing. *Science*, **234**, 842–844.

OLSEN, P. & KENT, D. 1996. Milankovitch climate forcing in the tropics of Pangaea during the late Triassic. *Palaeogeography, Palaeoclimatology, Palaeoecology*, **122**, 1–26.

OTTO-BLEISNER, B. & UPCHURCH, G. 1997. Vegetation-induced warming of high-latitude regions during the Late Cretaceous period. *Nature*, **385**, 804–807.

PAGANI, M., ARTHUR, M. & FREEMAN, K. 1999. Miocene evolution of atmospheric carbon dioxide. *Paleoceanography*, **14**, 273–292.

PARK, J., HERBERT, T. D. & D'HONDT, S. L. 1993. Late Cretaceous precessional cycles in double time: a warm-Earth Milankovitch response. *Science*, **261**, 1431–1434.

PEARSON, P. & PALMER, M. 1999. Middle Eocene seawater pH and atmospheric carbon dioxide concentrations. *Science*, **284**, 1824–1826.

PEDLOSKY, J. 1996. *Ocean Circulation Theory*. Springer, New York.

POLLARD, D., BERGENGREN, J., STILLWELL-SOLLER, L., FELZER, B. & THOMPSON, S. 1998. Climate simulations for 10,000 and 6,000 years BP using the GENESIS global climate model. *Palaeoclimates*, **2**, 183–218.

PRELL, W. & KUTZBACH, J. 1992. Sensitivity of the Indian monsoon to forcing parameters and implications for its evolution. *Nature*, **360**, 647–652.

RAYMO, M., RUDDIMAN, W. & FROELICH, P. 1988. Influence of late Cenozoic mountain building on ocean geochemical cycles. *Geology*, **16**, 649–653.

ROBERT, C. & CHAMLEY, H. 1991. Development of early Eocene warm climates, as inferred from clay mineral variations in oceanic sediments. *Palaeogeography, Palaeoclimatology, Palaeoecology*, **89**, 315–331.

ROBERT, C. & KENNETT, J. 1994. Antarctic subtropical humid episode at the Palaeocene–Eocene boundary: clay-mineral evidence. *Geology*, **22**, 211–214.

ROEHLER, H. W. 1993. *Eocene Climates, Depositional Environments, and Geography, Greater Green River Basin, Wyoming, Utah, and Colorado*. US Geological Survey, Professional Papers, **1506-F**.

SAVIN, S., DOUGLAS, R. & STEHLI, F. 1975. Tertiary marine palaeotemperatures. *Geological Society of America Bulletin*, **163**, 49–82.

SHORT, D., MENGEL, J., CROWLEY, T., HYDE, W. & NORTH, G. 1991. Filtering of Milankovitch cycles by Earth's geography. *Quaternary Research*, **157**, 157–173.

SLOAN, L. C., HUBER, M. & EWING, A. 1999. Polar stratospheric cloud forcing in a greenhouse world: a climate modeling sensitivity study. *In*: ABRANTES, F. & MIX, A. (eds) *Reconstructing Ocean History: a Window into the Future*. Kluwer Academic, Dordrecht; Plenum, New York.

SLOAN, L. C. & POLLARD, D. 1998. Polar stratospheric clouds: a high latitude winter warming mechanism in an ancient greenhouse world. *Geophysical Research Letters*, **25**, 3517–3520.

SLOAN, L. C. & REA, D. K. 1995. Atmospheric CO_2 of the Early Eocene: a general circulation modeling sensitivity study. *Palaeogeography, Palaeoclimatology, Palaeoecology*, **119**, 275–292.

SLOAN, L. C., WALKER, J. C. G., MOORE, T. C., Jr, REA, D. K. & ZACHOS, J. C. 1992. Possible methane-induced polar warming in the early Eocene. *Nature*, **357**, 320–322.

STALLARD, R. & EDMOND, J. 1983. Geochemistry of the Amazon, 2, the Influence of Geology and Weathering Environment on the Dissolved Load. *Journal of Geophysical Research*, **88**, 9671–9688.

THOMAS, E. & ZACHOS, J. 1999. Isotopic, palaeontologic, and other evidence for multiple transient thermal maxima in the Palaeocene and Eocene. *EOS Transactions, American Geophysical Union*.

THOMPSON, S. L. & POLLARD, D. 1997. Greenland and Antarctic mass balances for present and doubled atmospheric CO_2 from the GENESIS Version-2 global climate model. *Journal of Climate*, **10**, 871–900.

TRENBERTH, K., LARGE, W. & OLSON, J. 1990. The mean annual cycle in global ocean wind stress. *Journal of Physical Oceanography*, **20**, 1742–1760.

VERBITSKY, M. & OGLESBY, R. 1992. The effect of atmospheric carbon dioxide concentration on continental glaciation of the Northern Hemisphere. *Journal of Geophysical Research*, **97**, 5895–5909.

WADE, B. & KROON, D. 1999, High resolution stable isotope stratigraphy of the late middle Eocene (Leg 171B). *In*: ANDREASSON, F., SCHMITZ, B. & THOMPSON, E. (eds) *Early Palaeogene Warm Climates and Biosphere Dynamics, Abstracts Volume*. GFF, Geologiska Fóreningen, Goteborg.

WALLACE, J. & HOBBS, P. 1977. *Atmospheric Science*. Academic Press, New York.

WILF, P. & LABANDEIRA, C. 1999. Response of plant–insect associations to Palaeocene–Eocene warming. *Science*, **284**, 2153–2156.

WING, S. L., BAO, H. & KOCH, P. L. 2000. An early Eocene cool period? Evidence for continental cooling during the warmest part of the Cenozoic. *In*: HUBER, B. T., MACLEOD, K. & WING, S. L. (eds) *Warm Climates in Earth History*. Cambridge University Press, Cambridge, 197–138.

ZACHOS, J. C., FLOWER, B. P. & PAUL, H. 1997. Orbitally paced climate oscillations across the Oligocene/Miocene boundary. *Nature*, **388**, 567–570.

ZACHOS, J. C., LOHMANN, K., WALKER, J. & WISE, S. 1993. Abrupt climate change and transient climates during the Palaeogene: a marine perspective. *Journal of Geology*, **101**, 191–213.

ZACHOS, J. C., STOTT, L. D. & LOHMANN, K. C. 1994. Evolution of early Cenozoic marine temperatures. *Paleoceanography*, **9**, 353–387.

Orbitally forced climate change in late mid-Eocene time at Blake Nose (Leg 171B): evidence from stable isotopes in foraminifera

BRIDGET S. WADE[1], DICK KROON[1] & RICHARD D. NORRIS[2]

[1]*Department of Geology and Geophysics, University of Edinburgh, Grant Institute, West Mains Road, Edinburgh, EH9 3JW, UK (e-mail: B.Wade@glg.ed.ac.uk)*

[2]*Department of Geology and Geophysics, Woods Hole Oceanographic Institution, MS-23, Woods Hole, MA 02543-1541, USA*

Abstract: Previous stable oxygen isotopic data from surface-dwelling foraminifera indicate that Eocene tropical sea surface temperatures (SSTs) were significantly lower than at present. Here we show that stable isotopic analyses ($\delta^{18}O$, $\delta^{13}C$) of the late mid-Eocene mixed-layer dweller *Morozovella spinulosa* are consistent with mid-Eocene mid-latitude SSTs close to, or slightly lower than modern temperatures at Blake Nose, western North Atlantic. In contrast, isotopic analyses of the benthic foraminifer, *Nuttalides truempyi* reveal a gradual fall in mean bottom-water temperatures from 8 to 7 °C over *c.* 500 ka years. These deep intermediate-water temperatures are significantly higher than modern ones and are similar to intermediate- and bottom-water temperatures recorded from earlier in Palaeogene and late Cretaceous time.

Large shifts are seen in the $\delta^{18}O$ and $\delta^{13}C$ values of the planktonic foraminifers, of up to 1‰ and 2.6‰, respectively, that probably reflect temperature and nutrient fluctuations controlled by regional changes in upwelling intensity and runoff. The surface to benthos $\delta^{18}O$ gradient decreases from 3‰ PDB to a minimum of *c.* 0.5‰ PDB over 400 ka, which could relate to the intensity of upwelling. Spectral analysis reveals precessional forcing in the foraminiferal $\delta^{18}O$ records, which shows the direct influence of low-latitude insolation on surface-water stratification. Monsoonal wind systems may have forced the upwelling cycles and/or freshwater input. The benthic foraminifer $\delta^{18}O$ record also contains the obliquity cycle, in addition to the precessional cycles, indicating the inheritance of mid- and high-latitude forcing to subtropical deep waters.

Blake Nose, or Blake Spur, is a salient in the western North Atlantic, located due east of northern Florida, on the eastern margin of the Blake Plateau (Fig. 1). Here the Palaeogene and Cretaceous deposits have never been deeply buried by younger sediments and are thus little affected by diagenesis. A thin veneer of manganese-rich sand and nodules shields these sediments from modern-day erosional processes. Five sites were drilled along a transect of Blake Nose by the Ocean Drilling Program (ODP) Leg 171B (Norris *et al.* 1998).

Site 1051 (30°03′N, 76°21′W) is located at a water depth of 1980 m. This site comprises an expanded and virtually complete Palaeogene section of pelagic and hemipelagic sediments, from Palaeocene to early late Eocene age. The sediments consist predominantly of a siliceous nannofossil and foraminifer ooze (Norris *et al.* 1998). The mid-Eocene foraminifera recovered at this site are well preserved and therefore highly suitable for isotopic examination.

High-resolution stable isotopic investigations ($\delta^{18}O$, $\delta^{13}C$) were conducted on late mid-Eocene planktonic and benthic foraminifera from the upper 30 m of Hole 1051B. The primary aim of this investigation was to examine the stability of subtropical sea surface temperatures (SSTs) at a high temporal resolution in the Milankovitch frequency band. The examination of both planktonic and benthic foraminifers also allows the reconstruction of the surface-to-benthos thermal and trophic gradients in the late mid-Eocene ocean.

The Eocene climate

Abundant palaeontological and geochemical data indicate that the Eocene climate was extremely different from that at present. Early

From: KROON, D., NORRIS, R. D. & KLAUS, A. (eds). 2001 *Western North Atlantic Palaeogene and Cretaceous Palaeoceanography*. Geological Society, London, Special Publications, **183**, 273–291.
0305-8719/01/$15.00

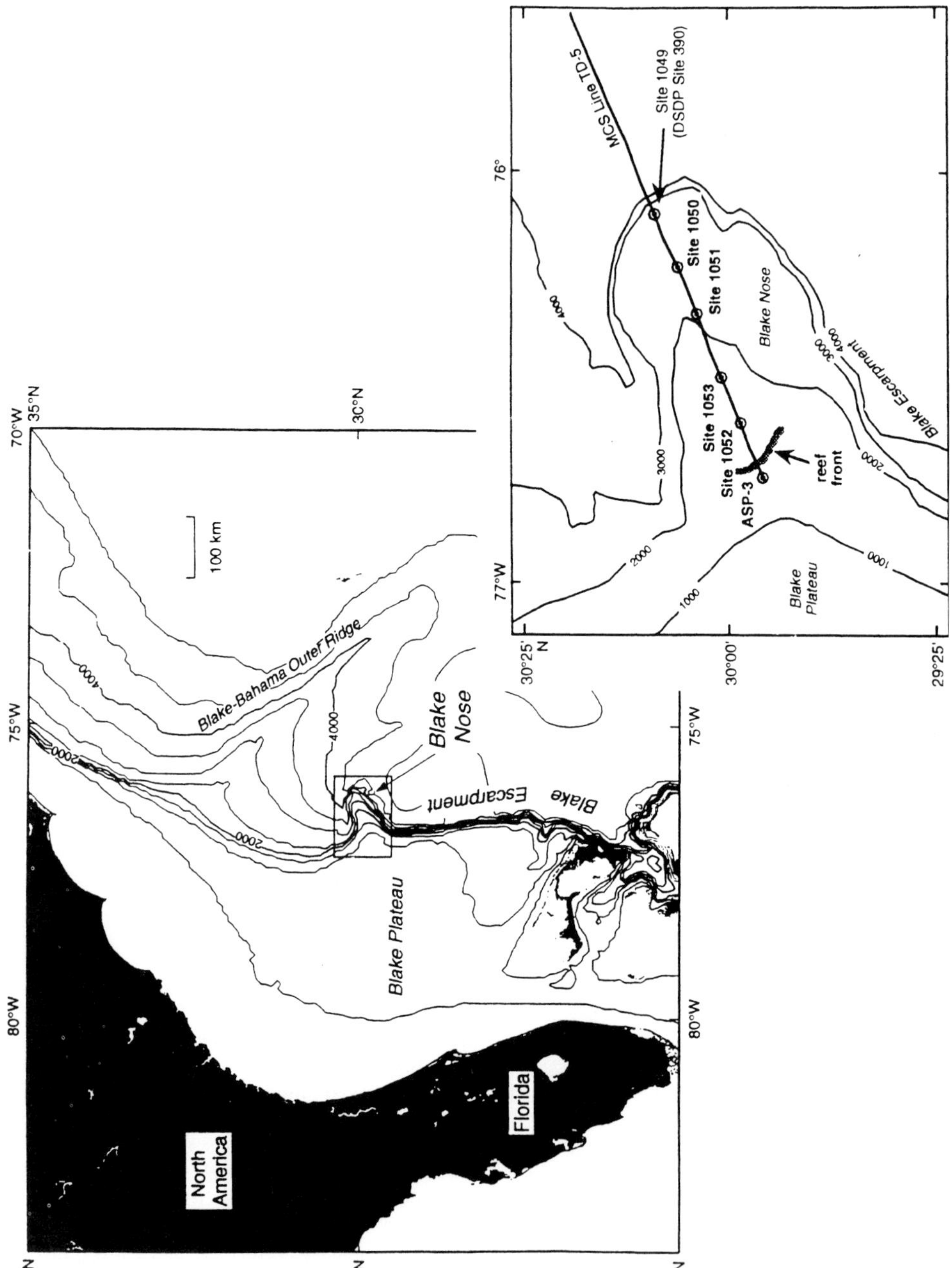

Fig. 1. Location map of the Blake Nose and Blake Plateau. Inset: ODP Leg 171B drill locations and bathymetry (Norris *et al.* 1998).

Eocene marine and terrestrial records reveal that the Earth was warmer than at any time during the last 65 Ma, representing a fundamentally different climate state from any other interval in the Cenozoic era. The Early Eocene climate is characterized by a reduced meridional surface temperature gradient, increased global mean temperature and the absence of large-scale continental glaciation (Zachos *et al.* 1994).

The early stages of the Cenozoic cooling trend began with a step-like climatic deterioration during early mid-Eocene time. This heralded the onset of global cooling, ultimately leading to the Quaternary ice age. Mid-Eocene time saw two modest increases in $\delta^{18}O$ values of planktonic foraminifera, at the early to mid-Eocene boundary and during the late mid-Eocene time (Shackleton & Boersma 1981). Benthic foraminifera also illustrate the cooling trend, indicating a synchronous cooling in both bottom and surface ocean waters in the mid- and high-latitudes (Kennett & Shackleton 1976). The increases in $\delta^{18}O$ have been explained as reductions in the global ocean temperature either with or without ice formation. There may have been seasonal winter ice in the Arctic region and continental ice in West Antarctica, though there is no confirmation of a major ice cap at this time (Wise *et al.* 1991).

Although the climatic shifts during Eocene time are understood in general terms, the mid-Eocene climate is inadequately documented in terms of its variability, timing of cooling and effect on oceanographic structure. This is because high-resolution, low-latitude Palaeogene records are scarce and previously acquired Eocene records are incomplete, intermittently cored or disturbed by drilling through Eocene chert (Stott & Zachos 1991). Complete records either lack the necessary resolution, or their microfossils are not sufficiently well preserved to record the rapid palaeoceanographic changes that are associated with the transition period from a non-glacial to glacial climate.

Tropical sea surface temperatures have been shown to be essential to verify the character of forcing and feedback of warm climatic states (Manabe & Bryan 1985; Covey & Thompson 1989; Horrell 1990; Crowley 1991; Zachos *et al.* 1994). However, a disparity has previously prevailed between proxy temperature interpretations from isotopic and other palaeontological indices for the mid-Eocene tropics. Previous stable oxygen isotopic data from surface-dwelling foraminifera indicate that Eocene tropical SSTs were significantly lower than present values (e.g. Boersma *et al.* 1987; Zachos *et al.* 1994; Bralower *et al.* 1995), whereas other palaeontological proxies give rise to SSTs equal to present-day temperatures (e.g. Adams *et al.* 1990; Graham 1994; Andreasson & Schmitz 1998). This discrepancy in palaeoclimate proxies may result from the diagenetic alteration of planktonic foraminifera, foraminifer palaeoecology or errors in estimating the $\delta^{18}O$ of Eocene seawater. The advantage of Site 1051 is that the sediments have never been deeply buried and there is a clearly defined biostratigraphy and magnetostratigraphy with no known unconformities throughout the upper middle Eocene sequence.

The record from Site 1051 allows the examination of high-frequency climatic variability and accurate documentation of the timing and scale of relatively short-term climatic changes associated with the switch from the early Eocene greenhouse to the late Eocene icehouse world in the Atlantic Ocean. The results from the Shipboard Scientific Party (Norris *et al.* 1998) indicate clear high-frequency variability in the colour reflectance data (Fig. 2), which are thought to be driven by Milankovitch-scale climate oscillations. The examination of both planktonic and benthic foraminifer stable isotopes in association with the colour record allows climatic dynamics in late mid-Eocene time to be examined much more fully than in previous studies. We explore the effect of orbital forcing (Milankovitch cycles) upon the late mid-Eocene climate. In addition, the comparison of isotopic records derived from planktonic and benthic foraminifera permits the reconstruction of the surface-to-benthos thermal and trophic gradients in the western North Atlantic.

Methods and procedures

Sample preparation

For isotopic analyses, 2.8 cm^3 samples from Hole 1051B were examined at 10 cm intervals. All samples were dried, weighed, soaked in a Calgon–peroxide solution overnight, and wet sieved on a 63 μm mesh. The >63 μm size fraction was then oven dried at <50 °C and weighed again to obtain a measurement of percentage coarse fraction (>63 μm). For planktonic foraminiferal analyses the >63 μm fraction was dry sieved into three size fractions: >355 μm, 250–355 μm and <250 μm. Isotopic measurements were conducted on planktonic foraminifera from the 250–355 μm size fraction. A narrow size fraction was selected to constrain vital and ontogenetic effects on stable isotopic interpretation (Shackleton *et al.* 1985; Corfield & Cartlidge 1991; Pearson *et al.* 1993; Norris 1998).

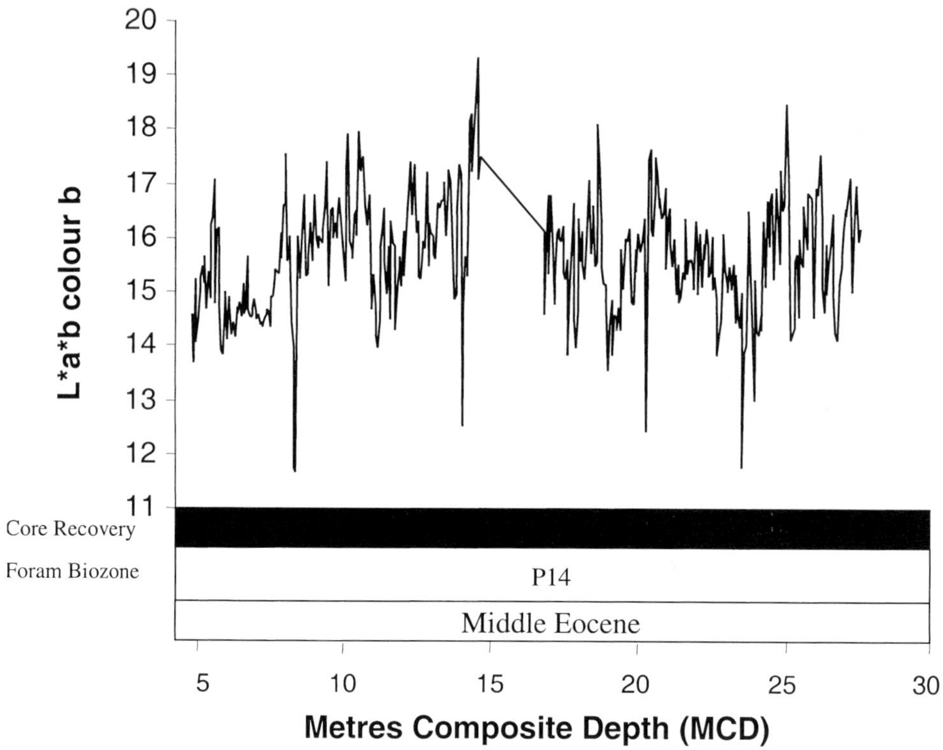

Fig. 2. Spliced L×a×b colour code b from 5 to 28 m composite depth at Site 1051 after Norris *et al.* (1998). Core recovery is shown in black along the *x*-axis with planktonic foraminifer biozone.

Isotopic examination

For planktonic foraminiferal isotopic investigation, multiple specimens of the mixed layer dweller *Morozovella spinulosa* (Fig. 3) were picked for each sample studied; normally 5–20 individuals depending on abundance. This species was selected for analysis because of its surface-dwelling habitat, ease of identification and abundance within the samples, although *M. spinulosa* was not present in all samples. Analysis of multiple specimens provides results that are nearer to the species' mean stable isotopic value than analyses conducted on singular specimens; however, using this method, data on intraspecific deviation are lost (Pearson & Shackleton 1995). Sample weights were usually 0.17 ± 0.03 mg. Before analysis, specimens were placed in methanol and cleaned ultrasonically to dislodge attached fine calcite particles. Ultrasonic cleaning was repeated when visual examination proved this to be required.

All planktonic foraminifer samples were analysed isotopically using a VG Isogas Prism III mass spectrometer at the University of Edinburgh. Normal corrections were employed and results of stable isotope measurements are expressed in ‰ relative to the Pee Dee Belemnite (PDB) standard reference carbonate of zero (Craig 1957). Silver Mine (SM) calcite powdered standard was measured concurrently (mean = 0.20 mg) to record analytical precision and instrument calibration. Replicate analyses of standards gave rise to standard deviations of 0.09‰ for $\delta^{18}O$ and 0.05‰ for $\delta^{13}C$.

The benthic foraminifer *Nuttalides truempyi* was picked from the >150 μm fraction and analysed isotopically with a Finnigan MAT252 mass spectrometer and associated automated carbonate device ('Kiel Device') in the Department of Geology and Geophysics at the Woods Hole Oceanographic Institution. Samples (mean = 60 μg) were lightly crushed in the reaction vessels to ensure complete reaction with phosphoric acid at 70 °C. Six standards were run with each set of 40 unknowns; standards included Carrara Marble, *Atlantis II* deep-sea coral and B-1 marine carbonate. Results were corrected to VPDB (PDB estimated by analysis of NBS-19), followed by a second correction for sample gas volume. Replicate analyses of all three standards yields standard

Fig. 3. Representative planktonic foraminifera from the Middle Eocene sequence at Blake Nose. All specimens are from sample 171B, 1051B, 2H-1, 50–52 cm, except (**a**) and (**d**), which are from sample 171B, 1051B, 4H-2, 90–92 cm. Scale bars represent 100 μm, except in (**d**) and (**h**) where they represent 20 μm. (**a**) *Morozovella spinulosa* umbilical view. (**b**) *Morozovella spinulosa* spiral view. (**c**) *Morozovella spinulosa* edge view. (**d**) *Morozovella spinulosa*, view of test surface. (3e) *Morozovella crassata* edge view. (**f**) *Acarinina praetopilensis* umbilical view. (**g**) *Acarinina praetopilensis* spiral view. (**h**) *Acarinina praetopilensis*, view of test surface.

errors of 0.08‰ for $\delta^{18}O$ and 0.04‰ for $\delta^{13}C$ as averages for reproducibility on the A and B lines in the Kiel Device. All isotopic data have been given by Wade *et al.* (2000).

Age model

The combination of abundant and well-preserved microfossils and nannofossils with the clear magnetostratigraphy by the Shipboard Scientific Party (Norris *et al.* 1998) provides an excellent chronostratigraphic framework for Site 1051. The age model (Table 1) calibrated for Site 1051 is based on the magnetostratigraphy and the age scale of Berggren *et al.* (1995), assuming a constant sedimentation rate between tie points and the best-fit line extrapolated to the top of the hole (Fig. 4). There is considerable scatter in the biostratigraphic datum levels; this is thought to be due to the lower sampling resolution of the biostratigraphy (every 1.5 m) compared with the magnetostratigraphy (every 5 cm), or inaccuracies in the current calibration of some of the bio-events. Although there are no biostratigraphic or magnetostratigraphic datum levels for the upper 60 m of the hole, the high abundance of *Morozovella spinulosa* indicates that the top of Hole 1051B cannot be younger than the extinction of this species at 38.1 Ma (Berggren *et al.* 1995).

This study focuses on *c.* 500 ka in late mid-Eocene time, equivalent to planktonic foraminiferal Biozone P14 and Magnetochron C18n.1n. Sampling was at 10 cm intervals and therefore the time span between each sample is *c.* 2500 years. Bioturbation will undoubtedly smooth the results to some extent. However, the large fluctuations in the colour and isotopic records (see Results) suggest that the smoothing effect is minimal over this interval. We ran spectral analysis on the colour record and $\delta^{18}O$ records to obtain information on the cyclicity in the records.

Results

Foraminifera preservation

Leg 171B material has never been deeply buried, resulting in the good preservation of foraminifera. Light and scanning electron microscopy shows the foraminifera to be devoid of carbonate infilling and visible dissolution (Fig. 3). Pores on the outer and inner test walls are plainly visible with no surficial overgrowth. Preservation of primary calcite is shown by cross-sections through the test walls, where pores are open and smooth. Minor amounts of fine carbonate debris (mainly coccoliths) are seen attached to the test surfaces. Good preservation is also confirmed by the difference in $\delta^{18}O$ between planktonic and benthic foraminiferal species, whereas uniform $\delta^{18}O$ values are forecast from models of bulk carbonate diagenetic alteration (Killingley 1983; Schrag *et al.* 1995). Our results will contribute to the debate on the mechanisms of Eocene climate change by establishing subtropical SSTs from well-preserved and complete material, without the obvious consequences of diagenetic alteration.

Stable isotope records

Morozovella spinulosa. The surface-water record provided by the planktonic foraminifer *Morozovella spinulosa* indicates large oscillations in $\delta^{18}O$ from −1.7‰ to 0‰ PDB (Fig. 5a). Between 24 and 28 m composite depth (mcd) the record varies by *c.* 0.4‰ PDB. However, the fluctuations increase to as much as 0.9‰ PDB at 18 mcd.

Table 1. *Magnetostratigraphic and biostratigraphic age datum levels for Hole 1051B*

	Datum	Age (Ma)	Maximum depth (m)	Minimum depth (m)	Mean depth (m)
B	C18n.1n	39.55	64.90	67.90	66.40
T	C18n.2n	39.63	68.50	71.50	70.00
B	C18n.2n	40.13	86.00	87.20	86.60
T	C19	41.26	135.80	139.50	137.65
B	C19	41.52	139.50	145.50	142.50
FO	*D. bisecta*	38.00	86.00	91.00	88.50
LO	*A. primitiva*	39.00	101.50	111.00	106.25
LO	*G. beckmanni*	40.10	72.00	82.00	77.00
FO	*G. beckmanni*	40.50	91.50	101.00	96.25
LO	*C. solitus*	40.40	114.00	115.00	114.50
FO	*T. pomeroli*	42.40	140.00	149.00	144.50

All depths are in metres composite depth (mcd). B, base; T, top; FO, first occurrence; LO, last occurrence. All ages from Berggren *et al.* (1995).

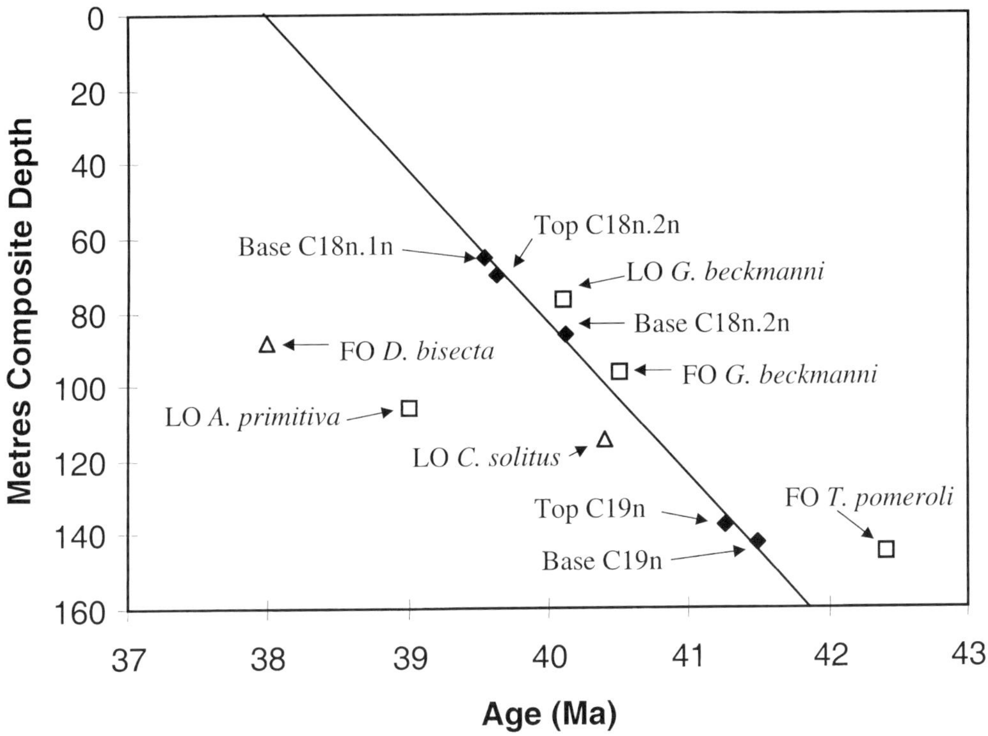

Fig. 4. Age model for Site 1051 as listed in Table 1. This is based on the magnetostratigraphy from the Shipboard Scientific Party (Norris *et al.* 1998) and the age scale of Berggren *et al.* (1995), assuming a constant sedimentation rate between tie points. Points represent the mean metres composite depth (mcd) of each datum. ◆, magnetostratigraphic datum levels; △, nannofossil datum levels; □, foraminifer datum levels. FO, first occurrence; LO, last occurrence.

There are several intervals (19, 9.5 and 5.5 mcd) where $\delta^{18}O$ values are lighter than −1.2‰ PDB.

The carbon isotope values of *M. spinulosa* range from 3.4‰ to 1.6‰ PDB (Fig. 5b). Between 24 and 28 mcd, values of $\delta^{13}C$ fluctuate between 3.4‰ and 2.9‰ PDB. This is followed by a sharp decrease in the $\delta^{13}C$ record at 22 mcd, when values fall from 3.3‰ to 1.9‰ PDB over *c.* 100 ka. The record between 18 and 5 mcd shows large fluctuations in $\delta^{13}C$ between 3.1‰ and 1.6‰ PDB.

Nuttalides truempyi. The oxygen isotope record of the benthic foraminifer *N. truempyi* mainly fluctuates between 0.5‰ and 1.1‰ PDB (Fig. 6a). There appears to be a small step in the record at 9 mcd when average values increase from 0.7‰ to 0.9‰ PDB. The benthic carbon isotope record (Fig. 6b) is very similar to the oxygen isotope record, with values fluctuating between 0.2‰ and 1.2‰ PDB. The same step is seen at 9 mcd when mean carbon values shift from 0.8‰ to 1.2‰ PDB; however, this step appears more gradual in the carbon record than in the oxygen record.

Palaeotemperature reconstruction

Ancient marine temperatures were calculated using the calcite–water $\delta^{18}O$ temperature equation of Erez & Luz (1983):

$$T\,^{\circ}C = 16{\cdot}998 - 4{\cdot}52(\delta^{18}O_{cc} - \delta^{18}O_{sw}) + 0{\cdot}028(\delta^{18}O_{cc} - \delta^{18}O_{sw})^2 \quad (1)$$

where $\delta^{18}O_{cc}$ is the oxygen isotope composition of the sample calcite relative to PDB, and $\delta^{18}O_{sw}$ is the oxygen isotope composition of the ambient seawater relative to Standard Mean Ocean Water (SMOW).

Sea surface temperatures. It is imperative that the mean oxygen isotope composition of the

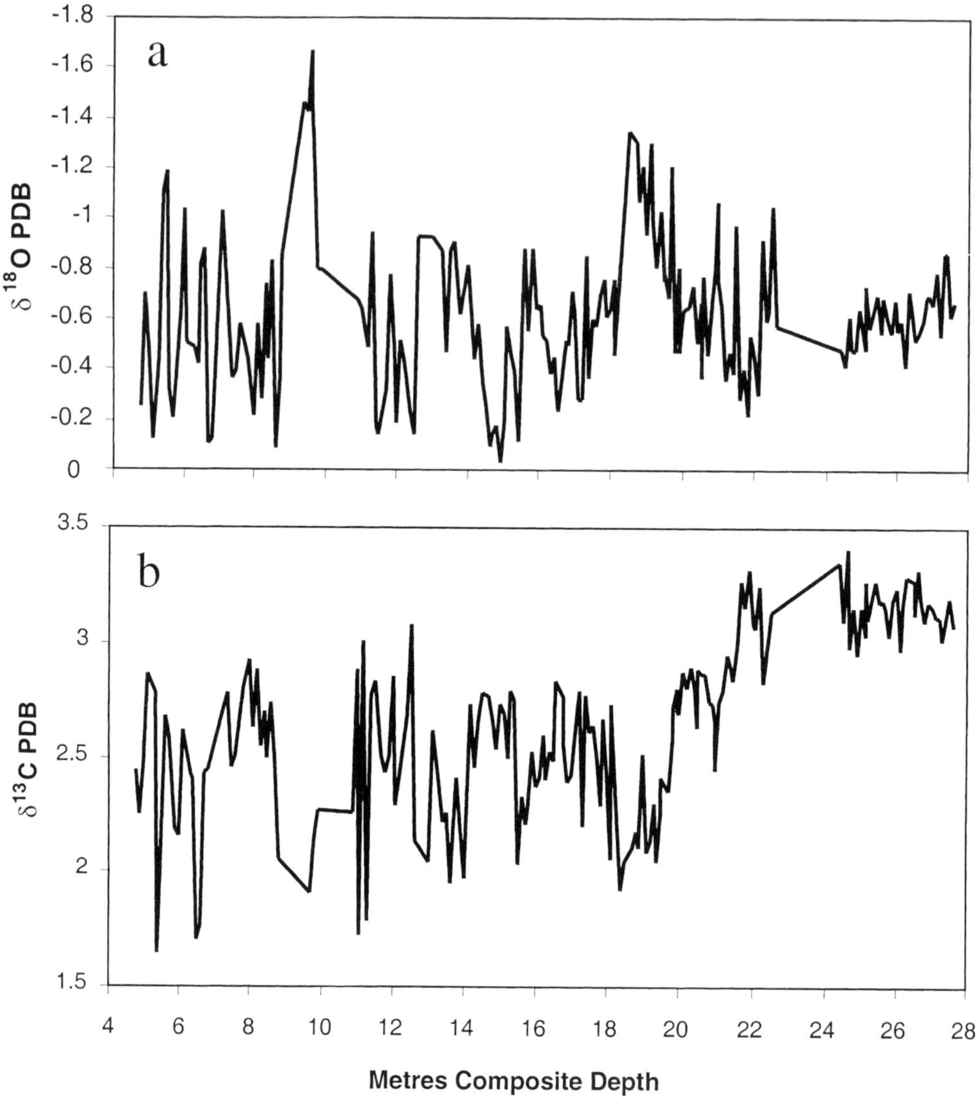

Fig. 5. Planktonic foraminifera (*Morozovella spinulosa*) stable isotope record at Hole 1051B. (**a**) Oxygen isotope record. (**b**) Carbon isotope record.

seawater in which the calcite was precipitated ($\delta^{18}O_{sw}$) is constrained in the calculation of palaeotemperatures from equation (1). During Eocene time, in the absence of significant continental ice, the estimated mean composition of seawater is −1‰ SMOW (Shackleton & Kennett 1975). Instead of applying a mean $\delta^{18}O$ value for the entire ocean when calculating SSTs from planktonic foraminifera, it is desirable to account for local variations in sea surface $\delta^{18}O$ as a function of latitude. This value is dependent on evaporation, precipitation and atmospheric vapour transport in the open ocean. There is at present no direct measure of palaeoseawater $\delta^{18}O_{sw}$. Palaeotemperatures were reconstructed using the assumption that local salinity did not severely fluctuate and thus temperature is the primary influence on the stable oxygen isotope record. It is possible that our assumption of constant salinity is incorrect (see below) but we have no direct means of quantifying the salinity effect on surface-water $\delta^{18}O$ at this stage.

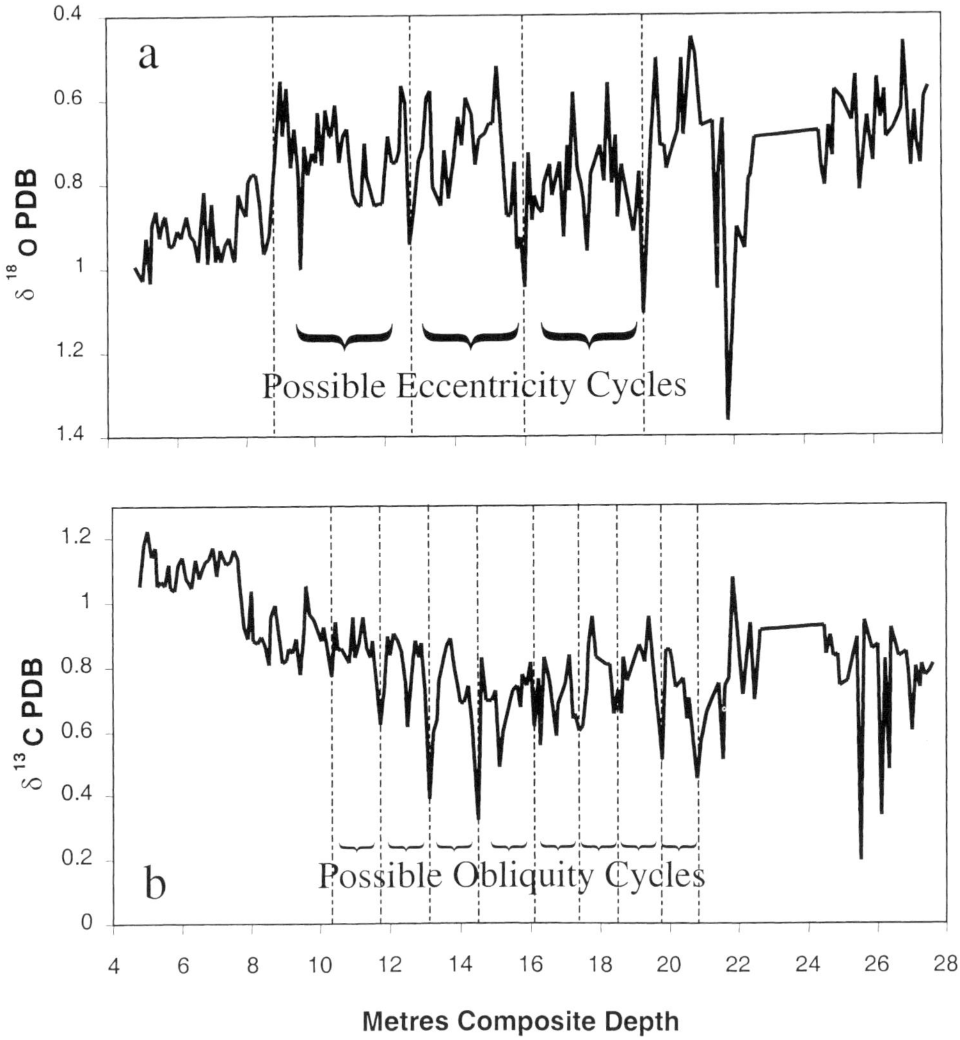

Fig. 6. Benthic foraminifera (*Nuttalides truempyi*) stable isotope record at Hole 1051B. (**a**) Oxygen isotope record, indicating possible eccentricity cycles. (**b**) Carbon isotope record, indicating possible obliquity cycles (discussed in text).

Zachos *et al.* (1994) introduced an equation to calculate the influence of atmospheric moisture transport and evaporation on regional $\delta^{18}O$:

$$y = 0{\cdot}576 + 0{\cdot}041x - 0{\cdot}0017x^2 + 1{\cdot}35 \times 10^{-5}x^3 \qquad (2)$$

where y is the oxygen isotopic composition of the seawater ($\delta^{18}O$, SMOW) and x is the absolute latitude in the range of 0°–70°. This equation is applied to decrease the effect of the surface-water $\delta^{18}O_{sw}$ gradient on palaeotemperature reconstruction. The equation implies that the zonally averaged meridional surface water $\delta^{18}O$ gradient was similar to that of the present day. Although it is conceivable that the latitudinal $\delta^{18}O$ gradient was different in mid-Eocene time, there is no indication that the amplitude or direction of this gradient was significantly dissimilar to that at present, as supported by Eocene climatic modelling studies (e.g. Sloan & Rea 1995).

As the actual isotopic composition of the mid-Eocene seawater cannot be determined, two constant values of $\delta^{18}O_{sw}$ were considered in

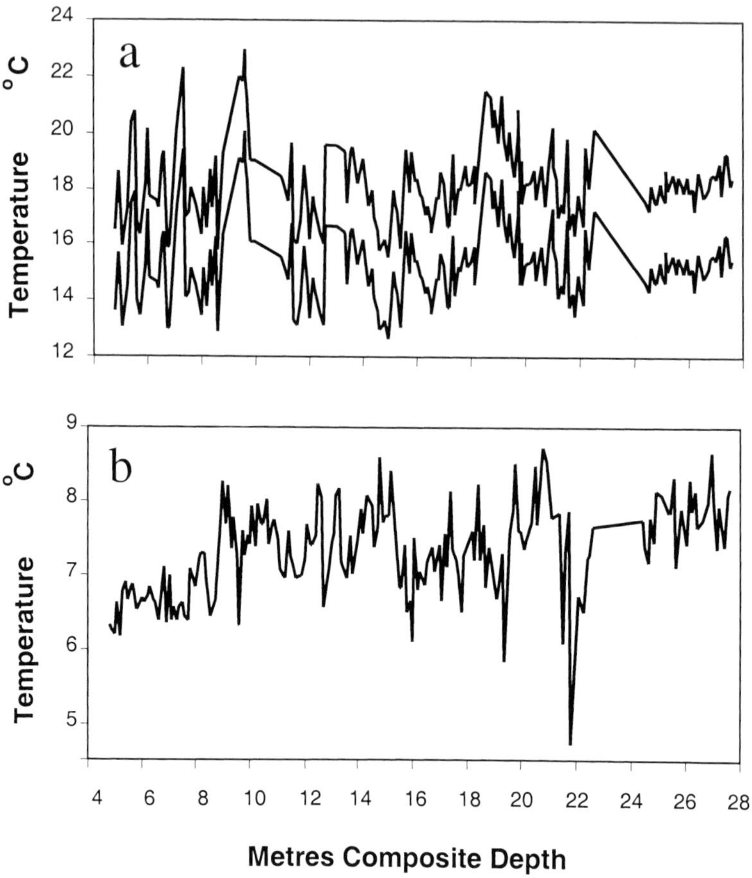

Fig. 7. (**a**) Surface-water temperatures calculated from the oxygen isotope record of *Morozovella spinulosa*. Lower line, $\delta^{18}O_{sw}$ of −1‰ (SMOW) (Shackleton & Kennett 1975); upper line, $\delta^{18}O_{sw}$ of −0.36‰ (SMOW) applying equation (2) (Zachos *et al.* 1994). (**b**) Bottom-water temperatures calculated from the oxygen isotope record of *Nuttalides truempyi*.

the calculation of sea surface temperatures. These were: (1) −1‰ (SMOW) where the net evaporation effect on $\delta^{18}O_{sw}$ is not considered and (2) −0.36‰ (SMOW) where the palaeolatitude dependent effect on $\delta^{18}O_{sw}$ was determined from equation (2) (Zachos *et al.* 1994), and the resulting value reduced by 1‰ to account for ice-free global conditions (Shackleton & Kennett 1975).

The calculated sea surface temperatures at Site 1051 are shown in Fig. 7a. When applying a $\delta^{18}O_{sw}$ value of −1‰ (SMOW), temperatures range between 12.5 °C and 20 °C. These values are raised by 3 °C when the equation of Zachos *et al.* (1994) (equation (2)) is employed to estimate palaeotemperatures between 15.5 °C and 23 °C in our data.

Bottom-water temperatures. In modern benthic foraminifera it has been demonstrated that in certain species the stable isotopic composition is not in equilibrium with the surrounding seawater (Duplessy *et al.* 1970; Shackleton 1974; Woodruff *et al.* 1980; Belanger *et al.* 1981; Graham *et al.* 1981; Shackleton *et al.* 1984; Keigwin & Corliss 1986; Zachos *et al.* 1992). The recorded stable isotopic values of oxygen are reduced compared with the forecast equilibrium values (Boersma *et al.* 1979). In many studies the measured oxygen isotope values are adjusted to account for this disparity. The proposed 'correction factors' range from +0.35‰ to +0.6‰ (Shackleton *et al.* 1984; Keigwin & Corliss 1986; Kennett & Stott 1990; Barrera & Huber 1991; Zachos *et al.* 1992, 1994). We have applied an adjustment of +0.4‰ (Shackleton *et al.* 1984) to the measured $\delta^{18}O$ values of *Nuttalides* to account for the suspected species-specific non-equilibrium fractionation, and allow comparison

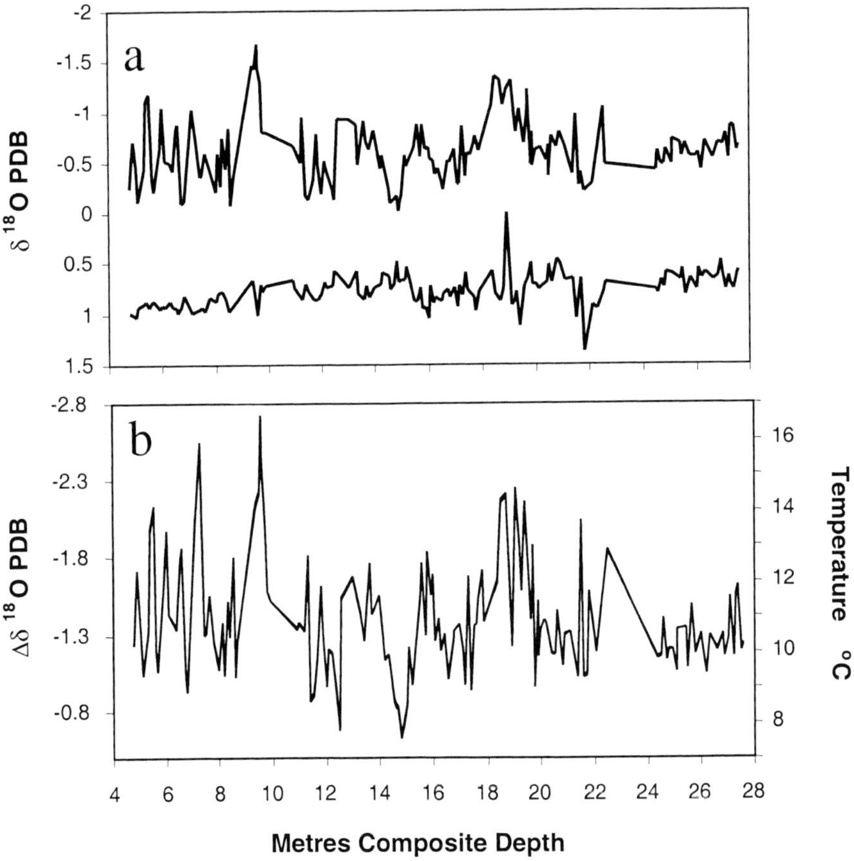

Fig. 8. (**a**) Planktonic and benthic foraminifera oxygen isotope record at Hole 1051B. Upper line, *Morozovella spinulosa*; lower line, *Nuttalides truempyi*. (**b**) Variation in the surface to benthos thermal gradient. The primary *y*-axis indicates the change in temperature (°C); the secondary *y*-axis indicates $\Delta\delta^{18}O$ PDB.

with previous work. Benthic foraminifera $\delta^{13}C$ were not adjusted.

Beneath the surface water the isotopic composition of seawater can be inferred to be the same as the global mean value. In calculations of benthic palaeotemperature, the $\delta^{18}O_{sw}$ value is thus taken to be −1‰ (SMOW) (Shackleton & Kennett 1975). Benthic foraminiferal temperatures (Fig. 7b) generally range between 6 and 9 °C. There is a decrease in the benthic foraminiferal temperatures over time, from a mean of 8 °C at 28 mcd to 6.5 °C at 5 mcd. These bottom-water temperatures are significantly higher than those of today and are similar to those recorded in Palaeocene and late Cretaceous time (e.g. Corfield & Norris (1996, 1998), and references therein).

Surface to benthos thermal gradient. When palaeotemperatures deduced from benthic foraminiferal results are subtracted from sea surface temperatures, the large oscillations in the surface to benthos temperature gradient can be seen (Fig. 8). This fluctuates between maximum values of 17 °C (−2.7‰ PDB) to a minimum of 7.5 °C (−0.65‰ PDB).

Surface to benthos carbon gradient

The surface to benthos carbon gradient decreases over time (Fig. 9a). At 22 mcd there is a reduction in surface carbon isotope values from 3.4‰ to 2.0‰ PDB. There is an increase in the benthic carbon isotope values at 8 mcd when mean carbon values shift from 0.8‰ to 1.2‰ PDB. The decline in planktonic $\delta^{13}C$ throughout the record and increase in benthic $\delta^{13}C$ at 8 mcd serve to reduce the surface to benthos $\delta^{13}C$ gradient from a maximum of *c.* 3‰ PDB at 25.5 mcd to a minimum of *c.* 0.5‰ PDB at

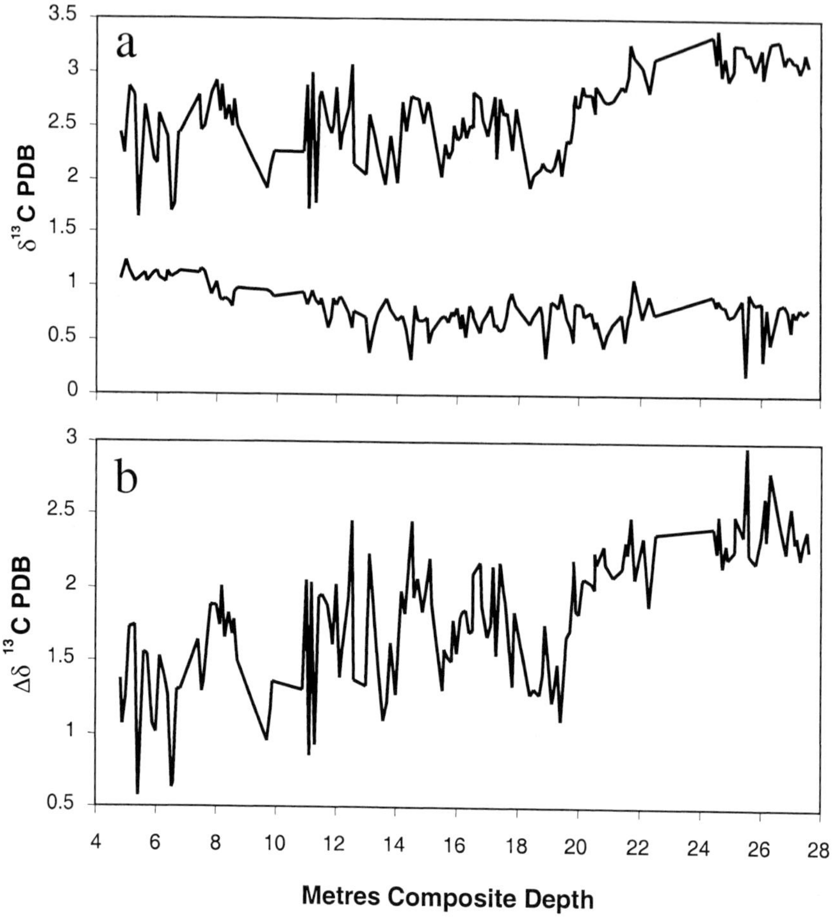

Fig. 9. (**a**) Planktonic and benthic foraminifera carbon isotope record at Hole 1051B. Upper line, *Morozovella spinulosa*; lower line, *Nuttalides truempyi*. (**b**) Variation in the surface to benthos trophic gradient.

5.5 mcd (Fig. 9b). This is a 2.5‰ PDB decrease in the δ^{13}C gradient over *c.* 450 ka.

A decrease in surface to benthos carbon gradient has also been seen (to a lesser magnitude) in previous studies of late Mid-Eocene time; for example, by Vergnaud-Grazzini & Saliege (1985) and Boersma *et al.* (1987). Those workers noted that a 1‰ decrease in surface-water δ^{13}C values occurred like that observed at Site 1051, in combination with a gradual increase in benthic foraminifer values.

Orbital forcing

It is thought that a significant fraction of the climatic variance is driven in some way by insolation changes caused by orbital forcing (Milankovitch cyclicity). The amount of solar radiation received and its geographical distribution are influenced by three variable elements of the Earth's orbit. Periodic variations in the insolation patterns of the Earth are produced by the cycles of equinox precession (19–23 ka), obliquity of the Earth's rotation axis (41–54 ka), and eccentricity of the orbit around the Sun (97–123 ka for short-term and 413 ka for long-term components) (Milankovitch 1941).

Over a 97 ka cycle, the Earth's orbit changes from being circular to an elliptical shape. This eccentricity of orbit causes seasonal variations in the amount of solar insolation, by modulating the precessional cycles. Every 41 ka, the tilt of the Earth's axis changes between 24.5° and 22.1°; the greater the tilt the more marked the seasons. The time of year at which the Earth is nearest the Sun (perihelion) also varies; this alternates with an average 21 ka cycle. It is these changes that appear to be responsible for remarkable variations in the complex oceanic and atmospheric systems. Changes in orbital parameters give rise to altered seasonal distributions and intensities of sunlight that are thought to lead to changes in marine productivity, runoff, sediment erosion and transport, and carbonate rain rates, among other things. The question arises how variations in insolation as a function of orbital cycles propagated in the climate system during late mid-Eocene time.

Evidence for orbital forcing in the Palaeogene time has been provided by spectral analysis of climatic records, which has led to recognition of orbital frequencies in foraminiferal assemblages, stable isotope data and physical property records. For example, Herbert & D'Hondt (1990) recognized precessional climate cyclicity in the South Atlantic during Danian time. Low-frequency (*c.* 400 ka) oscillations in climate and carbon cycle were found by Zahn & Diester-Haass (1995) at ODP Site 689 (Weddell Sea, Antarctica) through Eocene and Oligocene time, and Norris & Röhl (1999) recognized precessional frequencies in the late Palaeocene record at Site 1051.

Spectral analysis was conducted on the isotopic and colour records from Site 1051 using the Blackman–Tukey method in the ANALYSERIES software package (Paillard *et al.* 1996). Spectral analysis of the spliced L×a×b colour record (Fig. 10) reveals cycles at wavelengths of 1.0 and 1.4 cycles m^{-1}. This is calculated as 24 and 18 ka, respectively, taking the average sedimentation rate of 4 cm ka^{-1} determined from the bio-magnetostratigraphy (Fig. 4). These are thought to represent the characteristic double peaks of the precessional period (23 and 19 ka). This indicates that the colour record is primarily controlled by low-latitude changes in solar insolation. The variations in the colour record most probably represent fluctuations in the comparable alteration of biogenic to lithogenic sediment components. These could derive from modifications in factors such as wind strength, productivity and lithogenic input. The slight offset in the cycles is probably due to variations in sedimentation rate.

Spectral analysis of the planktonic foraminifera oxygen isotope record (Fig. 11) and the calculated surface to benthos oxygen isotope gradient (Fig. 12) reveal power peaks at 0.15, 0.3, 1.05 and 1.25 cycles m^{-1}. These are thought to represent the eccentricity cycles and the precessional double peaks, respectively. There is also the half-precessional cycle (1.7 cycles m^{-1}), which has also been recorded in previous studies (e.g. Crowley *et al.* 1992; Hagelberg *et al.* 1994). Precessional forcing indicates that surface water $\delta^{18}O$ is responding somehow to regional variations in solar insolation. This could be Ekman driven upwelling or changes in the strength of the thermocline that are sensitive to precessional forcing, as forecast from climatic modelling studies (e.g. Bice *et al.* 2000; Sloan & Huber 2000) or variations in continental runoff or evaporation–precipitation. Power spectra for the planktonic foraminifer carbon isotopic results are not well defined (not shown). This is

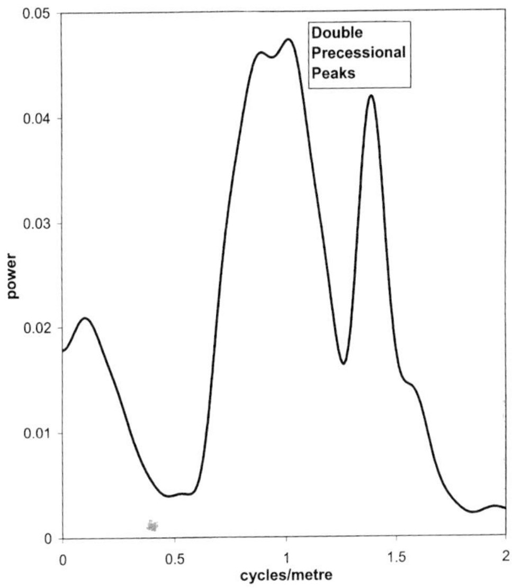

Fig. 10. Power spectra at Site 1051 for the spliced L×a×b colour code b record, illustrated in Fig. 2.

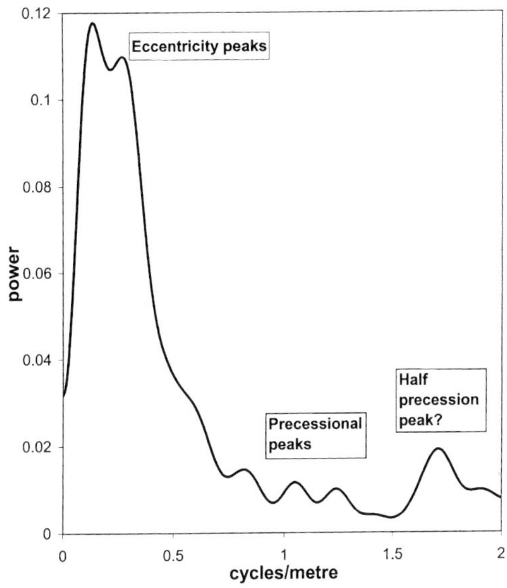

Fig. 11. Power spectra for Site 1051 planktonic foraminifer oxygen isotope record.

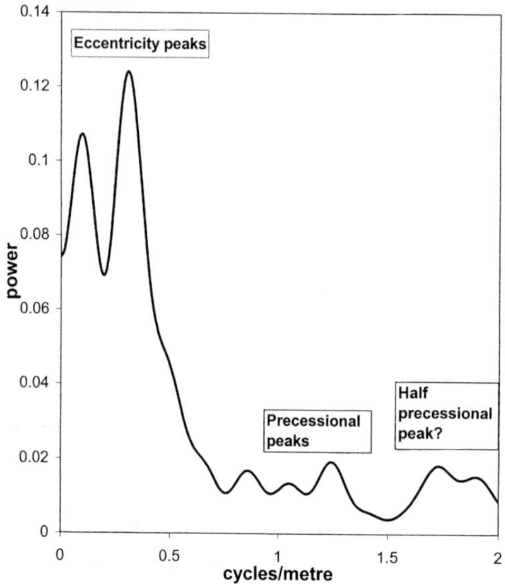

Fig. 12. Power spectra for Site 1051 surface to benthos oxygen isotope gradient.

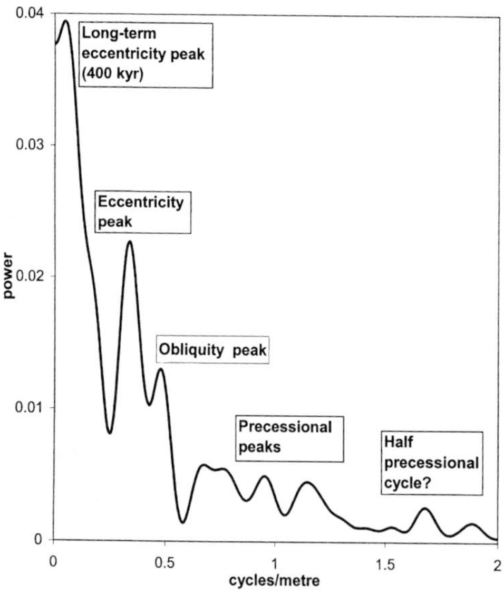

Fig. 13. Power spectra for Site 1051 benthic foraminifer oxygen isotope record.

thought to be due to vital effects within the planktonic carbon isotope record, such as changes in photosymbiotic effects or non-cyclic variation in surface-water nutrient status.

The benthic foraminifer oxygen isotope record (Fig. 13) reveals spectral peaks at 0.06, 0.3, 0.5, 0.95 and 1.15 cycles m^{-1}. These wavelengths of cycle are consistent with the long- and short-term components of the eccentricity, obliquity and precession respectively. The possible eccentricity and obliquity cycles are visible in the isotope records and are indicated in Fig. 6. The obliquity signal within the benthic record is highly significant, indicating that mid- to high latitudes influenced subtropical deep waters. It is of interest that both the long- and short-term components of eccentricity are seen in the benthic foraminifer oxygen isotope record. This suggests that the benthic foraminiferal oxygen isotope record is more affected by long-term changes in seawater $\delta^{18}O$, whereas the sea surface oxygen isotope variations are more influenced by short-term changes in solar insolation.

The results of all spectral analysis show that orbital forcing clearly propagates through the system and was an important short-term and long-term component of late mid-Eocene climate change at Blake Nose.

Comparison with stable isotope record from other sites

The main interests in this site were to allow the vertical structure of the Palaeogene oceans to be interpreted and to provide sea surface temperatures by utilizing well-preserved calcareous microfossils. Despite observing only 500 ka years during the late mid-Eocene period, the high-resolution study of Site 1051 presents many significant attributes that were not determined in low-resolution studies and presents a significant new understanding of subtropical palaeoceanography in the western North Atlantic.

The short-term, high-amplitude (*c.* 1‰ PDB) variability in planktonic foraminifera $\delta^{18}O$ and $\delta^{13}C$ values at Blake Nose has not been previously described. Several reasons may account for this; the record here is at a much higher resolution than in many other Eocene studies, with *c.* 2500 years between data points, allowing the high-amplitude oscillations at this site to be documented. The preservation of foraminifera at Blake Nose is also excellent and thus the record has not been smoothed by diagenetic alteration. The large changes in upwelling intensity, productivity or salinity at Site 1051 may also account for the greater $\delta^{18}O$ variability at this locality compared with previous studies. The generally reduced surface to benthos temperature gradient in Eocene time compared with the present day gradient may facilitate the upwelling of deep water to the surface ocean.

The maximum sea surface temperatures calculated for late mid-Eocene time were between

15.5 and 23 °C, with a mean value of 18 °C when palaeolatitude-dependent effects on $\delta^{18}O_{sw}$ are considered. Present-day temperatures at this site range between 20 °C and 28.5 °C with a mean value of 24 °C (Bottomley *et al.* 1990). The temperatures reconstructed for late mid-Eocene time at Blake Nose are therefore slightly lower than at present, by no more than 5–8 °C depending on the $\delta^{18}O_{sw}$ value used.

Although the sea surface temperatures reconstructed for late mid-Eocene time at Blake Nose are lower than modern-day values at this site, they are periodically significantly higher than those recorded in previous planktonic foraminiferal studies (e.g. Boersma *et al.* 1987; Keigwin & Corliss 1986; Zachos *et al.* 1994; Bralower *et al.* 1995). Minimal $\delta^{18}O$ values for *Morozovella spinulosa* are regularly greater than −1‰ PDB. These values are more typical in early mid-Eocene time (e.g. Zachos *et al.* 1994; Bralower *et al.* 1995), a period of pronounced warming. Our planktonic foraminifer oxygen isotopic values lighter than −1‰ PDB suggest that there may have been several warming intervals during mid-Eocene time that were forced at Milankovitch periodicities.

There are three possible reasons why these values are not recorded in other studies of late mid-Eocene time. A first explanation is that previous low-resolution studies missed the negative $\delta^{18}O$ shifts recorded in this study. By virtue of sampling the record at a resolution of *c.* 2500 years between data points we have identified short-term events in the surface-water isotopic record. The average sampling spacing for previous studies (e.g. Keigwin & Corliss 1986; Bralower *et al.* 1995) was *c.* 175 ka or greater, sufficient to document the long-term trends but not the short-term fluctuations. A second explanation is that recrystallization or dissolution caused a preservational bias towards heavier oxygen isotope values in previous records. Tropical mixed-layer foraminifera have more porous tests than thermocline-dwellers or benthic species, therefore mixed-layer species will frequently be removed first by partial dissolution rather than the heavier calcified forms (Erez 1979; Thunell & Honjo 1981; Wu & Berger 1989). This could potentially lower estimates of SSTs at low latitudes by several degrees (Paull *et al.* 1988; Wu *et al.* 1990). However, it seems a remote possibility that all previous planktonic foraminifer records from low and mid-latitudes are altered by dissolution. A third explanation is that the results here do not reflect temperature effects, but instead decreases in surface-water salinity that give rise to isotopically light $\delta^{18}O$ values.

Discussion

Late mid-Eocene time was previously thought to be a time of stable climatic conditions. However, planktonic foraminifer results reveal large shifts in the $\delta^{18}O$ values of the surface waters between −2 and 0‰ PDB (Fig. 5a). These, in combination with large variations in the surface to benthic thermal gradient (Fig. 8b), indicate that the palaeoceanographic conditions at Blake Nose were highly variable and very different from those of the present day. As ice volume effects in mid-Eocene time are generally believed to be modest, the changes in $\delta^{18}O$ in the tests of the foraminifera are thought to primarily reflect either temperature or salinity effects on the ambient seawater ($\delta^{18}O_{sw}$).

Climatic modelling results for early Eocene time (e.g. Bice *et al.* 2000; Sloan & Huber 2000) predict Ekman driven upwelling along the eastern North American margin, which was sensitive to precessional forcing. If the $\delta^{18}O$ profile of *M. spinulosa* reflects temperature alone, then the rapid fluctuations in surface-water $\delta^{18}O$ and the surface to benthos thermal gradient could be attributed to variations in upwelling intensity. The reduction in carbon isotope gradient is mainly due to decreases in the surface-water $\delta^{13}C$. This supports an increase in upwelling intensity over the course of the record, as a result of the transport of ^{12}C from deep to surface waters. The high abundance of radiolarians at this site is also indicative of a high-productivity environment.

There is a disparity in the relationship between oxygen and carbon isotopic values within the planktonic foraminifer record. Although the fluctuations in the oxygen isotope record are consistent with an upwelling system, this contradicts the isotopic shifts seen in the carbon data. If the variations in $\delta^{18}O$ of *Morozovella* reflected variations in upwelling, then we would expect to see positive shifts in $\delta^{18}O$ to be associated with negative shifts in $\delta^{13}C$, as deep water is upwelled to the surface. However, in this record we see positive shifts in $\delta^{18}O$ associated with positive shifts in $\delta^{13}C$. This would suggest that upwelling is not the obvious cause of the $\delta^{18}O$ fluctuations and that some other factor could be held responsible.

One possibility is that the heavy $\delta^{13}C$ values could be brought about by the increased nutrient availability, giving rise to phytoplankton blooms. Increased photosynthesis would subsequently remove the ^{12}C and leave the surface waters heavy with ^{13}C, explaining the heavy carbon and heavy oxygen events. This relationship has been seen in present-day upwelling systems, for instance in the Arabian Sea (Kroon & Ganssen 1989). Temperature could also have an effect on the metabolic rates of planktonic foraminifera (Bemis *et al.* 2000), or some species could have unexplained vital effects.

There also exists the possibility that the fluctuations in $\delta^{18}O$ reflect changes in salinity, rather than temperature. Perhaps at periodic intervals large amounts of warm, low-salinity water, possibly from the Mississippi outflow entrained in the proto Gulf Stream, were transported from the Gulf of Mexico, north over Blake Nose. Pinet & Popenoe (1985) found evidence that during Cenozoic time the position of the Gulf Stream periodically shifted across the Blake Plateau. The temporal climatically forced variations in the strength and location of the Gulf Stream, associated with freshwater input, could result in the fluctuations in planktonic foraminifer $\delta^{18}O$ observed at Blake Nose. This would also explain the shifts in $\delta^{13}C$, as freshwater tends to be depleted in ^{13}C relative to marine water, although interpretation of the carbon isotope record is complicated by vital effects. It should be noted that if we assume a constant temperature, a 4‰ shift in salinity (Broecker 1989) would be required for the maximum 2‰ PDB shift in $\delta^{18}O$ seen here; this change is considered too large to be consistent with the open-ocean setting for Site 1051. Salinity alone can therefore not explain the record, and temperature must play a significant role. It is unlikely that temperature shifts occurred without variations in salinity as a result of evaporation and precipitation. Therefore the shifts in $\delta^{18}O$ are most probably a combination of temperature and salinity changes. Fourier analysis shows the prodominance of the Milankovitch frequencies in the $\delta^{18}O$ records. Orbitally induced variations in solar insolation somehow account for the variations in surface-water properties, although it is not yet known exactly how.

Conclusions

Stable isotopic examinations conducted on well-preserved planktonic foraminifera of late mid-Eocene age from Blake Nose indicate temperatures equal to or slightly lower than modern-day temperatures, with maximum SSTs of 23 °C. The evidence that upwelling occurred at Blake Nose during late mid-Eocene time is the large oscillations in subtropical SSTs with surface-water temperatures periodically approaching deep-water temperatures and the high abundance of radiolarians. The greater decrease in surface $\delta^{13}C$ as the surface to benthos carbon isotopic gradient declines and recent climatic modelling results also support the upwelling hypothesis. We conclude that the oscillation in the stable isotopic profiles at Hole 1051B is due to climatic control on the intensity of upwelling of deep water to the surface ocean.

Warm climatic intervals were previously thought to be times of relative stability. However, the stable isotopic results suggest a highly unstable climate with variability like that seen in Plio-Pleistocene time. Milankovitch cyclicity consistent with the wavelengths of eccentricity, obliquity and precession is seen in the planktonic, benthic and colour records, indicating that orbital forcing was an important factor of climatic change during late mid-Eocene time.

Bottom-water temperatures range between 6 and 9 °C and gradually decline over the 500 ka studied. Spectral analysis reveals the influence of the obliquity cycle in the benthic foraminifera record, confirming intermediate- or deep-water formation at mid- to high latitudes during this period.

We would like to thank C. Chilcott for assistance with the mass spectrometer at the University of Edinburgh. S. Prew helped with picking benthic foraminifera and their analysis in the Woods Hole Oceanographic Institution's mass spectrometry facility. We are grateful to P. Pearson for useful discussions on Eocene planktonic foraminifer taxonomy and constructive review of the manuscript. This research was supported by UK Natural Environment Research Council reference number GT 04/97/93/ES. Analyses at Woods Hole and sample preparation were supported by a grant from JOI–USSSP to R.D.N.

References

ADAMS, C. G., LEE, D. E. & ROSEN, B. R. 1990. Conflicting isotopic and biotic evidence for tropical sea-surface temperatures during the Tertiary. *Palaeogeography, Palaeoclimatology, Palaeoecology*, **77**, 289–313.

ANDREASSON, F. P. & SCHMITZ, B. 1998. Tropical Atlantic seasonal dynamics in the early mid-Eocene from stable oxygen and carbon isotope profiles of mollusk shells. *Palaeoceaography*, **13**, 183–192.

BARRERA, E. & HUBER, B. T. 1991. Palaeogene and early Neogene oceanography of the southern Indian Ocean: Leg 119 foraminifer stable isotope results. *In:* BARRON, J., LARSEN, B. *et al.* (eds) *Proceedings of the Ocean Drilling Program, Scientific Results*, **119**. Ocean Drilling Program, College Station, TX, 693–717.

BELANGER, P. E., CURRY, W. B. & MATTHEWS, R. K. 1981. Core-top evaluation of benthic foraminiferal isotopic ratios for palaeo-oceanographic interpretations. *Palaeogeography, Palaeoclimatology, Palaeoecology*, **33**, 205–220.

BEMIS, B. E., SPERO, H. J., LEA, D. W. & BIJMA, J. 2000. Temperature influence on the carbon isotope composition of *Globigerina bulloides* and *Orbulina universa* (planktonic foraminifera). *Marine Micropalaeontology*, **38**, 213–228.

BERGGREN, W. A., KENT, D. V., SWISHER, C. C. III & AUBRY, M.-P. 1995. A revised Cenozoic geochronology and chronostratigraphy. *In*: BERGGREN, W. A., KENT, D. V., AUBRY, M.-P. & HARDENBOL, J. (eds) *Geochronology, Time Scales and Global Stratigraphic Correlation: a Unified Temporal Framework for an Historical Geology*. SEPM Special Publications, **54**, 129–212.

BICE, K. L., SLOAN, L. C. & BARRON, E. J. 2000. Comparison of early Eocene isotopic palaeotemperatures and the three-dimensional OGCM temperature field: the potential for use of model-derived surface water $\delta^{18}O$. *In*: HUBER, B. T., MACLEOD, K. G. & WING, S. L. (eds) *Warm Climates in Earth History*. Cambridge University Press, Cambridge, 79–131.

BOERSMA, A., PREMOLI SILVA, I. & SHACKLETON, N. J. 1987. Atlantic Eocene planktonic foraminiferal palaeohydrographic indicators and stable isotope palaeoceanography. *Palaeoceanography*, **2**, 287–331.

BOERSMA, A., SHACKLETON, N. J., HALL, M. & GIVEN, Q. 1979. Carbon and oxygen isotope records at DSDP Site 384 (North Atlantic) and some Palaeocene palaeotemperatures and carbon isotope variations in the Atlantic Ocean. *In*: TUCHOLKE, B. E., VOGT, P. R., MURDMAA, I. O. *et al.* (eds). *Initial Reports of the Deep Sea Drilling Project*, **43**. US Government Printing Office, Washington, DC, 695–717.

BOTTOMLEY, M., FOLLAND, C. K., HSILING, J., NEWELL, R. E. & PARKER, P. E. 1990. *Global Ocean Surface Temperature Atlas*. A Joint Project of the Meteorological Office and Massachusetts Institute of Technology, Meteorological Office, Bracknell.

BRALOWER, T. J., ZACHOS, J. C., THOMAS, E. *et al.* 1995. Late Palaeocene to Eocene palaeoceanography of the equatorial Pacific Ocean: stable isotopes recorded at Ocean Drilling Program Site 865, Allison Guyot. *Palaeoceanography*, **10**, 841–865.

BROECKER, W. S. 1989. The salinity contrast between the Atlantic and Pacific Oceans during glacial time. *Palaeoceanography*, **4**, 207–212.

CORFIELD, R. M. & CARTLIDGE, J. E. 1991. Isotopic evidence for the depth stratification of fossil and Recent *Globigerinina*: a review. *Historical Biology*, **5**, 37–63.

CORFIELD, R. M. & NORRIS, R. D. 1996. Deep water circulation in the Palaeocene Ocean. *In*: KNOX, R. W., CORFIELD, R. M. & DUNAY, R. E. (eds) *Correlation of the Early Palaeogene in Northwest Europe*. Geological Society, London, Special Publications, **101**, 443–456.

CORFIELD, R. M. & NORRIS, R. D. 1998. The oxygen and carbon isotopic context of the Palaeocene–Eocene Epoch boundary. *In*: AUBRY, M.-P., LUCAS, S. & BERGGREN, W. A. (eds) *Late Palaeocene–Early Eocene Climatic and Biotic Events in the Marine and Terrestrial Records*. Columbia University Press, New York, 124–137.

COVEY, C. & THOMPSON, S. L. 1989. Testing the effects of ocean heat transport on climate. *Global Planetary Change*, **1**, 331–341.

CRAIG, H. 1957. Isotopic standards for carbon and oxygen correction factors for mass spectrometric analysis of CO_2. *Geochimica et Cosmochimica Acta*, **12**, 133–149.

CROWLEY, T. J. 1991. Past CO_2 changes and tropical sea surface temperatures. *Palaeoceanography*, **6**, 387–394.

CROWLEY, T. J., KIM, K.-Y., MENGEL, J. & SHORT, D. 1992. Modeling 100,000-year climate fluctuations in pre-Pleistocene time series. *Science*, **255**, 705–707.

DUPLESSY, J. C., LALOU, C. & VINOT, A. C. 1970. Differential isotopic fractionation in benthic foraminifera and palaeotemperatures reassessed. *Science*, **168**, 250–251.

EREZ, J. 1979. Modification of the oxygen isotope records in deep sea cores by Pleistocene dissolution cycles. *Nature*, **281**, 535–538.

EREZ, J. & LUZ, B. 1983. Experimental palaeotemperature equation for planktonic foraminifera. *Geochimica et Cosmochimica Acta*, **47**, 1025–1031.

GRAHAM, A. 1994. Neotropical Eocene coastal floras and $^{18}O/^{16}O$-estimated warmer vs. cooler equatorial waters. *American Journal of Botany*, **81**, 301–306.

GRAHAM, D. W., CORLISS, B. H., BENDER, M. L. & KEIGWIN, L. D., Jr 1981. Carbon and oxygen isotopic disequilibria of recent deep-sea benthic foraminifera. *Marine Micropalaeontology*, **6**, 483–497.

HAGELBERG, T. K., BOND, G. & DEMENOCAL, P. 1994. Milankovitch band forcing of sub-Milankovitch climate variability during the Pleistocene. *Palaeoceanography*, **9**, 545–558.

HERBERT, T. D. & D'HONDT, S. L. 1990. Precessional climate cyclicity in Late Cretaceous–Early Tertiary marine sediments: a high resolution chronometer of Cretaceous–Tertiary boundary events. *Earth and Planetary Science Letters*, **99**, 263–275.

HORRELL, M. A. 1990. Energy balance constraints on ^{18}O based palaeo-sea surface temperature estimates. *Palaeoceanography*, **5**, 339–348.

KEIGWIN, L. D. & CORLISS, B. H. 1986. Stable Isotopes in late mid-eocene to Oligocene Foraminifera. *Geological Society of America Bulletin*, **97**, 335–345.

KENNETT, J. P. & SHACKLETON, N. J. 1976. Oxygen isotopic evidence for the development of the psychrosphere 38 m.y. ago. *Nature*, **260**, 513–515.

KENNETT, J. P. & STOTT, L. D. 1990. Proteus and proto-oceanus: ancestral Palaeogene oceans as revealed from Antarctic stable isotopic results: ODP Leg 113. *In*: BARKER, P. F., KENNETT, J. P. *et al.* (eds) *Proceedings of the Ocean Drilling*

Program, Scientific Results, **113**. Ocean Drilling Program, College Station, TX, 865–880.

KILLINGLEY, J. S. 1983. Effects of diagenetic recrystallization on $^{18}O/^{16}O$ values of deep-sea sediments. *Nature*, **301**, 594–597.

KROON, D. & GANSSEN, G. 1989. Northern Indian Ocean upwelling cells and the stable isotope composition of living planktonic foraminifers. *Deep-Sea Research*, **36**, 1219–1236.

MANABE, S. & BRYAN, K. 1985. CO_2-induced change in a coupled ocean–atmosphere model and its palaeoclimatic implications. *Journal of Geophysical Research*, **90**, 11689–11708.

MILANKOVITCH, M. M. 1941. *Kanon der erdbest rahlling und seine anwendung auf das eiszeiten problem.* Academic Royale Serbe, edition speciale **333**.

NORRIS, R. D. 1998. Recognition and macroevolutionary significance of photosymbiosis in molluscs, corals and foraminifera. *In*: NORRIS, R. D. & CORFIELD, R. M. (eds) *Isotope Palaeobiology and Palaeoecology*. Palaeontological Society Papers, **4**, 68–100.

NORRIS, R. D. & RÖHL, U. 1999. Carbon cycling and chronology of climate warming during the Palaeocene/Eocene transition. *Nature*, **401**, 775–778.

NORRIS, R. D., KROON, D., KLAUS, A. *et al.* (eds) 1998. *Proceedings of the Ocean Drilling Program, Initial Results*, **171**. Ocean Drilling Program, College Station, TX.

PAILLARD, D., LABEYRIE, L. & YIOU, P. 1996. Macintosh program performs time-series analysis. *Eos Transactions, American Geophysical Union*, **77**, 379.

PAULL, C. K., HILLS, S. J. & THIERSTEIN, H. R. 1988. Progressive dissolution of fine carbonate particles in pelagic sediments. *Marine Geology*, **81**, 27–40.

PEARSON, P. N. & SHACKLETON, N. J. 1995. Neogene multispecies planktonic foraminifer stable isotope record, site 871, Limalok Guyot. *In*: HAGGERTY, J. A., PREMOLI SILVA, I., RACK, F. & MCNUTT, M. K. (eds) *Proceedings of the Ocean Drilling Program, Scientific Results*, **144**. Ocean Drilling Program, College Station, TX, 401–410.

PEARSON, P. N., SHACKLETON, N. J. & HALL, M. A. 1993. Stable isotope palaeoecology of mid-Eocene planktonic foraminifera and multi-species isotope stratigraphy, DSDP Site 523, South Atlantic. *Journal of Foraminiferal Research*, **23**, 123–140.

PINET, P. R. & POPENOE, P. 1985. A scenario of Mesozoic–Cenozoic ocean circulation over the Blake Plateau and its environs. *Geological Society of America Bulletin*, **96**, 618–626.

SCHRAG, D. P., DEPAOLO, D. J. & RICHTER, F. M. 1995. Reconstructing past sea surface temperatures: correcting for diagenesis of bulk marine carbonate. *Geochimica et Cosmochimica Acta*, **59**, 2265–2278.

SHACKLETON, N. J. 1974. Attainment of isotopic equilibrium between ocean water and benthonic foraminifera genus *Uvigerina*: isotopic changes in the ocean during the last glacial. Colloques Internationaux du CNRS, **219**, 203–209.

SHACKLETON, N. J. & BOERSMA, A. 1981. The climate of the Eocene ocean. *Journal of the Geological Society London*, **138**, 153–157.

SHACKLETON, N. J. & KENNETT, J. P. 1975. Palaeotemperature history of the Cenozoic and the initiation of Antarctic glaciation: oxygen and carbon isotope analyses in DSDP Sites 277, 279, and 281. *In*: KENNETT, J. P., HOULTZ, R. E. *et al.* (eds) *Initial Reports of the Deep Sea Drilling Project*, **29**. US Government Printing Office, Washington, DC, 743–755.

SHACKLETON, N. J., CORFIELD, R. M. & HALL, M. A. 1985. Stable isotope data and the ontogeny of Palaeocene planktonic foraminifera. *Journal of Foraminiferal Research*, **15**, 321–336.

SHACKLETON, N. J., HALL, M. A. & BOERSMA, A. 1984. Oxygen and carbon isotope data from Leg 74 foraminifers. *In*: MOORE, T. C., Jr, RABINOWITZ, P. D. *et al.* (eds) *Initial Reports of the Deep Sea Drilling Project*, **74**. US Government Printing Office, Washington, DC, 599–612.

SLOAN, L. C. & HUBER, M. 2000. North Atlantic climate variability in early Palaeogene time: a climate modelling sensitivity study. *This volume.*

SLOAN, L. C. & REA, D. K. 1995. Atmospheric carbon dioxide and early Eocene climate: a general circulation modeling sensitivity study. *Palaeogeography, Palaeoclimatology, Palaeoecology*, **119**, 275–292.

STOTT, L. D. & ZACHOS, J. C. 1991. Paleogene paleoceanography workshop report, *JOI-USSAC.*

THUNELL, R. C. & HONJO, S. 1981. Calcite dissolution and the modification of planktonic foraminifera assemblages. *Marine Micropalaeontology*, **6**, 169–182.

VERGNAUD-GRAZZINI, C. & SALIEGE, J. F. 1985. Les événements isotopiques en milieu océanique à la transition Eocène/Oligocène dans le Pacifique et l'Atlantique; paléocirculations profondes en Atlantique sud. *Bulletin de la Société Géologique de France*, **1**, 441–455.

WADE, B. S., NORRIS, R. D. & KROON, D. 2000. Data report: high-resolution stable isotope stratigraphy of the late mid-Eocene at Site 1051, Blake Nose. *In*: KROON, D., NORRIS, R. D., KLAUS, A. (eds) *Proceedings of the Ocean Drilling Program, Scientific Results*, **171**. Ocean Drilling Program, College Station, TX, in press. [Available on world wide web: http://www.ODP.TAMU.EDU/PUBLICATIONS/MIB-SR/CHAP_05/CHAP_05.HTM].

WISE, S. W. Jr., BREZA, J. R., HARWOOD, D. M. & WEI, W. 1991. Palaeogene glacial history of Antarctica. *In*: MUELLER, D. W., MCKENZIE, J. A. & WEISSERT, H. (eds) *Controversies in Modern Geology*. Academic Press, San Diego, CA, 133–171.

WOODRUFF, F., SAVIN, S. M. & DOUGLAS, R. G. 1980. Biological fractionation of oxygen and carbon isotopes by Recent benthic foraminifera. *Marine Micropalaeontology*, **5**, 3–11.

WU, G. & BERGER, W. H. 1989. Planktonic foraminifer: differential dissolution and the

Quaternary stable isotope record in the west equatorial Pacific. *Palaeoceanography*, **4**, 181–198.

WU, G., HERGUERA, C. & BERGER, W. H. 1990. Differential dissolution, modification of late Pleistocene oxygen isotope records in the western equatorial Pacific. *Palaeoceanography*, **5**, 581–594.

ZACHOS, J. C., REA, D. K., SETO, K., NOMURA, R. & NIITSUMA, N. 1992. Palaeogene and Early Neogene deep water palaeoceanography of the Indian Ocean as determined from benthic foraminifer stable carbon and oxygen isotope records. *In*: DUNCAN, R. A., REA, D. K. *et al.* (eds) *Synthesis of Results of Scientific Drilling in the Indian Ocean*. Geophysical Monograph American Geophysical Union, **70**, 351–385.

ZACHOS, J. C., STOTT, L. D. & LOHMANN, K. C. 1994: Evolution of early Cenozoic marine temperatures. *Palaeoceanography*, **9**, 353–387.

ZAHN, R. & DIESTER-HAASS, L. 1995. Orbital forcing of Eocene–Oligocene climate: high resolution records of benthic isotopes from ODP Site 689, Maud Rise. ICP V Programme Abstracts, 5th International Conference on Paleoceanography, Halifax, 177.

Carbon addition and removal during the Late Palaeocene Thermal Maximum: basic theory with a preliminary treatment of the isotope record at ODP Site 1051, Blake Nose

GERALD R. DICKENS

School of Earth Sciences, James Cook University, Townsville, Qld 4811, Australia

Present address: Department of Geology & Geophysics, Rice University, Houston, TX 77251, USA (e-mail: Jerry@geophysics.rice.edu)

Abstract: The late Palaeocene Thermal Maximum (LPTM) was a brief interval at *c.* 55 Ma characterized by a −2.5 to −3‰ shift in the $\delta^{13}C$ of global carbon reservoirs. The geochemical perturbation probably represents a massive input of ^{12}C-rich carbon to the exogenic carbon cycle. Largely unresolved issues concerning this carbon injection during the LPTM are the rates of carbon input and removal. Simple expressions are developed here to describe a $\delta^{13}C$ excursion in the exogenic carbon cycle after carbon input. A change in global $\delta^{13}C$ ($d\delta_{Ex}/dt$) can be explained to a first approximation by a set of parameters: the initial mass and isotopic composition of the global carbon cycle ($M_{Ex(o)}$, $\delta_{Ex(o)}$), and the fluxes and isotopic compositions of external carbon inputs, outputs and injected carbon (F_{In}, δ_{In}, F_{Out}, δ_{Out}, F_{Add}, δ_{Add}). In general, for a given exogenic carbon cycle, a large F_{Add} or low δ_{Add} results in a larger $\delta^{13}C$ excursion. Likewise, for a given negative $\delta^{13}C$ excursion, a large M_{Ex} or low δ_{Ex} requires a greater input of ^{12}C. Differences in F_{In}, δ_{In}, F_{Out} and δ_{Out} cause changes in the response of δ_{Ex} over time. For a negative $\delta^{13}C$ excursion of given magnitude, a greater F_{In} requires a greater input of ^{12}C and lessens the time for δ_{Ex} to return to initial conditions. A decrease in δ_{Out} (caused by an increase in the relative output of organic matter and carbonate) has a similar effect. Variable dM_{Add}/dt produces transients in δ_{Ex} that are related to the source function but modified by carbon removal. In theory, a well-dated and representative global $\delta^{13}C$ excursion could be used to derive the carbon inputs and ouputs. Ocean Drilling Program (ODP) Site 1051 has an expanded early Palaeogene section, and recent work at this location has provided a well-dated $\delta^{13}C$ record across the LPTM. This $\delta^{13}C$ record contains transient variations of apparently global nature. These observed transients are best explained by a pulsed injection of CH_4 into an exogenic carbon cycle with a greater carbon throughput or enhanced burial of organic matter after carbon addition.

Superimposed on general early Palaeogene warmth was a brief event at *c.* 55 Ma when deep-ocean and high-latitude surface temperatures soared by at least 4–7°C (Kennett & Stott 1991; Bralower *et al.* 1995; Fricke *et al.* 1998; Katz *et al.* 2000). This event, termed the Late Palaeocene Thermal Maximum (LPTM) (Zachos *et al.* 1993), coincided with profound global environmental change and an extraordinary decrease in the $^{13}C/^{12}C$ ratio ($\delta^{13}C$) of all carbon in the ocean and atmosphere (e.g. Koch *et al.* 1992; Thomas & Shackleton 1996; Katz *et al.* 1999). In marine records examined to date, from both shallow and deep water, the isotope excursion is an abrupt drop in $\delta^{13}C$ of −2.5 to −3‰ over 5–20 cm followed by a roughly logarithmic return to near-initial values over 1 to 4 m (Fig. 1a). The main drop and gradual return in $\delta^{13}C$ across the LPTM probably spanned less than 10 ka and *c.* 120 ka, respectively (Kennett & Stott 1991; Bralower *et al.* 1998; Norris & Röhl 1999).

The magnitude, timing and widespread nature of the $\delta^{13}C$ excursion strongly suggest a sudden and massive injection of ^{12}C-rich carbon to the atmosphere or ocean during the LPTM (Dickens *et al.* 1995; Thomas & Shackleton 1996). This is an important conclusion because it makes the LPTM our best (and perhaps only) past time interval to understand the long-term fate of future anthropogenic carbon addition to the atmosphere (Dickens 1999). Within this context, rates of carbon injection and carbon removal become critical issues. In theory, these rates could be extracted from a global $\delta^{13}C$ record if the source of carbon addition, the timing of isotope perturbations, and the composition of

From: KROON, D., NORRIS, R. D. & KLAUS, A. (eds). 2001 *Western North Atlantic Palaeogene and Cretaceous Palaeoceanography*. Geological Society, London, Special Publications, **183**, 293–305.
0305-8719/01/$15.00

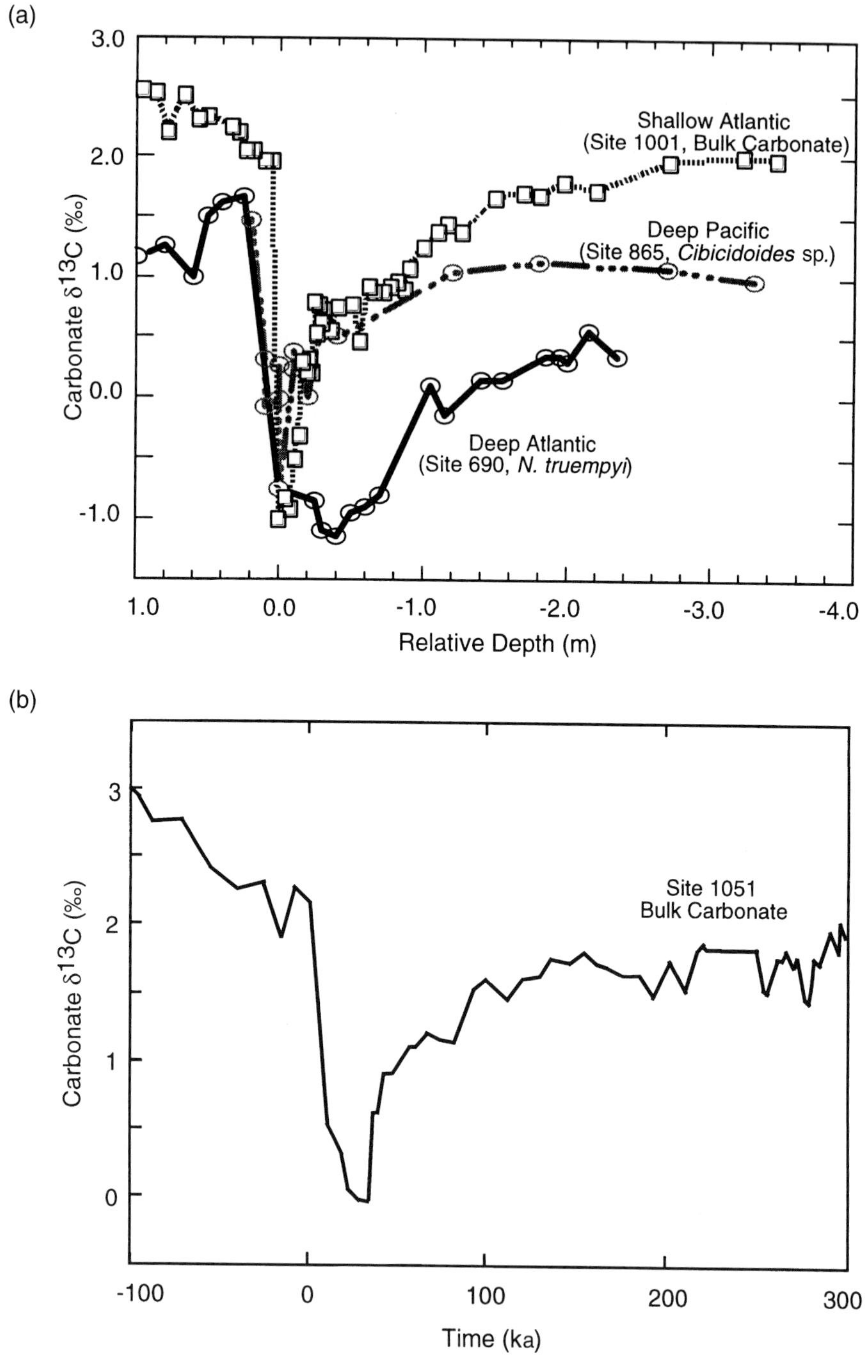

Fig. 1. High-resolution carbon isotope records at four widely separated Ocean Drilling Program (ODP) sites. (**a**) The benthic foraminifera records at Site 690 (Maud Rise, South Atlantic; Kennett & Stott, 1991) and Site 865 (Allison Guyot, Equatorial Pacific; Bralower *et al.* 1995), and the bulk carbonate record at Site 1001 (Nicaraguan Rise, Caribbean; Bralower *et al.* 1997) placed on a common depth scale where the $\delta^{13}C$ minimum is at 0.0 m. (Note the abrupt drop in $\delta^{13}C$ of −2.5 to −3‰ over 5–20 cm followed by a roughly logarithmic return to near-initial values over 1–4 m.) (**b**) The recently constructed bulk carbonate record at Site 1051 (Blake Nose, North Atlantic) placed on an orbitally tuned time scale (Norris & Röhl 1999).

the ocean and atmosphere were known (Berger & Vincent 1986).

Most current workers agree that sea-floor CH_4 is the probable source of carbon addition during the LPTM (e.g. Dickens *et al.* 1995, 1997*a*; Bralower *et al.* 1997, 1998; Thomas 1998; Bains *et al.* 1999; Dickens 2000; Katz *et al.* 1999; Norris & Röhl 1999). On the other hand, the timing of short-term $\delta^{13}C$ variations across the LPTM and parameters of the Palaeogene carbon cycle have been unconstrained. Initial modelling exercises of the LPTM therefore assumed a simple one-step injection of CH_4 carbon into the pre-industrial Holocene atmosphere (Dickens *et al.* 1997*a*) or ocean (Dickens 2000) over 10 ka followed by a sequestering of carbon according to poorly quantified global weathering parameters (Walker & Kasting 1992).

Recent work at Ocean Drilling Program (ODP) Site 1051 (Bains *et al.* 1999; Katz *et al.* 1999; Norris & Röhl 1999) has constructed an astronomically tuned and apparently continuous $\delta^{13}C$ record across the LPTM. In bulk carbonate, this record (Fig. 1b) contains short-term variations in $\delta^{13}C$ that cannot be explained by simple one-step injection of CH_4 to the atmosphere or ocean (Bains *et al.* 1999). Moreover, numerous papers have suggested that carbon masses in the pre-Neogene ocean and atmosphere were generally higher than at the present day (e.g. Kump & Arthur 1999). The diverse scientific community studying the early Palaeogene age thus needs models to deal with complex release of carbon to a Palaeogene carbon cycle and its subsequent sequestering over time. In this study I offer the simplest useful mathematical expression describing this behaviour. I then present general solutions to the expression to show how variations in carbon injection and a set of fundamental parameters will affect global $\delta^{13}C$ over time. Finally, I provide a very basic carbon input and removal model to explain transient $\delta^{13}C$ variations in an isotope record at Site 1051. The model for Site 1051 is imperfect in detail because of its simplicity, and because the $\delta^{13}C$ record was made from analyses of bulk carbonate. Nevertheless, the treatment presented here provides crucial insight for more sophisticated work concerning the extraordinary injection of carbon during the LPTM.

Theoretical background

The exogenic carbon cycle includes all carbon stored in the ocean, atmosphere and biomass (Fig. 2a). Because carbon cycles through these reservoirs in less than 2000–3000 years at the present day (and presumably in the past), the mass and isotope composition of the exogenic carbon cycle can be considered a single entity (Fig. 2b) over relatively short time scales (e.g. Sundquist 1986; Walker & Kasting 1992). All papers cited above concerning the LPTM agree that the observed $\delta^{13}C$ anomaly represents a sudden and massive injection of ^{12}C-rich carbon to an early Palaeogene exogenic carbon cycle.

The effect of a sudden and massive carbon injection upon the $\delta^{13}C$ composition of the exogenic carbon cycle can be evaluated through relatively straightforward mass balance relationships. Let us consider the entire exogenic carbon cycle as a reservoir of mass M_{Ex} and isotope composition δ_{Ex} (Fig. 2b). To this reservoir there is an external carbon input of flux F_{In} and isotope composition δ_{In} and an external carbon output of flux F_{Out} and isotope composition δ_{Out}. These inputs and outputs primarily represent delivery from rivers, weathering and volcanoes, and burial of carbonate and organic matter (e.g. Walker & Kasting 1992; Compton & Mallison 1996; Kump & Arthur 1999). Now let us consider a carbon addition of flux F_{Add} and isotope composition δ_{Add} from an external reservoir to the exogenic carbon cycle (Fig. 2b). The appropriate mathematical expressions for such carbon exchange are:

$$M_{Ex} = F_{In}t + F_{Add}t - F_{Out}t + M_{Ex(o)} \tag{1}$$

and

$$M_{Ex}\delta_{Ex} = F_{In}\delta_{In}t + F_{Add}\delta_{Add}t - F_{Out}\delta_{Out}t + M_{Ex(o)}\delta_{Ex(o)} \tag{2}$$

where $M_{Ex(o)}$ and $\delta_{Ex(o)}$ are the mass and isotope composition of the exogenic carbon cycle before carbon addition and t is time. Both equations can be differentiated with respect to t to render:

$$\frac{dM_{Ex}}{dt} = F_{In} + F_{Add} - F_{Out} \tag{3}$$

and

$$\frac{dM_{Ex}\delta_{Ex}}{dt} = F_{In}\delta_{In} + F_{Add}\delta_{Add} - F_{Out}\delta_{Out} \tag{4}$$

Applying the product rule and combining these two equations gives a fundamental expression for the change in $\delta^{13}C$ of the exogenic carbon cycle over time resulting from changes in

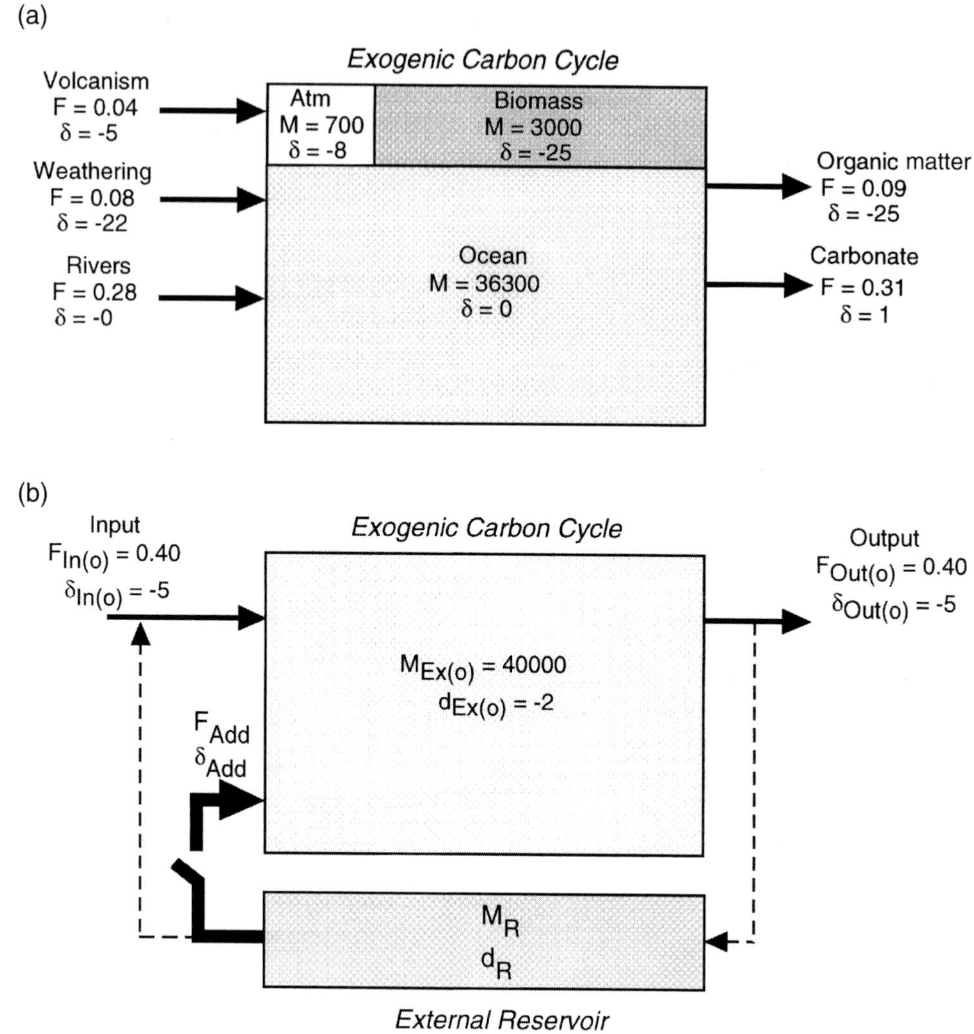

Fig. 2. The Holocene exogenic carbon cycle. (**a**) Carbon enters from volcanism, weathering and rivers, cycles through the ocean, atmosphere and biomass, and leaves through carbonate and organic matter. (**b**) Schematic representation of sudden and massive carbon addition from an external reservoir where all internal reservoirs of the exogenic carbon cycle experience a similar change in $\delta^{13}C$. Masses in Gt; fluxes in Gt a^{-1}; composition in ‰.

carbon inputs and outputs (adapted from Kump & Arthur 1999):

$$\frac{d\delta_{Ex}}{dt} = \frac{F_{Add}}{M_{Ex}}(\delta_{Add}-\delta_{Ex})+\frac{F_{In}}{M_{Ex}}(\delta_{In}-\delta_{Ex}) - \frac{F_{Out}}{M_{Ex}}(\delta_{Out}-\delta_{Ex}) \quad (5)$$

The first term of this equation accounts for the change in $\delta^{13}C$ caused by carbon addition whereas the second and third terms account for the change in $\delta^{13}C$ caused by external carbon ‘throughput’. The external input and output terms can be expanded to include constituent fluxes of different isotope composition (e.g. Compton & Mallison 1996; Kump & Arthur 1999). In the special case where δ_{Ex} is examined immediately after instantaneous carbon addition ($F_{In}t = F_{Out}t = 0$; $F_{Add}t = M_{Add}$), the above expression reduces to (Dickens *et al.* 1995):

$$\delta_{Ex} = \frac{M_{Add}\delta_{Add}+M_{Ex(o)}\delta_{Ex(o)}}{(M_{Add}+M_{Ex(o)})} \quad (6)$$

General solutions relevant to the LPTM carbon isotope excursion

At the most basic level, three fluxes (F_{Add}, F_{In}, F_{Out}), each with an isotope composition (δ_{Add}, δ_{In}, δ_{Out}), are needed to describe the evolution of the mass and $\delta^{13}C$ of the exogenic carbon cycle upon carbon addition from an external reservoir (equation (5)). The mass and composition of the exogenic carbon cycle before carbon injection ($M_{Ex(o)}$, $\delta_{Ex(o)}$) also need definition. Equation (5) can be solved by stepwise integration if the eight parameters are known.

Carbon injection to the Holocene carbon cycle

As a convenient starting point, the exogenic carbon cycle before carbon injection is assumed to be at steady-state conditions ($F_{Add} = 0$, $F_{In} = F_{Out}$, and $\delta_{In} = \delta_{Out}$) with six key parameters similar to those for pre-industrial Holocene time. The approximate values for these six parameters are (see Broecker 1974; Berger & Vincent 1986; Walker & Kasting 1992; Broecker & Peng 1993; Siegenthaler 1993; Dickens *et al.* 1995): $M_{Ex(o)} = 40\,000$ Gt (10^{15} g), $\delta_{Ex(o)} = -2.0‰$, $F_{In} = F_{Out} = 0.4$ Gt a^{-1}, $\delta_{In} = \delta_{Out} = -5.0‰$. It should be noted that $\delta_{Ex(o)}$ is negative because the isotope reference (0‰) is defined by the composition of marine inorganic carbon but the exogenic carbon cycle contains significant amounts (>3000 Gt) of organic matter with a $\delta^{13}C$ of about −25‰.

Carbonate and organic carbon are formed from carbon within the exogenic carbon cycle. Thus, F_{Out} and δ_{Out} are functions of M_{Ex} and δ_{Ex} rather than functions of F_{In} and δ_{In}. For initial modelling, the assumed relationships are

$$F_{Out} = kM_{Ex} \quad (7)$$

and

$$\delta_{Out} = \delta_{Ex} + f \quad (8)$$

where k and f are constants. The Holocene exogenic carbon is at steady-state conditions when $1/k$ is the e-folding or residence time (100 ka), and f is the carbon isotope fractionation during net carbon output (−3‰).

The infinite range of potential carbon inputs can be restricted greatly by considering a key factor: if the direction and magnitude of a global $\delta^{13}C$ excursion are known, there are a limited number of possible combinations for F_{Add} and δ_{Add}, especially if t is small ($t \ll k$). Figure 3 shows subsets of these combinations that can explain −2.5‰ and −5.0‰ shifts in the $\delta^{13}C$ of the Holocene exogenic carbon cycle given a simple one-step release of carbon over 10 ka. As seen in Fig. 3, and in general, the amount of carbon required to cause a global negative $\delta^{13}C$ excursion is directly related to the magnitude of the $\delta^{13}C$ excursion but inversely related to the absolute value of its isotopic composition. Moreover, if f is negative, disproportionately large masses of carbon with less negative $\delta^{13}C$ are needed to cause a given negative excursion (Fig. 3). Thus, a relatively small carbon input with a $\delta^{13}C$ of −60‰ can have the same effect on

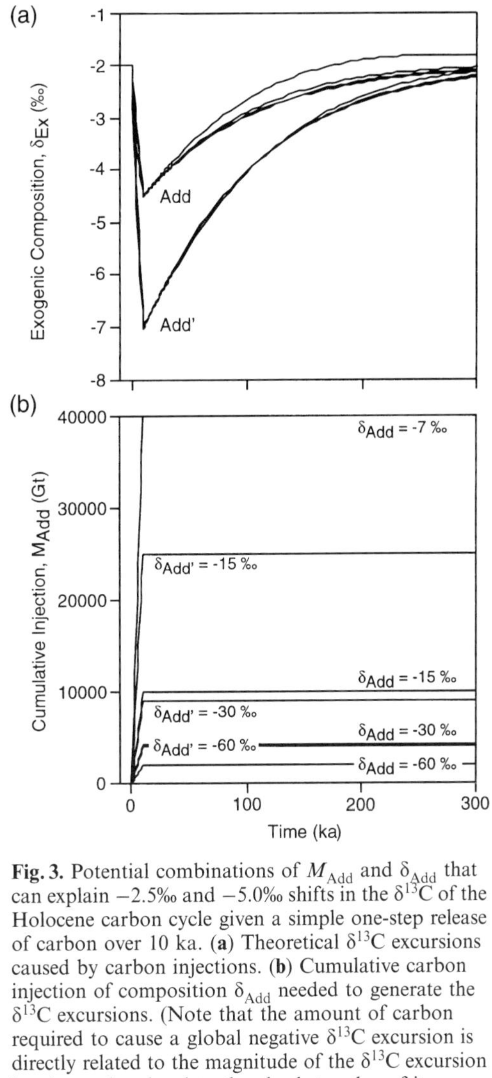

Fig. 3. Potential combinations of M_{Add} and δ_{Add} that can explain −2.5‰ and −5.0‰ shifts in the $\delta^{13}C$ of the Holocene carbon cycle given a simple one-step release of carbon over 10 ka. (**a**) Theoretical $\delta^{13}C$ excursions caused by carbon injections. (**b**) Cumulative carbon injection of composition δ_{Add} needed to generate the $\delta^{13}C$ excursions. (Note that the amount of carbon required to cause a global negative $\delta^{13}C$ excursion is directly related to the magnitude of the $\delta^{13}C$ excursion but inversely related to the absolute value of its isotopic composition. Both relationships are non-linear.)

the isotope composition of the ocean, atmosphere and biomass as an immensely large carbon input with a $\delta^{13}C$ of −7‰ (Dickens *et al.* 1995; Fig. 3).

Importantly, most theoretical carbon inputs are implausible causes for a −2.5‰ excursion in the $\delta^{13}C$ of the exogenic carbon cycle over 10 ka. First, there are only a limited number of significant external carbon reservoirs (M_R), each with a specific average δ_R (Fig. 2). Second, most carbon reservoirs of δ_R cannot supply a sufficient F_{ADD} over short time scales (Dickens *et al.* 1995; Thomas & Shackleton 1996). For example, the mantle, a large external carbon reservoir with a $\delta^{13}C$ of −5‰ (at least in this model, Fig. 2a), would have to release >40 000 Gt of carbon to explain a −2.5‰ excursion within the context of the Holocene exogenic carbon cycle (Fig. 3). Such carbon release seems impossible over 10 ka because it would signify a sudden doubling of M_{Ex} and an abrupt increase in mean Phanerozoic volcanic inputs (<0.1 Gt a^{-1}) by 1–2 orders of magnitude (Dickens *et al.* 1995; Thomas & Shackleton 1996). It should be noted that the absolute magnitude of the LPTM $\delta^{13}C$ excursion is at least 2.5‰ but remains unconstrained because of carbonate dissolution or sediment disturbance at the onset of the event in most deep-ocean records (e.g. Bralower *et al.* 1997; Thomas 1998; Katz *et al.* 1999).

On the basis of mass balance calculations (equation (2); Fig. 3), Dickens *et al.* (1995) have argued that a global $\delta^{13}C$ excursion of at least −2.5‰ must be associated with massive release of CH_4 from marine gas hydrate deposits. Gas hydrates are ice-like solids composed of gas and water that are stable at high pressure, low temperature and high gas concentration (Kvenvolden 1993; Dickens & Quinby-Hunt 1994). At the present day (and presumably in the Palaeocene time), about 7000–15 000 Gt of exceptionally ^{12}C-rich carbon ($\delta^{13}C$ = −60‰) are stored as methane hydrate and associated free gas in sediment on continental slopes (Kvenvolden 1993; Gornitz & Fung 1994; Dickens *et al.* 1997*b*). Large portions of this global reservoir may have transferred to the exogenic carbon cycle through sediment failure during the sudden bottom-water warming on slopes at the onset of the LPTM (Dickens *et al.* 1995; Katz *et al.* 1999). According to equation (5), a simple one-step release of 1880 Gt of carbon with a $\delta^{13}C$ of −60‰ over 10 ka would cause a global $\delta^{13}C$ excursion of −2.5‰ in the present-day exogenic carbon cycle (Fig. 3). This predicted amount of carbon is significantly greater than suggested by Dickens *et al.* (1997*a*) for several reasons, some of which are discussed below. None the less, the quantity is plausible within our current understanding of bottom-water warming during the LPTM and the global gas hydrate reservoir (Dickens *et al.* 1995).

Single-step methane injection to other carbon cycles

The general warmth and shallow carbonate compensation depth of early Palaeogene time (Van Andel 1975; Zachos *et al.* 1994) suggest that pCO_2 and $M_{Ex(o)}$ were larger in late Palaeocene time than at the present day. Long-term carbon isotope records also show that δ_{Ex} was between 1 and 2‰ greater in late Palaeocene time (Shackleton & Hall 1984). The effect of different initial values for M_{Ex} and δ_{Ex} upon $d\delta_{Ex}/dt$ (equation (5)) can be assessed by comparing theoretical $\delta^{13}C$ excursions caused by the same carbon injection. Figure 4 shows $\delta^{13}C$ excursions generated by a simple one-step release of 1880 Gt of CH_4 carbon over 10 ka into steady-state exogenic carbon cycles where F_{In} is constant at 0.4 Gt a^{-1} but initial values of M_{Ex} and δ_{Ex} range between 28 000 and 72 000 Gt and −4 and 0‰, respectively. As seen in Fig. 4, and in general, the magnitude of a global negative $\delta^{13}C$ excursion for a given injection is inversely related to the size of M_{Ex}, but directly related to δ_{Ex}. Interestingly, carbon sources for the LPTM excursion other than gas hydrate become even less appealing if one considers a more massive early Palaeogene exogenic carbon cycle.

The logarithmic return of δ_{Ex} to initial conditions depends on how fast the injected carbon is flushed through the system. In the above scenarios, steady-state conditions before carbon addition are maintained by adjusting k so that the e-folding time increases with increasing $M_{Ex(o)}$. Thus, for the same carbon addition at constant F_{In}, $d\delta_{Ex}/dt$ is greater for a less massive exogenic carbon cycle (Fig. 4). However, a Palaeogene carbon cycle with a greater carbon throughput (shorter residence time) than at present is plausible (e.g. Compton & Mallinson 1996; Kump & Arthur 1999). As before, the effect of different F_{In} and k upon $d\delta_{Ex}/dt$ can be assessed by comparing theoretical $\delta^{13}C$ excursions caused by the same carbon injection. Figure 5 shows $\delta^{13}C$ excursions generated by the above one-step release of CH_4 carbon into a steady-state carbon cycle (Fig. 2) where F_{In} varies between 0.2 and 0.8 Gt a^{-1} ($1/k$ between 200 ka and 50 ka). As seen in Fig. 5, an increase in F_{In} both shortens the time for δ_{Ex} to return to the initial value $\delta_{Ex(o)}$ and slightly reduces

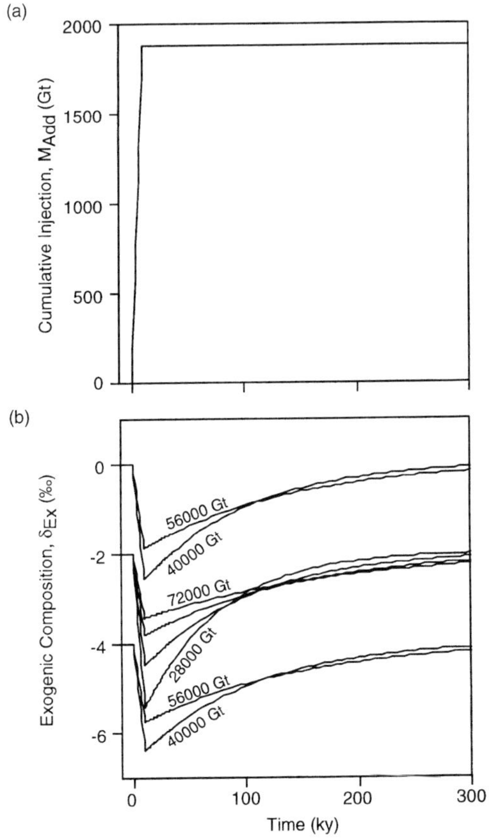

Fig. 4. Theoretical $\delta^{13}C$ excursions generated by a simple one-step release of CH_4 carbon into a steady-state carbon cycle where initial values of M_{Ex} and δ_{Ex} are different. (**a**) Cumulative injection of 1880 Gt of carbon with $\delta^{13}C$ of −60‰ over 10 ka. (**b**) The excursions caused by this injection where M_{Ex} and δ_{Ex} range between 28 000 and 72 000 Gt, and −4 and 0‰, respectively. (Note that the magnitude of a global negative $\delta^{13}C$ excursion for a given injection is inversely related to the size of M_{Ex} but directly related to δ_{Ex}.)

the magnitude of a global negative $\delta^{13}C$ excursion for a given carbon input. The latter effect is caused by the removal of excess carbon to the rock cycle during protracted carbon injection.

Carbonate and organic matter each fractionate carbon isotopes by a different amount. Thus, their compositions are different (Fig. 2a), and the net isotope fractionation during carbon output depends on the relative proportion of carbonate and organic matter removal (e.g. Compton & Mallison 1996; Kump & Arthur 1999). To address the effect of different carbonate/organic

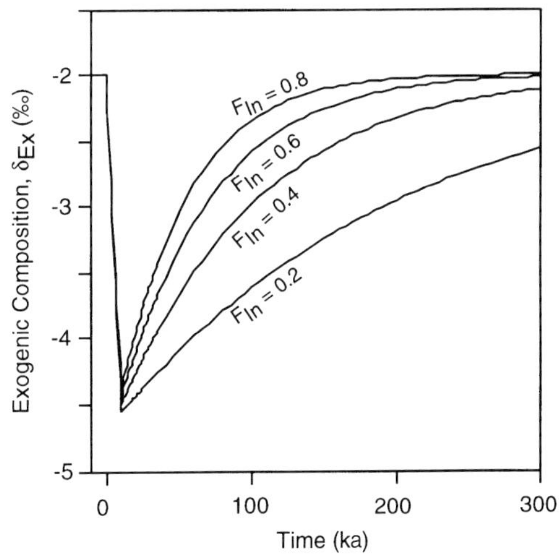

Fig. 5. Theoretical $\delta^{13}C$ excursions generated by a simple one-step release of CH_4 carbon (Fig. 4a) into a steady-state exogenic carbon cycle where F_{In} ranges between 0.2 and 0.8 Gt a^{-1}. (Note that an increase in F_{In} shortens the time for δ_{Ex} to return to its initial value and decreases the magnitude of the $\delta^{13}C$ excursion for a given injection.)

carbon output ratios upon $d\delta_{Ex}/dt$, equation (8) must be expanded as follows:

$$\delta_{Out} = \frac{F_{Carb}(\delta_{Ex}+f_{Carb})+F_{Org}(\delta_{Ex}+f_{Org})}{F_{Out}} \quad (9)$$

where F_{Carb} and f_{Carb} and F_{Org} and f_{Org} are the fluxes and fractionations of the separate carbonate and organic constituents of the output flux, respectively. Figure 6 shows $\delta^{13}C$ excursions generated by the aforementioned one-step release of CH_4 carbon into a steady-state carbon cycle (Fig. 2) where F_{Org}/F_{Carb} varies after injection in one case. In both cases, f_{Carb} and f_{Org} are constant at 3‰ and −23‰, respectively. As seen in Fig. 6, the logarithmic return of δ_{Ex} to initial conditions is shortened with an increase in F_{Org}/F_{Carb} after injection.

Variable methane injection

It is highly unlikely that the LPTM was marked by a single blast of carbon to the exogenic carbon cycle. For example, assuming that the LPTM $\delta^{13}C$ excursion signifies massive release of CH_4 by thermal dissociation of marine gas hydrate (Dickens *et al.* 1995), carbon might be added in pulses as different oceans or depth horizons containing gas hydrate were warmed (Bains *et al.* 1999; Dickens 2000). The effect of variable carbon input (dM_{Add}/dt) upon $d\delta_{Ex}/dt$ can be

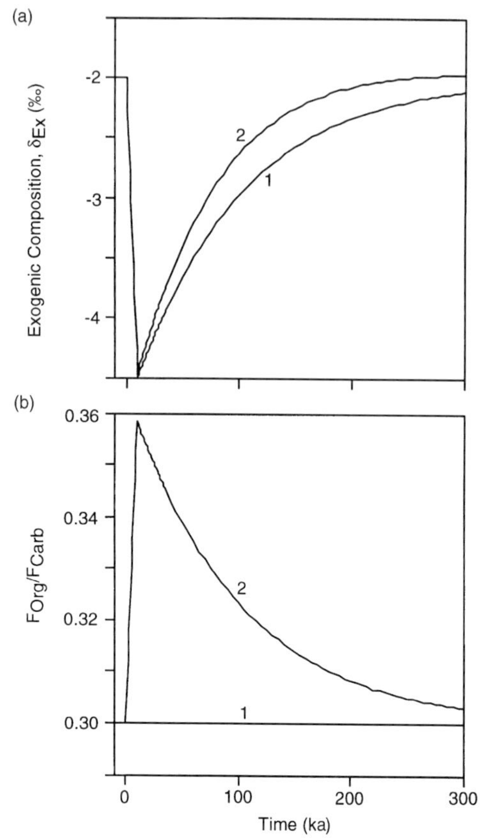

Fig. 6. Theoretical $\delta^{13}C$ excursions of *c.* – 2.5‰ generated by the same carbon injection into a steady-state carbon cycle (Figs 4 and 5) with different organic carbon to carbonate output ratios over time. (**a**) Modelled excursions. (**b**) Ratios of F_{Org}/F_{Out} ratio over time where (1) is constant and (2) is proportion to total carbon mass (M_{Ex}). (Note that an increase in the F_{Org}/F_{Out} ratio shortens the time for δ_{Ex} to return to its initial value.)

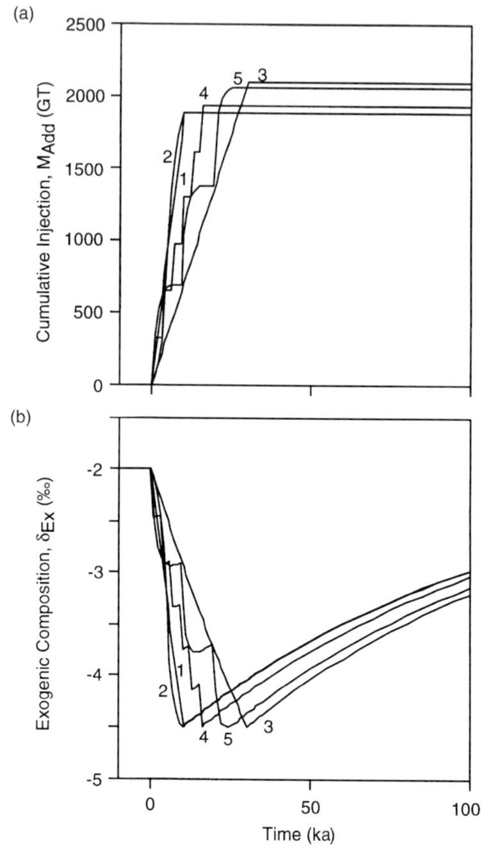

Fig. 7. Theoretical $\delta^{13}C$ excursions of −2.5‰ generated by different carbon inputs (dM_{Add}/dt) into the Holocene steady-state carbon cycle. (**a**) Five inputs of carbon including (1) a simple one-step release of 1880 Gt over 10 ka, (2) a 'smooth' one-step release of 1879 Gt over 10 ka, (3) a protracted one-step release of 2088 Gt over 30 ka, (4) a 'sawtooth' six-step release of 1933 Gt over 16 ka, and (5) a 'pulsed' three-step release of 2059 Gt over 26 ka. All carbon inputs have a $\delta^{13}C = -60$‰. (**b**) The excursions caused by these various carbon inputs. (Note that variations in carbon input over time produce transients in δ_{Ex} that are related to dM_{Add}/dt but modified by an increase in $\delta^{13}C$.)

assessed by comparing theoretical $\delta^{13}C$ excursions caused by different inputs into the same exogenic carbon cycle. Figure 7 shows five $\delta^{13}C$ excursions of −2.5‰ generated by different carbon inputs into the present-day exogenic carbon cycle, including simple one-step, 'smooth' one-step, gradual one-step, pulsed three-step and 'sawtooth' six-step additions. A very interesting aspect of rapid carbon injection to the exogenic carbon cycle becomes apparent in Fig. 7: variations in carbon input over time produce transients in δ_{Ex} that are related to dM_{Add}/dt but modified by carbon removal. Thus, multi-stepped or pulsed inputs of ^{12}C-rich carbon are characterized by multi-stepped or pulsed changes in δ_{Ex} separated by intervals of increasing δ_{Ex} when excess carbon is sequestered to the rock cycle (Fig. 7). The rate of increase in δ_{Ex} during the transients with lower carbon input is related to the flushing time (F_{IN} and k).

A preliminary treatment of the carbon isotope record at Site 1051, Blake Nose

Ocean Drilling Program (ODP) Leg 171B recovered an apparently continuous sediment sequence across the LPTM at Site 1051 in the western Atlantic Ocean (Norris *et al.* 1998). This record is exceptional because it is greatly expanded and contains obvious sediment cycles (Norris *et al.* 1998; Katz *et al.* 1999; Norris & Röhl 1999). Spectral analysis of these cycles shows that they have a regular frequency consistent with the Earth's precessional periods of about 20 ka (Norris & Röhl 1999). Using these sediment cycles, an astronomically calibrated time scale has been constructed across the LPTM with a precision on relative age to within 2000 years (Norris & Röhl 1999).

Carbon isotope compositions have been determined on bulk carbonate (Bains *et al.* 1999; Norris & Röhl 1999) and single species benthic foraminifera (Katz *et al.* 1999) at Site 1051. Although there are differences between the isotope records, all of the records show several similar features as exemplified by the bulk carbonate record (Fig. 2b). There are at least two abrupt drops in $\delta^{13}C$, each occurring within 5000–7000 years, separated by at least one transitional interval of relatively slowly decreasing $\delta^{13}C$ over about 8000 years. Following these changes in $\delta^{13}C$ are a second transitional interval of about 10 ka, and a gradual return to near-initial conditions over at least 120 ka. The transient features in the $\delta^{13}C$ record appear to represent global changes in δ_{Ex} because they also exist in isotope records of bulk carbonate and benthic foraminifera at Site 690 in the South Atlantic (Kennett & Stott 1991; Bains *et al.* 1999).

The combination of a high-resolution age model and a detailed $\delta^{13}C$ record of global significance at Site 1051 provides an unprecedented opportunity to investigate rates of carbon injection to and removal from the exogenic carbon cycle during the LPTM. A change in the $\delta^{13}C$ of the exogenic carbon cycle after carbon addition ($d\delta_{Ex}/dt$) can be explained to a first approximation by an injection function and a set of masses, fluxes and compositions for the carbon reservoirs, inputs and outputs (equation (5)). In theory, given a global $\delta^{13}C$ excursion, one can solve equation (5) by numerical integration to derive the carbon source function or estimate unknown parameters (Berger & Vincent 1986). Of the three available $\delta^{13}C$ records at Site 1051, I take the bulk carbonate record presented by Norris & Röhl (Fig. 2a) as representing $d\delta_{Ex}/dt$ for the purpose of a preliminary treatment of carbon injection and removal during the LPTM. It should be noted, however, that the small magnitude of the bulk carbonate excursion at Site 1051 (−1.6‰) suggests that this record is not entirely representative of the global signal. Unfortunately, at this juncture, the benthic foraminifera record at Site 1051 (Katz *et al.* 1999) lacks sufficient detail, and the detailed bulk record at Site 690 (Bains *et al.* 1999) lacks an independent age model.

The masses, fluxes and isotopic compositions of the late Palaeocene exogenic carbon cycle before carbon injection have not been estimated previously. For this paper, I assume a modified 'Phanerozoic' carbon cycle by Kump & Arthur (1999). This steady-state carbon cycle (Fig. 8) has an M_{Ex} of 56 000 Gt with an atmospheric pCO_2 about twice that of the present day and a biomass 1.23 greater than at the present day. The input and output fluxes at steady-state conditions are 1.5 times greater than in Holocene time, but are the same isotopic composition. However, the ratio of F_{Org}/F_{Carb} is 0.28 at steady-state conditions. Assuming that CH_4 is the correct source function, M_{Add} is between 1500 and 2200 Gt (Fig. 7), and δ_{Add} is −60‰.

Transients in $d\delta_{Ex}/dt$ at the onset of the LPTM suggest variations in dM_{Add}/dt (Fig. 7). These variations in carbon input can be determined for a given steady-state carbon cycle (equation (5)). Figure 9 shows two $\delta^{13}C$ excursions generated by massive CH_4 injection to a 'Palaeogene' exogenic carbon cycle (Fig. 8) where f is constant (case 1) or variable (case 2). In both cases, a somewhat complicated two-step pulsed input reproduces the observed transients in $d\delta_{Ex}/dt$ at the onset of the LPTM. However, in the first case the predicted $d\delta_{Ex}/dt$ does not return to initial conditions fast enough over the first 50 ka after carbon injection. If the age model for Site 1051 and the carbon cycle shown in Fig. 8 are correct, apparently ^{12}C was removed from the Palaeogene carbon cycle at an enhanced rate following carbon injection. One possibility is that the carbon throughput increased (Fig. 5). An alternative explanation is that organic matter output increased (Fig. 9).

The carbon cycle perturbation during the LPTM, as exemplified by the $\delta^{13}C$ record at Site 1051, is best explained by a pulsed input of exceptionally ^{12}C-rich carbon into an exogenic carbon cycle where ^{12}C output is enhanced significantly after carbon input (i.e. beyond expectations set by equations (7) and (8)). The complicated pulsed carbon input is consistent with non-linear carbon release from an external reservoir. Thermal dissociation of gas hydrates and release of CH_4 from multiple locations at different water depths remains a plausible scenario. The rapid increase in $\delta^{13}C$ after carbon injection is consistent with a greater throughput of carbon. Atmospheric pCO_2 should rise after a massive input of carbon to the exogenic carbon cycle. Expected effects of this higher pCO_2 are elevated Earth surface temperatures, greater

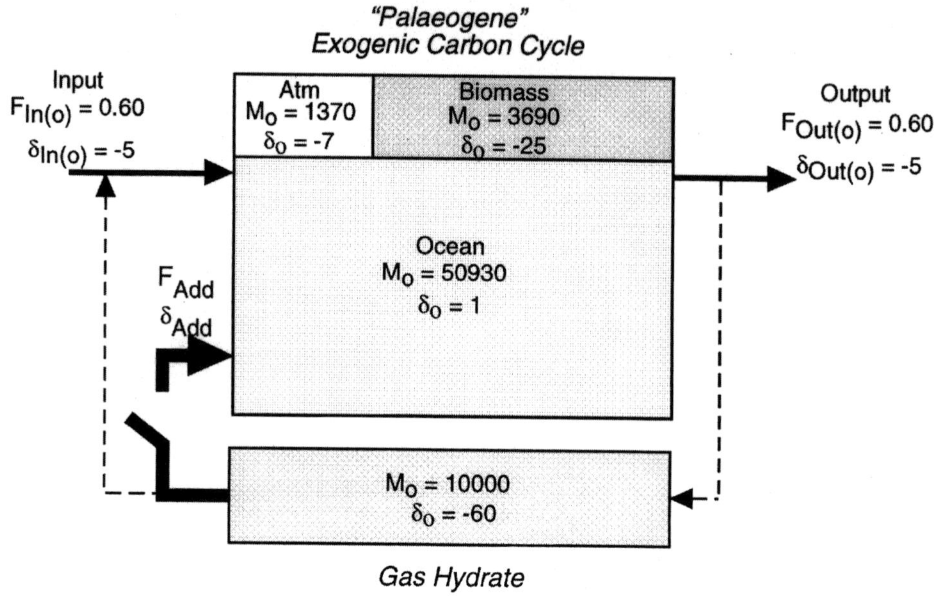

Fig. 8. A plausible steady-state Palaeogene exogenic carbon cycle. Units the same as in Fig. 2.

weathering, and increased delivery of riverine loads, including HCO_3^- (Walker *et al.* 1981). This enhanced carbon input with an isotopic composition of −5‰ would increase $d\delta_{Ex}/dt$ after carbon injection. However, on the basis of the $\delta^{13}C$ record alone, a sudden increase in biomass and burial of ^{12}C-rich organic matter after carbon injection cannot be excluded. The alternative explanation could represent CO_2 fertilization of biomass (e.g. Esser 1987).

Problems and future research

The expression presented here for describing the evolution of the global carbon cycle after carbon input is exceptionally simple. Alternatively, there exist moderately complex numerical models that explicitly address massive carbon input to the present-day exogenic carbon cycle (e.g. Sundquist 1986; Walker & Kasting 1992). In essence, these models are great extensions of equation (5) with many additional terms and parameters. The problem with these models is that the importance of basic parameters and the significance of transients are not immediately clear. To a large degree it is hoped that a diverse range of researchers can use and modify equation (5) to understand and test various aspects of carbon cycle behaviour during the LPTM.

Although a simple expression for carbon input to and removal from the exogenic carbon cycle is convenient, we have already encountered one problem with its use. There must be at least one feedback so that F_{In} or f is a function of M_{Ex}. In fact, both are probably true (Walker *et al.* 1981). The numerical model by Walker & Kasting (1992) has a global flux of riverine carbon linked to atmospheric pCO_2 by a set of weathering constants. Carbon sequestering in their model after a one-step release of CH_4 does result in an increase of $\delta^{13}C$ over 140 ka that is somewhat similar to the observed rise in $\delta^{13}C$ at Site 1051 (Dickens *et al.* 1997*a*; Katz *et al.* 1999; Norris & Röhl 1999).

Carbonate dissolution presents a second major complication for strict use of equation (5). Deep water in the oceans is undersaturated with respect to $CaCO_3$ at some depth. Assuming constant pressure, temperature and abundant Ca^{2+}, this saturation depth will primarily depend on the concentration of CO_3^{2-} over moderate (<1 Ma) time scales. An input of carbon to the exogenic carbon cycle will result in an increase in ΣCO_2 to all deep ocean reservoirs. Such a ΣCO_2 increase should cause the CO_3^{2-} concentration to decrease at fixed alkalinity in the deep ocean, resulting in $CaCO_3$ dissolution (Broecker & Takahashi 1977; Sundquist 1986; Walker & Kasting 1992). The carbonate dissolution will provide additional carbon input with a different $\delta^{13}C$ composition from the primary external carbon input. There is very good evidence that

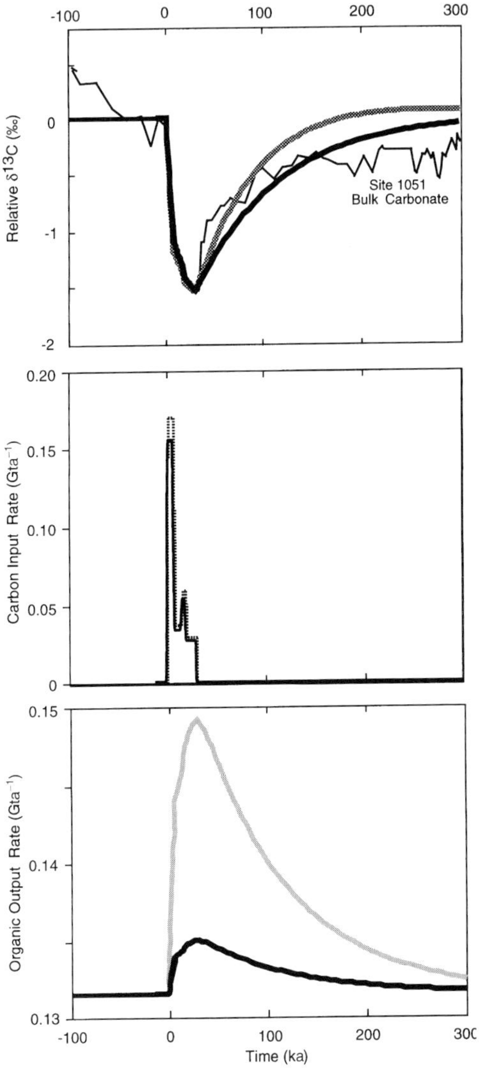

Fig. 9. Observed $\delta^{13}C$ excursion in bulk carbonate at Ocean Drilling Program (ODP) Site 1051 (Norris & Röhl 1999) and two theoretical $\delta^{13}C$ excursions generated by slightly different CH_4 carbon inputs (dM_{Add}/dt) into 'Palaeogene' exogenic carbon cycle (Fig. 8). The first model (black line) assumes a constant ratio of organic matter and carbonate outputs; the second model (grey line)has an output ratio (F_{Org}/F_{Carb}) that increases with increasing mass of the exogenic carbon cycle. The second model implies a greatly enhanced burial of organic carbon after CH_4 injection.

such carbonate dissolution occurred during the LPTM (Bralower *et al.* 1997; Thomas 1998; Katz *et al.* 1999), although the amount has not been quantified.

Conclusions

A simple mathematical expression (equation (5)) is developed to describe a $\delta^{13}C$ excursion in the exogenic carbon cycle after massive injection of carbon. In general, for a given exogenic carbon cycle: (1) a more massive carbon injection will cause a greater $\delta^{13}C$ excursion, and (2) a greater difference between the $\delta^{13}C$ of the carbon input and the exogenic carbon cycle will cause a greater $\delta^{13}C$ excursion. For a given negative $\delta^{13}C$ excursion: (3) a more massive exogenic carbon cycle requires a greater input of ^{12}C, and (4) a lower $\delta^{13}C$ of exogenic carbon cycle requires a greater input of ^{12}C. For a negative $\delta^{13}C$ excursion in an exogenic carbon cycle of known mass and composition: (5) a faster throughput requires a greater input of ^{12}C and shortens the timing of the overall excursion, and (6) an increase in the ratio of organic matter flux to carbonate flux shortens the timing of the overall excursion.

Lastly, a variable injection of carbon to an exogenic carbon cycle produces: (7)transients in global $\delta^{13}C$ that are related to the source function but modified by carbon removal.

All of these effects should be independent of a chosen model for carbon injection and removal during the LPTM or other times (e.g. the Early Toarcian, Hesselbo *et al.* 2000). If the well-dated bulk $\delta^{13}C$ record at ODP Site 1051 was representative of global $\delta^{13}C$, carbon cycle perturbations across the LPTM are best explained by a pulsed input of CH_4 into an exogenic carbon cycle where the carbon throughput or organic matter flux increases after carbon injection.

I am greatly indebted to many colleagues who have helped my efforts at understanding the LPTM. For this particular work, I thank R. Norris, M. Katz and D. Pak for sharing data, and J. C. G. Walker for providing an annotated computer output of his numerical model. This paper has also benefited from an insightful review by L. Kump and general queries by J. Zachos.

References

BAINS, S., CORFIELD, R. M. & NORRIS, R. D. 1999. Mechanisms of climate warming at the end of the Palaeocene. *Science*, **285**, 724–727.

BERGER, W. H. & VINCENT, E. 1986. Deep-sea carbonates: reading the carbon-isotope signal. *Geologische Rundschau*, **75**, 249–269.

BRALOWER, T. J., THOMAS, D. J., THOMAS, E. & ZACHOS, J. C. 1998. High-resolution records of the late Palaeocene thermal maximum and circum-Caribbean volcanism: is there a causal link?: reply. *Geology*, **26**, 671.

BRALOWER, T. J., THOMAS, D. J., ZACHOS, J. C. *et al.* 1997. High-resolution records of the late Palaeocene thermal maximum and circum-Caribbean volcanism: is there a causal link? *Geology*, **25**, 963–966.

BRALOWER, T. J., ZACHOS, J. C., THOMAS, E. *et al.* 1995. Late Palaeocene to Eocene palaeoceanography of the equatorial Pacific Ocean: stable isotopes recorded at Ocean Drilling Program Site 865, Allison Guyot. *Palaeoceanography*, **10**, 841–865.

BROECKER, W. S. 1974. *Chemical Oceanography*. Harcourt Brace, New York.

BROECKER, W. S. & PENG, T.-H. 1993. What caused the glacial to interglacial CO_2 change? *In*: HEIMANN, M. (ed.) *The Global Carbon Cycle*. NATO ASI Series I, 15. Kluwer, Dordrecht, 95–115.

BROECKER, W. S. & TAKAHASHI, T. 1977. Neutralization of fossil fuel CO_2 by marine calcium carbonate. *In*: ANDERSEN, N. & MALAHOF, A. (eds) *Fate of Fossil Fuel CO_2 in the Oceans*. Plenum, New York, 213–241.

COMPTON, J. S. & MALLINSON, D. J. 1996. Geochemical consequences of increased late Cenozoic weathering rates and the global CO_2 balance since 100 Ma. *Paleoceanography*, **11**, 431–446.

DICKENS, G. R. 1999. The blast in the past. *Nature*, **401**, 752–755.

DICKENS, G. R. 2000. Methane oxidation during the Late Palaeocene Thermal Maximum. *Bulletin de la Société Géologique de France*, **171**, 37–49.

DICKENS, G. R. & QUINBY-HUNT, M. S. 1994. Methane hydrate stability in seawater. *Geophysical Research Letters*, **21**, 2115–2118.

DICKENS, G. R., CASTILLO, M. M. & WALKER, J. C. G. 1997*a*. A blast of gas in the latest Palaeocene: simulating first-order effects of massive dissociation of methane hydrate. *Geology*, **25**, 259–262.

DICKENS, G. R., O'NEIL, J. R., REA, D. K. & OWEN, R.M. 1995. Dissociation of oceanic methane hydrate as a cause of the carbon isotope excursion at the end of the Palaeocene. *Paleoceanography*, **10**, 965–971.

DICKENS, G. R., PAULL, C. K., WALLACE, P. & ODP LEG 164 SCIENTIFIC PARTY 1997*b*. Direct measurement of *in situ* methane quantities in a large gas hydrate reservoir. *Nature*, **385**, 426–428.

ESSER, G. 1987. Sensitivity of global carbon cycle pools and fluxes to human and potential climatic impacts. *Tellus*, **39**, 245–260.

FRICKE, H. C., CLYDE, W. C., O'NEIL, J. R. & GINGERICH, P. D. 1998. Evidence for rapid climate change in North America during the Latest Palaeocene thermal maximum: oxygen isotope compositions of biogenic phosphate from the Bighorn Basin (Wyoming). *Earth and Planetary Science Letters*, **160** 193–208.

GORNITZ, V. & FUNG, I. 1994. Potential distribution of methane hydrates in the world's oceans. *Global Biogeochemical Cycles*, **8**, 335–347.

HESSELBO, S. P., GRÖCKE, D. R., JENKYNS, H. C. *et al.* 2000. Massive dissociation of gas hydrate during a Jurassic oceanic anoxic event. *Nature*, **406**, 392–395.

KATZ, M. E., PAK, D. K., DICKENS, G. R. & MILLER, K. G. 1999. The source and fate of massive carbon input during the Latest Palaeocene Thermal Maximum: new evidence from the North Atlantic Ocean. *Science*, **286**, 1531–1533.

KENNETT, J. P. & STOTT, L. D. 1991. Abrupt deep sea warming, palaeoceanographic changes and benthic extinctions at the end of the Palaeocene. *Nature*, **353**, 225–229.

KOCH, P. L., ZACHOS, J. C. & GINGERICH, P. 1992. Correlation between isotope records in marine and continental carbon reservoirs near the Palaeocene/Eoceneboundary. *Nature*, **358**, 319–322.

KUMP, L. R. & ARTHUR, M. A. 1999. Interpreting carbon-isotope excursions: carbonates and organic matter. *Chemical Geology*, **161**, 181–198.

KVENVOLDEN, K. A. 1993. Gas hydrates: geological perspective and global change. *Reviews of Geophysics*, **31**, 173–187.

NORRIS, R. D. & RÖHL, U. 1999. Carbon cycling and chronology of climate warming during the Palaeocene/Eocene transition. *Nature*, **401**, 775–778.

NORRIS, R. D., KROON, D., KLAUS, A. *et al.* (eds) 1998. *Proceedings of the Ocean Drilling Program, Initial Reports*, **171B**. Ocean Drilling Program, College Station, TX.

SHACKLETON, N. J. & HALL, M. A. 1984. Carbon isotope data from Leg 74 sediments. *In*: *Initial Reports of the Deep Sea Drilling Project*, **74**, US Government Printing Office, Washington, DC, 613–620.

SIEGENTHALER, U. 1993. Modelling the present-day oceanic carbon cycle. *In*: HEIMANN, M. (ed.) *The Global Carbon Cycle*. NATO ASI Series I, 15. Kluwer, Dordrecht, 367–395.

SUNDQUIST, E. T. 1986. Geologic analogs: their value and limitations in carbon dioxide research. *In*: TRABALKA, J. R. & REICHLE, D. E. (eds) *The Changing Carbon Cycle: a Global Analysis*. Springer, New York, 371–402.

THOMAS, E. 1998. Biogeography of the late Palaeocene benthic foraminiferal extinction. *In*: AUBRY, M.-P., LUCAS, S. & BERGGREN, W. A. (eds) *Late Palaeocene–Early Eocene Climatic and Biotic Events*. Columbia University Press, New York, 214–243.

THOMAS, E. & SHACKLETON, N. J. 1996. The Palaeocene–Eocene benthic foraminiferal extinction and stable isotope anomalies. *In*: KNOX, R. W. O., CORFIELD, R. M. & DUNAY, R. E. (eds) *Correlations of the Early Palaeogene in Northwest Europe*. Geological Society, London, Special Publications, **101**, 401–411.

VAN ANDEL, T. H. 1975. Mesozoic/Cenozoic calcite compensation depth and the global distribution of calcareous sediments. *Earth and Planetary Science Letters*, **26**, 187–194.

WALKER, J. C. G. & KASTING, J. F. 1992. Effects of fuel and forest conservation on future levels of atmospheric carbon dioxide. *Palaeogeography, Palaeoclimatology, Palaeoecology*, **97**, 151–189.

WALKER, J. C. G., HAYS, P. B. & KASTING, J. F. 1981. A negative feedback mechanism for the long-term stabilization of Earth's surface temperature. *Journal of Geophysical Research*, **86**, 9776–9782.

ZACHOS, J. C., LOHMANN, K. C., WALKER, J. C. G. & WISE, S. W. 1993. Abrupt climate change and transient climates during the Palaeogene: a marine perspective. *Journal of Geology*, **101**, 191–213.

ZACHOS, J. C., STOTT, L. D. & LOHMANN, K. C. 1994. Evolution of early Cenozoic temperatures. *Palaeoceanography*, **9**, 353–387.

Palaeoenvironmental implications of palygorskite clays in Eocene deep-water sediments from the western Central Atlantic

THOMAS PLETSCH

Institut für Geowissenschaften, Universität Kiel, Olshausenstr. 40–60, 24118 Kiel, Germany
Present address: Geologisches Institut, Universität Köln, Zülpicher Str. 49a, 50674 Köln, Germany

Abstract: Clay mineral analyses were performed on Eocene sediments from drill sites in the western Central Atlantic. The investigated sites cover the full range of early Palaeogene deep waters above and below the calcite compensation depth (CCD), but otherwise represent different depositional and hydrographic regimes. Palygorskite clays with authigenic microstructures were discovered in Lower Eocene hemipelagic sediments from the distal end of the Blake Nose depth transect and in pelagic clays of the same age from the distal Nares Abyssal Plain, where terrigenous input was reduced. Palygorskite clays were not detected in coeval sediments from a distal near-CCD setting on Bermuda Rise that received major terrigenous input. The distribution of palygorskite clays at these sites, the microstructures of the constituent minerals, their absence from contemporaneous deposits on the American margin, and the position of the northerly sites outside the range of a potential African aeolian supply strongly suggest an authigenic origin of these clays at the early Eocene sea floor. Palygorskite clays are widely distributed in lower Eocene sediments from about 50° N to 50° S palaeolatitude. The most widespread distributions and peak abundances in Atlantic oceanic sediments are reported from shelf to deep-water sites of the palaeo-tropical and -subtropical belt and correlate with the Early Eocene period of extreme warmth. Marine authigenic palygorskite clay may provide an indication of the localities and the time periods that were characterized by high bottom-water temperatures, by elevated alkalinity, silica and magnesium concentrations, and by reduced sediment accumulation rates.

Lower Palaeogene sediments in general, and Lower Eocene deep-sea deposits from the equatorial to mid-latitude Atlantic in particular are often characterized by a peculiar clay mineral composition. They frequently contain the magnesium-rich, fibrous minerals palygorskite and (usually minor) sepiolite, collectively called palygorskite clay here (Kastner 1981; Callen 1984). Marine palygorskite clays have often been attributed to detrital input from land or marginal marine environments, by analogy with modern, net-evaporative settings where palygorskite is a common detrital component (Chamley *et al.* 1977; Coudé-Gaussen & Blanc 1985). As terrestrial palygorskite is formed only in arid areas (Paquet 1970; Singer 1984), detrital palygorskite in deep-marine sediments provides an excellent proxy for arid climates on land and has been used in numerous palaeoclimatic studies (Chamley 1989).

Other major occurrences of palygorskite clay are difficult, if not impossible, to explain by terrestrial formation and detrital supply to the oceans. Alternatively, it has been proposed that these palygorskite clays formed at the deep sea floor, probably in areas with reduced sediment accumulation rates, elevated ambient temperatures, high magnesium and silica contents, and high alkalinity (Berger & von Rad 1972; Couture 1977; Kastner 1981; López Galindo 1987; Thiry & Jacquin 1993). Acceptance of the authigenic mechanism has been hampered, however, by the absence of modern examples for marine palygorskite formation, and by the lack of a satisfactory thermodynamic model (Kastner 1981; Jones & Galán 1988; Chamley 1989).

As the genesis of marine palygorskite clays has fundamental implications for their palaeoclimatic interpretation, the existence and the relevance of marine authigenic palygorskite is a matter of continuing debate (Couture 1977, 1978; Weaver & Beck 1977; Beck & Weaver 1978; Singer 1979; Callen 1981; Chamley 1989; Thiry & Jacquin 1993). The discovery of massive, authigenic Lower Eocene palygorskite clay deposits in the Gulf of Guinea has recently reanimated this debate. These deposits were interpreted as a result of *in situ* alteration at, or close to, the sea

From: KROON, D., NORRIS, R. D. & KLAUS, A. (eds). 2001 *Western North Atlantic Palaeogene and Cretaceous Palaeoceanography*. Geological Society, London, Special Publications, **183**, 307–316.
0305-8719/01/$15.00

floor. It was proposed that the frequent occurrence of palygorskite clay in Lower Eocene deposits is related to warm saline bottom waters (WSBW) that presumably bathed much of the Palaeocene–Eocene ocean floors (Kennett & Stott 1990; Pletsch 1998). Clearly, an authigenic mode of formation at the sea floor would compromise the straightforward application of palygorskite clays as a proxy for aridity on land, although the basic palaeoclimatic interpretation of net evaporation would remain.

In this paper, mineralogical and microstructural data on Palaeocene and Eocene palygorskite clays from Deep Sea Drilling Program (DSDP) and Ocean Drilling Program (ODP) holes in the western Central Atlantic are presented and compared with data on deposits of similar age in the Gulf of Guinea. The western Central Atlantic is particularly well suited to monitor the origin of palygorskite clay as contemporaneous terrigenous supply of palygorskite clay from both adjacent and remote source areas was either inexistent or negligible, which reduces the number of potential formation mechanisms. A causal connection is proposed here between the occurrence of deep-marine authigenic palygorskite and the times and locations of drastic changes in sea-water temperature and composition that characterized the Early Eocene ocean.

Materials and methods

This work is based on a larger reconnaissance X-ray diffraction (XRD) study of the clay size-fraction of about 300 samples from three locations in the western Central Atlantic (Pletsch, unpub. data). Samples were taken onboard *JOIDES Resolution* or provided by repositories and individual researchers. A selection of these samples with ages ranging from Late Cretaceous to mid-Eocene time were studied by scanning electron microscopy (SEM) and coupled energy dispersive analysis (EDX). The latter yields semiquantitative chemical results. Samples for XRD were disaggregated in water and the sand size fraction and carbonates were removed by sieving and gentle acid attack. Most samples were prepared as smeared mounts and measured at Sédimentologie et Géodynamique, Université de Lille, France, according to routine procedures of this laboratory (Pletsch 1998). Aliquots and additional samples were measured as filter-peel mounts on a diffractometer at Institut für Geowissenschaften, University of Kiel, Germany. No mineral percentages were calculated, because routine XRD analyses of palygorskite are difficult to evaluate quantitatively as a result of its fibrous habit and interferences with diffraction peaks of other common clay minerals (Pletsch 1998). In spite of these difficulties, replicate analyses at the two laboratories yielded similar intensity ratios for diffraction peaks attributed to palygorskite and those of other minerals.

Geological setting and previous studies

The three locations investigated represent different depositional and hydrographic regimes. Site 417 is at presently at 5470 m water depth on the top of a volcanic basement hill on Nares Abyssal Plain, at the southern end of the Bermuda Rise. During Eocene time, Site 417 was located in a distal position outside the reach of turbidites and well below the CCD, at about 4500–5000 m water depth (Chenet & Francheteau 1979). Abundant palygorskite clays were detected in a previous study (Mann & Müller 1979) at the boundary between dark brown radiolarian ooze and clay and underlying multicoloured, carbonate-free zeolite clay (Donelly *et al.* 1979). Both units were deposited at rates characteristic of pelagic clay sedimentation (*c.* 1 m Ma^{-1}). A more precise figure is not available, because the complete dissolution of calcareous marker fossils makes conventional biostratigraphic dating of this interval problematic. A preliminary Palaeocene to lower Eocene age for Cores 417A-16 to -20 is based on phosphatic fish remains (Kozarek & Orr 1979). Inspection of the washed residues from samples discussed here has slightly modified this age assignment (W. Kuhnt, pers. commun.): agglutinated foraminifers indicate a Late Cretaceous age for Core 18 and below, and an Eocene age for Core 17 and above. The abundance of radiolarians at the top of Core 17 points to an early Eocene age. Although it is not possible to calculate reliable sedimentation rates for this interval, it is obvious that sediment accumulation was extremely slow, as is also suggested by the enrichment in phosphatic debris and the lack of turbidites.

DSDP Site 386 lies on the Bermuda Rise, about 140 km southeast of the island of Bermuda and at 4780 m water depth. The sediments studied here are Lower Eocene greenish grey claystones with variable, but generally low carbonate contents (0–7 %) and with minor porcellanite separated from the underlying Lower Eocene to Upper Palaeocene dark olive grey radiolarian mudstones with minor carbonate, porcellanite and chert by a suspect erosional contact (Thierstein & Okada 1978; Tucholke & Vogt 1979). The depositional environment is assumed to have been below the CCD during Late Palaeocene and Eocene time, with linear sedimentation rates of the order of 25 m Ma^{-1}. This relatively rapid deposition rate was attributed to fine-grained turbidity current input from sites on the American lower continental rise near the CCD (Tucholke & Mountain 1986). Thus, in spite of a similar distance to the American continent, Site 386 differs markedly from Site 417 by the rapid

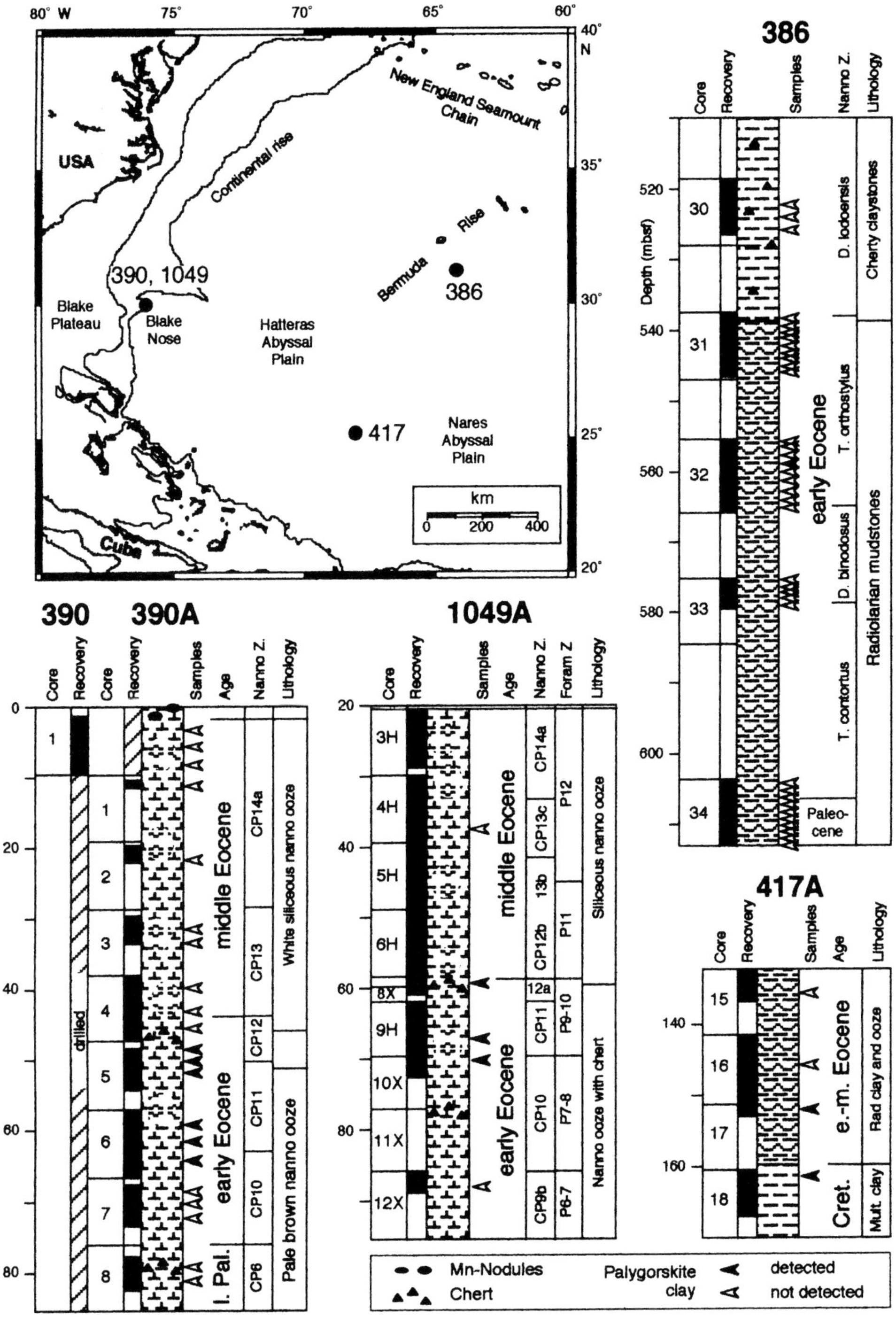

Fig. 1. Locations of the studied DSDP and ODP Holes and simplified stratigraphy and lithofacies of the palygorskite clay intervals in Holes 386, 390A, 417A and 1049A. Numbers and black bars to the right of lithology columns indicate core number and recovery, respectively. Stratigraphic data from sources discussed in the text.

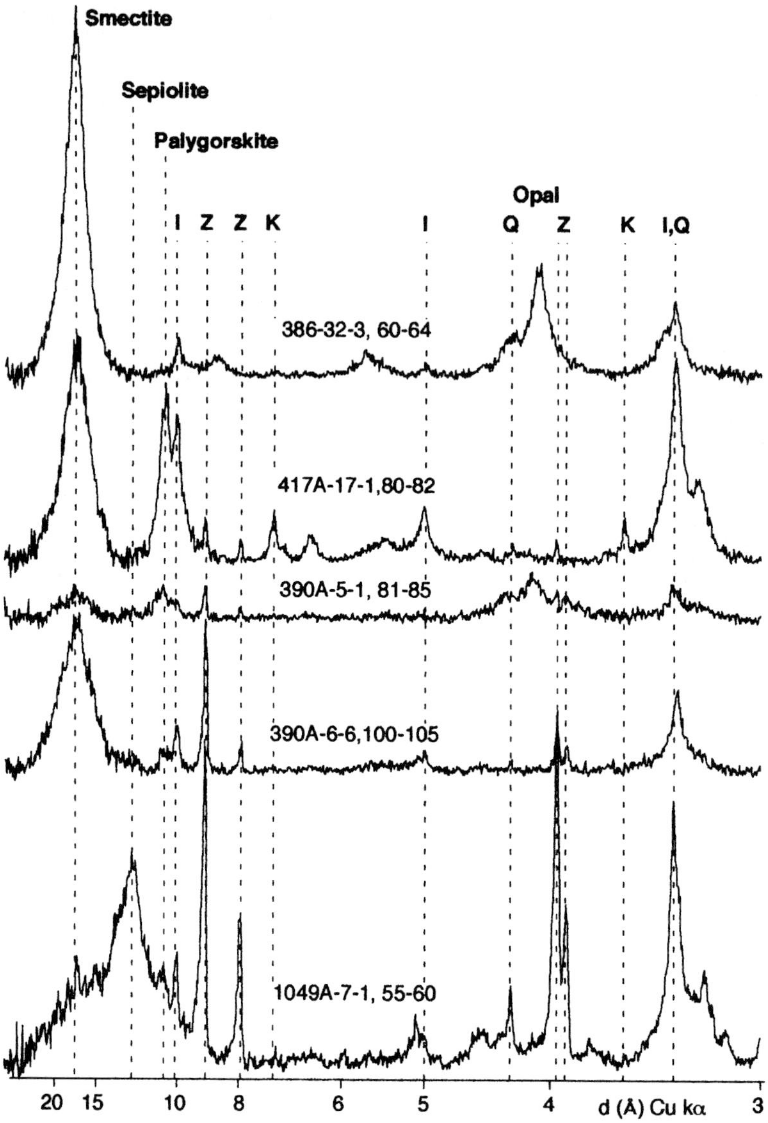

Fig. 2. X-ray diffractograms (ethylene glycol solvated mounts) of selected Lower Eocene samples from Sites 386, 390, 417 and 1049. Diffractogram from Site 1049 has been converted to the 1° divergence slit-setting used for the other samples. I, illite; Z, zeolite (probably clinoptilolite); K, kaolinite; Q, quartz. A noteworthy feature is the clear diffraction peaks attributed to presence of palygorskite and/or sepiolite, except for diffractogram from Site 386. The latter bears the strongest (yet inconclusive) indication for the presence of palygorskite out of 38 samples from the Lower Eocene interval of that site. Further discussion is given in the text.

accumulation of fine-grained terrigenous sediments. Non-carbonate sedimentation rates were of the order of 18–20 m Ma^{-1}, assuming a thickness of 124–133 m, an average carbonate content of 5 % (McCave 1979), and a duration of 6 Ma for Early Eocene time (Berggren *et al.* 1995). Findings of palygorskite clays reported previously from the interval studied here (Rothe 1989) could only be reproduced with significant uncertainty for one sample (Figs 1, 2).

The Leg 171B drill sites, including a reoccupation of DSDP Site 390, were drilled along the crest of Blake Nose, a spur that projects into the western Central Atlantic. Site 390/1049, in

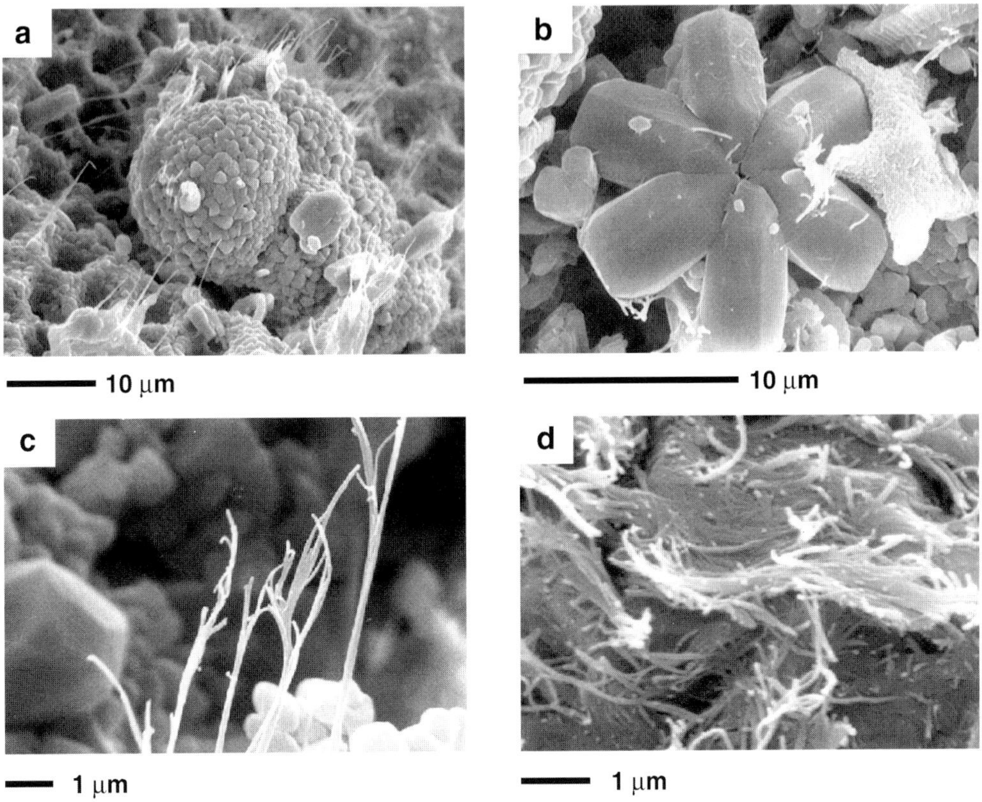

Fig. 3. Scanning electron micrographs of Lower Eocene sediments. (**a**) Sample 390A-6-6, 100–105 cm. Overview of open void inside foraminifer test with delicate palygorskite clay fibres. (**b**) Same sample. Palygorskite clay fibres on *Discoaster* sp. (**c**) Detail of (**a**) revealing biomorphic strands of palygorskite clays. Element mapping mode reveals Mg- and Al-rich composition. (**d**) Sample 417A-17-1, 80–82 cm. Interwoven mats of palygorskite fibres constitute much of the sample.

2670 m of water, is the deepest site of the Blake Nose depth transect. The interval studied comprises siliceous nannofossil ooze and nannofossil ooze with chert (Benson *et al.* 1978; Norris *et al.* 1998). The calcareous nannofossil zonation for Site 390 (Schmidt 1978) was converted to the one used during Leg 171B with reference to Berggren *et al.* (1995). According to shipboard biostratigraphic results, the lower Eocene sequence at Site 390/1049 is relatively complete, although the basal calcareous nannofossil zone CP9a was not recognized (Schmidt 1978; Norris *et al.* 1998). Lower Eocene sediments at Site 390/1049 record hemipelagic deposition with carbonate contents between 60 and 80%. Bulk sedimentation rates were moderately low (11 m Ma^{-1}; Norris *et al.* 1998), and terrigenous input was strongly reduced by the distal position of these sites, seaward of the Blake Plateau. This becomes clear when the carbonate-free sedimentation rates are compared with those for Site 386. Using a conservative estimate of 70% carbonate (which does not account for a significant component of biogenic opal) for the interval comprising nannofossil zones CP9b–CP12a (4.6 Ma), the non-carbonate sedimentation rate at Site 1049 amounts to only 2.6 m Ma^{-1}, or less than a seventh that at Site 386. The stratigraphic range of palygorskite clays at Site 390/1049 is entirely within early Eocene calcareous nannofossil zones CP10–CP12. Palygorskite was not detected in a previous study of material from Site 390, perhaps as a result of a wider sample spacing (Pastouret *et al.* 1978).

Results

Eocene sediments from Holes 390A and 1049A contain major smectite, some kaolinite and minor illite. Non-clay minerals in the <2 μm size fraction include quartz, opal-CT, and zeolites of the clinoptilolite–heulandite group.

This is a common mineral assemblage in Upper Cretaceous and Palaeogene sediments throughout the Atlantic (Chamley 1989). Lower Eocene samples, all from calcareous nannofossil zones CP10–CP12, are characterized by minor, but significant reflections attributed to palygorskite and sepiolite, and by abundant zeolite. Kaolinite and smectite peaks are conspicuously low when compared with samples from above and below zones CP10–CP12 (Figs 1 and 2). SEM observations and EDX analyses reveal filamentous silicates with elevated Al and Mg contents in samples from this interval (Fig. 3). The curved fibres and filaments occur both in the open pore space of the sediment and as coatings on calcareous microfossils. Fragile bundles of silicate strands are attached to the inner walls of foraminifer tests, with their pointed and bent edges towards the centre of the void (Fig. 3a). Such an arrangement strongly suggests that the silicate fibres grew from the microfossil wall into the open space, rather than being depositional infill. It is important to mention that no palygorskite clay was detected in coeval sediments from the more proximal sites of the Blake Nose transect, i.e. Sites 1050–1052, or in any other stratigraphic interval of Site 390/1049. In addition, neither XRD nor SEM analyses provided evidence of palygorskite clays in the discrete ash layers that are scattered throughout the Eocene sections of the Leg 171B drill sites (Norris *et al.* 1998; Reicherter & Pletsch 1998).

The clay mineral assemblage throughout the studied interval of Hole 386 is very uniform, with strongly dominant smectite, common illite and occasional minor kaolinite. Other minerals in the clay size fraction are minor quartz, zeolite and opal-CT. Zeolite is restricted to those samples that have no kaolinite. Palygorskite could not be identified reliably, but may be present in trace amounts in one sample (Fig. 2). Although minor palygorskite or sepiolite peaks may have been overlooked as a result of overlapping diffraction peaks of illite and smectite, it is unlikely that palygorskite clays are present as significant amounts in any of the samples studied.

Clay minerals of the interval studied from Hole 417A comprise major smectite and palygorskite, variable but generally minor illite, and kaolinite, which is abundant in those samples where palygorskite is not present. Other minerals are zeolite and minor quartz. Palygorskite occurs in two out of four samples analysed, which have been tentatively assigned to latest Cretaceous and to early Eocene time, respectively (Fig. 2; Kuhnt, pers. comm.). The lower Eocene sample has the highest relative intensity of the palygorskite diffraction peaks in any of the studied samples from the four sites studied. In SEM images, this sample shows a dense meshwork of interwoven mats that are composed of flexed palygorskite fibres (Fig. 3d). No detrital grains or structures were observed.

Discussion

The distribution of palygorskite clay shows a correlation with varying depositional conditions, most notably with terrigenous sedimentation rates. Concentrations of palygorskite clay are highest where terrigenous detrital supply is lowest (Sites 390/1049 and 417) and concentrations are nil or below the detection limit where terrigenous input is high (Site 386). This pattern suggests that the occurrence of palygorskite required moderate to low terrigenous accumulation rates. Thus a detrital input of palygorskite clays from the American continent can be ruled out, unless mixing of sediments from multiple sources on the adjacent margin is invoked. However, contemporaneous terrestrial or shallow-water palygorskite deposits are unknown from the US seaboard (Callen 1984; Riggs & Manheim 1988; Weaver 1989; and references therein). Long-distance aeolian transport of palygorskite clay from arid sources on the African continent, where extensive terrestrial and shallow marine palygorskite deposits do occur (Millot 1970; Daoudi *et al.* 1995), is equally unlikely, except at Site 417, which, according to a General Circulation Model for late Palaeocene time, was in the palaeo-NE trade wind belt (O'Connell *et al.* 1996). However, pervasive authigenic microstructures in samples from Site 417 indicate that *in situ* formation was dominant, if not the only source of palygorskite at that site. The influence of aeolian input from Africa was probably even smaller at the other drill sites. Their Eocene positions at about 25° N (Norris *et al.* 1998) places them even further outside the reach of the palaeo-trade winds than Site 417, which is about 5° southward. Although an aeolian supply of palygorskite clay to Site 417 cannot be excluded on the basis of the available data, the more northerly sites in the western Central Atlantic were probably unaffected by detrital palygorskite supply from both proximal and distal source areas.

The negligible influence of detrital palygorskite clay is corroborated by the observation of authigenic micromorphologies in samples from Site 390/1049, which indicate post-depositional growth in an open pore space of hemipelagic sediments. Formation must have been related to peculiar and transient environmental factors, rather than to burial diagenetic conditions because the occurrence of palygorskite clays is restricted to a narrow stratigraphic interval, but not to a particular depth or lithology. The most striking aspect of these palygorskite clays in the western Central Atlantic is their correlation with a world-wide maximum of palygorskite abundance and distribution during early Eocene time.

Sediments containing palygorskite clays have been described from many DSDP or ODP drill sites from all oceans, from shallow-marine and terrestrial basins, and from palaeo-latitudes ranging to 50° north and south, including palaeo-equatorial sites (Callen 1984; Pletsch 1998). Of these occurrences, those from marine basins are of special interest, because they can often be dated at a level of precision that allows comparison with available data on early Eocene depositional processes, oceanographic conditions and climate.

Early Eocene time was a period of extreme warmth, both on the surface of the Earth and at the deep sea floor (Kennett & Stott 1990; Corfield 1994; Zachos *et al.* 1994). It is widely accepted that warming was triggered initially through the massive input of greenhouse gases during a late Palaeocene–early Eocene period of increased volcanic and hydrothermal activity, although the ultimate causes of global warming are not fully understood (Owen & Rea 1985; Rea *et al.* 1990; Sloan *et al.* 1992; Eldholm & Thomas 1993; Sloan & Rea 1995; Norris & Röhl 1999). On the basis of stable isotope evidence, it has been suggested that oceanic circulation during this period was driven, at least in part, by warm saline waters that originated from the low-latitude shelves and marginal seas of the Atlantic and Tethys oceans (Miller *et al.* 1987; Kennett & Stott 1990; Pak & Miller 1992; Sloan *et al.* 1995; Charisi & Schmitz 1996; Corfield & Norris 1996). As the maximum long-term warming of bottom waters falls within early Eocene time (Zachos *et al.* 1994; Bralower *et al.* 1995; Corfield & Norris 1996), it is tempting to infer a causal relationship with the widespread occurrence of palygorskite clays in the tropical to mid-latitude oceans, the underlying implication being that many of those occurrences previously interpreted as detrital palygorskite clays were formed authigenically at the sea floor.

In this scenario, evaporation over low-latitude shallow seas produced warm and ion-enriched waters, notably during periods of high sea level. Eventually, these saline waters became dense enough to travel down the continental slopes in spite of their elevated temperature. These waters replaced or mixed with the existing intermediate and deep waters to form the warm saline bottom waters (WSBW) that have been proposed by several workers (e.g. Brass *et al.* 1982; Kennett & Stott 1990). Higher temperatures increased the dissolved silica concentrations and reaction rates, and favoured the formation of palygorskite clays and other silica-rich minerals (Couture 1977). Given the relatively shallow CCD during early Eocene time (Tucholke &] 1979), bottom-water alkalinity and pH would have been high enough for the precipitation of palygorskite and sepiolite (Morse & Mackenzie 1990). Elevated temperatures, enrichment in silica and magnesium, reduced terrigenous input, and an 'alkaline chemical environment' are among the conditions that are classically cited to promote the formation of palygorskite clays (Millot 1970; Couture 1978; Kastner 1981). Although the details of this process are poorly understood, it seems plausible that lower Eocene authigenic palygorskite clays formed from the interaction of Early Eocene WSBW with pre-existing clayey sediments.

Similar models for the authigenic growth of palygorskite clays under elevated water temperatures at the sea floor have been proposed by several workers (e.g. Couture 1977; Kastner 1981; Thiry & Jacquin 1993), but it was not until the discovery of Lower Eocene palygorskite clays in the Gulf of Guinea by ODP Leg 159 that the stratigraphic correlation and the potential causal relationship with the early Eocene period of extreme warmth became established (Pletsch 1998). Thick intervals of palygorskite clays, often with very minor admixture of minerals such as smectite, quartz or zeolite, and without indications of detrital input, hydrothermal influence or deep burial diagenesis, were recovered at Sites 960 and 961 in palaeo-water depths of more than 1000 m. Formation of palygorskite at these sites was explained as an early diagenetic transformation of pre-existing clay minerals, probably smectite (Pletsch 1998). A recent geochemical and stable isotope study on palygorskite and smectite from these sites supports the hypothesis of a mineral transformation, accompanied by an isotopic exchange with ambient seawater (Pletsch & Botz 1997; Pletsch *et al.* 2000). Although the stratigraphic range of palygorskite occurrence at Sites 960 and 961 overlaps with that observed at Site 390/1049, the maximum abundance in the Gulf of Guinea is in calcareous nannofossil zones CP9b–CP11 (Pletsch 1998), or 1–3 Ma earlier than in the western Central Atlantic. This difference may reflect variations in the intensity and/or dispersal of WSBW, and would indicate that peak WSBW conditions in the western Central Atlantic were reached significantly later than in the Equatorial Atlantic.

Future mapping and stratigraphic dating of authigenic marine palygorskite clays may identify where and when WSBW bathed the sea floor during proposed periods of extreme warmth, such as Early Eocene time. Other promising intervals are the middle and upper Cretaceous sequences, which also host abundant and extensive palygorskite clay deposits (Callen 1984). Although this approach is apparently limited to areas of low terrigenous input, palygorskite clays may provide a tracer for WSBW in deep-sea sediments, because these silicates, once formed, are resistant to carbonate dissolution that plague conventional stable isotope investigations.

It should be noted, however, that the proposed mechanism of deep-marine authigenic palygorskite clay formation remains speculative in the absence of modern analogues, crystallization experiments or thermodynamic models. No satisfactory explanation has yet been given for the assumed transformation from a phyllosilicate precursor to the chain or ribbon structure characteristic of palygorskite clays (see reviews by Jones & Galán (1988) and Singer (1989)). In addition, the postulated enrichments of bottom waters in Mg, Si and alkalinity required for the growth of palygorskite clays await rigorous geochemical testing. These caveats notwithstanding, the available evidence suggests that deep-marine authigenic palygorskite clays are more common than previously thought, that some of them formed during periods of warm ocean bottom waters, and that the same conditions that are required to form WSBW would promote the formation of palygorskite according to existing models.

Conclusions

Significant amounts of palygorskite clays with authigenic microstructures have been detected in Lower Eocene sediments from Sites 390/1049 and 417 in the western Central Atlantic. Terrigenous sediment accumulation at these sites was moderate to very slow. In contrast, little or no palygorskite clay is found in coeval sediments at Site 386, which received abundant terrigenous material from the American continental rise. Coeval sediments in more shoreward positions also do not contain palygorskite clays, so that a detrital origin from the American margin is unlikely. Aeolian supply from African sources may have contributed minor palygorskite to Site 417, but the northerly drill sites were probably unaffected by this aeolian supply from the east.

The mineralogical, microstructural and stratigraphic data presented here suggest that palygorskite clays at Sites 390/1049 and 417 formed at, or just below, the early Eocene sea floor. The widespread occurrence and maximum abundance of palygorskite clays at the same time as the long-term negative excursion in oxygen isotopes of benthic foraminifers and the proposed formation of warm saline bottom waters in Early Eocene time suggests that these phenomena are genetically linked. The higher temperatures and elevated ionic concentrations of Early Eocene deep waters may have given rise to processes of silicate formation that have no modern analogue.

H. Chamley, J. F. Deconinck, W. Kuhnt and C. Samtleben are thanked for permission to use the laboratory facilities at Lille and Kiel, as well as for discussions about the origin of palygorskite over the years. Constructive reviews by E. Arnold, L. Krissek and S. Hovan helped significantly to improve the manuscript. P. Recourt, P. Vanderaveroet and D. Malengros (Lille), and U. Stender, O. Krüger, W. Reimann and U. Schuldt (Kiel) helped with sample preparation and measurements. Shore-based sampling benefited from the assistance of W. Hale and A. Wülbers (Bremen). W. Kuhnt (Kiel) and R. D. Norris (Woods Hole) provided DSDP samples and stratigraphic information. Participation on Leg 171B and shore-based research was funded by the German Research Council (DFG), Project Ku 649/7.

References

BECK, K. C. & WEAVER, C. E. 1978. Miocene of the S.E. United States: a model for chemical sedimentation in a peri-marine environment—reply. *Sedimentary Geology*, **21**, 154–157.

BENSON, W. E., SHERIDAN, R. E. *et al.* (eds) 1978. *Initial Reports of the Deep Sea Drilling Project*, **44**. US Government Printing Office, Washington, DC.

BERGER, W. H. & VON RAD, U. 1972. Cretaceous and Cenozoic sediments from the Atlantic Ocean. *In*: HAYES, D. E., PIMM, A. C. *et al.* (eds) *Initial Reports of the Deep Sea Drilling Project*, **14**. US Government Printing Office, Washington, DC, 787–954.

BERGGREN, W. A., KENT, D. V., SWISHER, C. C. & AUBRY, M.-P. 1995. A revised Cenozoic geochronology and chronostratigraphy. *In*: BERGGREN, W. A., KENT, D. V., AUBRY, M. P. & HARDENBOL, J. (eds) *Geochronology, Time Scales and Global Stratigraphic Correlation: a Unified Temporal Framework for a Historial Geology*. SEPM, Special Publications, **54**, 129–212.

BRALOWER, T. J., ZACHOS, J. C., THOMAS, E. *et al.* 1995. Late Palaeocene to Eocene oceanography of the equatorial Pacific Ocean: stable isotopes recorded at Ocean Drilling Program Site 865, Allison Guyot. *Palaeoceanography*, **10**, 841–865.

BRASS, G. W., SOUTHAM, J. R. & PETERSON, W. H. 1982. Warm saline bottom water in the ancient ocean. *Nature*, **296**, 620–623.

CALLEN, R. A. 1981. Palygorskite in sediments: detrital, diagenetic or neoformed—a critical review. Discussion. *Geologische Rundschau*, **70**, 1303–1305.

CALLEN, R. A. 1984. Clays of the palygorskite–sepiolite group: depositional environment, age and distribution. *In*: SINGER, A. & GALÁN, E. (eds) *Palygorskite—Sepiolite. Occurrence, Genesis and Uses*. Developments in Sedimentology, **37**. Elsevier, Amsterdam, 1–37.

CHAMLEY, H. 1989. *Clay Sedimentology*. Springer, Berlin.

CHAMLEY, H., DIESTER-HAASS, L. & LANGE, H. 1977. Terrigenous material in East Atlantic sediment cores as an indicator of NW African climates. *'Meteor' Forschungsergebnisse, C*, **26**, 44–59.

CHARISI, S. D. & SCHMITZ, B. 1996. Early Eocene palaeoceanography and palaeoclimatology of the eastern North Atlantic: stable isotope results for DSDP Hole 550. *In*: KNOX, R. W. O., CORFIELD, R. M. & DUNAY, R. E. (eds) *Correlation of the Early Palaeogene in Northwest Europe*. Geological Society, London, Special Publications, **101**, 457–472.

CHENET, P. Y. & FRANCHETEAU, J. 1979. Bathymetric reconstruction method: application to the Central Atlantic basin between 10° N and 40° N. *In*: DONELLY, T., FRANCHETEAU, J., BRYAN, W. *et al.* (eds) *Initial Reports of the Deep Sea Drilling Project*, **51–53**. US Government Printing Office, Washington, DC, 1501–1513.

CORFIELD, R. M. 1994. Palaeocene oceans and climate: an isotope perspective. *Earth-Science Reviews*, **37**, 225–252.

CORFIELD, R. M. & NORRIS, R. D. 1996. Deep water circulation in the Palaeocene Ocean. *In*: KNOX, R. W. O., CORFIELD, R. M. & DUNAY, R. E. (eds) *Correlation of the Early Palaeogene in Northwest Europe*. Geological Society, London, Special Publications, **101**, 443–456.

COUDÉ-GAUSSEN, G. & BLANC, P. 1985. Présence de grains éolisés de palygorskite dans les poussières actuelles et les sédiments récents d'origine désertique. *Bulletin de la Société Géologique de France*, **8**, 571–579.

COUTURE, R. A. 1977. Composition and origin of palygorskite-rich and montmorillonite-rich zeolite-containing sediments from the Pacific Ocean. *Chemical Geology*, **19**, 113–130.

COUTURE, R. A. 1978. Miocene of the S.E. United States: a model for chemical sedimentation in a peri-marine environment—comments. *Sedimentary Geology*, **21**, 149–157.

DAOUDI, L., CHARROUD, M., DECONINCK, J. F. & BOUABDELLI, M. 1995. Distribution et origine des minéraux argileux des formations Crétacé–Éocene du Moyen Atlas sud-occidental (Maroc): signification paléogéographique. *Annales de la Société Géologique du Nord, (2)*, **4**, 31–40.

DONELLY, T., FRANCHETEAU, J., BRYAN, W. *et al.* 1979. *Initial Reports of the Deep Sea Drilling Project*, US Government Printing Office, Washington, DC, 51–53.

ELDHOLM, O. & THOMAS, E. 1993. Environmental impact of volcanic margin formation. *Earth and Planetary Science Letters*, **117**, 319–329.

JONES, B. F. & GALÁN, E. 1988. Sepiolite and palygorskite. *In*: BAILEY, S. W. (ed.) *Hydrous Phyllosilicates (Exclusive of Micas)*. Mineralogical Society of America, Reviews in Mineralogy, **19**, 631–674.

KASTNER, M. 1981. Authigenic silicates in deep-sea sediments: formation and diagenesis. *In*: EMILIANI, C. (ed.) *The Sea*, **7**. Wiley, New York, 915–980.

KENNETT, J. P. & STOTT, L. D. 1990. Proteus and Proto-Oceanus: ancestral Palaeogene oceans as revealed from Antarctic stable isotope results. *In*: BARKER, P. F. & KENNETT, J. P. (eds) *Proceedings of the Ocean Drilling Program, Scientific Results*, **113**. Ocean Drilling Program, College Station, TX, 829–848.

KOZAREK, R. J. & ORR, W .N. 1979. Ichthyoliths, Deep Sea Drilling Project Legs 51 through 53. *In*: DONELLY, T., FRANCHETEAU, J., BRYAN, W. *et al.* (eds) *Initial Reports of the Deep Sea Drilling Project*, **51–53**. US Government Printing Office, Washington, DC, 857–895.

LÓPEZ GALINDO, A. 1987. Paligorskita en sedimentos cretácicos de la Zona Subbetica. Origen. *Boletín de la Sociedad Española de Mineralogía*, **10**, 131–139.

MANN, U. & MÜLLER, G. 1979. X-ray mineralogy of Deep-Sea Drilling Project Legs 51 through 53, western North Atlantic. *In*: DONELLY, T., FRANCHETEAU, J., BRYAN, W. *et al.* (eds) *Initial Reports of the Deep Sea Drilling Project*, **51–53**. US Government Printing Office, Washington, DC, 721–729.

MILLER, K. G., JANECEK, T. R., KATZ, M. E. & KEIL, D. J. 1987. Abyssal circulation and benthic foraminiferal changes near the Palaeocene/Eocene boundary. *Palaeoceanography*, **2**, 741–761.

MILLOT, G. 1970. *Geology of Clays*. Springer, Berlin.

MORSE, J. W. & MACKENZIE, F. T. 1990. *Geochemistry of Sedimentary Carbonates*. Developments in Sedimentology, **48**. Elsevier, Amsterdam.

MCCAVE, I. N. 1979. Diagnosis of turbidites at Sites 386 and 387 by particle-counter size analysis of the silt (2–40 μm) fraction. *In*: TUCHOLKE, B. E., VOGT, P. R. *et al.* (eds) *Initial Reports of the Deep Sea Drilling Project*, **43**. US Government Printing Office, Washington, DC, 395–405.

NORRIS, R. D. & RÖHL, U. 1999. Carbon cycling and chronology of climate warming during the Paleocene/Eocene transition. *Nature*, **401**, 775–778.

NORRIS, R. D., KROON, D., KLAUS, A. *et al.* (eds) 1998 *Proceedings of the Ocean Drilling Program, Initial Reports*, **171B**. Ocean Drilling Program, College Station, TX.

O'CONNELL, S., CHANDLER, M. A. & RUEDY, R. 1996. Implications for the creation of warm saline deep water: Late Palaeogene reconstructions and global climate model simulations. *Geological Society of America Bulletin*, **108**, 270–284.

OWEN, R. M. & REA, D. K. 1985. Sea-floor hydrothermal avtivity links climate to tectonics: the

Eocene carbon dioxide greenhouse. *Science*, **227**, 16–169.

PAK, D. K. & MILLER, K. G. 1992. Palaeocene to Eocene benthic foraminiferal isotopes and assemblages: implications for deepwater circulation. *Palaeoceanography*, **7**, 405–422.

PAQUET, H. 1970. Évolution géochimique des minéraux argileux dans les altérations et les sols des climats méditerranéens et tropicaux à saisons contrastées. *Mémoires du Service de la Carte Géologique d'Alsace–Lorraine*, **30**, 1–212.

PASTOURET, L., AUFFRET, G. A. & CHAMLEY, H. 1978. Microfacies of some sediments from the western North Atlantic: palaeoceanographic implications (Leg 44 DSDP). *In*: BENSON, W. E. & SHERIDAN, R. E. (eds) *Initial Reports, Deep Sea Drilling Project*, **44**, US Government Printing Office, Washington, DC, 477–501.

PLETSCH, T. 1998. Origin of Lower Eocene palygorskite claystones on the Côte d'Ivoire–Ghana margin (Leg 159, eastern Equatorial Atlantic). *In*: MASCLE, J., LOHMANN, G. P., MOULLADE, M. *et al.* (eds) *Proceedings of the Ocean Drilling Program, Scientific Results*, **159**. Ocean Drilling Program, College Station, TX, 137–152.

PLETSCH, T. & BOTZ, R. 1997. *Sedimentary facies, clay mineral and oxygen isotope composition of deep marine palygorskite clay from ODP Leg 159.* International Clay Conference, Ottawa, Canada, 15–21 June 1997.

PLETSCH, T., BOTZ, R., BARRERA, E., HISADA, K., KAJIWARA, Y. & WRAY, D. S. in prep. Palygorskite clays and their relation to the Palaeocene–Eocene period of extreme warmth.

PLETSCH, T., BARRERA, E., BOTZ, R., HISADA, K. & KAJIWARA, Y. 2000. Marine palygorskite clay and early Paleogene oceanography. *Mitteilungen der Gesellschaft für Geologie und Bergbaustudenten in Österreich*, **43**, 106–107.

REA, D. K., ZACHOS, J. C., OWEN, R. M. & GINGERICH, P. D. 1990. Global change at the Palaeocene–Eocene boundary: climatic and evolutionary effects of tectonic events. *Palaeogeography, Palaeoclimatology, Palaeoecology*, **79**, 117–128.

RIGGS, S. R. & MANHEIM, F. T. 1988. Mineral resources of the U.S. Atlantic continental margin. *In*: SHERIDAN, R. E. & GROW, J. A. (eds) *The Atlantic Continental Margin: U.S. The Geology of North America*, **I-2**. Geological Society of America, Boulder, CO, 501–520.

ROTHE, P. 1989. Mineral composition of sedimentary formations in the North Atlantic Ocean. *Geologische Rundschau*, **78**, 903–942.

SCHMIDT, R. R. 1978. Calcareous nannoplankton from the western North Atlantic, DSDP Leg 44. *In*: BENSON, W. E. & SHERIDAN, R. E. (eds) *Initial Reports of the Deep Sea Drilling Project*, **44**. US Government Printing Office, Washington, DC, 703–718.

SINGER, A. 1979. Palygorskite in sediments: detrital, diagenetic or neoformed—a critical review. *Geologische Rundschau*, **68**, 996–1008.

SINGER, A. 1984. The palaeoclimatic interpretation of clay minerals in sediments—a review. *Earth-Science Reviews*, **21**, 251–293.

SINGER, A. 1989. Palygorskite and sepiolite group minerals. *In*: *Minerals in Soil Environments.* SSSA Book Series 1, Soil Science Society of America, Madison, WI, 996–1008.

SLOAN, L. C. & REA, D. K. 1995. Atmospheric carbon dioxide and early Eocene climate: a general circulation modeling sensitivity study. *Palaeogeography, Palaeoclimatology, Palaeoecology*, **119**, 275–292.

SLOAN, L. C., WALKER, J. C. G. & MOORE, T. C. 1995. Possible role of heat transport in early Eocene climate. *Paleoceanography*, **10**, 347–356.

SLOAN, L. C., WALKER, J. C. G., MOORE, T. C., REA, D. K. & ZACHOS, J. Z. 1992. Possible methane-induced polar warming in the early Eocene. *Nature*, 357, 320–322.

THIERSTEIN, H. R. & OKADA, H. 1979. The Cretaceous/Tertiary boundary event in the North Atlantic. *In*: TUCHOLKE, B. E., VOGT, P. R. *et al.* (eds) *Initial Reports of the Deep Sea Drilling Project*, **43**. US Government Printing Office, Washington, DC, 601–616.

THIRY, M. & JACQUIN, T. 1993. Clay mineral distribution related to rift activity, sea-level changes and oceanography in the Cretaceous of the Atlantic Ocean. *Clay Minerals*, **28**, 61–84.

TUCHOLKE, B. E. & MOUNTAIN, G. S. 1986. Tertiary palaeoceanography of the Western North Atlantic Ocean. *In*: VOGT, P. R. & TUCHOLKE, B. E. (eds) *The Western North Atlantic Region. The Geology of North America*, **M**. Geological Society of America, Boulder, CO, 631–650.

TUCHOLKE, B. E. & VOGT, P. R. 1979. Western North Atlantic: sedimentary evolution and aspects of tectonic history. *In*: TUCHOLKE, B. E., VOGT, P. R. *et al.* (eds) *Initial Reports of the Deep Sea Drilling Project*, **43**. US Government Printing Office, Washington, DC, 791–825.

WEAVER, C. E. 1989. *Clays, Muds, and Shales.* Developments in Sedimentology, **44**. Elsevier, Amsterdam.

WEAVER, C. E. & BECK, K. C. 1977. Miocene of the S.E. United States: a model for chemical sedimentation in a peri-marine environment. *Sedimentary Geology*, **17**, 1–234.

ZACHOS, J. C., STOTT, L. D. & LOHMANN, K. C. 1994. Evolution of early Cenozoic marine temperatures. *Palaeoceanography*, **9**, 353–387.

Index

Note: Page numbers of in *italics* refer to tables and those in **bold** refer to figures.

acoustic reflectors
  Bermuda Rise 23–48
    locations **25**
Albian black shale facies, Blake Nose 5–7, 49–72
  Oceanic Anoxic Events 49–72, 73–91
Albian–Cenomanian boundary events 1–19
Allison Guyot (ODP Site 865), carbon isotope records **204**
Aptian–Albian benthic foraminiferal record, ODP Leg 171B 73–91
  correspondence analysis **80**
  evolutionary change, factors 83–6
  stratigraphic distribution **82**
astronomical calibration of Danian time scale 163–83
Atlantic Coastal Plain, sediment biostratigraphic subdivision and correlation 93–108
atmospheric general circulation model (AGCM), climate modelling, Palaeogene 254

Bermuda Rise sites
  acoustic reflectors **25**, 23–48
  palygorskite clays in deep-water sediments 307–14
  stratigraphy 37–40
Blake Nose sites (ODP Leg 171B) 1–19
  acoustic reflectors 23–48
  Albian black shale facies 5–7, 49–72
  Aptian–Albian benthic foraminiferal record 73–91
  correlation with South Carolina Coastal Plain coreholes 93–108
  Cretaceous–Tertiary (K–T) boundary 9–10, 35–7
    Chicxulub ejecta deposits, geochemistry 131–47
    climate change in subtropical North Atlantic 2–13
    spherules as record of Chicxulub ejecta deposits 149–61
  Eocene deep-water sediments 32–4
    palygorskite clays 307–14
  geochemical data **134–6**
  isotope records, comparison with other sites 286–7
  Late Palaeocene Thermal Maximum 10
    carbon addition and removal 293–305
  lithostratigraphy and seismic stratigraphy 2–4, 28–32
  Maastrichtian, implications for global change 111–26
    extinctions and palaeoceanographic events 9
    setting and importance 114–26
  Maastrichtian, Upper, sediment biostratigraphic subdivision and correlation 93–108
  Mid-Cretaceous sea surface temperatures and OAE (1d and 2) 7–9
  Mid-to Late Eocene organic walled dinoflagellate cysts, offshore Florida 225–50
  orbitally forced climate change, stable isotopes in foraminifera 273–91
  palaeobathymetry 27–8
  previous work 25–6
  seismic profile **5**
  Site-1049 **51**, 53–60
    geochemical data *134–6*, *139–41*
    palygorskite clays 307–14
  Site-1050 **51**, 62–3, 94, **95–6**
    bio and magnetostratigraphy **170**
    Danian time scale 163–83
    geochemical data *137*, *142–3*
  Site-1051
    age model **279**
    carbon isotope records **204**, 300–302
    Late Palaeocene Thermal Maximum carbon addition/removal 293–305
    magnetostratigraphic and biostratigraphic datum levels *278*
    mid-latitude Palaeocene–Eocene radiolarian faunas 185–224
  Site-1052 **51**, 60–2, 94, **95–6**
    geochemical data *137*, *142–3*
  Site-1053, dinoflagellate cysts 225–50
  sites, maps and 3-D 3, **26–7**, **112**, **186–7**
  spherule bed 151–57
    comparisons with other K–T ejecta deposits 157–9
  see also Ocean Drilling Program
Blake–Bahama Basin, burial history 53

Campanian–Maastrichtian refrigeration, stable isotope records 16
carbon dioxide, climate variability in early Palaeogene 253–70
carbon isotopes
  Blake Nose sites
    Maastrichtian 118
    Site-1051 **204**, 300–302
  carbon addition/removal, Late Palaeocene Thermal Maximum 293–305
  composition of organic matter 71–2
  Holocene carbon cycle 297–8
  Pacific, Atlantic and Caribbean **204**
  theoreial background 295–6
Caribbean Sea (ODP Leg 165: Site 1001A) 163–83
  bio and magnetostratigraphy **174**, **176**
  carbon isotope records **204**
  XRF Fe record **171**
Ceratolithoides taxa, Upper Maastrichtian 99–102
Chicxulub ejecta deposits 9–10, 131–47, 149–61
  comparison with other deposits 146
  geochemistry 131–47
  iridium anomaly **36**, 131
  spherules 149–61
clay mineral analyses, Eocene 307–14
climate change in subtropical North Atlantic
  Cretaceous–Tertiary/Palaeogene (K–T/C–P) boundary 1–19
  lithostratigraphy and seismic stratigraphy 2–5

orbital forcing, stable isotopes in foraminifera 273–91
climate variability in early Palaeogene 253–70
sensitivity study
model and methods 256–7
results 257–63
continental runoff 261–2, **264–5**, 268
continental slope mass wasting, C–P boundary sites 35–7, **39**
correspondence analysis, Aptian–Albian benthic foraminiferal record **80**
Cretaceous
climate modelling sensitivity study 253–70
magnetic polarity time scale, calibration array **179**
Cretaceous sea surface temperatures, and OAE [(1d and 2)] 7–9
Cretaceous–Tertiary/Palaeogene (K–T/C–P) boundary
climate change in subtropical North Atlantic 1–19
continental slope mass wasting 35–7, **39**
element stratigraphy 138–46
detrital elements 145–6
redox-sensitive elements 138–45
Sr and Mg 138–9
geochemistry 131–47
comparison with other K–T ejecta deposits 146
geochemical data *134–7*, *139–43*
samples and analytical methods 132–8
spherules as record of Chicxulub ejecta deposits 9–10, 131–47, 149–61
stable isotope records 13–18
Campanian–Maastrichtian refrigeration 16
Cretaceous climate optimum 15–16
Danian climate 16–17
Palaeocene–Eocene climate trends 17–18

Danian
climate, stable isotope records, K–T boundary 16–17
foraminiferal zone P-alpha 132
Danian time scale 163–83
astronomical calibration 177–81
Palaeogene time scale and Danian stage 164–6
spectral analysis and direct cycle counts 168–76
Upper Danian stratigraphy at ODP Sites 1050 and 1001! 166–7
X-ray fluorescence (XRF) scanning 167–8
Dinoflagellate cysts from ODP Leg 171B (Site 1053A), offshore Florida 225–50
absolute ages *235*
first and last occurrences *232–3*, **234**
material and methods 226–9
neritic ratios **237**, *238*
palynomorph counts *230–1*, **236**
previous studies 228–9
results 229–35
systematic palynology 239–43
taxonomic appendix 243–50

Ekman transport divergence 263
Eocene
climate and foraminiferal record 273–5
climate modelling sensitivity study 253–70
Palaeocene–Eocene climate trends 10–13, 17–18
palygorskite clays in deep-water sediments 307–14
precessional cycle, orbital forcing 256
Eocene deep water, Late Palaeocene Thermal Maximum and continental slope mass wasting during C–P impact 23–45
Bermuda Rise, deposit stratigraphy 37–40

foraminiferal record
excursion fauna 34
isotopes, evidence for orbitally forced climate change 273–91
ODP Leg 171B, Aptian–Albian 73–91
Palaeocene–Eocene transition 34–5

greenhouse gases, climate variability in early Palaeogene 253–70

high-resolution sampling 70–1

inoceramid extinction, Blake Nose sites, Maastrichtian 117
iridium anomaly, Chicxulub ejecta deposits **36**, 131
isotope records
Blake Nose sites, comparison with other sites 286–7
Cretaceous–Tertiary/Palaeogene (K–T/C–P) boundary 13–18

JOIDES project 23–48, 225–50
Joint Time–Frequency Analysis (JTFA) Tool 168–76

Late Palaeocene Thermal Maximum 10
carbon addition/removal, Site-1051, Blake Nose 293–305
carbon isotope excursion, general solutions 297–300
climate modelling 255–6
and continental slope mass wasting during C–P impact 23–45
Bermuda Rise, deposit stratigraphy 37–40
layer lamination and thickness 71
Lithraphidites taxa, Upper Maastrichtian 102–3

Maastrichtian
Blake Nose sites 114–15
implications for global palaeoceanographic and biotic changes 111–26
benthic foraminifera 120
inoceramid extinction 117
Maastrichtian models 123–6
models 123–6
other environments 119–26
palaeogeographical map **113**
sea levels 122–3
Maastrichtian, Lower, disconformity 108
Maastrichtian, Upper
Atlantic Coastal Plain and Blake Nose, sediment biostratigraphic subdivision and correlation 93–108
extinctions 9–10
maceral analysis 71
magnesium, Cretaceous–Tertiary/Palaeogene (K–T/C–P) boundary 138
Maud Rise (ODP Site 690), carbon isotope records **204**

methane
  climate variability in early Palaeogene 253–70
  injection into carbon cycles 298–300
Micula taxa, Upper Maastrichtian 103–6
Milankovitch cyclicity *see* orbital forcing

North Atlantic
  continental runoff 261–2, **264–5**, 268
  net moisture balance 260–1, 267–8
  sea ice, modelling 260, 267
  upwelling 263, **266**
  see also sea surface temperatures

Ocean Drilling Program
  Bermuda Rise sites 23–48, 307–14
  Blake Nose sites 2–13, **5**, **50**, **51**, **166**
    Early Albian black shale OAEs 5–7, 49–72
    lithostratigraphy and seismic stratigraphy 2–4, 28–32
    see also Blake Nose sites (ODP Leg 171B)
  Caribbean Sea site (ODP Leg 165: Site 1001A) 163–83
  Pacific, Allison Guyot (ODP Site 865) **204**
  South Atlantic, Maud Rise (ODP Site 690) **204**
Oceanic Anoxic Events 49–72, 73–91
  1b, Aptian–Albian benthic foraminiferal record 73–91
orbital forcing
  Eocene
    climate change 284–6
    precessional cycle 256
orbitally forced climate change, stable isotopes in foraminifera 273–91
organic matter *see* carbon isotopes
oscillations
  obliquity vs precession 163–4, 175–6
  spectral analysis and direct cycle counts 168–76
oxygen isotopes, Blake Nose sites, Maastrichtian 117–18

Pacific, Allison Guyot (ODP Site 865), carbon isotope records **204**
Palaeocene–Eocene radiolarian faunas 185–224
  see also radiolarian faunas
Palaeocene–Eocene transition 34–5
  climate trends 10–13, 17–18
  see also Late Palaeocene Thermal Maximum
Palaeogene, magnetic polarity time scale
  biochronology **165**
  calibration array **179**
  Danian stage 164–6
palygorskite clays, Eocene deep-water sediments, Blake Nose sites 307–14
Podorhabdus? elkefensis, Upper Maastrichtian 106–7

polar stratospheric clouds (PSCs) 254–6

radiolarian faunas, Palaeocene–Eocene 185–224
  across P–E boundary and LPTM interval 199–201
  biostratigraphy and biochronology 188–91
  first and last occurrences *188–90*
  hiatuses 197–9
  lithostratigraphy 187–8
  RP6, *Bekoma campechensis* Interval Zone 196
  RP7, *Bekoma bidartensis* Interval Zone 196
  RP8, *Buryella clinata* Interval Zone 196
  RP9, *Phormocyrtis striata* Interval Zone 195
  RP10, *Theocotyle cryptocephala* Interval Zone 195
  RP11, *Dictyoprora mongolfieri* Interval Zone 195
  RP12, *Thyrsocyrtis (Pentalcorys) triacantha* Interval Zone 194–5
  RP13, *Podocyrtis (Podocyrtoges) ampla* Lineage Zone 194
  RP14, *Podocyrtis (Lampterium) mitra* Lineage Zone 194
  RP15, *Podocyrtis (Lampterium) chalara* Lineage Zone 193–4
  RP16, *Podocyrtis (Lampterium) goetheana* Interval Zone 191
  species list and taxonomic notes 208–220
  systematics 202–8
Rock–Eval analysis 53

sea surface temperatures
  Cretaceous 7–9
  Mid-Eocene 279–87
  modelling 257–60
  responses to forcing 263–8
sedimentary organic matter (SOM) *see* Albian black shale facies; carbon
seismic profiling 23–48
  horizon A* 43–5
  reflectors A^u, A^b, A^c 40–5
South Atlantic
  Maud Rise (ODP Site 690), carbon isotope records **204**
  spreading rates **180**
South Carolina Coastal Plain coreholes, correlation with Blake Nose sites (ODP Leg 171B) 93–108
spherules *see* Chicxulub ejecta deposits
strontium, Cretaceous–Tertiary/Palaeogene (K–T/C–P) boundary 138

upwelling 263, **266**

wetlands, climate variability in early Palaeogene 254

X-ray fluorescence (XRF) scanning, Danian 167–8